Contemporary Quantitative Finance

Carl Chiarella · Alexander Novikov
Editors

Contemporary Quantitative Finance

Essays in Honour of Eckhard Platen

 Springer

Editors
Carl Chiarella
Alexander Novikov
School of Finance and Economics
Department of Mathematical Sciences
The University of Technology, Sydney
P.O. Box 123, Broadway 2007
NSW Australia
carl.chiarella@uts.edu.au
alex.novikov@uts.edu.au

ISBN 978-3-642-43858-5   ISBN 978-3-642-03479-4 (eBook)
DOI 10.1007/978-3-642-03479-4
Springer Heidelberg Dordrecht London New York

Mathematics Subject Classification (2000): 49Kxx, 49Nxx, 60G35, 62M05, 60G70, 65Mxx

Cover design: WMXDesign GmbH, Heidelberg

Printed on acid-free paper

Springer is part of Springer Science+Business Media (www.springer.com)

Preface

This volume contains a collection of papers dedicated to Professor Eckhard Platen to celebrate his 60th birthday, which occurred in 2009. The contributions have been written by a number of his colleagues and co-authors. All papers have been reviewed and presented as keynote talks at the international conference "Quantitative Methods in Finance" (QMF) in Sydney in December 2009. The QMF Conference Series was initiated by Eckhard Platen in 1993 when he was at the Australian National University (ANU) in Canberra. Since joining UTS in 1997 the conference came to be organised on a much larger scale and has grown to become a significant international event in quantitative finance.

Professor Platen has held the Chair of Quantitative Finance at the University of Technology, Sydney (UTS) jointly in the Faculties of Business and Science since 1997. Prior to this appointment, he was the Founding Head of the Centre for Financial Mathematics at the Institute of Advanced Studies at ANU, a position to which he was appointed in 1994.

Eckhard completed a PhD in Mathematics at the Technical University in Dresden in 1975 and in 1985 obtained his Doctor of Science degree (Habilitation degree in the German system) from the Academy of Sciences in Berlin where he headed the Stochastics group at the Weierstrass Institute.

Eckhard has served on the editorial boards of many distinguished journals, and he is currently a member of the following Editorial Boards: Mathematical Finance, Acta Applicandae Mathematicae, Communications on Stochastic Analysis, Monte Carlo Methods and Applications, Quantitative Finance, the new Bocconi-Springer Book Series and an adviser to Asia-Pacific Financial Markets.

Eckhard worked initially on the development of the theory and applications of numerical methods for stochastic differential equations, at first in the context of engineering applications the 1970's and 1980's and then continuing this work in 1990–2000's with a strong focus on finance applications. He uncovered the key role in the derivation of high order weak and strong approximations for stochastic differential equations (both with or without jump components) played by a stochastic analogue of the Taylor expansion, also called the Wagner-Platen (or, also Platen-Wagner) formula. This result has been generalised and extended in numerous papers, the best reference being the monographs Kloeden P. and Platen E. (1999) [2]

and Kloeden P., Platen E. and Shurz H. (2003) [3]; both of which have so far had three editions at Springer.

From the beginning of 1990's Eckhard turned the focus of his research interests to different aspects of quantitative finance. At first his work was devoted to developing general principles for the modelling of incomplete financial markets. Over the last decade he has worked on essential problems with the existing methodologies for derivative pricing and risk measurement. His benchmark approach, described in the monograph Platen & Heath (2006) [4], models the dynamics of the financial market by using the growth optimal portfolio (GOP). In the context of portfolio optimisation the GOP serves as a benchmark, and for the purpose of pricing as well as hedging it can be taken as the numeraire. Under this approach the real world measure is used as the pricing measure and so the approach can be used also in those cases in which a risk neutral measure does not exist. One of the advantages of this approach is the fact that it allows one to establish a link between modern mathematical finance methodologies and the more classical ones that are still in wide use, especially among practitioners in both the finance and insurance industries.

The benchmark approach is furthermore in line with some current investigations that take into account the fact that in various financial applications a local martingale may not be a martingale (see for example the article by Hulley in this volume). There are in fact market models for which an equivalent risk neutral measure does not exist because the corresponding Radon-Nikodým derivative is a strict supermartingale. A discrete time version of the benchmark approach with a view to insurance applications was discussed in Bühlmann & Platen (2003) [1]. As in the case of financial pricing the standard actuarial pricing rule is obtained as a particular case of fair pricing when the contingent claim is independent of the growth optimal portfolio.

In addition to the above, Eckhard has published a number of papers on diverse subjects including stochastic control, modelling of stochastic vibrations, modelling of interacting stochastic particle systems and modelling of charge transport. During his career Eckhard has published four books and about 150 academic articles. His most recent book with the late Nicola Bruti-Liberati [5] contains his recent research on the numerical solution of stochastic differential equations with jumps.

His current research includes estimation of discretely observed financial markets (with Michael Sørenson), study of transition densities for diffusion processes (with Mark Craddock), and work on estimation techniques in search of the most appropriate financial market model (with Renata Rendek).

The papers in this collection fall broadly under three headings. The first group of papers of D. Fernholz and I. Karatzas; C. Kardaras; H. Hulley and M. Schweizer; H. Hulley; W. Runggaldier and G. Galesso deal with various current issues relating to martingale measures in the context of the arbitrage approach to financial markets; the second group of papers by Bao, Delbaen and Hu; Cohen and Elliott, and Imkeller, Reis and Zhang deal with certain recent advances in backward stochastic differential equations for the purposes of utility maximisation and hedging. The third group of eleven papers contains various applications, five on interest rate modelling and the rest on diverse applications from stochastic partial differential equations and portfolio choice to maximum likelihood estimation for integrated diffusion processes. Many of the authors are amongst the leading figures internationally in mathematical and quantitative finance.

The editors would like to thank all of the authors for their contributions and the anonymous reviewers for their prompt and helpful reports. We would particularly thank both authors and reviewers for responding in such a timely manner to the extreme dead-line pressure. We would like to thank Ms. Stephanie Ji-Won Ough for all her work in preparing the final draft of the book in the Springer layout. Finally we would like to thank the staff at the publisher Springer Verlag for assisting the publishing process in many ways.

Sydney, Australia   Carl Chiarella
                                                                              Alexander Novikov

References

1. Bühlmann, H., Platen, E.: A discrete time benchmark approach for insurance and finance. ASTIN Bull. **33**(2), 153–172 (2003)
2. Kloeden, P.E., Platen, E.: Numerical Solution of Stochastic Differential Equations. *Applications of Mathematics* 23. Springer, Berlin, 632 pp., ISBN 3-540-54062-8, (1992), (1995), Third extensively revised printing (1999)
3. Kloeden, P.E., Platen, E., Schurz, H.: Numerical Solution of SDE Through Computer Experiments. *UNIVERSITEXT-Series* (with diskette). Springer, Berlin, 294 pp., ISBN 3-540-57074-8, (1994), Second revised printing (1997), Corrected third printing (2003)
4. Platen, E., Heath, D.: A Benchmark Approach to Quantitative Finance. *Springer Finance*. Springer, Berlin, 700 pp., ISBN-10 3-540-26212-1 (2006), Corrected second printing (2010)
5. Platen, E., Bruti-Liberati, N.: Numerical Solution of SDEs with Jumps in Finance. *Applications of Mathematics*. Springer, Berlin, 882 pp., (2010)

Contents

Probabilistic Aspects of Arbitrage 1
Daniel Fernholz and Ioannis Karatzas

Finitely Additive Probabilities and the Fundamental Theorem of Asset Pricing 19
Constantinos Kardaras

M^6—On Minimal Market Models and Minimal Martingale Measures . . 35
Hardy Hulley and Martin Schweizer

The Economic Plausibility of Strict Local Martingales in Financial Modelling 53
Hardy Hulley

A Remarkable σ-finite Measure Associated with Last Passage Times and Penalisation Problems 77
Joseph Najnudel and Ashkan Nikeghbali

Pricing Without Equivalent Martingale Measures Under Complete and Incomplete Observation 99
Wolfgang J. Runggaldier and Giorgia Galesso

Existence and Non-uniqueness of Solutions for BSDE 123
Xiaobo Bao, Freddy Delbaen, and Ying Hu

Comparison Theorems for Finite State Backward Stochastic Differential Equations 135
Samuel N. Cohen and Robert J. Elliott

Results on Numerics for FBSDE with Drivers of Quadratic Growth . . . 159
Peter Imkeller, Gonçalo Dos Reis, and Jianing Zhang

Variance Swap Portfolio Theory 183
Dilip B. Madan

Stochastic Partial Differential Equations and Portfolio Choice 195
Marek Musiela and Thaleia Zariphopoulou

**Issuers' Commitments Would Add More Value than Any Rating Scheme
Could Ever Do** 217
Carlos Veiga and Uwe Wystup

Pricing and Hedging of CDOs: A Top Down Approach 231
Damir Filipović and Thorsten Schmidt

**Constructing Random Times with Given Survival Processes and
Applications to Valuation of Credit Derivatives** 255
Pavel V. Gapeev, Monique Jeanblanc, Libo Li, and Marek Rutkowski

**Representation of American Option Prices Under Heston Stochastic
Volatility Dynamics Using Integral Transforms** 281
Carl Chiarella, Andrew Ziogas, and Jonathan Ziveyi

Buy Low and Sell High 317
Min Dai, Hanqing Jin, Yifei Zhong, and Xun Yu Zhou

**Continuity Theorems in Boundary Crossing Problems for Diffusion
Processes** 335
Konstantin A. Borovkov, Andrew N. Downes, and Alexander A. Novikov

Binomial Models for Interest Rates 353
John van der Hoek

**Lognormal Forward Market Model (LFM) Volatility Function
Approximation** 369
In-Hwan Chung, Tim Dun, and Erik Schlögl

Maximum Likelihood Estimation for Integrated Diffusion Processes . . . 407
Fernando Baltazar-Larios and Michael Sørensen

Probabilistic Aspects of Arbitrage

Daniel Fernholz and Ioannis Karatzas

Abstract Consider the logarithm $\log(1/U(T, \mathbf{z}))$ of the highest return on investment that can be achieved relative to a market with Markovian weights, over a given time-horizon $[0, T]$ and with given initial market weight configuration $\mathscr{X}(0) = \mathbf{z}$. We characterize this quantity (i) as the smallest amount of relative entropy with respect to the *Föllmer exit measure*, under which the market weight process $\mathscr{X}(\cdot)$ is a diffusion with values in the unit simplex Δ and the same covariance structure but zero drift; and (ii) as the smallest "total energy" expended during $[0, T]$ by the respective drift, over a class of probability measures which are absolutely continuous with respect to the exit measure and under which $\mathscr{X}(\cdot)$ stays in the interior Δ^o of the unit simplex at all times, almost surely. The smallest relative entropy, or total energy, corresponds to the conditioning of the exit measure on the event $\{\mathscr{X}(t) \in \Delta^o, \ \forall \ 0 \le t \le T\}$; whereas, under this "minimal energy" measure, the portfolio $\widehat{\pi}(\cdot)$ generated by the function $U(\cdot, \cdot)$ has the numéraire and relative log-optimality properties. This same portfolio $\widehat{\pi}(\cdot)$ also attains the highest possible relative return on investment with respect to the market.

1 Introduction

The pioneering work of Fernholz [6] demonstrated that, under appropriate conditions, it is possible systematically to outperform a market portfolio over sufficiently

I. Karatzas is on leave from the Department of Mathematics, Columbia University, Mail Code 4438, New York, NY 10027. e-mail: ik@math.columbia.edu

D. Fernholz
Daniel Fernholz LLC, 603 West 13th Street, Suite 1A-440, Austin, TX 78701, USA
e-mail: df@danielfernholz.com

I. Karatzas (✉)
INTECH Investment Management LLC, One Palmer Square, Suite 441, Princeton, NJ 08542, USA
e-mail: ik@enhanced.com

C. Chiarella, A. Novikov (eds.), *Contemporary Quantitative Finance*,
DOI 10.1007/978-3-642-03479-4_1, © Springer-Verlag Berlin Heidelberg 2010

long time horizons. Since then there has been an effort to understand various aspects of such "relative arbitrage": its implementation on arbitrary time horizons, the nature and behavior of portfolios that implement it, pricing and hedging in the context of strict local martingale deflators, etcetera. These efforts are summarized in the survey paper [7].

The fairly recent article [5] addresses an issue that arises naturally in this context: if arbitrage exists on a given time-horizon relative to a given market, what is the "best possible" arbitrage of this type? how can it be characterized? what portfolio(s) implement it? Within a Markovian weight model and under suitable regularity conditions, it was shown in [5] that the reciprocal $U(T, \mathbf{z}) \in (0, 1]$ of the highest relative return on investment, that can be achieved with respect to the market over a given time-horizon $[0, T]$ and with initial market weight configuration $\mathscr{L}(0) = \mathbf{z}$ in the interior Δ^o of the unit simplex Δ, is equal to the probability under an auxiliary measure $\widehat{\mathbb{Q}}$ that the process of relative market weights $\mathscr{L}(\cdot)$ stays in Δ^o throughout the time interval $[0, T]$. The probability measure $\widehat{\mathbb{Q}}$ is the celebrated *Föllmer exit measure* [9, 10]: the original measure $\mathbb{P}$ is absolutely continuous with respect to it; it renders $\mathscr{L}(\cdot)$ a Δ-valued diffusion with the same covariance structure but zero drift, thus a *martingale*; and it anoints $\mathscr{L}(\cdot)$ with the "numéraire property", of a portfolio that cannot be outperformed. It was also shown in [5] that $U(\cdot, \cdot)$ satisfies a linear, second-order partial differential equation on $(0, \infty) \times \Delta^o$, determined entirely on the basis of the covariance structure of the underlying model. The function $U(\cdot, \cdot)$ is the smallest nonnegative supersolution of this equation subject to $U(0, \cdot) \equiv 1$ on Δ^o, and the portfolio $\widehat{\pi}(\cdot)$ which attains this best possible relative arbitrage is the one generated by this function in the sense of Fernholz [6, p. 46].

We show here that this portfolio $\widehat{\pi}(\cdot)$ has *itself* the numéraire and log-optimality properties, but under *yet another* probability measure $\mathbb{P}_\star(\cdot) = \widehat{\mathbb{Q}}(\cdot \,|\, \mathscr{L}(t) \in \Delta^o, \forall\, 0 \le t \le T)$, the conditioning of the Föllmer measure on the event that the boundary of the simplex is not attained by the process $\mathscr{L}(\cdot)$ during $[0, T]$. This $\mathbb{P}_\star$ has a further characterization as the probability measure whose associated drift $\vartheta^{\mathbb{P}_\star}(\cdot)$ keeps the process $\mathscr{L}(\cdot)$ inside of Δ^o throughout $[0, T]$ with the smallest amount of "total energy", or effort, $(1/2)\, \mathbb{E}^{\mathbb{P}} \int_0^T \|\vartheta^{\mathbb{P}}(t)\|^2\, dt$, over all probability measures $\mathbb{P} \ll \widehat{\mathbb{Q}}$ with $\mathbb{P}(\mathscr{L}(t) \in \Delta^o, \forall\, 0 \le t \le T) = 1$; it is also the measure with the smallest relative entropy $H_T(\mathbb{P} \,|\, \widehat{\mathbb{Q}})$ over this same class, and this smallest value is the logarithm $\log(1/U(T, \mathbf{z}))$ of the highest achievable return on investment relative to the market. We call this measure $\mathbb{P}_\star$ "minimal energy measure".

The paper is organized as follows. Sections 2 and 3 provide necessary background material and set up the model. Section 4 introduces the notions of relative arbitrage and of the minimal capital necessary to implement it, as well as of the Föllmer exit measure and its properties. We also describe in Sect. 4 the analytical properties of the function $U(\cdot, \cdot)$, and of the "optimal arbitrage" portfolio $\widehat{\pi}(\cdot)$ generated by it. The conditional measure $\mathbb{P}_\star$ is introduced in Sect. 5, and under it the numéraire and relative log-optimality properties of $\widehat{\pi}(\cdot)$ are developed. Section 6 establishes the minimum relative entropy and stochastic control characterizations of $\mathbb{P}_\star$, building on a Schrödinger-type equation satisfied by the function $\log(1/U(\cdot, \cdot))$ and on the relation of this equation to one of the Hamilton-Jacobi-Bellman (HJB) type. We conclude in Sect. 7 with several additional characterizations of the optimal arbitrage function, including one that involves a zero-sum stochastic game, and with a compilation of results.

2 Preliminaries

On a given filtered probability space $(\Omega, \mathscr{F}, \mathbb{P})$, $\mathbb{F} = \{\mathscr{F}(t)\}_{0 \leq t < \infty}$ we consider a vector $\mathfrak{X}(\cdot) = (X_1(\cdot), \ldots, X_n(\cdot))'$ of strictly positive semimartingales which represent the capitalizations in an equity market with $n \geq 2$ assets. We denote by $X(\cdot) := X_1(\cdot) + \cdots + X_n(\cdot)$ the total market capitalization, and by $Z_1(\cdot) := X_1(\cdot)/X(\cdot), \ldots,$ $Z_n(\cdot) := X_n(\cdot)/X(\cdot)$ the corresponding relative weights of the individual assets. The vector $\mathscr{Z}(\cdot) = (Z_1(\cdot), \ldots, Z_n(\cdot))'$ of these weights is a semimartingale with values in the interior Δ^o of the simplex $\Delta := \{\mathbf{z} = (z_1, \ldots, z_n)' \in [0, 1]^n : \sum_{i=1}^n z_i = 1\}$ in $n - 1$ dimensions; we shall let $\Gamma := \Delta \setminus \Delta^o$ denote the boundary of Δ. We shall take throughout $\mathscr{F}(0) = \{\emptyset, \Omega\}$ mod. $\mathbb{P}$.

One way to think of investment in such a market is in terms of selecting a *portfolio* $\pi(\cdot) = (\pi_1(\cdot), \ldots, \pi_n(\cdot))'$, i.e., an $\mathbb{F}$-predictable process with values in $\{\mathbf{z} \in \mathbb{R}^n : \sum_{i=1}^n z_i = 1\}$, such that $\pi_i(\cdot)/X_i(\cdot)$ is an admissible integrand for the semimartingale $X_i(\cdot)$; then $\pi_i(t)$ stands for the proportion of wealth that gets invested at time $t > 0$ in the ith asset, for each $i = 1, \ldots, n$. With this interpretation, the dynamics of the *wealth process* $V^{w,\pi}(\cdot)$, corresponding to portfolio $\pi(\cdot)$ and initial wealth $w \in (0, \infty)$, are given by

$$\frac{d V^{w,\pi}(t)}{V^{w,\pi}(t)} = \sum_{i=1}^n \pi_i(t) \frac{dX_i(t)}{X_i(t)}, \qquad V^{w,\pi}(0) = w. \tag{1}$$

The collection of all portfolios will be denoted by Π. Portfolios that take values in Δ will be called "long-only"; the most conspicuous of these is the *market portfolio* $\mathscr{Z}(\cdot)$, which actually takes values in Δ^o and generates wealth $V^{w,\mathscr{Z}}(\cdot) = wX(\cdot)/X(0)$ proportional to the total market capitalization at all times.

If one is interested in performance relative to the market, it makes sense to consider for any given portfolio $\pi(\cdot) \in \Pi$ the ratio, or relative performance,

$$Y^{q,\pi}(\cdot) := V^{qX(0),\pi}(\cdot)/X(\cdot) = V^{qX(0),\pi}(\cdot)/V^{X(0),\mathscr{Z}}(\cdot); \tag{2}$$

here the scalar $q > 0$ measures initial wealth $w = qX(0)$ as a proportion of total market capitalization at time $t = 0$. The dynamics of this relative performance are shown to be

$$\frac{d Y^{q,\pi}(t)}{Y^{q,\pi}(t)} = \sum_{i=1}^n \pi_i(t) \frac{dZ_i(t)}{Z_i(t)}, \qquad Y^{q,\pi}(0) = q. \tag{3}$$

2.1 Change of Variables

It is occasionally convenient to describe a portfolio $\pi(\cdot) = (\pi_1(\cdot), \ldots, \pi_n(\cdot))'$ in terms of the vector $\psi(\cdot) = (\psi_1(\cdot), \ldots, \psi_n(\cdot))'$ of its *scaled relative weights*

$$\psi_i(\cdot) := \pi_i(\cdot) / Z_i(\cdot), \quad i = 1, \ldots, n.$$

This allows us to rewrite the relative performance dynamics (3) simply as

$$\frac{d\, Y^{q,\pi}(t)}{Y^{q,\pi}(t)} = \sum_{i=1}^{n} \psi_i(t)\, dZ_i(t) = \psi'(t)\, d\mathscr{Z}(t), \quad Y^{q,\pi}(0) = q. \tag{4}$$

Since $\sum_{i=1}^{n} dZ_i(t) = 0$, the vector $\psi(t)$ of scaled portfolio weights need be specified only modulo a scalar factor; then $\psi_1(t), \ldots, \psi_n(t)$ are recovered from the ordinary portfolio weights

$$\pi_i(t) = Z_i(t)\left(\psi_i(t) + 1 - \sum_{j=1}^{n} Z_j(t)\psi_j(t)\right), \quad i = 1, \ldots, n. \tag{5}$$

3 The Model

We shall postulate from now onwards an *Itô process model* for the Δ^o-valued relative market weights $\mathscr{Z}(\cdot) = (Z_1(\cdot), \ldots, Z_n(\cdot))'$ process, of the form

$$d\mathscr{Z}(t) = \mathrm{s}\big(\mathscr{Z}(t)\big)\big(dW(t) + \vartheta(t)\, dt\big), \quad \mathscr{Z}(0) = \mathbf{z} \in \Delta^o. \tag{6}$$

Here $W(\cdot)$ is an n-dimensional $\mathbb{P}$-Brownian motion; $\vartheta(\cdot)$ is an n-dimensional, progressively measurable and locally square-integrable process, i.e., $\int_0^T \|\vartheta(t)\|^2 dt < \infty$ holds $\mathbb{P}$-a.s. for every $T \in (0, \infty)$; and the volatility structure is characterized in terms of $\mathrm{s}(\cdot) = (\mathrm{s}_{i\ell}(\cdot))_{1 \leq i, \ell \leq n}$, a matrix-valued function with continuous components $\mathrm{s}_{i\ell} : \Delta \to \mathbb{R}$ that satisfy the condition $\sum_{i=1}^{n} \mathrm{s}_{i\ell}(\cdot) \equiv 0$ for all $\ell = 1, \ldots, n$.

We shall assume that the corresponding *covariance matrix*

$$\mathrm{a}(\mathbf{z}) := \mathrm{s}(\mathbf{z})\, \mathrm{s}'(\mathbf{z}), \quad \mathbf{z} \in \Delta \tag{7}$$

has rank $n - 1$ for every $\mathbf{z} \in \Delta^o$ (cf. Lemma 3.1 in [7]), as well as rank k in the interior $\mathfrak{d}^o$ of every k-dimensional sub-simplex $\mathfrak{d} \subset \Gamma, k = 1, \ldots, n - 2$.

If, on some given time-horizon $[0, T]$ of finite length, we can write $\vartheta(t) = \Theta(T - t, \mathscr{Z}(t)), 0 \leq t \leq T$ for some continuous function $\Theta : [0, T] \times \Delta \to \mathbb{R}^n$, we shall say that on this time-horizon $[0, T]$ the resulting diffusion

$$d\mathscr{Z}(t) = \mathrm{b}\big(T - t, \mathscr{Z}(t)\big)\, dt + \mathrm{s}(\mathscr{Z}(t))\, dW(t), \quad \mathscr{Z}(0) = \mathbf{z} \in \Delta^o, \tag{8}$$

with $\mathrm{b}(\tau, \mathbf{z}) := \mathrm{s}(\mathbf{z})\Theta(\tau, \mathbf{z})$, constitutes a *Markovian Market Weight (MMW)* model.

3.1 Numéraire and Log-Optimality Properties

Consider now two portfolios $\pi(\cdot)$, $v(\cdot)$ with corresponding scaled relative weights $\psi_i(\cdot) = \pi_i(\cdot)/Z_i(\cdot)$ and $\varphi_i(\cdot) = v_i(\cdot)/Z_i(\cdot)$, $i = 1, \ldots, n$. An application of Itô's rule in conjunction with (4) gives

$$d\left(\frac{Y^{q,\pi}(t)}{Y^{q,v}(t)}\right) = \left(\frac{Y^{q,\pi}(t)}{Y^{q,v}(t)}\right)(\psi(t) - \varphi(t))'\left[d\mathscr{Z}(t) - \mathrm{a}(\mathscr{Z}(t))\varphi(t)\,dt\right]; \quad (9)$$

whereas, on the strength of (6), the last term (in brackets) is

$$\mathrm{s}(\mathscr{Z}(t))\,dW(t) + \mathrm{s}(\mathscr{Z}(t))\left(\vartheta(t) - (\mathrm{s}(\mathscr{Z}(t)))'\varphi(t)\right)dt.$$

The finite variation part of this expression vanishes, if we select the portfolio $v(\cdot)$ as in (5) with scaled relative weights $\varphi(\cdot) = (\varphi_1(\cdot), \ldots, \varphi_n(\cdot))'$ that satisfy

$$(\mathrm{s}(\mathscr{Z}(\cdot)))'\varphi(\cdot) = \vartheta(\cdot). \quad (10)$$

Thus, for $v(\cdot) \equiv v^{\mathbb{P}}(\cdot)$ selected this way, *the ratio $Y^{q,\pi}(\cdot)/Y^{q,v^{\mathbb{P}}}(\cdot)$ is, for any portfolio $\pi(\cdot) \in \Pi$, a positive local martingale, thus also a supermartingale.*

We express this by saying that the portfolio $v^{\mathbb{P}}(\cdot)$ has the "numéraire property" (e.g., [18, 21, 24] and the references there). As was observed in [7, 18], no arbitrage relative to such a portfolio is possible over any finite time-horizon. Note also that, if $\vartheta(\cdot) \equiv 0$, we can select $\varphi(\cdot) \equiv 0$ in (10), so the resulting market portfolio $\mathscr{Z}(\cdot)$ from (5) has then the numéraire property.

With $v^{\mathbb{P}}(\cdot)$ selected as in (5) from scaled relative weights $\varphi(\cdot) = (\varphi_1(\cdot), \ldots, \varphi_n(\cdot))'$ that satisfy (10), the expression of (9) becomes

$$d\left(Y^{q,\pi}(t)/Y^{q,v^{\mathbb{P}}}(t)\right) = \left(Y^{q,\pi}(t)/Y^{q,v^{\mathbb{P}}}(t)\right)(\psi(t) - \varphi(t))'\mathrm{s}(\mathscr{Z}(t))\,dW(t)$$

and yields

$$d\log\left(\frac{Y^{q,\pi}(t)}{Y^{q,v^{\mathbb{P}}}(t)}\right) = (\psi(t) - \varphi(t))'\mathrm{s}(\mathscr{Z}(t))dW(t)$$
$$-\frac{1}{2}(\psi(t) - \varphi(t))'\mathrm{a}(\mathscr{Z}(t))(\psi(t) - \varphi(t))dt.$$

We deduce as in [18, 19] (see also [1, 2]) the *relative log-optimality* property of $v^{\mathbb{P}}(\cdot)$: for every portfolio $\pi(\cdot) \in \Pi$ and $(T, q) \in (0, \infty)^2$, we have

$$\mathbb{E}^{\mathbb{P}}\left[\log\frac{Y^{q,\pi}(T)}{q}\right] \leq \mathbb{E}^{\mathbb{P}}\left[\log\frac{Y^{q,v^{\mathbb{P}}}(T)}{q}\right] = \frac{1}{2}\mathbb{E}^{\mathbb{P}}\int_0^T \|\vartheta(t)\|^2\,dt. \quad (11)$$

4 Relative Arbitrage

Let us recall briefly some results from [5]. Within a market model as in the previous section, and for a given time-horizon $[0, T]$ and initial market weight configuration $\mathscr{Z}(0) = \mathbf{z} \in \Delta^o$, we introduce the *relative arbitrage function*

$$U(T; \mathbf{z}) := \inf\left\{q > 0 : \exists\,\pi(\cdot) \in \Pi \ \text{s.t.}\ \mathbb{P}(Y^{q,\pi}(T) \geq 1) = 1\right\}. \quad (12)$$

To wit, $U(T, \mathbf{z}) \in (0, 1]$ is the smallest relative initial wealth required at $t = 0$, in order to attain at time $t = T$ relative wealth of (at least) 1 with respect to the market, $\mathbb{P}$-a.s. Equivalently, the quantity $(1/U(T, \mathbf{z}))$ gives the maximal relative amount by which the market portfolio can be outperformed over the time horizon $[0, T]$.

If $U(T, \mathbf{z}) = 1$, it is not possible to outperform the market over $[0, T]$; if, on the other hand, $U(T, \mathbf{z}) < 1$, then for every $q \in (U(T, \mathbf{z}), 1)$ (and even for $q = U(T, \mathbf{z})$, when the infimum in (12) is attained) there exists a portfolio $\pi_q(\cdot) \in \Pi$ such that

$$\frac{V^{1, \pi_q}(T)}{V^{1, \mathscr{Z}}(T)} \geq \frac{1}{q} > 1 \quad \text{holds } \mathbb{P}\text{-a.s.}$$

In the terminology of [7], each such $\pi_q(\cdot)$ is then *strong arbitrage relative to the market portfolio* $\mathscr{Z}(\cdot)$ over the time-horizon $[0, T]$.

Remark 1 It is shown in [7, Sect. 11] (see also [5]) that a sufficient condition leading to $U(T, \mathbf{z}) < 1$, is that there exist a real constant $h > 0$ for which

$$\sum_{i=1}^{n} z_i \frac{a_{ii}(\mathbf{z})}{z_i^2} \geq h, \quad \forall \mathbf{z} \in \Delta^o. \tag{13}$$

The expression on the left-hand side of this inequality is the average, weighted by market weight z_i, of the variances $a_{ii}(\mathbf{z})/z_i^2$ of individual asset log-returns relative to the market and, as such, a measure of the market's "intrinsic volatility". The condition (13) posits a positive lower bound on this quantity as sufficient for $U(T, \mathbf{z}) < 1$, i.e., for the possibility of outperforming the market.

Here, we shall characterize $U(T, \mathbf{z})$ in various equivalent ways, thus providing necessary and sufficient conditions for this possibility.

Remark 2 A concrete example where (13) is satisfied concerns the model

$$d \log X_i(t) = (\kappa/Z_i(t)) \, dt + \left(1/\sqrt{Z_i(t)}\right) dW_i(t)$$

for the log-capitalizations $\log X_i(t)$, $i = 1, \ldots, n$ with some constant $\kappa \geq 0$, or equivalently

$$dZ_i(t) = 2\kappa\left(1 - n Z_i(t)\right)dt + \sqrt{Z_i(t)} \, dW_i(t) - Z_i(t) \sum_{k=1}^{n} \sqrt{Z_k(t)} \, dW_k(t)$$

for the market weight process $\mathscr{Z}(\cdot)$. It is checked that the variances are of the Wright-Fisher type $a_{ii}(\mathbf{z}) = z_i(1 - z_i)$ in this case, so (13) holds as an equality with $h = n - 1 \geq 1$. For the details and analysis of this model see Sect. 12 in [5], as well as [15, 22].

4.1 The Föllmer "Exit Measure"

Under appropriate "canonical" conditions on the filtered measurable space $(\Omega, \mathscr{F})$, $\mathbb{F} = \{\mathscr{F}(t)\}_{0 \leq t < \infty}$ there exists on it a probability measure $\widehat{\mathbb{Q}}$, the so-called *Föllmer exit measure*, with respect to which the original probability measure $\mathbb{P}$ is absolutely continuous and which has the following properties: The process

$$\widehat{W}(t) := W(t) + \int_0^t \vartheta(s)\,ds, \quad 0 \leq t < \infty,$$

whose differential appears in (6), is a $\widehat{\mathbb{Q}}$-Brownian motion; the exponential process

$$\Lambda(t) := \exp\left\{ \int_0^t \vartheta'(s)\,d\widehat{W}(s) - \frac{1}{2}\int_0^t \|\vartheta(s)\|^2\,ds \right\}, \quad 0 \leq t < \infty \quad (14)$$

is a $\widehat{\mathbb{Q}}$-martingale, indeed we have $\mathbb{P}(A) = \int_A \Lambda(t)\,d\widehat{\mathbb{Q}}$ for $A \in \mathscr{F}(t)$; whereas the vector $\mathscr{Z}(\cdot) = (Z_1(\cdot), \ldots, Z_n(\cdot))'$ of relative market weights is a $\widehat{\mathbb{Q}}$-martingale and a Markov process, with values in Δ and "purely diffusive" dynamics of the form

$$d\mathscr{Z}(t) = \mathrm{s}(\mathscr{Z}(t))\,d\widehat{W}(t), \quad \mathscr{Z}(0) = \mathbf{z} \in \Delta^o. \quad (15)$$

In particular, viewed as a portfolio, $\mathscr{Z}(\cdot)$ has the numéraire property under $\widehat{\mathbb{Q}}$; i.e., we have $\mathscr{Z}(\cdot) \equiv \nu^{\widehat{\mathbb{Q}}}(\cdot)$ in the notation of Sect. 3.1. We shall let

$$\mathscr{T} := \inf\left\{ t \geq 0 : \mathscr{Z}(t) \in \Gamma \right\} = \min_{1 \leq i \leq n} \mathscr{T}_i, \quad \mathscr{T}_i := \inf\left\{ t \geq 0 : Z_i(t) = 0 \right\} \quad (16)$$

be the first time the process $\mathscr{Z}(\cdot)$ reaches the boundary of the simplex Δ, with the usual convention $\inf \emptyset = \infty$. We have then the representation

$$U(T, \mathbf{z}) = \widehat{\mathbb{Q}}^{\mathbf{z}}(\mathscr{T} > T), \quad (T, \mathbf{z}) \in (0, \infty) \times \Delta^o \quad (17)$$

for the relative arbitrage function of (12), as the probability under the Föllmer measure (which we index here by the starting position $\mathscr{Z}(0) = \mathbf{z}$ of the diffusion in (15)) that $\mathscr{Z}(\cdot)$ has not reached the boundary Γ of the simplex by time $t = T$. Since each $Z_i(\cdot)$ is a nonnegative $\widehat{\mathbb{Q}}$-martingale, it is clear that on $\{\mathscr{T}_i < \infty\}$ we have $Z_i(\mathscr{T}_i + u) = 0$, $\forall u \geq 0$, $\widehat{\mathbb{Q}}$-a.s. Thus, the process $\mathscr{Z}(\cdot)$ stays $\widehat{\mathbb{Q}}$-a.s. on Γ once it gets there; and so on by induction, regarding the boundaries $\gamma = \eth \setminus \eth^o$ of lower-dimensional subsimplices $\eth$ of Γ.

For proofs of these claims we refer the reader to [5], as well as to [3, 4, 23, 25] and of course to the original work of Föllmer [9, 10]. It is also seen in [5] that $\mathscr{T} = \inf\{t \geq 0 : \Lambda(t) = 0\}$ holds $\widehat{\mathbb{Q}}$-a.s., and that the relative arbitrage function of (12) admits the representation

$$U(T, \mathbf{z}) = \mathbb{E}^{\mathbb{P}^{\mathbf{z}}}\left[\frac{1}{\Lambda(T)} \right], \quad (18)$$

where now we index by the starting position $\mathscr{Z}(0) = \mathbf{z}$ of the process $\mathscr{Z}(\cdot)$ in (6) the original probability measure $\mathbb{P}$. Under this measure, the *deflator* process

$$\frac{1}{\Lambda(\cdot)} = \exp\left\{ -\int_0^{\cdot} \vartheta'(t)\,dW(t) - \frac{1}{2}\int_0^{\cdot} \|\vartheta(t)\|^2\,dt \right\} \equiv \frac{q}{Y^{q,\nu^{\mathbb{P}}}(\cdot)} \quad (19)$$

appearing in (18) is, of course, a strictly positive local martingale and supermartingale; here $q > 0$ is a constant, and $\nu^{\mathbb{P}}(\cdot)$ the numéraire portfolio of Sect. 3.1.

It may be helpful to think of the passage from the original measure $\mathbb{P}$ to the Föllmer measure $\widehat{\mathbb{Q}}$, as a generalized Girsanov-like change of probability that "removes the drift" in (6), when the deflator process $1/\Lambda(\cdot)$ of (19) is a strict local martingale under $\mathbb{P}$, i.e., when $U(T, \mathbf{z}) < 1$. As $\Lambda(\cdot)$ can reach the origin with positive $\widehat{\mathbb{Q}}$-probability, this is in general *not* an equivalent change of measure. Nonetheless, the market weight process $\mathscr{X}(\cdot)$ is a $\widehat{\mathbb{Q}}$-martingale, so the Föllmer measure $\widehat{\mathbb{Q}}$ can be thought of as of a "martingale measure" for the model under consideration.

4.2 The Functionally-Generated Portfolio

It is shown in [5] that, under appropriate regularity conditions on the covariance structure $a(\cdot)$ and on the relative drift $\vartheta(\cdot)$, the function $U(\cdot, \cdot)$ defined in (12) is of class $\mathscr{C}^{1,2}$ on $(0, \infty) \times \Delta^o$ and satisfies on this domain the equation

$$D_\tau U(\tau, \mathbf{z}) = \frac{1}{2} \sum_{i=1}^{n} \sum_{j=1}^{n} a_{ij}(\mathbf{z}) \, D_{ij}^2 U(\tau, \mathbf{z}), \tag{20}$$

or equivalently $D_\tau U = \frac{1}{2}\mathrm{Tr}(a \, D^2 U)$; and that $U(\cdot, \cdot)$ is also the smallest nonnegative supersolution of this equation, subject to the initial condition

$$U(0, \mathbf{z}) \equiv 1, \quad \forall \, \mathbf{z} \in \Delta^o. \tag{21}$$

We shall use throughout the notation $D_\tau f = \partial f/\partial \tau$, $D_i f = \partial f/\partial x_i$, $Df = (\partial f/\partial x_1, \ldots, \partial f/\partial x_n)'$ for the gradient, and $D^2 f = (D_{ij}^2 f)_{1 \le i,j \le n}$ with $D_{ij}^2 f = \partial^2 f/\partial x_i \partial x_j$ for the Hessian. Let us consider the process

$$\widehat{Y}(t) := U\big(T - t, \mathscr{X}(t)\big) \, 1_{\{\mathscr{T} > t\}}, \quad 0 \le t \le T; \tag{22}$$

it satisfies $\widehat{Y}(0) = U(T, \mathbf{z})$ and $\widehat{Y}(T) = 1$, $\mathbb{P}$-a.s. An application of Itô's rule leads now, in conjunction with (20) and (6), to the $\mathbb{P}$-dynamics

$$d\widehat{Y}(t) = \widehat{Y}(t) \big(\widehat{\psi}(t)\big)' d\mathscr{X}(t) \tag{23}$$

as in (4), with

$$\widehat{\psi}_i(t) := \left(\frac{D_i U}{U}\right)\big(T - t, \mathscr{X}(t)\big) = D_i \log U\big(T - t, \mathscr{X}(t)\big), \quad i = 1, \ldots, n \tag{24}$$

and, via (5), to the *functionally-generated* (in the terminology of [6, p. 46]) portfolio

$$\widehat{\pi}_i(t) := Z_i(t) \Bigg[D_i \log U\big(T - t, \mathscr{X}(t)\big) + 1$$

$$- \sum_{j=1}^{n} Z_j(t) \, D_j \log U\big(T - t, \mathscr{X}(t)\big) \Bigg]. \tag{25}$$

If $U(T, \mathbf{z}) < 1$, this portfolio implements the optimal arbitrage under the original probability measure, in the sense that with $q = U(T, \mathbf{z})$ we have

$$\widehat{Y}(\cdot) \equiv Y^{q, \widehat{\pi}}(\cdot), \quad \mathbb{P}\text{-a.s.} \tag{26}$$

as in (2), (3). In particular, we see that the infimum in (12) is then attained.

5 Induced Drifts

One potentially intriguing aspect of the arbitrage function in (12), and of the portfolio (25) which is generated by it and implements the optimal arbitrage, is that they appear not to depend at all on the drift vector $s(\mathscr{L}(\cdot)) \vartheta(\cdot)$ of the Itô process $\mathscr{L}(\cdot)$ in (6). That is, as long as we know the covariance structure of (7), we can construct the function $U(\cdot, \cdot)$, at least in principle, as the smallest nonnegative (super)solution of the Cauchy problem (20), (21), and from $U(\cdot, \cdot)$ the portfolio $\widehat{\pi}(\cdot)$ of (25) that attains the prescribed amount of arbitrage. Even if we were given the drift as well, we would not, in general, be able to improve upon this optimal arbitrage.

Of course, the drift is not entirely irrelevant; we have assumed, for one, that it satisfies certain regularity conditions. More significantly, we have posited that under the measure $\mathbb{P}$ the market weight process $\mathscr{L}(\cdot)$ never hits the boundary $\varGamma$ of the simplex; this assumption alone places non-trivial restrictions on the drift term or, equivalently, on the "relative market price of risk" process $\vartheta(\cdot)$ of (6); cf. [13, Chaps. 9 and 11].

5.1 Conditioning

One way to construct such a drift is as follows. We reverse the above procedure and *start* with a probability measure $\widehat{\mathbb{Q}}$ on the canonical filtered measurable space $(\varOmega, \mathscr{F})$, $\mathbb{F} = \{\mathscr{F}(t)\}_{0 \le t < \infty}$, under which the relative weight process $\mathscr{L}(\cdot)$ is a Markovian diffusion and a martingale on the state space Δ with dynamics as in (15), and with $\widehat{W}(\cdot)$ an n-dimensional Brownian motion. For the given time-horizon $[0, T]$, we construct then a *new* probability measure $\mathbb{P}_\star$ on $\mathscr{F}(T)$ via the recipe

$$\mathbb{P}_\star(A) := \widehat{\mathbb{Q}}\left(A \mid \mathscr{T} > T\right), \quad A \in \mathscr{F}(T) \tag{27}$$

in the notation of (16). In other words, the distribution of the market weight process $\mathscr{L}(\cdot)$ under $\mathbb{P}_\star$ on $\mathscr{F}(T)$ is the same as the conditional distribution under $\widehat{\mathbb{Q}}$ of the process $\mathscr{L}(\cdot)$, conditioned on the event that this process has not reached the boundary $\varGamma$ of the simplex by time T.

An elementary computation gives

$$\frac{d\mathbb{P}_\star}{d\widehat{\mathbb{Q}}}\bigg|_{\mathscr{F}(t)} = \frac{U\left(T - t, \mathscr{L}(t)\right)}{U(T, \mathbf{z})} 1_{\{\mathscr{T} > t\}} = \frac{\widehat{Y}(t)}{\widehat{Y}(0)} =: \Lambda^{\mathbb{P}_\star}(t), \quad 0 \le t \le T, \tag{28}$$

$\widehat{\mathbb{Q}}$-a.s. In particular, the process $\widehat{Y}(\cdot)$ of (22) is a $\widehat{\mathbb{Q}}$-martingale, so the representation (17) follows then from optional sampling and (21).

Repeating the steps of our analysis in Sect. 4.2, now in reverse, we see that the $\widehat{\mathbb{Q}}$-martingale $\Lambda^{\mathbb{P}_\star}(\cdot)$ of (28) can be cast in the manner of (14), with $\vartheta(\cdot)$ replaced by the n-dimensional process

$$\vartheta^{\mathbb{P}_\star}(\cdot) = \Theta_\star\big(T - \cdot, \mathscr{L}(\cdot)\big), \quad \text{where } \Theta_\star(\tau, \mathbf{z}) := \big(\mathbf{s}(\mathbf{z})\big)' D \log U(\tau, \mathbf{z}). \quad (29)$$

Moreover, (6) takes now the form of an *MMW model*

$$d\mathscr{L}(t) = \mathbf{a}\big(\mathscr{L}(t)\big) D \log U\big(T - t, \mathscr{L}(t)\big)dt + \mathbf{s}\big(\mathscr{L}(t)\big) dW^{\mathbb{P}_\star}(t),$$
$$\mathscr{L}(0) = \mathbf{z} \in \Delta^o \qquad\qquad\qquad (30)$$

with $W^{\mathbb{P}_\star}(\cdot) := \widehat{W}(\cdot) - \int_0^\cdot \vartheta^{\mathbb{P}_\star}(t) \, dt$ an n-dimensional $\mathbb{P}_\star$-Brownian motion.

5.2 Numéraire and Log-Optimality Properties of $\widehat{\pi}(\cdot)$

It is clear from (29), (30), (10) and (24) that the portfolio $\widehat{\pi}(\cdot)$ of (25) has the numéraire and log-optimality properties *under the probability measure* $\mathbb{P}_\star$ of (27); i.e.,

the process $Y^{q,\pi}(t)/Y^{q,\widehat{\pi}}(t)$, $0 \le t \le T$ is a $\mathbb{P}_\star$-supermartingale, and

$$\mathbb{E}^{\mathbb{P}_\star}\left[\log \frac{Y^{q,\pi}(t)}{q}\right] \le \mathbb{E}^{\mathbb{P}_\star}\left[\log \frac{Y^{q,\widehat{\pi}}(t)}{q}\right] = \frac{1}{2} \mathbb{E}^{\mathbb{P}_\star} \int_0^t \|\vartheta^{\mathbb{P}_\star}(u)\|^2 \, du \quad (31)$$

holds for $q > 0$, $t \in [0, T]$ on the strength of (11), for every $\pi(\cdot) \in \Pi$. In the notation of Sect. 3.1, we can thus write $\widehat{\pi}(\cdot) \equiv v^{\mathbb{P}_\star}(\cdot)$.

We may characterize the strategy that achieves optimal arbitrage over $[0, T]$ thus: *Start with a probability measure* $\widehat{\mathbb{Q}}$ *that generates* driftless, Markovian *market weights as in* (15); *then construct the measure* $\mathbb{P}_\star$ *by conditioning* $\widehat{\mathbb{Q}}$ *on the event* $\{\mathscr{T} > T\}$; *finally, find a portfolio* $\widehat{\pi}(\cdot)$ *that maximizes expected logarithmic relative returns under* $\mathbb{P}_\star$.

Let us also note that, in the context of the MMW model (30) with $\vartheta(\cdot) \equiv \vartheta^{\mathbb{P}_\star}(\cdot)$ as in (29), the representation (18), (19) can be cast on the strength of (28), (26) as

$$U(T, \mathbf{z}) = \frac{1}{\Lambda^{\mathbb{P}_\star}(T)} = \frac{q}{Y^{q,\widehat{\pi}}(T)}, \quad \mathbb{P}_\star\text{-a.s.} \qquad (32)$$

6 Relative Entropy

The definitions in (14) and (28) suggest a re-writing of (31) for $t = T$, $q > 0$ as

$$\mathbb{E}^{\mathbb{P}_\star}\left[\log \frac{Y^{q,\pi}(T)}{q}\right] \le \mathbb{E}^{\mathbb{P}_\star}\left[\log \frac{Y^{q,\widehat{\pi}}(T)}{q}\right] = \mathbb{E}^{\mathbb{P}_\star}\big[\log \Lambda^{\mathbb{P}_\star}(T)\big],$$
$$\forall \, \pi(\cdot) \in \Pi, \qquad\qquad\qquad (33)$$

and thence a further interpretation of the arbitrage function and of the portfolio $\widehat{\pi}(\cdot)$ generated by it, this time in terms of *relative entropy*. In particular, we claim that with $L := \log(1/U)$ we have the string of equalities

$$
\begin{aligned}
H_T(\mathbb{P}_\star | \widehat{\mathbb{Q}}) &:= \mathbb{E}^{\mathbb{P}_\star}\Big[\log\big((d\mathbb{P}_\star/d\widehat{\mathbb{Q}})|_{\mathscr{F}(T)}\big)\Big] \\
&= \mathbb{E}^{\mathbb{P}_\star}\big[\log \Lambda^{\mathbb{P}_\star}(T)\big] = \frac{1}{2}\,\mathbb{E}^{\mathbb{P}_\star}\int_0^T \|\vartheta^{\,\mathbb{P}_\star}(t)\|^2 dt \\
&= \frac{1}{2}\,\mathbb{E}^{\mathbb{P}_\star}\int_0^T \big(DL(T-t,\mathscr{L}(t))\big)'a\big(\mathscr{L}(t)\big)\,DL\big(T-t,\mathscr{L}(t)\big)dt \\
&= \log\big(1/U(T,\mathbf{z})\big).
\end{aligned}
\tag{34}
$$

Proof of (34), (33) Let us start by observing from the Cauchy problem of (20), (21) that the function $L = \log(1/U)$ satisfies the semilinear *Schrödinger-type equation*

$$
D_\tau L(\tau,\mathbf{z}) = \frac{1}{2}\sum_{i=1}^n \sum_{j=1}^n a_{ij}(\mathbf{z})\big(D_{ij}^2 L - D_i L \cdot D_j L\big)(\tau,\mathbf{z})
\tag{35}
$$

or equivalently $D_\tau L = \frac{1}{2}\mathrm{Tr}(a\,D^2 L) - \frac{1}{2}(DL)'a\,(DL)$ on $(0,\infty)\times\Delta^o$, and the initial condition $L(0,\cdot)\equiv 0$ on Δ^o.

In conjunction with (30) and the notation of (29), this implies that the process

$$
\frac{1}{2}\int_0^\cdot \big\|\Theta_\star(T-t,\mathscr{L}(t))\big\|^2 dt + L\big(T-\cdot,\mathscr{L}(\cdot)\big) = L(T,\mathbf{z}) - M_\star(\cdot),
\tag{36}
$$

with $M_\star(\cdot) := \int_0^\cdot(\Theta_\star(T-t,\mathscr{L}(t)))'\,dW^{\mathbb{P}_\star}(t)$, is a nonnegative $\mathbb{P}_\star$-local martingale, thus a $\mathbb{P}_\star$-supermartingale; in particular,

$$
\mathbb{E}^{\mathbb{P}_\star}\int_0^T \big\|\Theta_\star(T-t,\mathscr{L}(t))\big\|^2 dt \le 2\,L(T,\mathbf{z}).
\tag{37}
$$

We have used here the identity $L(0,\cdot)\equiv 0$ on Δ^o, and the fact that the market weight process $\mathscr{L}(\cdot)$ takes values in Δ^o, $\mathbb{P}_\star$-a.s.

The quadratic variation of the $\mathbb{P}_\star$-local martingale $M_\star(\cdot)$ of (36) is given by $\langle M_\star\rangle(\cdot) = \int_0^\cdot \|\Theta_\star(T-t,\mathscr{L}(t))\|^2 dt$, so (37) implies that $M_\star(\cdot)$ is actually a (square-integrable) martingale under $\mathbb{P}_\star$. In particular, (37) holds as equality and the last equation in (34) holds.

The only other claim in (34) that needs discussion, is its third equality; but now this follows from the work of H. Föllmer, namely, Proposition 2.11 in [12] and Lemma 2.6 in [11]. As for (33), this is now just a restatement of (31). $\qquad\square$

6.1 Stochastic Control

Emboldened by all this, let us consider on the filtered measurable space $(\Omega,\mathscr{F})$, $\mathbb{F} = \{\mathscr{F}(t)\}_{0\le t\le T}$ the collection $\mathfrak{P}$ of all probability measures that satisfy $\mathbb{P}\ll\widehat{\mathbb{Q}}$ on $\mathscr{F}(T)$ and $\mathbb{P}(\mathscr{L}(t)\in\Delta^o,\ \forall\ 0\le t\le T) = 1$.

The measure $\mathbb{P}_\star$ of Sect. 5.1 belongs to $\mathfrak{P}$, as does the generic probability measure of our model introduced in Sect. 3. It is not hard to see that $\mathfrak{P}$ consists of precisely those probability measures $\mathbb{P}$ on $(\Omega, \mathscr{F})$ which are absolutely continuous with respect to $\mathbb{P}_\star$ on $\mathscr{F}(T)$. The elements of $\mathfrak{P}$ are absolutely continuous with respect to $\widehat{\mathbb{Q}}$ but, in general, *not* equivalent: the case $\widehat{\mathbb{Q}}^z(\mathscr{X}(t) \in \Delta^o, \ \forall \ 0 \le t \le T) = U(T, \mathbf{z}) < 1$ is the most interesting.

For every $\mathbb{P} \in \mathfrak{P}$, the nonnegative $\widehat{\mathbb{Q}}$-martingale $(d\mathbb{P}/d\widehat{\mathbb{Q}})|_{\mathscr{F}(t)} = \Lambda^{\mathbb{P}}(t)$, $0 \le t \le T$ admits a representation of the form (14) for an appropriate n-dimensional process $\vartheta(\cdot) \equiv \vartheta^{\mathbb{P}}(\cdot)$ which: is $\mathbb{F}$-progressively measurable; is square-integrable on $[0, T]$, $\mathbb{P}$-a.s.; and satisfies $\int_0^T \|\vartheta(t)\|^2 dt < \int_0^{\mathscr{T}} \|\vartheta(t)\|^2 dt = \infty$, $\widehat{\mathbb{Q}}$-a.s. on $\{\mathscr{T} > T\}$. The vector $\mathscr{X}(\cdot)$ of relative market weights is under $\mathbb{P}$ an Itô process of the form $d\mathscr{X}(t) = \mathrm{s}(\mathscr{X}(t))(dW^{\mathbb{P}}(t) + \vartheta^{\mathbb{P}}(t)\,dt)$, $\mathscr{X}(0) = \mathbf{z} \in \Delta^o$ as in (6), with values in Δ^o and $W^{\mathbb{P}}(\cdot) \equiv W(\cdot)$ an n-dimensional $\mathbb{P}$-Brownian motion. Finally, just as before, the relative entropy of $\mathbb{P}$ with respect to the probability measure $\widehat{\mathbb{Q}}$ on $\mathscr{F}(T)$ is given as

$$H_T(\mathbb{P} \,|\, \widehat{\mathbb{Q}}) := \mathbb{E}^{\mathbb{P}}\Big[\log\big((d\mathbb{P}/d\widehat{\mathbb{Q}})|_{\mathscr{F}(T)}\big)\Big] = \mathbb{E}^{\mathbb{P}}\big[\log \Lambda^{\mathbb{P}}(T)\big]$$

$$= \frac{1}{2}\,\mathbb{E}^{\mathbb{P}}\int_0^T \|\vartheta^{\mathbb{P}}(t)\|^2 dt\,. \tag{38}$$

The crucial next step is a simple observation that goes back at least to Holland [17] and Fleming [8], namely, that the Schrödinger equation (35) can be cast in the *Hamilton-Jacobi-Bellman (HJB)* form

$$D_\tau L = \frac{1}{2}\,\mathrm{Tr}\big(\mathbf{a}\,D^2 L\big) + \min_{\theta \in \mathbb{R}^n}\Big[(DL)'\mathrm{s}\theta + \frac{1}{2}\,\|\theta\|^2\Big];$$

and that the minimization is attained at $\theta_\star = -\mathrm{s}'DL$, just as in (29).

Of course, every time an HJB equation makes its appearance, an associated stochastic control problem cannot lag very far behind. In our context, this problem amounts to minimizing, over all probability measures $\mathbb{P} \in \mathfrak{P}$, the "total energy" $(1/2)\,\mathbb{E}^{\mathbb{P}}\int_0^T \|\vartheta^{\mathbb{P}}(t)\|^2 dt$ during $[0, T]$, of the drift term in (6) that keeps the relative weight process $\mathscr{X}(\cdot)$ inside of Δ^o during this time-horizon. Equivalently, from (38), this amounts to finding a probability measure that minimizes the relative entropy $H_T(\mathbb{P} \,|\, \widehat{\mathbb{Q}})$ over all $\mathbb{P} \in \mathfrak{P}$. The answer to (both incarnations of) this question should now be obvious; it is the subject of the proposition that follows.

It is instructive to consider first the subclass $\mathfrak{P}^\dagger \subseteq \mathfrak{P}$ of probability measures

$$\mathfrak{P}^\dagger := \left\{ \mathbb{P} \in \mathfrak{P} \ : \ \mathbb{E}^{\mathbb{P}}\left(\int_0^T \big\|\Theta_\star(T - t, \mathscr{X}(t))\big\|^2 dt\right)^{1/2} < \infty \right\}. \tag{39}$$

Clearly $\mathbb{P}_\star$ belongs to $\mathfrak{P}^\dagger$, because of the property $\mathbb{E}^{\mathbb{P}_\star}\int_0^T \|\Theta_\star(T-t, \mathscr{X}(t))\|^2 dt < \infty$ from (37). This implies, furthermore, that every probability measure $\mathbb{P}$ with $\mathbb{P} \ll \mathbb{P}_\star$ on $\mathscr{F}(T)$ and $\mathbb{E}^{\mathbb{P}_\star}[((d\mathbb{P}/d\mathbb{P}_\star)|_{\mathscr{F}(T)})^2] < \infty$ also belongs to $\mathfrak{P}^\dagger$.

Let us observe that $\mathbb{P}_\star$ minimizes the "total energy" $\frac{1}{2}\,\mathbb{E}^{\mathbb{P}}\int_0^T \|\vartheta^{\mathbb{P}}(t)\|^2 dt$ over all probability measures in $\mathfrak{P}^\dagger$, i.e.,

$$\min_{\mathbb{P} \in \mathfrak{P}^\dagger} \frac{1}{2} \mathbb{E}^{\mathbb{P}} \int_0^T \| \vartheta^{\mathbb{P}}(t) \|^2 \, dt = \frac{1}{2} \mathbb{E}^{\mathbb{P}_*} \int_0^T \| \vartheta^{\mathbb{P}_*}(t) \|^2 \, dt$$

$$= \log \left(1 / U(T, \mathbf{z}) \right). \tag{40}$$

Proof of (40) This amounts to a standard verification argument, one part of which has already been established in the proof of (34). There, we obtained in (36) the $\mathbb{P}_*$-semimartingale decomposition of the process $L(T - \cdot, \mathscr{Z}(\cdot))$; now we obtain for this process its $\mathbb{P}$-semimartingale decomposition

$$\frac{1}{2} \int_0^{\cdot} \| \vartheta^{\mathbb{P}}(t) \|^2 dt + L\left(T - \cdot, \mathscr{Z}(\cdot)\right)$$

$$= L(T, \mathbf{z}) + \frac{1}{2} \int_0^{\cdot} \| \vartheta^{\mathbb{P}}(t) - \vartheta^{\mathbb{P}_*}(t) \|^2 \, dt - M(\cdot), \tag{41}$$

where the $\mathbb{P}$-local martingale $M(\cdot) := \int_0^{\cdot} (\vartheta^{\mathbb{P}^*}(t))' dW^{\mathbb{P}}(t)$ has quadratic variation $\langle M \rangle(\cdot) = \int_0^{\cdot} \| \Theta_*(T - t, \mathscr{Z}(t)) \|^2 \, dt$. But the condition $\mathbb{E}^{\mathbb{P}} (\sqrt{\langle M \rangle(T)}) < \infty$ of (39) implies that $M(\cdot)$ is $\mathbb{P}$-uniformly integrable, by the Burkholder-Davis-Gundy inequalities (e.g., Theorem 3.3.28 in [20]), thus a bona-fide $\mathbb{P}$-martingale (ibid., Problem 1.5.19(i)). This makes the right-hand side of (41) a $\mathbb{P}$-submartingale, and leads to the comparison $(1/2) \mathbb{E}^{\mathbb{P}} \int_0^T \| \vartheta^{\mathbb{P}}(t) \|^2 \, dt \geq L(T, \mathbf{z}) = - \log U(T, \mathbf{z})$, completing the verification argument.

We have used here yet again the identity $L(0, \cdot) \equiv 0$ on Δ^o in conjunction with the fact that $\mathscr{Z}(\cdot)$ takes values in Δ^o, $\mathbb{P}$-a.s. $\square$

It might well be possible to strengthen this argument, and show that $\mathfrak{P}^\dagger$ can be replaced in (40) by the larger set $\mathfrak{P}$. This would be the case if, for instance, one could show that the function $\Theta_*(\cdot, \cdot)$ of (29) is bounded uniformly over $[0, T] \times \Delta^o$; then $\mathfrak{P}^\dagger = \mathfrak{P}$ would follow from (39). Here is, however, a totally different and elementary argument based on first principles, that establishes this stronger result.

Proposition 1 *The measure $\mathbb{P}_*$ of (27) has the smallest relative entropy with respect to the Föllmer measure $\widehat{\mathbb{Q}}$ of Sect. 4.1, among all probability measures $\mathbb{P} \in \mathfrak{P}$; whereas its associated drift $\vartheta^{\mathbb{P}_*}(\cdot)$ in (29) keeps the process $\mathscr{Z}(\cdot)$ inside Δ^o during this time-horizon $[0, T]$ with the smallest amount of total energy. These two minima are equal, and*

$$\log \left(1 / U(T, \mathbf{z}) \right) = \min_{\mathbb{P} \in \mathfrak{P}} H_T(\mathbb{P} | \widehat{\mathbb{Q}}) = H_T(\mathbb{P}_* | \widehat{\mathbb{Q}})$$

$$= \min_{\mathbb{P} \in \mathfrak{P}^\dagger} \frac{1}{2} \mathbb{E}^{\mathbb{P}} \int_0^T \| \vartheta^{\mathbb{P}}(t) \|^2 \, dt = \frac{1}{2} \mathbb{E}^{\mathbb{P}_*} \int_0^T \| \vartheta^{\mathbb{P}_*}(t) \|^2 \, dt$$

$$= \mathbb{E}^{\mathbb{P}_*} \left[\log \frac{Y^{q, \widehat{\pi}}(T)}{q} \right]$$

$$= \max_{\pi(\cdot) \in \Pi} \mathbb{E}^{\mathbb{P}_*} \left[\log \frac{Y^{q, \pi}(T)}{q} \right], \quad q > 0. \tag{42}$$

Proof In light of (33) and (34), only the second equation needs justification. For an arbitrary $\mathbb{P} \in \mathfrak{P}$ we have $\mathbb{P}(\mathscr{T} > T) = 1$, so

$$H_T(\mathbb{P} \,|\, \widehat{\mathbb{Q}}) = \mathbb{E}^{\mathbb{P}}\Big[\log \big((d\mathbb{P}/d\widehat{\mathbb{Q}}) |_{\mathscr{F}(T)} \big) \Big] = \int_{\{\mathscr{T} > T\}} \log \big((d\mathbb{P}/d\widehat{\mathbb{Q}}) |_{\mathscr{F}(T)} \big) \, d\mathbb{P}.$$

But (28) gives $(d\mathbb{P}_\star / d\widehat{\mathbb{Q}})|_{\mathscr{F}(T)} = 1/U(T, \mathbf{z})$, $\widehat{\mathbb{Q}}$-a.s. on the event $\{\mathscr{T} > T\}$, thus

$$H_T(\mathbb{P} \,|\, \widehat{\mathbb{Q}}) = \int_{\{\mathscr{T} > T\}} \Big(\log (d\mathbb{P}/d\mathbb{P}_\star)|_{\mathscr{F}(T)} + \log \big(1/U(T, \mathbf{z}) \big) \Big) d\mathbb{P}$$

$$= H_T(\mathbb{P} \,|\, \mathbb{P}_\star) + \log \big(1/U(T, \mathbf{z}) \big) \geq \log \big(1/U(T, \mathbf{z}) \big) = H_T(\mathbb{P}_\star \,|\, \widehat{\mathbb{Q}}).$$

$\square$

According to this result, the quantity $\log(1/U(T, \mathbf{z}))$ – logarithm of the maximal relative amount by which the market portfolio can be outperformed over the horizon $[0, T]$, $\mathbb{P}$-a.s. for every $\mathbb{P} \in \mathfrak{P}$ – is also the maximal $\mathbb{P}_\star$-expected-log-outperformance of the market during $[0, T]$ that can be achieved over all $\pi(\cdot) \in \Pi$.

On the other hand, we have shown that the probability measure $\mathbb{P}_\star$ of (27) has the smallest relative entropy with respect to the Föllmer measure $\widehat{\mathbb{Q}}$, over the collection $\mathfrak{P}$; and that its associated drift $\vartheta^{\mathbb{P}_\star}(\cdot)$ keeps the relative weight process $\mathscr{Z}(\cdot)$ inside Δ^o with the smallest amount of total energy. For this reason we shall call $\mathbb{P}_\star$ *minimal energy measure* for the process $\mathscr{Z}(\cdot)$.

This terminology echoes that of Frittelli [14] (see also [16]), where the term "minimal entropy martingale measure" is used in a different context (the valuation of contingent claims in incomplete markets) and with an eye towards banishing arbitrage rather than implementing it in the best way when it exists. In that context, the Minimal Entropy *Martingale* (MEM) measure renders the underlying price process a martingale, which our $\mathbb{P}_\star$ in general does not.

6.2 Stochastic Game

Having come thus far, let us also note that (33) can be written as

$$\mathbb{E}^{\mathbb{P}_\star}\left[\log \frac{Y^{q,\pi}(T)}{q} \right] \leq \mathbb{E}^{\mathbb{P}_\star}\left[\log \frac{Y^{q,\widehat{\pi}}(T)}{q} \right]$$

$$= \mathbb{E}^{\mathbb{P}_\star}\big[\log \Lambda^{\mathbb{P}_\star}(T) \big] = H_T(\mathbb{P}_\star \,|\, \widehat{\mathbb{Q}}) = \log \big(1/U(T, \mathbf{z}) \big)$$

$$= \mathbb{E}^{\mathbb{P}}\left[\log \frac{Y^{q,\widehat{\pi}}(T)}{q} \right], \quad \forall \, (\mathbb{P}, \pi(\cdot)) \in \mathfrak{P} \times \Pi. \tag{43}$$

For this last equality we have used (28) in the form

$$\Lambda^{\mathbb{P}_\star}(T) = Y^{q,\widehat{\pi}}(T)/q = \widehat{Y}(T)/\widehat{Y}(0) = \big(U(0, \mathscr{Z}(T))/U(T, \mathbf{z}) \big) 1_{\{\mathscr{T} > T\}}, \quad \mathbb{P}\text{-a.s.}$$

and yet again (21), (26) as well as the fact $\mathbb{P}(\mathscr{L}(t) \in \Delta^o, \ \forall \ 0 \le t \le T) = 1$, to obtain that $\Lambda^{\mathbb{P}_*}(T) = Y^{q,\widehat{\pi}}(T)/q = 1/U(T, \mathbf{z})$ holds $\mathbb{P}$-a.s.

This gives also the following strengthening of (32), (18): for every $q > 0$, $\mathbb{P} \in \mathfrak{P}$ we have

$$U(T, \mathbf{z}) = \mathbb{E}^{\mathbb{P}}\left[\frac{1}{\Lambda^{\mathbb{P}}(T)}\right] = \mathbb{E}^{\mathbb{P}}\left[\frac{q}{Y^{q,\upsilon^{\mathbb{P}}}(\cdot)}\right] = \frac{1}{\Lambda^{\mathbb{P}_*}(T)}$$

$$= \frac{q}{Y^{q,\widehat{\pi}}(T)}, \quad \mathbb{P}\text{-a.s.} \tag{44}$$

- We conclude from (43) that the pair $(\mathbb{P}_*, \widehat{\pi}(\cdot))$ is a saddle point for the *Stochastic Game* with value

$$\log\left(1/U(T, \mathbf{z})\right) = \min_{\mathbb{P} \in \mathfrak{P}} \ \max_{\pi(\cdot) \in \Pi} \ \mathbb{E}^{\mathbb{P}}\left[\log \frac{Y^{q,\pi}(T)}{q}\right]$$

$$= \max_{\pi(\cdot) \in \Pi} \ \min_{\mathbb{P} \in \mathfrak{P}} \ \mathbb{E}^{\mathbb{P}}\left[\log \frac{Y^{q,\pi}(T)}{q}\right].$$

The "maximizer" in this zero-sum game is an investor who tries to outperform the market by his choice of portfolio $\pi(\cdot) \in \Pi$; whereas we can think of the "minimizer" as Nature, or the goddess *Tyche* herself, who tries to thwart the investor's efforts by choosing the probability measure $\mathbb{P} \in \mathfrak{P}$, or equivalently its induced drift $\vartheta^{\mathbb{P}}(\cdot) \equiv \vartheta(\cdot)$, to the investor's detriment.

7 Conclusion

We recapitulate and summarize by listing together in a single proposition the various interpretations of the arbitrage function we have discussed.

Proposition 2 *Consider a Markovian Market Weight model as in (8), and subject to the regularity assumptions of [5]. Then the relative arbitrage function*

$$U(T, \mathbf{z}) := \inf\left\{q > 0 : \exists \ \pi(\cdot) \in \Pi \ \text{ s.t. } \ \mathbb{P}\left(Y^{q,\pi}(T) \ge 1\right) = 1\right\},$$

$$(T, \mathbf{z}) \in (0, \infty) \times \Delta^o$$

of (12) admits the following representations, in terms of:

1. *The Cauchy problem of (20)–(21), as its smallest nonnegative classical solution;*
2. *The probability under the Föllmer exit measure $\widehat{\mathbb{Q}}$ (which makes the relative market weight process $\mathscr{L}(\cdot)$ a diffusion with zero drift, thus a martingale) that $\mathscr{L}(\cdot)$ has not reached the boundary of the simplex by time T, i.e.,*

$$U(T, \mathbf{z}) = \widehat{\mathbb{Q}}\left(\mathscr{L}(t) \in \Delta^o, \ \forall \ 0 \le t \le T\right);$$

3. *The expected log-relative-return of $\widehat{\pi}(\cdot) \equiv \upsilon^{\mathbb{P}_*}(\cdot)$, the numéraire portfolio under the probability measure $\mathbb{P}_*(\cdot) := \widehat{\mathbb{Q}}(\cdot \mid \mathscr{L}(t) \in \Delta^o, \ \forall \ 0 \le t \le T)$ of (27); this*

quantity is also the maximal $\mathbb{P}_\star$-expected log-relative-return over all portfolios, to wit, for $q > 0$,

$$\log\left(1/U(T, \mathbf{z})\right) = \mathbb{E}^{\mathbb{P}_\star}\left[\log\left(Y^{q,\widehat{\pi}}(T)/q\right)\right] = \max_{\pi(\cdot) \in \Pi} \mathbb{E}^{\mathbb{P}_\star}\left[\log\left(Y^{q,\pi}(T)/q\right)\right];$$

4. *The relative return of the portfolio $\widehat{\pi}(\cdot)$ in item 3, which is also the maximal achievable relative return over $[0, T]$, in the sense*

$$\frac{1}{U(T, \mathbf{z})} = \sup\left\{b > 0 : \exists\, \pi(\cdot) \in \Pi \quad \text{s.t.} \quad \mathbb{P}\left(Y^{q,\pi}(T) \geq q\, b\right) = 1\right\}$$

$$= \left(\mathbb{E}^{\mathbb{P}}\left[\left(\Lambda^{\mathbb{P}}(T)\right)^{-1}\right]\right)^{-1} = \left(\mathbb{E}^{\mathbb{P}}\left[\left(Y^{q,\nu^{\mathbb{P}}}(T)/q\right)^{-1}\right]\right)^{-1}$$

$$= \frac{Y^{q,\widehat{\pi}}(T)}{q}, \quad \mathbb{P}\text{-}a.s.;$$

5. *The value of the zero-sum stochastic game*

$$\log\left(1/U(T, \mathbf{z})\right) = \min_{\mathbb{P} \in \mathfrak{P}} \max_{\pi(\cdot) \in \Pi} \mathbb{E}^{\mathbb{P}}\left[\log\frac{Y^{q,\pi}(T)}{q}\right]$$

$$= \max_{\pi(\cdot) \in \Pi} \min_{\mathbb{P} \in \mathfrak{P}} \mathbb{E}^{\mathbb{P}}\left[\log\frac{Y^{q,\pi}(T)}{q}\right],$$

which has the pair $(\mathbb{P}_\star, \widehat{\pi}(\cdot))$ as saddle point in $\mathfrak{P} \times \Pi$;

6. *The relative entropy $H_T(\mathbb{P}_\star | \widehat{\mathbb{Q}})$ of $\mathbb{P}_\star$ with respect to the Föllmer exit measure, which is also the smallest such relative entropy attainable among probability measures in $\mathfrak{P}$, namely*

$$\log\left(1/U(T, \mathbf{z})\right) = H_T(\mathbb{P}_\star | \widehat{\mathbb{Q}}) = \min_{\mathbb{P} \in \mathfrak{P}} H_T(\mathbb{P} | \widehat{\mathbb{Q}}); \quad \text{as well as}$$

7. *The "total energy" $(1/2)\,\mathbb{E}^{\mathbb{P}_\star} \int_0^T \|\vartheta^{\mathbb{P}_\star}(t)\|^2\, dt$ expended over $[0, T]$ by the $\mathbb{P}_\star$-induced drift $\vartheta^{\mathbb{P}_\star}(\cdot)$ as in (29) to keep the relative market weight diffusion process $\mathscr{L}(\cdot)$ in the interior of the simplex, which is also the smallest such total energy attainable among probability measures in the set $\mathfrak{P}$, that is,*

$$\log\left(1/U(T, \mathbf{z})\right) = \frac{1}{2}\,\mathbb{E}^{\mathbb{P}_\star}\int_0^T \|\vartheta^{\mathbb{P}_\star}(t)\|^2\, dt = \min_{\mathbb{P} \in \mathfrak{P}} \frac{1}{2}\,\mathbb{E}^{\mathbb{P}}\int_0^T \|\vartheta^{\mathbb{P}}(t)\|^2\, dt.$$

In items (3)–(5) the portfolio $\widehat{\pi}(\cdot)$ is given by (25), and Π denotes the class of all portfolios; whereas in item (5) the set $\mathfrak{P}$ is the collection of all probability measures $\mathbb{P} \ll \widehat{\mathbb{Q}}$ which are absolutely continuous with respect to the Föllmer exit measure on $\mathscr{F}(T)$ and satisfy $\mathbb{P}(\mathscr{L}(t) \in \Delta^o, \forall\, 0 \leq t \leq T) = 1$.

Acknowledgements   Work supported in part by the National Science Foundation under Grant NSF-DMS-09-05754. We are grateful to Tomoyuki Ichiba and Johannes Ruf for their comments on an early version of this work.

References

1. Algoet, P., Cover, T.M.: Asymptotic optimality and asymptotic equipartition property of log-optimal investment. Ann. Probab. **16**, 876–898 (1988)
2. Cover, T.M., Thomas, J.A.: Elements of Information Theory, Second edn. Wiley, New York (2006)
3. Delbaen, F., Schachermayer, W.: The no-arbitrage property under a change of numéraire. Stoch. Stoch. Rep. **53**, 213–226 (1995)
4. Delbaen, F., Schachermayer, W.: Arbitrage possibilities in Bessel processes and their relations to local martingales. Probab. Theory Relat. Fields **102**, 357–366 (1995)
5. Fernholz, D., Karatzas, I.: On optimal arbitrage. Ann. Appl. Probab., to appear (2010)
6. Fernholz, E.R.: Stochastic Portfolio Theory. Springer, New York (2002)
7. Fernholz, E.R., Karatzas, I.: Stochastic portfolio theory: A survey. In: Bensoussan, A., Zhang, Q. (eds.) Mathematical Modeling and Numerical Methods in Finance. *Handbook of Numerical Analysis*, pp. 89–168. Elsevier, Amsterdam (2009)
8. Fleming, W.H.: Exit probabilities and optimal stochastic control. Appl. Math. Optim. **4**, 329–346 (1978)
9. Föllmer, H.: The exit measure of a supermartingale. Z. Wahrscheinlichkeitstheor. Verw. Geb. **21**, 154–166 (1972)
10. Föllmer, H.: On the representation of semimartingales. Ann. Probab. **1**, 580–589 (1973)
11. Föllmer, H.: An entropy approach to the time reversal of difusion processes. In: *Lecture Notes in Control and Information Systems*, vol. 69, pp. 156–163. Springer, New York (1985)
12. Föllmer, H.: Time reversal in Wiener space. In: *Lecture Notes in Mathematics*, vol. 1158, pp. 119–129. Springer, New York (1986)
13. Friedman, A.: Stochastic Differential Equations & Applications. Reprint of the 1975/76 edition, two volumes bound as one. Dover, New York (2006)
14. Frittelli, M.: The minimal entropy martingale measure and the valuation problem in incomplete markets. Math. Finance **10**, 39–52 (2000)
15. Goia, I.: Bessel and volatility-stabilized processes. Doctoral Dissertation, Columbia University (2009)
16. Grandits, P., Rheinländer, T.: On the minimal entropy martingale measure. Ann. Probab. **30**, 1003–1038 (2002)
17. Holland, C.J.: A new energy characterization of the smallest eigenvalue of the Schrödinger equation. Commun. Pure Appl. Math. **30**, 755–765 (1978)
18. Karatzas, I., Kardaras, C.: The numéraire portfolio in semimartingale models. Finance Stoch. **11**, 447–493 (2007)
19. Karatzas, I., Lehoczky, J.P., Shreve, S.E., Xu, G.-L.: Martingale and duality methods for utility maximization in an incomplete market. SIAM J. Control Optim. **29**, 707–730 (1991)
20. Karatzas, I., Shreve, S.E.: Brownian Motion and Stochastic Calculus, Second edn. Springer, New York (1991)
21. Long Jr., J.B.: The numéraire portfolio. J. Financ. Econ. **26**, 29–69 (1990)
22. Pal, S.: Analysis of market weights under volatility stabilized market models. Preprint, University of Washington, Seattle (2007)
23. Pal, S., Protter, Ph.: Strict local martingales, bubbles, and no early exercise. Preprint, Cornell University, Ithaca (2007)
24. Platen, E.A.: Benchmark approach to finance. Math. Finance **16**, 131–151 (2006)
25. Ruf, J.: Optimal trading strategies under arbitrage. Preprint, Columbia University, New York (2009)

Finitely Additive Probabilities and the Fundamental Theorem of Asset Pricing

Constantinos Kardaras

Abstract This work aims at a deeper understanding of the mathematical implications of the economically-sound condition of *absence of arbitrages of the first kind* in a financial market. In the spirit of the Fundamental Theorem of Asset Pricing (FTAP), it is shown here that the absence of arbitrages of the first kind in the market is equivalent to the existence of a finitely additive probability, weakly equivalent to the original and only locally countably additive, under which the discounted wealth processes become "local martingales". The aforementioned result is then used to obtain an independent proof of the classical FTAP, as it appears in Delbaen and Schachermayer (Math. Ann. 300:463–520, 1994). Finally, an elementary and short treatment of the previous discussion is presented for the case of continuous-path semimartingale asset-price processes.

1 Introduction

In the Quantitative Finance literature, the most common normative assumption placed on financial market models in the literature is the existence of an Equivalent Local Martingale Measure (ELMM), i.e., a probability, equivalent to the original one, that makes discounted asset-price processes local martingales. There is, of course, a very good reason for postulating the existence of an ELMM in the market: the Fundamental Theorem of Asset Pricing (FTAP) establishes[1] the equivalence between a precise market viability condition, coined "No Free Lunch with Vanishing Risk" (NFLVR) with the existence of an ELMM (see [7] and [9]).

C. Kardaras (✉)
Mathematics and Statistics Department, Boston University, 111 Cummington Street, Boston, MA 02215, USA
e-mail: kardaras@bu.edu

[1]At least in the case where asset-price processes are nonnegative semimartingales; see [9] for the case of general semimartingales, where σ-martingales, a generalization of local martingales, have to be utilized.

C. Chiarella, A. Novikov (eds.), *Contemporary Quantitative Finance*,
DOI 10.1007/978-3-642-03479-4_2, © Springer-Verlag Berlin Heidelberg 2010

The importance of condition NFLVR notwithstanding, there has lately been considerable interest in researching models where an ELMM might fail to exist. Major examples include the benchmark approach in financial modeling of [24], as well as the emergence of stochastic portfolio theory [10], a *descriptive* theory of financial markets. Even though the previous approaches allow for the existence of some form of arbitrage, they still deal with viable models of financial markets. In fact, the markets there satisfy a weaker version of the NFLVR condition; more precisely, there is *absence of arbitrages of the first kind*[2] (see Definition 1 of the present paper), which we abbreviate as condition NA_1. In the recent work [19], it was shown that condition NA_1 is equivalent to the existence of a *strictly positive local martingale deflator*, i.e., a strictly positive process with the property that every asset-price, when deflated by it, becomes a local martingale. The previous mathematical counterpart of the economic NA_1 condition is rather elegant; however, and in order to provide a closer comparison with the FTAP of [7], it is still natural to wish to equivalently express the NA_1 condition in terms of the existence of some measure that makes discounted asset-prices have some kind of martingale property.

In an effort to connect, expand, and simplify previous research, the purpose of this paper is threefold; in particular, we aim at:

1. presenting a weak version of the FTAP, stating the equivalence of the NA_1 condition with the existence of a "probability" that makes discounted nonnegative wealth processes "local martingales";
2. using the previous result as an intermediate step to obtain the FTAP as it appears in [7];
3. providing an elementary proof of the above weak version of the FTAP discussed in (1) above when the asset-prices are continuous-path semimartingales.

In order to tackle (1), we introduce the concept of a Weakly Equivalent Local Martingale Measure (WELMM). A WELMM is a *finitely additive* probability[3] that is *locally countably additive* and makes discounted asset-price processes behave like local martingales. Of course, the last local martingale property has to be carefully and rigorously defined, as only finitely additive probabilities are involved—see Definition 3 later on in the text. In Theorem 1, and in a general semimartingale market model, we obtain the equivalence between condition NA_1 and the existence of a WELMM.

Theorem 1 can be also seen as an intermediate step in proving the FTAP of [7]. Under the validity of Theorem 1, and using the very important optional decomposition theorem, this task becomes easier, as the proof of Theorem 2 of the present paper shows.

---

[2]The terminology "arbitrage of the first kind" was introduced in [14], although our definition, involving a limiting procedure, is closer in spirit to arbitrages of the first kind in the context of large financial markets, as appears in [15]. Here, one should also mention [22], where arbitrages of the first kind are called *cheap thrills*.

[3]Finitely additive measures have appeared quite often in economic theory in a financial equilibrium setting in cases of infinite horizon (see [13]) or even finite-time horizon with credit constraints on economic agents (see [22] and [23]).

We now come to the issue raised at (3) above. In order to establish our weak version of the FTAP, we need to invoke the main result from [19], which itself depends heavily upon results of [16]. The immense level of technicality in the proofs of the previous results render their presentation in graduate courses almost impossible. The same is true for the FTAP of [7]. Given the importance of such type of results, this is really discouraging. We provide here a partial resolution to this issue in the special case where the asset-prices are continuous-path semimartingales. As is shown in Theorem 4, proving of our main Theorem 1 becomes significantly easier; in fact, the only non-trivial result that is used in the course of the proof is the representation of a continuous-path local martingales as time-changed Brownian motion. Furthermore, in Theorem 4, condition NA_1 is shown to be equivalent to the existence and square-integrability of a risk-premium process, which has nice economic interpretation and can be easily checked once the model is specified.

The structure of the paper is as follows. In Sect. 2, the market is introduced, arbitrages of the first kind and the concept of a WELMM are defined, and Theorem 1, the weak version of the FTAP, is stated. Section 3 deals with a proof of the FTAP as it appears in [7]. Finally, Sect. 4 contains the statement and elementary proof of Theorem 4, which is a special case of Theorem 1 when the asset-price processes are continuous-path semimartingales.

2 Arbitrages of the First Kind and Weakly Equivalent Local Martingale Measures

2.1 General Probabilistic Remarks

All stochastic processes in the sequel are defined on a filtered probability space $(\Omega, \mathscr{F}, (\mathscr{F}_t)_{t \in \mathbb{R}_+}, \mathbb{P})$. Here, $\mathbb{P}$ is a probability on $(\Omega, \mathscr{F})$, where $\mathscr{F}$ is a σ-algebra that will make all involved random variables measurable. The filtration $(\mathscr{F}_t)_{t \in \mathbb{R}_+}$ is assumed to satisfy the usual hypotheses of right-continuity and saturation by $\mathbb{P}$-null sets. A finite financial planning horizon T will be assumed. Here, T is a $\mathbb{P}$-a.s. finite stopping time and all processes will be assumed to be constant, and equal to their value they have at T, after time T. It will be assumed throughout that $\mathscr{F}_0$ is trivial modulo $\mathbb{P}$ and that $\mathscr{F}_T = \mathscr{F}$.

2.2 The Market and Investing

Henceforth, S will be denoting the *discounted*, with respect to some baseline security, price process of a financial asset, satisfying:

$$S \text{ is a nonnegative semimartingale.} \tag{1}$$

Starting with capital $x \in \mathbb{R}_+$, and investing according to some predictable and S-integrable strategy ϑ, an economic agent's discounted wealth is given by the process

$$X^{x,\vartheta} := x + \int_0^\cdot \vartheta_t \, dS_t. \qquad (2)$$

In frictionless, continuous-time trading, credit constraints have to be imposed on investment in order to avoid doubling strategies. Define then $\mathscr{X}(x)$ to be the set of all wealth processes $X^{x,\vartheta}$ in the notation of (2) such that $X^{x,\vartheta} \geq 0$. Also, let $\mathscr{X} := \bigcup_{x \in \mathbb{R}_+} \mathscr{X}(x)$ denote the set of all nonnegative wealth processes.

2.3 Arbitrages of the First Kind

The market viability notion that will be introduced now will be of central importance in our discussion.

Definition 1 An $\mathscr{F}_T$-measurable random variable ξ will be called an *arbitrage of the first kind on* $[0, T]$ if $\mathbb{P}[\xi \geq 0] = 1$, $\mathbb{P}[\xi > 0] > 0$, and *for all $x > 0$ there exists* $X \in \mathscr{X}(x)$, *which may depend on x, such that* $\mathbb{P}[X_T \geq \xi] = 1$.

If there are no arbitrages of the first kind in the market, we say that condition NA_1 holds.

In view of Proposition 3.6 from [7], condition NA_1 is weaker than condition NFLVR. In fact, condition NA_1 is exactly the same as condition "No Unbounded Profit with Bounded Risk" (NUPBR) of [16], as we now show.

Proposition 1 *Condition NA_1 is equivalent to the requirement that the set of terminal outcomes starting from unit wealth $\{X_T \mid X \in \mathscr{X}(1)\}$ is bounded in probability.*

Proof Using the fact that $\mathscr{X}(x) = x\mathscr{X}(1)$ for all $x > 0$, it is straightforward to check that if an arbitrage of the first kind exists, then $\{X_T \mid X \in \mathscr{X}(1)\}$ is not bounded in probability. Conversely, assume that $\{X_T \mid X \in \mathscr{X}(1)\}$ is not bounded in probability. Since $\{X_T \mid X \in \mathscr{X}(1)\}$ is further convex, Lemma 2.3 of [4] implies the existence of $\Omega_u \in \mathscr{F}_T$ with $\mathbb{P}[\Omega_u] > 0$ such that, for all $n \in \mathbb{N}$, there exists $\widetilde{X}^n \in \mathscr{X}(1)$ with $\mathbb{P}[\{\widetilde{X}_T^n \leq n\} \cap \Omega_u] \leq \mathbb{P}[\Omega_u]/2^{n+1}$. For all $n \in \mathbb{N}$, let $A^n = \mathbb{I}_{\{\widetilde{X}_T^n > n\}} \cap \Omega_u \in \mathscr{F}_T$. Then, set $A := \bigcap_{n \in \mathbb{N}} A^n \in \mathscr{F}_T$ and $\xi := \mathbb{I}_A$. It is clear that ξ is $\mathscr{F}_T$-measurable and that $\mathbb{P}[\xi \geq 0] = 1$. Furthermore, since $A \subseteq \Omega_u$ and

$$\mathbb{P}\left[\Omega_u \setminus A\right] = \mathbb{P}\left[\bigcup_{n \in \mathbb{N}} (\Omega_u \setminus A^n)\right] \leq \sum_{n \in \mathbb{N}} \mathbb{P}\left[\Omega_u \setminus A^n\right] = \sum_{n \in \mathbb{N}} \mathbb{P}\left[\{\widetilde{X}_T^n \leq n\} \cap \Omega_u\right]$$

$$\leq \sum_{n \in \mathbb{N}} \frac{\mathbb{P}[\Omega_u]}{2^{n+1}} = \frac{\mathbb{P}[\Omega_u]}{2},$$

we obtain $\mathbb{P}[A] > 0$, i.e., $\mathbb{P}[\xi > 0] > 0$. For all $n \in \mathbb{N}$ set $X^n := (1/n)\widetilde{X}^n$, and observe that $X^n \in \mathscr{X}(1/n)$ and $\xi = \mathbb{I}_A \leq \mathbb{I}_{A^n} \leq X_T^n$ hold for all $n \in \mathbb{N}$. It follows that ξ is and arbitrage of the first kind, which finishes the proof. $\square$

2.4 Weakly Equivalent Local Martingale Measures

The mathematical counterpart of the economical NA_1 condition involves a weakening of the concept of an ELMM. The appropriate notion turns out to involve measures that behave like probabilities, but are *finitely additive* and only *locally countably additive*.

In what follows, a *localizing sequence* will refer to a *nondecreasing* sequence $(\tau^n)_{n \in \mathbb{N}}$ of stopping times such that $\uparrow \lim_{n \to \infty} \mathbb{P}[\tau^n \geq T] = 1$.

2.4.1 Local Probabilities Weakly Equivalent to $\mathbb{P}$

The concept that will be introduced below in Definition 2 is essentially a localization of countably additive probabilities.

Definition 2 A mapping $Q : \mathscr{F} \mapsto [0, 1]$ is a *local probability weakly equivalent to* $\mathbb{P}$ if:

1. $Q[\emptyset] = 0$, $Q[\Omega] = 1$, and Q is (finitely) additive: $Q[A \cup B] = Q[A] + Q[B]$ whenever $A \in \mathscr{F}$ and $B \in \mathscr{F}$ satisfy $A \cap B = \emptyset$;
2. for $A \in \mathscr{F}$, $\mathbb{P}[A] = 0$ implies $Q[A] = 0$;
3. there exists a localizing sequence $(\tau_n)_{n \in \mathbb{N}}$ such that, when restricted on $\mathscr{F}_{\tau_n}$, Q is countably additive and equivalent to $\mathbb{P}$, for all $n \in \mathbb{N}$. (Such sequence of stopping times will be called a *localizing sequence for* Q.)

Conditions (1) and (2) above imply that Q is a positive element of the dual of $\mathbb{L}^\infty$, the space of (equivalence classes modulo $\mathbb{P}$ of) $\mathscr{F}$-measurable random variable that are bounded modulo $\mathbb{P}$ equipped with the essential-sup norm. The theory of finitely additive measures is developed in great detail in [3]; for our purposes here, mostly results from the Appendix of [6], as well as some material from [18], will be needed.

To facilitate the understanding, finitely additive positive measures that are not necessarily countably additive will be denoted using sans-serif typeface (like "Q"), while for countably additive probabilities the blackboard bold typeface (like "$\mathbb{Q}$") will be used. As Q will be in the dual of $\mathbb{L}^\infty$, $\langle Q, \xi \rangle$ will denote the action of Q on $\xi \in \mathbb{L}^\infty$. The fact that Q is a positive functional enables to extend the definition of $\langle Q, \xi \rangle$ for $\xi \in \mathbb{L}^0$ with $\mathbb{P}[\xi \geq 0] = 1$, via $\langle Q, \xi \rangle := \lim_{n \to \infty} \langle Q, \xi \mathbb{I}_{\{\xi \leq n\}} \rangle \in [0, \infty]$. ($\mathbb{L}^0$ denotes the set of all $\mathbb{P}$-a.s. finitely-valued random variables modulo $\mathbb{P}$-equivalence equipped with the topology of convergence in probability.)

Remark 1 In general, a finitely additive probability $Q : \mathscr{F} \mapsto [0, 1]$ is called weakly absolutely continuous with respect to $\mathbb{P}$ if for each $A \in \mathscr{F}$ with $\mathbb{P}[A] = 0$ we have $Q[A] = 0$. Furthermore, Q is called strongly absolutely continuous with respect to $\mathbb{P}$ if for any $\varepsilon > 0$ there exists $\delta = \delta(\varepsilon) > 0$ such that $E \in \mathscr{F}$ and $\mathbb{P}[E] < \delta$ implies $Q[E] < \varepsilon$. It is clear that strong absolute continuity of Q with respect to $\mathbb{P}$ is a stronger requirement than weak absolutely continuity of Q with respect to $\mathbb{P}$.

Actually, the two notions coincide when Q is countable additive. Of course, similar definitions can be made with the roles of $\mathbb{P}$ and Q reversed. Then, $\mathbb{P}$ and Q are called weakly (respectively, strongly) equivalent if Q is weakly (respectively, strongly) absolutely continuous with respect to $\mathbb{P}$ and $\mathbb{P}$ is weakly (respectively, strongly) absolutely continuous with respect to Q.

In Definition 2, Q we called a local probability "weakly equivalent to $\mathbb{P}$"; however, condition (2) only implies that Q is weakly absolutely continuous with respect to $\mathbb{P}$. We claim that $\mathbb{P}$ is also weakly absolutely continuous with respect to Q. Indeed, let Q satisfy (1) and (3) of Definition 2. Pick any $A \in \mathscr{F}$ with $Q[A] = 0$. Since $A \cap \{\tau_n \geq T\} \in \mathscr{F}_{\tau_n}$ for all $n \in \mathbb{N}$, $Q[A \cap \{\tau_n \geq T\}] = 0$ implies that $\mathbb{P}[A \cap \{\tau_n \geq T\}] = 0$ by (3). Then, $\mathbb{P}[A] = \uparrow \lim_{n \to \infty} \mathbb{P}[A \cap \{\tau_n \geq T\}] = 0$.

Let Q be a local probability weakly equivalent to $\mathbb{P}$. When Q is only finitely, but not countably, additive, $\mathbb{P}$ and Q are not strongly equivalent, as we now explain. Write $Q = Q^r + Q^s$ for the unique decomposition of Q in its regular and singular part. (The regular part Q^r is countably additive, while the singular part Q^s is purely finitely additive, meaning that there is no nonzero countably additive measure that is dominated by Q^s. One can check [3] for more information.) According to Lemma A.1 in [6], for all $\varepsilon > 0$ one can find a set $A_\varepsilon \in \mathscr{F}$ with $\mathbb{P}[A_\varepsilon] < \varepsilon$ and $Q^s[A_\varepsilon] = Q^s[\Omega]$; therefore $Q[A_\varepsilon] \geq Q^s[\Omega]$. In other words, if Q^s is nontrivial, then Q is *not* strongly absolutely continuous with respect to $\mathbb{P}$. Note, however, that $\mathbb{P}$ *is* strongly absolutely continuous with respect to Q in view of condition (3) of Definition 2.

We briefly digress from our main topic to give a simple criterion that connects the countable additivity of Q, a local probability weakly equivalent to $\mathbb{P}$, with the strong equivalence between Q and $\mathbb{P}$, as the latter notion was introduced in Remark 1 above.

Proposition 2 *Let* Q *be a local probability weakly equivalent to* $\mathbb{P}$. *The following are equivalent*:

1. Q *is countably additive, i.e., a true probability.*
2. Q *is strongly absolutely continuous with respect to* $\mathbb{P}$.
3. $\uparrow \lim_{n \to \infty} Q[\tau^n \geq T] = 1$ *holds for any localizing sequence* $(\tau^n)_{n \in \mathbb{N}}$ *for* Q.
4. $\uparrow \lim_{n \to \infty} Q[\tau^n \geq T] = 1$ *holds for some localizing sequence* $(\tau^n)_{n \in \mathbb{N}}$ *for* Q.

Proof The implications (1) $\Rightarrow$ (2) $\Rightarrow$ (3) $\Rightarrow$ (4) are straightforward, so we only focus on the implication (4) $\Rightarrow$ (1). Let $(E^k)_{k \in \mathbb{N}}$ be a decreasing sequence of $\mathscr{F}$-measurable sets such that $\bigcap_{k \in \mathbb{N}} E^k = \emptyset$. We need show that $\downarrow \lim_{k \to \infty} Q[E^k] = 0$. Consider the Q-localizing sequence $(\tau^n)_{n \in \mathbb{N}}$ of statement (4). For each $n \in \mathbb{N}$ and $k \in \mathbb{N}$ we have $E^k \cap \{\tau^n \geq T\} \in \mathscr{F}_{\tau^n}$. (Here, remember that $\mathscr{F} = \mathscr{F}_T$.) This means that $\limsup_{k \to \infty} Q[E^k] \leq Q[\tau^n < T] + \limsup_{k \to \infty} Q[E^k \cap \{\tau^n \geq T\}] = Q[\tau^n < T]$, the last equality holding because Q is countably additive on $\mathscr{F}_{\tau^n}$, for all $n \in \mathbb{N}$. Sending n to infinity and using (4) we get the result. $\qquad\square$

2.4.2 Density Processes

For a local probability weakly equivalent to $\mathbb{P}$ as in Definition 2, one can associate a strictly positive local $\mathbb{P}$-martingale Y^Q, as will be now described. For all $n \in \mathbb{N}$, consider the $\mathbb{P}$-martingale $Y^{Q,n}$ defined by setting

$$Y^{Q,n}_\infty \equiv Y^{Q,n}_T := \frac{d(Q|_{\mathscr{F}_{\tau_n}})}{d(\mathbb{P}|_{\mathscr{F}_{\tau_n}})}.$$

It is clear that, $\mathbb{P}$-a.s., $Y^{Q,n}_0 = 1$ and $Y^{Q,n}_T > 0$. Furthermore, for all $n \in \mathbb{N} \setminus \{0\}$, $Y^{Q,n} = Y^{Q,n-1}$ on the stochastic interval $[\![0, \tau_{n-1}]\!]$. Therefore, patching the processes $(Y^{Q,n})_{n \in \mathbb{N}}$ together, one can define a local $\mathbb{P}$-martingale Y^Q such that, $\mathbb{P}$-a.s., $Y^Q_0 = 1$ and $Y^Q_T > 0$.

Remark 2 A general result in [18] shows that a *supermartingale* Y^Q can be associated to a finitely additive measure Q that satisfies (1) and (2) of Definition 2, but not necessarily (3). The construction of Y^Q in [18] is messier than the one provided above, exactly because condition (3) of Definition 2 is not assumed to hold. In the special case described here, the two constructions coincide.

A partial converse of the above construction is also possible. To wit, start with some local $\mathbb{P}$-martingale Y such that, $\mathbb{P}$-a.s., $Y_0 = 1$ and $Y_T > 0$. If $(\tau_n)_{n \in \mathbb{N}}$ is a localizing sequence for Y, one can define for each $n \in \mathbb{N}$ a probability $\mathbb{Q}^n$, equivalent to $\mathbb{P}$ on $\mathscr{F}$, via the recipe $d\mathbb{Q}^n := Y_{\tau_n} d\mathbb{P}$. By Alaoglu's Theorem (see, for example, Theorem 6.25, page 250 of [1]), the sequence $(\mathbb{Q}^n)_{n \in \mathbb{N}}$ has some cluster point Q for the weak* topology on the dual of $\mathbb{L}^\infty$, which will be a finitely-additive probability. Proposition A.1 of [6] gives that $d Q^r / d\mathbb{P} = Y_T$. It is easy to see that Q is a local probability weakly equivalent to $\mathbb{P}$, as well as that $Y^Q = Y$. (Note that, again by Proposition A.1 of [6], the sequence $(\mathbb{Q}^n)_{n \in \mathbb{N}}$ might have several cluster points, but all will have the same regular part. Therefore, Q is not uniquely defined, but it is always the case that $Y^Q = Y$.)

2.4.3 Local Martingales

When Q is a local probability weakly equivalent to $\mathbb{P}$ and fails to be countably additive, the concept of a Q-martingale, and therefore also of a local Q-martingale, is tricky to state. The reason is that existence of conditional expectations requires Q to be countably additive in order to invoke the Radon-Nikodým Theorem. To overcome this difficulty, we follow an alternative route. Let $\mathbb{Q}$ be a probability measure, equivalent to $\mathbb{P}$. According to the optional sampling theorem (see, for example, Sect. 1.3.C in [17]), a càdlàg process X is a local $\mathbb{Q}$-martingale if and only if there exists a localizing sequence $(\tau_n)_{n \in \mathbb{N}}$ such that $\langle \mathbb{Q}, X_{\tau^n \wedge \tau} \rangle = X_0$ for all $n \in \mathbb{N}$ and all stopping times τ. This characterization makes the following Definition 3 plausible.

Definition 3 Let Q be a local probability weakly equivalent to $\mathbb{P}$. A nonnegative càdlàg process X will be called a *local Q-martingale* if there exists a localizing sequence $(\tau_n)_{n \in \mathbb{N}}$ such that $\langle Q, X_{\tau^n \wedge \tau} \rangle = X_0$ for all $n \in \mathbb{N}$ and all stopping times τ.

Now, a characterization of local Q-martingales in terms of density processes will be given. This extends the analogous result in the case where Q is countably additive.

Proposition 3 *Let* Q *be a local probability weakly equivalent to* $\mathbb{P}$ *and let* Y^Q *be defined as in Sect. 2.4.2. A nonnegative process* X *is a local* Q*-martingale if and only if* $Y^Q X$ *is a local* $\mathbb{P}$*-martingale.*

Proof Start by assuming that X is a local Q-martingale. Since $\langle Q, X_{\tau^n \wedge \tau} \rangle = X_0$ for all $n \in \mathbb{N}$ and *all* stopping times τ, where $(\tau_n)_{n \in \mathbb{N}}$ is a localizing sequence, $(\tau_n)_{n \in \mathbb{N}}$ can be assumed to also localize Q. Then, since $X_{\tau^n \wedge \tau} \in \mathscr{F}_{\tau^n}$ for all $n \in \mathbb{N}$ and all stopping times τ, and since $\mathbb{Q}^n := Q|_{\mathscr{F}_{\tau^n}}$ is countably additive with $d\mathbb{Q}^n / (d\mathbb{P}|_{\mathscr{F}_{\tau^n}}) = Y^Q_{\tau^n}$, it follows that

$$Y^Q_0 X_0 = X_0 = \langle Q, X_{\tau^n \wedge \tau} \rangle = \mathbb{E}[Y^Q_{\tau^n} X_{\tau^n \wedge \tau}] = \mathbb{E}[\mathbb{E}[Y^Q_{\tau^n} \mid \mathscr{F}_{\tau^n \wedge \tau}] X_{\tau^n \wedge \tau}]$$
$$= \mathbb{E}[Y^Q_{\tau^n \wedge \tau} X_{\tau^n \wedge \tau}]$$

for all $n \in \mathbb{N}$ and all stopping times τ. This means that $Y^Q X$ is a local $\mathbb{P}$-martingale.

Conversely, suppose that $Y^Q X$ is a local $\mathbb{P}$-martingale. Let $(\tau_n)_{n \in \mathbb{N}}$ be a localizing sequence for both $Y^Q X$ and Q. Then, for all $n \in \mathbb{N}$ and all stopping times τ,

$$X_0 = Y^Q_0 X_0 = \mathbb{E}[Y^Q_{\tau^n \wedge \tau} X_{\tau^n \wedge \tau}] = \mathbb{E}[\mathbb{E}[Y^Q_{\tau^n} \mid \mathscr{F}_{\tau^n \wedge \tau}] X_{\tau^n \wedge \tau}] = \mathbb{E}[Y^Q_{\tau^n} X_{\tau^n \wedge \tau}]$$
$$= \langle Q, X_{\tau^n \wedge \tau} \rangle.$$

Therefore, X is a local Q-martingale.   □

2.4.4 Weakly Equivalent Local Martingale Measures

As will be shown in Theorem 1, the following definition gives the mathematical counterpart of the market viability condition NA_1.

Definition 4 A *weakly equivalent local martingale measure* (WELMM) Q is a local probability weakly equivalent to $\mathbb{P}$ such that S is a local Q-martingale.

Remark 3 (On the semimartingale property of S) Under the assumption that S is nonnegative, the existence of a WELMM *enforces* the semimartingale property on S. Indeed, write $S = (1/Y^Q)(Y^Q S)$, where Q is a WELMM and Y^Q is the density defined in Sect. 2.4.2. Since Y^Q is a local $\mathbb{P}$-martingale with $Y^Q_T > 0$, $\mathbb{P}$-a.s., and $Y^Q S$ is also a local $\mathbb{P}$-martingale, both $1/Y^Q$ and $Y^Q S$ are semimartingales, which gives that S is a semimartingale.

Semimartingales are essential in frictionless financial modeling. This has been made clear in Theorem 7.1 of [7], where it was shown that if S is locally bounded and *not* a semimartingale, condition NFLVR using only simple trading strategies fails. Furthermore, from the treatment in [20] it follows that, if S is nonnegative and *not* a semimartingale, one can construct an arbitrage of the first kind, *even* if one uses only *no-short-sale* and *simple* strategies.

If S satisfies (1), it is straightforward to check that a probability $\mathbb{Q}$ equivalent to $\mathbb{P}$ is an ELMM if and only if each $X \in \mathscr{X}$ is a local Q-martingale. The following result extends the last equivalence in the case of a WELMM.

Proposition 4 *Let* Q *be a local probability weakly equivalent to* $\mathbb{P}$. *If* S *satisfies* (1), *then* S *is a local* Q-*martingale if and only if every process* $X \in \mathscr{X}$ *is a local* Q-*martingale.*

Proof Start by assuming that S is a local Q-martingale. For $x \in \mathbb{R}_+$, let $X^{x,\vartheta}$ in the notation of (2) be a wealth process in $\mathscr{X}(x)$. A use of the integration-by-parts formula gives

$$Y^Q X^{x,\vartheta} = x + \int_0^{\cdot} \left(X_{t-}^{x,\vartheta} - \vartheta_t S_{t-} \right) dY_t^Q + \int_0^{\cdot} \vartheta_t \, d(Y^Q S)_t.$$

It follows that $Y^Q X^{x,\vartheta}$ is a positive martingale transform under $\mathbb{P}$, and therefore a local $\mathbb{P}$-martingale by the Ansel-Stricker Theorem (see [2]).

Now, assume that every process in $\mathscr{X}$ is a local Q-martingale. Since $S \in \mathscr{X}$, S is a local Q-martingale. □

Remark 4 Let Q be a local probability weakly equivalent to $\mathbb{P}$. Proposition 3 combined with Proposition 4 imply that Q is a WELMM if and only if $Y^Q X$ is a local $\mathbb{P}$-martingale for all $X \in \mathscr{X}$. In other words, the process Y^Q is a *strict martingale density* in the terminology of [25].

2.5 The Main Result

After the preparation of the previous sections, it is possible to state Theorem 1 below, which can be seen as a weak version of the FTAP in [7].

Theorem 1 *Suppose that* S *satisfies* (1). *Then, there are no arbitrages of the first kind in the market if and only if a weakly equivalent local martingale measure exists.*

Proof By Theorem 1.1 in [19], condition NA$_1$ is equivalent to the existence of a nonnegative càdlàg process Y with $Y_0 = 1$, $Y_T > 0$, and such that YX is a local $\mathbb{P}$-martingale for all $X \in \mathscr{X}$. Then, using also the discussion in Sect. 2.4.2 and Proposition 3, NA$_1$ holds if and only if there exists a local probability Q, weakly equivalent to $\mathbb{P}$, such that X is a local Q-martingale for all $X \in \mathscr{X}$. Proposition 4 gives that Q is a WELMM, which completes the proof. □

Remark 5 *If* the statement of the FTAP of [9] is assumed, one can provide a proof of Theorem 1 using the "change of numéraire" technique of [8]; a similar approach has been taken up in [5]. We opt here to prove Theorem 1 *directly*, using the result of [19] that is not relying on previous heavy results. Then, the classical FTAP itself becomes a corollary, as we shall see in Sect. 3 below. There is no claim that the path followed here is shorter or less arduous than the one taken up in [9], but certainly it has different focus.

Remark 6 As can be seen from its proof, Theorem 1 still holds if the nonnegativity assumption on S is removed, as long as we agree to reformulate the notion of a WELMM Q, asking that each $X \in \mathscr{X}$ is a local Q-martingale.

Furthermore, Theorem 1 holds without the assumption that S is one-dimensional. Indeed, in Remark 5 above it was discussed that Theorem 1 can be seen as a consequence of the FTAP in [9], which does not require S to be one-dimensional. Unfortunately, in [19] the assumption that S is one-dimensional is being made, mostly in order to avoid immense technical difficulties in the proof of Theorem 1.1 there, which is used to prove Theorem 1 above.

Remark 7 Undoubtedly, the notion of a WELMM is more complicated than that of an ELMM. However, checking the existence of a WELMM is fundamentally easier than checking whether an ELMM exists for the market. Indeed, in view of Theorem 1, existence of a WELMM is equivalent to the existence of the numéraire portfolio in the market. For checking the existence of the latter, there exists a necessary and sufficient criterion in terms of the predictable characteristics of the discounted asset-price process, as was shown in [16]. The details are rather technical, but if the asset-price process has continuous paths the situation is very simple, as will be discussed in Sect. 4 later.

3 The FTAP of Delbaen and Schachermayer

In this subsection, a proof of the FTAP as appears in [7] is given using the already-developed tools. Also, the Q-supermartingale property of wealth processes in $\mathscr{X}$ when Q is a WELMM is examined, and it is shown that the latter property holds only under the existence of an ELMM.

3.1 Proving the FTAP

In the notation of the present paper, the main technical difficulty for proving the FTAP in [7] is showing that the set $\{g \in \mathbb{L}^0 \mid 0 \leq g \leq X_T \text{ for some } X \in \mathscr{X}(1)\}$ is closed in probability under the NFLVR condition. This implies the weak* closedness of the set of bounded superhedgeable claims starting from zero capital and therefore allows for the use of the Kreps-Yan separation theorem (see [21] and [26]) in order to conclude the existence of a separating measure.

There is a way to establish the aforementioned closedness in probability using Theorem 1 and some additional well-known results. In fact, a seemingly stronger statement than the one in [7] will now be stated and proved.

Theorem 2 *If no arbitrages of the first kind are present in the market, the set* $\{g \in \mathbb{L}^0 \mid 0 \leq g \leq X_T$ *for some* $X \in \mathscr{X}(1)\}$ *is closed in probability.*

Proof Define $\mathscr{V}^{\downarrow}(1)$ to be the class of nonnegative, adapted, càdlàg, *nonincreasing* processes with $V_0 \leq 1$. Then, set[4]

$$\mathscr{X}^{\times\times}(1) := \mathscr{X}(1)\mathscr{V}^{\downarrow}(1) = \{XV \mid X \in \mathscr{X}(1) \text{ and } V \in \mathscr{V}^{\downarrow}(1)\}.$$

The statement of the Theorem can be reformulated to say that the *convex* set $\{\xi_T \mid \xi \in \mathscr{X}^{\times\times}(1)\}$ is closed in $\mathbb{L}^0$. Consider therefore a sequence $(\xi^n)_{n \in \mathbb{N}}$ such that $\mathbb{L}^0\text{-}\lim_{n \to \infty} \xi_T^n = \zeta$. It will be shown below that there exists $\xi^\infty \in \mathscr{X}^{\times\times}(1)$ such that $\xi_T^\infty = \zeta$.

In what follows in the proof, the concept of Fatou-convergence is used, which will now be recalled. Define $\mathbb{D} := \{k/2^m \mid k \in \mathbb{N}, m \in \mathbb{N}\}$ to be the set of dyadic rational numbers in $\mathbb{R}_+$. A sequence $(Z^n)_{n \in \mathbb{N}}$ of nonnegative càdlàg processes *Fatou-converges* to Z^∞ if

$$Z_t^\infty = \limsup_{\mathbb{D} \ni s \downarrow t} \left(\limsup_{n \to \infty} Z_s^n \right) = \liminf_{\mathbb{D} \ni s \downarrow t} \left(\liminf_{n \to \infty} Z_s^n \right)$$

holds $\mathbb{P}$-a.s. for all $t \in \mathbb{R}_+$. Note that, since all processes are assumed to be constant after time T, for any $t \geq T$ the above relationship simply reads $Z_T^\infty = \lim_{n \to \infty} Z_T^n$, $\mathbb{P}$-a.s.

From Theorem 1 and Proposition 3, under absence of arbitrages of the first kind in the market, there exists some nonnegative process $\overline{Y}$ with $\overline{Y}_0 = 1$ and $\overline{Y}_T > 0$, $\mathbb{P}$-a.s., such that $\overline{Y}X$ is a local $\mathbb{P}$-martingale for all $X \in \mathscr{X}(1)$. Then, $\overline{Y}\xi$ is a nonnegative $\mathbb{P}$-supermartingale for all $\xi \in \mathscr{X}^{\times\times}(1)$. Since $(\overline{Y}\xi^n)_{n \in \mathbb{N}}$ is a sequence of nonnegative $\mathbb{P}$-supermartingales with $\overline{Y}_0\xi_0^n \leq 1$, Lemma 5.2(1) of [12] gives the existence of a sequence $(\overline{\xi}^n)_{n \in \mathbb{N}}$ such that $\overline{\xi}^n$ is a convex combination of $\xi^n, \xi^{n+1}, \ldots$, for each $n \in \mathbb{N}$ (and, therefore, $\overline{\xi}^n \in \mathscr{X}^{\times\times}(1)$ for all $n \in \mathbb{N}$, since $\mathscr{X}^{\times\times}(1)$ is convex), and such that $(\overline{Y}\overline{\xi}^n)_{n \in \mathbb{N}}$ Fatou-converges to some nonnegative $\mathbb{P}$-supermartingale Z. Obviously, $Z_0 \leq 1$. Also, since $\mathbb{L}^0\text{-}\lim_{n \to \infty}(\overline{Y}_T\overline{\xi}_T^n) = \overline{Y}_T\zeta$, one gets $Z_T = \overline{Y}_T\zeta$. Define $\xi^\infty := Z/\overline{Y}$. Then, $(\overline{\xi}^n)_{n \in \mathbb{N}}$ Fatou-converges to ξ^∞ and $\xi_T^\infty = \zeta$. The last line of business is to show that $\xi^\infty \in \mathscr{X}^{\times\times}(1)$.

First of all, $\xi_0^\infty \leq 1$ and ξ^∞ is nonnegative. Let $\mathscr{Y}(1)$ be the class of all nonnegative process Y with $Y_0 = 1$, $\mathbb{P}$-a.s., such that YX is a $\mathbb{P}$-supermartingale for

---

[4] The notation "$\mathscr{X}^{\times\times}(1)$" is borrowed from [27] since it is *suggestive* of the fact that $\mathscr{X}^{\times\times}(1)$ is the process-bipolar of $\mathscr{X}(1)$, as is defined in [27]. Note, however, that it actually remains to show that $\mathscr{X}(1)$ is closed in probability to actually have that bipolar relationship.

all $X \in \mathcal{X}(1)$. Of course, for all $Y \in \mathcal{Y}(1)$ and all $\xi \in \mathcal{X}^{\times\times}(1)$, $Y\xi$ is a $\mathbb{P}$-supermartingale. It follows that $Y\overline{\xi}^n$ is a nonnegative $\mathbb{P}$-supermartingale for all $n \in \mathbb{N}$. Since, for any $Y \in \mathcal{Y}(1)$, $(Y\overline{\xi}^n)_{n \in \mathbb{N}}$ Fatou-converges to $Y\xi^\infty$, using Fatou's lemma one gets that $Y\xi^\infty$ is also a $\mathbb{P}$-supermartingale for all $Y \in \mathcal{Y}(1)$. Since there exists a local $\mathbb{P}$-martingale in $\overline{Y} \in \mathcal{Y}(1)$ with $\overline{Y}_T > 0$, $\mathbb{P}$-a.s., the optional decomposition theorem as appears in [11] implies that $\xi^\infty \in \mathcal{X}^{\times\times}(1)$. □

3.2 NFLVR and the Supermartingale Property of Wealth Processes Under a WELMM

We now move to another characterization of the NFLVR condition using the concept of WELMMs. We start with a simple observation. If $\mathbb{Q}$ is a probability measure equivalent to $\mathbb{P}$, it is straightforward to check that all $X \in \mathcal{X}$ are $\mathbb{Q}$-supermartingales if and only if $\langle \mathbb{Q}, X_T \rangle \leq X_0$ for all $X \in \mathcal{X}$. Consider now an ELMM $\mathbb{Q}$. Since nonnegative local $\mathbb{Q}$-martingales are $\mathbb{Q}$-supermartingales, every $X \in \mathcal{X}$ is a $\mathbb{Q}$-supermartingale; therefore, $\langle \mathbb{Q}, X_T \rangle \leq X_0$ for all $X \in \mathcal{X}$. One wonders, *does the last property hold when $\mathbb{Q}$ is replaced by a WELMM $\mathbb{Q}$?*

Before we state and prove a result along the lines of the above discussion, some terminology will be introduced. A mapping $\mathbb{Q} : \mathcal{F} \mapsto [0, 1]$ will be called a *weakly equivalent finitely additive probability* if (1) and (2) of Definition 2 hold, as well as, $\mathbb{P}$-a.s., $d\mathbb{Q}^r / d\mathbb{P} > 0$. Obviously, a local probability weakly equivalent to $\mathbb{P}$ is a weakly equivalent finitely additive probability. A *separating weakly equivalent finitely additive probability* is a weakly equivalent finitely additive probability $\mathbb{Q}$ such that $\langle \mathbb{Q}, X_T \rangle \leq X_0$ for all $X \in \mathcal{X}$. We can then think of the processes $X \in \mathcal{X}$ as being $\mathbb{Q}$-supermartingales. In accordance to the discussion above, the natural question that comes into mind is: when can we find a separating WELMM separating? In loose terms: *can we find a WELMM $\mathbb{Q}$ such that all elements of $\mathcal{X}$ $\mathbb{Q}$-supermartingales?* The answer, given in Theorem 3 below, is that this *only* happens under the NFLVR condition.

Theorem 3 *The following are equivalent*:

1. *The market satisfies the NFLVR condition.*
2. *There exists an ELMM $\mathbb{Q}$.*
3. *There exists a separating weakly equivalent finitely additive probability.*

Proof We prove (1) $\Rightarrow$ (3), (3) $\Rightarrow$ (2), and (2) $\Rightarrow$ (1) below.

(1) $\Rightarrow$ (2). This is a consequence of [9] and the fact that nonnegative σ-martingales are local martingales—see [2].

(2) $\Rightarrow$ (3). An ELMM is a separating weakly equivalent finitely additive probability.

(3) $\Rightarrow$ (1). In view of Proposition 3.6 of [7] and Proposition 1.3 proved previously in the present paper, condition NFLVR is equivalent to showing that (a) $\{X_T \mid X \in$

$\mathscr{X}(1)\}$ is bounded in probability, and (b) If $\mathbb{P}[X_T \geq X_0] = 1$ for some $X \in \mathscr{X}$, then $\mathbb{P}[X_T > X_0] = 0$. For (a), observe that

$$\sup_{X \in \mathscr{X}(1)} \mathbb{E}\left[\left(\frac{\mathrm{d}\mathsf{Q}^r}{\mathrm{d}\mathbb{P}}\right) X_T\right] = \sup_{X \in \mathscr{X}(1)} \langle \mathsf{Q}^r, X_T \rangle \leq \sup_{X \in \mathscr{X}(1)} \langle \mathsf{Q}, X_T \rangle \leq 1;$$

in particular, $\{(\mathrm{d}\mathsf{Q}^r/\mathrm{d}\mathbb{P}) X_T \mid X \in \mathscr{X}(1)\}$ is bounded in probability. In view of the fact that $\mathbb{P}[(\mathrm{d}\mathsf{Q}^r/\mathrm{d}\mathbb{P}) > 0] = 1$, we obtain that $\{X_T \mid X \in \mathscr{X}(1)\}$ is bounded in probability as well. To show (b), note that, for any $\varepsilon > 0$ and $X \in \mathscr{X}$ with $\mathbb{P}[X_T \geq X_0] = 1$, we have

$$X_0 \geq \langle \mathsf{Q}, X_T \rangle \geq \langle \mathsf{Q}, X_0 \mathbb{I}_\Omega + \varepsilon \mathbb{I}_{\{X_T > X_0 + \varepsilon\}} \rangle = X_0 + \varepsilon \mathsf{Q}[X_T > X_0 + \varepsilon]$$
$$\geq X_0 + \varepsilon \mathsf{Q}^r[X_T > 1 + \varepsilon].$$

It follows that $\mathsf{Q}^r[X_T > X_0 + \varepsilon] = 0$; since $\mathbb{P}[(\mathrm{d}\mathsf{Q}^r/\mathrm{d}\mathbb{P}) > 0] = 1$, this is equivalent to $\mathbb{P}[X_T > X_0 + \varepsilon] = 0$. The latter holds for all $\varepsilon > 0$, so we get $\mathbb{P}[X_T > X_0] = 0$, which completes the argument. $\square$

4 The Case of Continuous-Path Semimartingales

In this section, we shall state and prove a result that implies Theorem 1 in the case where S is a d-dimensional continuous-path semimartingale. Note that Assumption (1) will *not* be in force here; in particular, there can be more than one traded security and the prices of securities do not have to be nonnegative. In fact, Theorem 4 that is presented below actually sharpens the conclusion of Theorem 1 by providing a further equivalence in terms of the local rates of return and local covariances of the discounted prices $S = (S^i)_{i=1,\dots,d}$.

We first introduce some notation. Since S is a continuous-path semimartingale, one has the decomposition $S = A + M$, where $A = (A^1, \dots, A^d)$ has continuous paths and is of finite variation, and $M = (M^1, \dots, M^d)$ is a continuous-path local martingale. Denote by $[M^i, M^k]$ the quadratic (co)variation of M^i and M^k. Also, let $[M, M]$ be the $d \times d$ nonnegative-definite symmetric matrix-valued process whose (i, k)-component is $[M^i, M^k]$. Call now $G := \mathsf{trace}[M, M]$, where trace is the operator returning the trace of a matrix. Observe that G is an increasing, adapted, continuous process and that there exists a $d \times d$ nonnegative-definite symmetric matrix-valued process c such that $[M^i, M^k] = \int_0^{\cdot} c_t^{i,k} \, \mathrm{d}G_t$; $[M, M] = \int_0^{\cdot} c_t \, \mathrm{d}G_t$ in short.

Theorem 4 *In the continuous-semimartingale market described above, the following statements are equivalent:*

1. *There are no arbitrages of the first kind in the market.*
2. *There exists a strictly positive local $\mathbb{P}$-martingale Y with $Y_0 = 1$ such that $Y S^i$ is a local $\mathbb{P}$-martingale for all $i \in \{1, \dots, d\}$.*
3. *There exists a d-dimensional, predictable process ρ such that $A = \int_0^{\cdot} (c_t \rho_t) \, \mathrm{d}G_t$, as well as $\int_0^T \langle \rho_t, c_t \rho_t \rangle \, \mathrm{d}G_t < \infty$.*

Proof We prove (1) $\Rightarrow$ (3), (3) $\Rightarrow$ (2), and (2) $\Rightarrow$ (1) below.

(1) $\Rightarrow$ (3). We shall show that if statement (3) of Theorem 4 is not valid, then $\{X_T \mid X \in \mathscr{X}(1)\}$ is not bounded in probability. In view of Proposition 1, (1) $\Rightarrow$ (3) will be established.

Suppose that one *cannot* find a predictable d-dimensional process ρ such that $A = \int_0^{\cdot} (c_t \rho_t) \, dG_t$. In that case, linear algebra combined with a simple measurable selection argument gives the existence of some bounded predictable process θ such that (a) $\int_0^T \theta_t \, dG_t = 0$, (b) $\int_0^{\cdot} \langle \theta_t, \, dA_t \rangle$ is a *nondecreasing* process, and (c) $\mathbb{P}[\int_0^T \langle \theta_t, \, dA_t \rangle > 0] > 0$. This means that $X^{1,\theta} \in \mathscr{X}(1)$, in the notation of (2), satisfies $X^{1,\theta} \geq 1$, $\mathbb{P}[X_T^{1,\theta} > 1] > 0$. Then, $X^{1,k\theta} \in \mathscr{X}(1)$ for all $k \in \mathbb{N}$ and $(X^{1,k\theta})_{k \in \mathbb{N}}$ is not bounded in probability.

Now, suppose that $A = \int_0^{\cdot} (c_t \rho_t) \, dG_t$ for some predictable d-dimensional process ρ, but that $\mathbb{P}[\int_0^T \langle \rho_t, c_t \rho_t \rangle \, dG_t = \infty] > 0$. Consider the sequence $\pi^k := \rho \mathbb{I}_{\{|\rho| \leq k\}}$ and let X^k be defined via $X_0^k = 1$ and satisfying $dX_t^k = X_t^k \pi_t^k \, dS_t$. Then, Itô's formula implies that

$$\log X_T^k = -\frac{E_T^k}{2} + \int_0^T \left(\rho_t \mathbb{I}_{\{|\rho_t| \leq k\}} \right) \, dM_t,$$

holds for all $k \in \mathbb{N}$, where $E_T^k := \int_0^T \langle \rho_t, c_t \rho_t \rangle \mathbb{I}_{\{|\rho_t| \leq k\}} \, dG_t$ coincides with the total quadratic variation of the local martingale $\int_0^{\cdot} (\rho_t \mathbb{I}_{\{|\rho_t| \leq k\}}) \, dM_t$. It follows that, for every $k \in \mathbb{N}$, one can find a one-dimensional standard Brownian motion β^k such that

$$\log X_T^k = -\frac{E_T^k}{2} + \beta_{E_T^k}^k.$$

The strong law of large numbers for Brownian motion will imply that

$$\lim_{k \to \infty} \mathbb{P}\left[\left| \frac{\beta_{E_T^k}^k}{E_T^k} \right| > \varepsilon, \; \int_0^T \langle \rho_t, c_t \rho_t \rangle \, dG_t = \infty \right] = 0, \; \text{for all } \varepsilon > 0,$$

so that

$$\lim_{k \to \infty} \mathbb{P}\left[\frac{\log X_T^k}{E_T^k} > \frac{1}{2} - \varepsilon \; \bigg| \; \int_0^T \langle \rho_t, c_t \rho_t \rangle \, dG_t = \infty \right] = 1, \; \text{for all } \varepsilon > 0.$$

Choosing $\varepsilon = 1/4$, it follows that if $\mathbb{P}[\int_0^T \langle \rho_t, c_t \rho_t \rangle \, dG_t = \infty] > 0$, the sequence $(X_T^k)_{k \in \mathbb{N}}$ is not bounded in probability.

(3) $\Rightarrow$ (2). With the data of condition (3) there, define the process

$$Y := \exp\left(-\int_0^{\cdot} \langle \rho_t, \, dS_t \rangle + \frac{1}{2} \int_0^{\cdot} \langle \rho_t, c_t \rho_t \rangle \, dG_t \right).$$

Condition (3) ensures that Y is well-defined (meaning that the two integrals above make sense). Itô's formula easily shows that Y is a local $\mathbb{P}$-martingale. Then, a simple use of integration-by-parts gives that YS^i is a local martingale for all $i \in \{1, \ldots, d\}$.

(2) ⇒ (1). The proof of this implication is somewhat classic, but will be presented anyhow for completeness. Start with a sequence $(X^k)_{k \in \mathbb{N}}$ of wealth processes such that $\lim_{k \to \infty} X_0^k = 0$ as well as $X_T^k \geq \xi$ for some $\mathbb{R}_+$-valued random variable ξ. Since YS^i is a local $\mathbb{P}$-martingale for all $i \in \{1, \ldots, d\}$, a straightforward multidimensional generalization of the proof of Proposition 3 shows that, for all $k \in \mathbb{N}$, YX^k is a local $\mathbb{P}$-martingale. As nonnegative local $\mathbb{P}$-martingales are $\mathbb{P}$-supermartingales, we have $\mathbb{E}[Y_T \xi] \leq \mathbb{E}[Y_T X_T^k] \leq X_0^k$ holding for all $k \in \mathbb{N}$. Therefore, since $\lim_{k \to \infty} X_0^k = 0$, we obtain $\mathbb{E}[Y_T \xi] = 0$. Since $Y_T > 0$ and $\xi \geq 0$, $\mathbb{P}$-a.s, the last inequality holds if only if $\mathbb{P}[\xi = 0] = 1$. Therefore, $(X^k)_{k \in \mathbb{N}}$ is not an arbitrage of the first kind. □

Remark 8 (Market price of risk and the numéraire portfolio) Condition (3) of Theorem 4 has some economic consequences. Assume for simplicity that G is absolutely continuous with respect to Lebesgue measure, i.e., that $G := \int_0^\cdot g_t \, dt$ for some predictable process g. Under condition NA₁, we also have $A := \int_0^\cdot a_t \, dt$ for some predictable process g, and that there exists a predictable process ρ such that $c\rho = a$. (In fact, the latter process ρ can be taken to be equal to $c^\dagger a$, where $c^\dagger$ is the Moore-Penrose pseudo-inverse of c.) Now, take $c^{1/2}$ to be any root of the nonnegative-definite matrix c (that can be chosen in a predictable way) and define $\sigma := c^{1/2} \sqrt{g}$. Then, we can write $dS_t = \sigma_t(\lambda_t \, dt + dW_t)$, where W is a standard d-dimensional Brownian motion[5] and $\lambda := \sigma^\top \rho$ is a *risk premium* process (in the one-dimensional case also commonly known as the *Sharpe ratio*), that has to satisfy $\int_0^T |\lambda_t|^2 \, dt < \infty$ for all $T \in \mathbb{R}_+$. We conclude that condition NA₁ is valid if and only if a risk-premium process exists and is locally square-integrable in a pathwise sense.

References

1. Aliprantis, C.D., Border, K.C.: Infinite-Dimensional Analysis – A Hitchhiker's Guide, Second edn. Springer, Berlin (1999)
2. Ansel, J.-P., Stricker, C.: Couverture des actifs contingents et prix maximum. Ann. Inst. Henri Poincaré Probab. Stat. **30**, 303–315 (1994)
3. Bhaskara Rao, K.P.S., Bhaskara Rao, M.: Theory of Charges: A Study of Finitely Additive Measures. *Pure and Applied Mathematics*, vol. 109. Academic Press Inc. [Harcourt Brace Jovanovich Publishers], New York, With a foreword by D.M. Stone (1983)
4. Brannath, W., Schachermayer, W.: A bipolar theorem for $L_+^0(\Omega, \mathscr{F}, \mathbf{P})$. In: Séminaire de Probabilités, XXXIII. *Lecture Notes in Math.*, vol. 1709, pp. 349–354. Springer, Berlin (1999)

---

[5]In the case where c is nonsingular for Lebesgue-almost every $t \in \mathbb{R}$, $\mathbb{P}$-almost surely, we have $W_\cdot := \int_0^\cdot c_t^{-1/2} \, dM_t$. If c fails to be nonsingular for Lebesgue-almost every $t \in \mathbb{R}$, $\mathbb{P}$-almost surely, one can still construct a Brownian motion W in order to have $M_\cdot = \int_0^\cdot c_t^{1/2} \, dW_t$ holding by enlarging the probability space—check for example [17], Theorem 4.2 of Sect. 3.4.

5. Christensen, M.M., Larsen, K.: No arbitrage and the growth optimal portfolio. Stoch. Anal. Appl. **25**, 255–280 (2007)
6. Cvitanić, J., Schachermayer, W., Wang, H.: Utility maximization in incomplete markets with random endowment. Finance Stoch. **5**, 259–272 (2001)
7. Delbaen, F., Schachermayer, W.: A general version of the fundamental theorem of asset pricing. Math. Ann. **300**, 463–520 (1994)
8. Delbaen, F., Schachermayer, W.: The no-arbitrage property under a change of numéraire. Stoch. Stoch. Rep. **53**, 213–226 (1995)
9. Delbaen, F., Schachermayer, W.: The fundamental theorem of asset pricing for unbounded stochastic processes. Math. Ann. **312**, 215–250 (1998)
10. Fernholz, E., Karatzas, I.: Stochastic Portfolio Theory: An Overview. *Handbook of Numerical Analysis*, vol. 15, pp. 89–167 (2009)
11. Föllmer, H., Kabanov, Y.M.: Optional decomposition and Lagrange multipliers. Finance Stoch. **2**, 69–81 (1998)
12. Föllmer, H., Kramkov, D.: Optional decompositions under constraints. Probab. Theory Relat. Fields **109**, 1–25 (1997)
13. Gilles, C., LeRoy, S.F.: Bubbles and charges. Int. Econ. Rev. **33**, 323–339 (1992)
14. Ingersoll, J.E.: Theory of Financial Decision Making. *Rowman and Littlefield Studies in Financial Economics*. Rowman & Littlefield, Totowa (1987)
15. Kabanov, Y.M., Kramkov, D.O.: Large financial markets: asymptotic arbitrage and contiguity. Teor. Veroyatn. I Primenen. **39**, 222–229 (1994)
16. Karatzas, I., Kardaras, C.: The numéraire portfolio in semimartingale financial models. Finance Stoch. **11**, 447–493 (2007)
17. Karatzas, I., Shreve, S.E.: Brownian Motion and Stochastic Calculus. *Graduate Texts in Mathematics*, vol. 113, Second edn. Springer, New York (1991)
18. Karatzas, I., Žitković, G.: Optimal consumption from investment and random endowment in incomplete semimartingale markets. Ann. Probab. **31**, 1821–1858 (2003)
19. Kardaras, C.: Market viability via absence of arbitrages of the first kind. Submitted for publication. Preprint available at http://arxiv.org/abs/0904.1798 (2009)
20. Kardaras, C., Platen, E.: On the semimartingale property of discounted asset-price processes. Submitted for publication. Preprint available at http://arxiv.org/abs/0803.1890 (2008)
21. Kreps, D.M.: Arbitrage and equilibrium in economies with infinitely many commodities. J. Math. Econ. **8**, 15–35 (1981)
22. Loewenstein, M., Willard, G.A.: Local martingales, arbitrage, and viability. Free snacks and cheap thrills. Econom. Theory **16**, 135–161 (2000)
23. Loewenstein, M., Willard, G.A.: Rational equilibrium asset-pricing bubbles in continuous trading models. J. Econ. Theory **91**, 17–58 (2000)
24. Platen, E., Heath, D.: A Benchmark Approach to Quantitative Finance. *Springer Finance*. Springer, Berlin (2006)
25. Schweizer, M.: On the minimal martingale measure and the Föllmer-Schweizer decomposition. Stoch. Anal. Appl. **13**, 573–599 (1995)
26. Yan, J.A.: Caractérisation d'une classe d'ensembles convexes de L^1 ou H^1. Seminar on Probability XIV. *Lecture Notes in Math.*, vol. 784 pp. 220–222. Springer Berlin (1980)
27. Žitković, G.: A filtered version of the bipolar theorem of Brannath and Schachermayer. J. Theor. Probab. **15**, 41–61 (2002)

M⁶—On Minimal Market Models and Minimal Martingale Measures

Hardy Hulley and Martin Schweizer

Abstract The well-known absence-of-arbitrage condition NFLVR from the fundamental theorem of asset pricing splits into two conditions, called NA and NUPBR. We give a literature overview of several equivalent reformulations of NUPBR; these include existence of a growth-optimal portfolio, existence of the numeraire portfolio, and for continuous asset prices the structure condition (SC). As a consequence, the minimal market model of E. Platen is seen to be directly linked to the minimal martingale measure. We then show that reciprocals of stochastic exponentials of continuous local martingales are time changes of a squared Bessel process of dimension 4. This directly gives a very specific probabilistic structure for minimal market models.

1 Introduction

Classical mathematical finance has been built on pillars of absence of arbitrage; this is epitomised by the celebrated fundamental theorem of asset pricing (FTAP), due in its most general form to F. Delbaen and W. Schachermayer. However, several recent directions of research have brought up the question whether one should not also study more general models that do not satisfy all the stringent requirements of the FTAP; see also [21] for an early contribution in that spirit. One such line of research is the recent work of R. Fernholz and I. Karatzas on *diverse markets*,

This paper is dedicated to Eckhard Platen on the occasion of his 60th birthday.

H. Hulley
School of Finance and Economics, University of Technology, Sydney, P.O. Box 123, Broadway, NSW 2007, Australia
e-mail: hardy.hulley@uts.edu.au

M. Schweizer (✉)
Departement Mathematik, ETH-Zentrum, ETH Zürich, HG G 51.2, CH–8092 Zürich, Switzerland
e-mail: martin.schweizer@math.ethz.ch

C. Chiarella, A. Novikov (eds.), *Contemporary Quantitative Finance*,
DOI 10.1007/978-3-642-03479-4_3, © Springer-Verlag Berlin Heidelberg 2010

of which an overview is given in [17]. Another is the *benchmark approach* and the idea of *minimal market models* proposed and propagated by E. Platen and co-authors in several recent publications; see [24] for a textbook account. Finally, also some approaches to *bubbles* go in a similar direction.

Our goal in this paper is twofold. We first give a neutral overview of several equivalent formulations of an L^0-boundedness property, called NUPBR, that makes up a part, but not all of the conditions for the FTAP. For continuous asset prices, we then show that the *minimal market model* of E. Platen is very directly linked to the *minimal martingale measure* introduced by H. Föllmer and M. Schweizer. As a consequence, we exhibit a very specific probabilistic structure for minimal market models: We show that they are *time changes of a squared Bessel process of dimension 4 (a BESQ^4)*, under very weak assumptions. This extends earlier work in [22] to the most general case of a continuous (semimartingale) financial market.

The paper is structured as follows. Section 2 considers general semimartingale models, introduces basic notations, and recalls that the well-known condition NFLVR underlying the FTAP consists of two parts: no arbitrage NA, and a certain boundedness condition in L^0, made more prominent through its recent labelling as NUPBR by C. Kardaras and co-authors. We collect from the literature several equivalent formulations of this property, the most important for subsequent purposes being the existence of a *growth-optimal portfolio*. Section 3 continues this overview under the additional assumption that the basic price process S is continuous; the main addition is that NUPBR is then also equivalent to the *structure condition (SC)* introduced by M. Schweizer, and that it entails the existence of the minimal martingale density for S.

Both Sects. 2 and 3 contain only known results from the literature; their main contribution is the effort made to present these results in a clear, concise and comprehensive form. The main probabilistic result in Sect. 4 shows that reciprocals of stochastic exponentials of continuous local martingales are automatically time changes of BESQ^4 processes. Combining this with Sect. 3 then immediately yields the above announced structural result for minimal market models.

2 General Financial Market Models

This section introduces basic notations and concepts and recalls a number of general known results. Loosely speaking, the main goal is to present an overview of the relations between absence of arbitrage and existence of a log-optimal portfolio strategy, in a frictionless financial market where asset prices can be general semimartingales. The only potential novelty in all of this section is that the presentation is hopefully clear and concise. We deliberately only give references to the literature instead of repeating proofs, in order not to clutter up the presentation.

We start with a probability space $(\Omega, \mathcal{F}, P)$ and a filtration $\mathbb{F} = (\mathcal{F}_t)_{0 \le t \le T}$ satisfying the usual conditions of right-continuity and P-completeness. To keep notations simple, we assume that the time horizon $T \in (0, \infty)$ is nonrandom and finite. All our basic processes will be defined on $[0, T]$ which frees us from worrying about their behaviour at "infinity" or "the end of time". Results from the literature on processes living on $[0, \infty)$ are used by applying them to the relevant processes stopped at T.

We consider a financial market with $d + 1$ assets. One of these is chosen as numeraire or unit of account, labelled with the number 0, and all subsequent quantities are expressed in terms of that. So we have an asset $S^0 \equiv 1$ and d "risky" assets whose price evolution is modelled by an $\mathbb{R}^d$-valued semimartingale $S = (S_t)_{0 \le t \le T}$, where S_t^i is the price at time t of asset $i \in \{1, \dots, d\}$, expressed in units of asset 0. To be able to use stochastic integration, we assume that S is a semimartingale.

Trading in our financial market is frictionless and must be done in a self-financing way. Strategies are then described by pairs (x, ϑ), where $x \in \mathbb{R}$ is the initial capital or initial wealth at time 0 and $\vartheta = (\vartheta_t)_{0 \le t \le T}$ is an $\mathbb{R}^d$-valued predictable S-integrable process; we write $\vartheta \in L(S)$ for short. The latter means that the (real-valued) stochastic integral process $\vartheta \cdot S := \int \vartheta \, dS$ is well defined and then again a semimartingale. We remark in passing that $\vartheta \cdot S$ must be understood as a vector stochastic integral, which may be different from the sum of the componentwise stochastic integrals; see [14] for the general theory and [4] for an amplification of the latter point. In financial terms, ϑ_t^i is the number of units of asset i that we hold in our dynamically varying portfolio at time t, and the self-financing condition means that our wealth at time t is given by

$$X_t^{x,\vartheta} := x + \vartheta \cdot S_t = x + \int_0^t \vartheta_u \, dS_u, \quad 0 \le t \le T.$$

Not every $\vartheta \in L(S)$ yields a decent trading strategy. To exclude unpleasant phenomena resulting from doubling-type strategies, one has to impose some lower bound on the trading gains/losses $\vartheta \cdot S$. We call $\vartheta \in L(S)$ *a-admissible* if $\vartheta \cdot S \ge -a$, where $a \ge 0$, and *admissible* if it is a-admissible for some $a \ge 0$. We then introduce for $x > 0$ the sets

$$\mathcal{X}^x := \left\{ X^{x,\vartheta} \mid \vartheta \in L(S) \text{ and } X^{x,\vartheta} \ge 0 \right\} = \{x + \vartheta \cdot S \mid \vartheta \in L(S) \text{ is } x\text{-admissible}\},$$

$$\mathcal{X}^{x,++} := \left\{ X^{x,\vartheta} \mid \vartheta \in L(S) \text{ and } X^{x,\vartheta} > 0 \text{ as well as } X_-^{x,\vartheta} > 0 \right\},$$

and we set $\mathcal{X}_T^x := \{X_T^{x,\vartheta} \mid X^{x,\vartheta} \in \mathcal{X}^x\}$, with $\mathcal{X}_T^{x,++}$ defined analogously. So every $f \in \mathcal{X}_T^x$ represents a terminal wealth position that one can generate out of initial wealth x by self-financing trading while keeping current wealth always nonnegative (and even strictly positive, if f is in $\mathcal{X}_T^{x,++}$). We remark that $X_-^{x,\vartheta} > 0$ does not follow from $X^{x,\vartheta} > 0$ since we only know that $X^{x,\vartheta}$ is a semimartingale; we have no local martingale or supermartingale property at this point. Note that all the ϑ appearing in the definition of $\mathcal{X}_T^x$ have the same uniform lower bound for $\vartheta \cdot S$, namely $-x$. Finally, we need the set

$$\mathcal{C} := \left\{ X_T^{0,\vartheta} - B \mid \vartheta \in L(S) \text{ is admissible, } B \in L_+^0(\mathcal{F}_T) \right\} \cap L^\infty$$

of all bounded time T positions that one can dominate by self-financing admissible trading even from initial wealth 0.

With the above notations, we can now recall from [9] and [18] the following concepts.

Definition Let S be a semimartingale. We say that S *satisfies NA, no arbitrage*, if $C \cap L_+^\infty = \{0\}$; in other words, C contains no nonnegative positions except 0. We say that S *satisfies NFLVR, no free lunch with vanishing risk*, if $\overline{C}^\infty \cap L_+^\infty = \{0\}$, where $\overline{C}^\infty$ denotes the closure of C in the norm topology of L^∞. Finally, we say that S *satisfies NUPBR, no unbounded profit with bounded risk*, if $\mathcal{X}_T^x$ is bounded in L^0 for some $x > 0$ (or, equivalently, for all $x > 0$ or for $x = 1$, because $\mathcal{X}_T^x = x\mathcal{X}_T^1$).

The condition NFLVR is a precise mathematical formulation of the natural economic idea that it should be impossible in a financial market to generate something out of nothing without risk. The meta-theorem that "absence of arbitrage is tantamount to the existence of an equivalent martingale measure" then takes the precise form that S satisfies NFLVR if and only if there exists a probability measure Q equivalent to P such that S is under Q a so-called σ-martingale. This is the celebrated *fundamental theorem of asset pricing (FTAP)* in the form due to F. Delbaen and W. Schachermayer; see [6, 8].

In the sequel, our interest is neither in the FTAP nor in equivalent σ-martingale measures Q as above; hence we do not explain these in more detail. Our focus is on the condition NUPBR and its ramifications. The connection to NFLVR is very simple and direct:

S satisfies NFLVR if and only if it satisfies both NA and NUPBR.

This result can be found either in Sect. 3 of [6] or more concisely in Lemma 2.2 of [15]. Moreover, neither of the conditions NA and NUPBR implies the other, nor of course NFLVR; see Chap. 1 of [12] for explicit counterexamples.

The next definition introduces strategies with certain optimality properties.

Definition An element $X^{np} = X^{1,\vartheta^{np}}$ of $\mathcal{X}^{1,++}$ is called a *numeraire portfolio* if the ratio $X^{1,\vartheta}/X^{np}$ is a P-supermartingale for every $X^{1,\vartheta} \in \mathcal{X}^{1,++}$. An element $X^{go} = X^{1,\vartheta^{go}}$ of $\mathcal{X}^{1,++}$ is called a *growth-optimal portfolio* or a *relatively log-optimal portfolio* if

$$E\big[\log\big(X_T^{1,\vartheta}/X_T^{go}\big)\big] \le 0$$

for all $X^{1,\vartheta} \in \mathcal{X}^{1,++}$ such that the above expectation is not $\infty - \infty$. Finally, an element $X^{lo} = X^{1,\vartheta^{lo}}$ of $\mathcal{X}^{1,++}$ with $E[\log X_T^{lo}] < \infty$ is called a *log-utility-optimal portfolio* if

$$E\big[\log X_T^{1,\vartheta}\big] \le E\big[\log X_T^{lo}\big]$$

for all $X^{1,\vartheta} \in \mathcal{X}^{1,++}$ such that $E[(\log X_T^{1,\vartheta})^-] < \infty$.

For all the above concepts, we start with initial wealth 1 and look at self-financing strategies whose wealth processes (together with their left limits) must remain strictly positive. In all cases, we also commit a slight abuse of terminology by calling "portfolio" what is actually the wealth process of a self-financing strategy. In words, the above three concepts can then be described as follows:

- The *numeraire portfolio* has the property that, when used for discounting, it turns every wealth process in $\mathcal{X}^{1,++}$ into a supermartingale. Loosely speaking, this means that it has the best "performance" in the class $\mathcal{X}^{1,++}$.
- The *growth-optimal portfolio* has, in relative terms, a higher expected growth rate (measured on a logarithmic scale) than any other wealth process in $\mathcal{X}^{1,++}$.
- The *log-utility-optimal portfolio* maximises the expected logarithmic utility of terminal wealth essentially over all wealth processes in $\mathcal{X}^{1,++}$.

The next result gives the first main connection between the notions introduced so far.

Proposition 1

1) X^{np}, X^{go} and X^{lo} are all unique.
2) X^{np}, X^{go} and X^{lo} coincide whenever they exist.
3) X^{np} exists if and only if X^{go} exists. This is also equivalent to existence of X^{lo} if in addition $\sup\{E[\log X_T] \mid X \in \mathcal{X}^{1,++}$ with $E[(\log X_T)^-] < \infty\} < \infty$.
4) X^{np} (or equivalently X^{go}) exists if and only if S satisfies NUPBR.

Proof This is a collection of well-known results; see [3, Propositions 3.3 and 3.5], [5, Theorem 4.1], and [18, Proposition 3.19 and Theorem 4.12].   □

Our next definition brings us closer again to equivalent σ-martingale measures for S.

Definition An *equivalent supermartingale deflator* (for $\mathcal{X}^{1,++}$) is an adapted RCLL process $Y = (Y_t)_{0 \le t \le T}$ with $Y_0 = 1$, $Y \ge 0$, $Y_T > 0$ P-a.s. and the property that $Y X^{1,\vartheta}$ is a P-supermartingale for all $X^{1,\vartheta} \in \mathcal{X}^{1,++}$. The set of all equivalent supermartingale deflators is denoted by $\mathcal{Y}$.

Because $\mathcal{X}^{1,++}$ contains the constant process 1, we immediately see that each $Y \in \mathcal{Y}$ is itself a supermartingale; and $Y_T > 0$ implies by the minimum principle for supermartingales that then also $Y > 0$ and $Y_- > 0$. To facilitate comparisons, we mention that the class $\mathcal{X}^{1,++}$ is called $\mathcal{N}$ in [3], and $\mathcal{Y}$ is called $\mathcal{SM}$ there.

Definition A *σ-martingale density* (or *local martingale density*) *for S* is a local P-martingale $Z = (Z_t)_{0 \le t \le T}$ with $Z_0 = 1$, $Z \ge 0$ and the property that $Z S^i$ is a P-σ-martingale (or P-local martingale, respectively) for each $i = 1, \ldots, d$. If $Z > 0$, we call Z in addition *strictly positive*. For later use, we denote by $\mathcal{D}_{loc}^{++}(S, P)$ the set of all strictly positive local P-martingale densities Z for S.

From the well-known Ansel–Stricker result (see Corollaire 3.5 of [2]), it is clear that $Z X^{1,\vartheta}$ is a P-supermartingale for all $X^{1,\vartheta} \in \mathcal{X}^{1,++}$ whenever Z is a σ- or local martingale density for S. Hence $\mathcal{Y}$ contains all strictly positive σ- and local martingale densities for S. On the other hand, if Q is an equivalent σ- or local martingale measure for S (as in the FTAP, in the sense that each S^i is a Q-σ-martingale or local

Q-martingale, respectively), then the density process Z^Q of Q with respect to P is by the Bayes rule a strictly positive σ- or local martingale density for S, if it has $Z_0^Q = 1$ (which means that $Q = P$ on $\mathcal{F}_0$). In that sense, supermartingale deflators can be viewed as a generalisation of equivalent σ- or local martingale measures for S. This important idea goes back to [20].

The second main connection between the concepts introduced in this section is provided by

Proposition 2 *The $\mathbb{R}^d$-valued semimartingale S satisfies NUPBR if and only if there exists an equivalent supermartingale deflator for $\mathcal{X}^{1,++}$. In short:*

$$NUPBR \Longleftrightarrow \mathcal{Y} \neq \emptyset.$$

Proof This is part of [18, Theorem 4.12].   □

Combining what we have seen so far, we directly obtain the main result of this section.

Theorem 3 *For an $\mathbb{R}^d$-valued semimartingale S, the following are equivalent:*

1) *S satisfies NUPBR.*
2) *The numeraire portfolio X^{np} exists.*
3) *The growth-optimal portfolio X^{go} exists.*
4) *There exists an equivalent supermartingale deflator for $\mathcal{X}^{1,++}$, i.e. $\mathcal{Y} \neq \emptyset$.*

In each of these cases, X^{np} and X^{go} are unique, and $X^{\mathrm{np}} = X^{\mathrm{go}}$. If in addition $\sup\{E[\log X_T] \mid X \in \mathcal{X}^{1,++} \text{ with } E[(\log X_T)^-] < \infty\} < \infty$, then 1)–4) are also equivalent to

5) *The log-utility-optimal portfolio X^{lo} exists.*

In that case, also X^{lo} is unique, and $X^{\mathrm{lo}} = X^{\mathrm{np}} = X^{\mathrm{go}}$.

Remarks

1) We emphasise once again that all these results are known. In the above most general form, they are due to [18], but variants and precursors can already be found in [3, 5, 21]. In particular, Theorem 5.1 in [5] shows that under the assumption NA, the existence of the growth-optimal portfolio X^{go} is equivalent to the existence of a strictly positive σ-martingale density for S.
2) It seems that the key importance of the condition NUPBR, albeit not under that name and in the more specialised setting of a complete Itô process model, has first been recognised in [21], who relate NUPBR to the absence of so-called cheap thrills; see Theorem 2 in [21].
3) If the numeraire portfolio X^{np} exists, then it lies in $\mathcal{X}^{1,++}$ and at the same time, $1/X^{\mathrm{np}}$ lies in $\mathcal{Y}$, by the definitions of X^{np} and $\mathcal{Y}$. So another property equivalent to 1)–4) in Theorem 3 would be that $\mathcal{X}^{1,++} \cap (1/\mathcal{Y}) \neq \emptyset$ or $\mathcal{Y} \cap (1/\mathcal{X}^{1,++}) \neq \emptyset$.

4) Since we work on the closed interval $[0, T]$, all our processes so far are defined up to and including T. Hence we need not worry about finiteness of X_T^{np}, which is in contrast to [18].

5) In view of the link to log-utility maximisation, it is no surprise that there are also dual aspects and results for the above connections. This is for instance presented in [3] and [18], but is not our main focus here.

6) For yet another property equivalent to NUPBR, see the recent work in [19].

The third important result in this section would be a more explicit description of the numeraire portfolio X^{np} or, more precisely, its generating strategy ϑ^{np}. Such a description can be found in [18, Theorem 3.15], or more generally in [11, Theorem 3.1 and Corollary 3.2]. In both cases, ϑ^{np} can be obtained (in principle) by pointwise maximisation of a function (called g or Λ, respectively, in the above references) that is given explicitly in terms of certain semimartingale characteristics. In [18], this involves the characteristics of the returns process R, where each $S^i = S_0^i \mathcal{E}(R^i)$ is assumed to be a stochastic exponential. By contrast, the authors of [11] take a general semimartingale S and allow in addition to trading also consumption with a possibly stochastic clock; they then need the (joint) characteristics of (S, M), where M is a certain process defined via the stochastic clock. In the general case where S can have jumps, neither of these descriptions unfortunately gives very explicit expressions for ϑ^{np} since the above pointwise maximiser is only defined implicitly. For this reason, we do not go into more detail here, and focus in the next section on the much simpler case where S is continuous.

3 Continuous Financial Market Models

In this section, we focus on the special case when S on $(\Omega, \mathcal{F}, \mathbb{F}, P)$ from Sect. 2 is *continuous*. We introduce some more concepts and link them to those of Sect. 2. Again, all the results given here are well known from the literature, and we at most claim credit for a hopefully clear and concise overview.

So let $S = (S_t)_{0 \leq t \leq T}$ be an $\mathbb{R}^d$-valued continuous semimartingale with canonical decomposition $S = S_0 + M + A$. The processes $M = (M_t)_{0 \leq t \leq T}$ and $A = (A_t)_{0 \leq t \leq T}$ are both $\mathbb{R}^d$-valued, continuous and null at 0. Moreover, M is a local P-martingale and A is adapted and of finite variation. The bracket process $\langle M \rangle$ of M is the adapted, continuous, $d \times d$-matrix-valued process with components $\langle M \rangle^{ij} = \langle M^i, M^j \rangle$ for $i, j = 1, \ldots, d$; it exists because M is continuous, hence locally square-integrable.

Definition We say that S satisfies the *weak structure condition (SC')* if A is absolutely continuous with respect to $\langle M \rangle$ in the sense that there exists an $\mathbb{R}^d$-valued predictable process $\widehat{\lambda} = (\widehat{\lambda}_t)_{0 \leq t \leq T}$ such that $A = \int d\langle M \rangle \widehat{\lambda}$, i.e.

$$A_t^i = \sum_{j=1}^d \int_0^t \widehat{\lambda}_u^j \, d\langle M \rangle_u^{ij} = \sum_{j=1}^d \int_0^t \widehat{\lambda}_u^j \, d\langle M^i, M^j \rangle_u \quad \text{for } i = 1, \ldots, d \text{ and } 0 \leq t \leq T.$$

We then call $\widehat{\lambda}$ the *(instantaneous) market price of risk* for S and sometimes infor-mally write $\widehat{\lambda} = dA/d\langle M \rangle$.

Definition If S satisfies the weak structure condition (SC′), we define

$$\widehat{K}_t := \int_0^t \widehat{\lambda}_u^{\mathrm{tr}} d\langle M \rangle \widehat{\lambda}_u = \sum_{i,j=1}^d \int_0^t \widehat{\lambda}_u^i \widehat{\lambda}_u^j d\langle M^i, M^j \rangle_u, \quad 0 \le t \le T$$

and call $\widehat{K} = (\widehat{K}_t)_{0 \le t \le T}$ the *mean-variance tradeoff process* of S. Because $\langle M \rangle$ is positive semidefinite, the process $\widehat{K}$ is increasing and null at 0; but note that it may take the value $+\infty$ in general. We say that S satisfies the *structure condition (SC)* if S satisfies (SC′) and $\widehat{K}_T < \infty$ P-a.s.

Remarks

1) There is some variability in the literature concerning the structure condition; some authors call (SC) what we label (SC′) here. The terminology we have cho-sen is consistent with [26]. For discussions of the difference between (SC′) and (SC), we refer to [7] and [16].
2) The weak structure condition (SC′) comes up very naturally via Girsanov's the-orem. In fact, suppose S is a local Q-martingale under some Q equivalent to P and $1/Z$ is the density process of P with respect to Q. Then the process $M := S - S_0 - \int Z_- d\langle S, \frac{1}{Z} \rangle$ is by Girsanov's theorem a local P-martingale null at 0 and continuous like S, and $A := \int Z_- d\langle S, \frac{1}{Z} \rangle = \int Z_- d\langle M, \frac{1}{Z} \rangle$ is abso-lutely continuous with respect to $\langle M \rangle$ by the Kunita–Watanabe inequality. These results are of course well known from stochastic calculus; but their relevance for mathematical finance was only discovered later around the time when the impor-tance of equivalent local martingale measures was highlighted by the FTAP.

If S satisfies (SC), the condition $\widehat{K}_T < \infty$ P-a.s. can equivalently be formu-lated as $\widehat{\lambda} \in L^2_{\mathrm{loc}}(M)$ (on $[0, T]$, to be accurate). This means that the stochastic integral process $\widehat{\lambda} \cdot M = \int \widehat{\lambda} dM$ is well defined and a real-valued continuous local P-martingale null at 0, and we have $\widehat{K} = \langle \widehat{\lambda} \cdot M \rangle$. The stochastic exponential

$$\widehat{Z}_t := \mathcal{E}(-\widehat{\lambda} \cdot M)_t = \exp\left(-\widehat{\lambda} \cdot M_t - \frac{1}{2} \widehat{K}_t\right), \quad 0 \le t \le T$$

is then also well defined and a strictly positive local P-martingale with $\widehat{Z}_0 = 1$. For reasons that will become clear presently, $\widehat{Z}$ is called the *minimal martingale density* for S.

Proposition 4 *Suppose S is an $\mathbb{R}^d$-valued continuous semimartingale. Then S sat-isfies the structure condition (SC) if and only if there exists a strictly positive local martingale density Z for S. In short:*

$$(SC) \Longleftrightarrow \mathcal{D}^{++}_{\mathrm{loc}}(S, P) \ne \emptyset.$$

Proof If S satisfies (SC), we have seen above that $\widehat{Z} = \mathcal{E}(-\widehat{\lambda} \cdot M)$ is a strictly positive local P-martingale with $\widehat{Z}_0 = 1$. Moreover, using (SC), it is a straightforward computation via the product rule to check that each $\widehat{Z} S^i$ is a local P-martingale. Hence we can take $Z = \widehat{Z}$.

Conversely, suppose that ZS and $Z > 0$ are both local P-martingales. If Z is a true P-martingale on $[0, T]$, it can be viewed as the density process of some Q equivalent to P such that, by the Bayes rule, S is a local Q-martingale. Then the Girsanov argument in Remark 2 above gives (SC'). In general, applying the product rule shows that ZS has the finite variation part $\int Z_- \, dA + \langle Z, S \rangle$, which must vanish because ZS is a local P-martingale. This gives

$$A = -\int \frac{1}{Z_-} d\langle S, Z \rangle = -\int \frac{1}{Z_-} d\langle M, Z \rangle,$$

hence again (SC'), and with some more work, one shows that even (SC) is satisfied. For the details, we refer to Theorem 1 of [26]. $\qquad \square$

Since S is continuous, Theorem 1 of [26] also shows, by an application of the Kunita–Watanabe decomposition with respect to M to the stochastic logarithm N of $Z = \mathcal{E}(N)$, that every strictly positive local martingale density Z for S can be written as $Z = \widehat{Z} \mathcal{E}(L)$ for some local P-martingale L null at 0 which is strongly P-orthogonal to M. From that perspective, $\widehat{Z}$ is minimal in that it is obtained for the simplest choice $L \equiv 0$. One can also exhibit other minimality properties of $\widehat{Z}$, but this is not our main focus here.

Remark While $\widehat{Z}$ is (for continuous S) always strictly positive, it is in general only a local, but not a true P-martingale. But if $\widehat{Z}$ happens to be a true P-martingale (on $[0, T]$), or equivalently if $E_P[\widehat{Z}_T] = 1$, we can define a probability measure $\widehat{P}$ equivalent to P via $d\widehat{P} := \widehat{Z}_T \, dP$. The local P-martingale property of $\widehat{Z}S$ is by the Bayes rule then equivalent to saying that S is a local $\widehat{P}$-martingale. This $\widehat{P}$, if it exists, is called the *minimal martingale measure*; see [10].

As we have already seen in Sect. 2, the family $\mathcal{Y}$ of all equivalent supermartingale deflators for $\mathcal{X}^{1,++}$ contains the family $\mathcal{D}_{\text{loc}}^{++}(S, P)$ of all strictly positive local martingale densities for S. Therefore $\mathcal{D}_{\text{loc}}^{++}(S, P) \neq \emptyset$ implies that $\mathcal{Y} \neq \emptyset$, and this already provides a strong first link between the results in this section and those in Sect. 2. A second link is given by the following connection between the minimal martingale density $\widehat{Z}$ and the numeraire portfolio X^{np}.

Lemma 5 *Suppose S is an $\mathbb{R}^d$-valued continuous semimartingale. If S satisfies (SC), then X^{np} exists and is given by $X^{\text{np}} = 1/\widehat{Z}$.*

Proof Since S satisfies (SC), $\widehat{Z} = \mathcal{E}(-\widehat{\lambda} \cdot M)$ exists and

$$1/\widehat{Z} = \exp\left(\widehat{\lambda} \cdot M + \frac{1}{2} \widehat{K}\right) = \exp\left(\widehat{\lambda} \cdot S - \frac{1}{2} \widehat{K}\right) = \mathcal{E}(\widehat{\lambda} \cdot S),$$

because $S = S_0 + M + \int d\langle M\rangle\widehat{\lambda}$. This shows that $1/\widehat{Z} = X^{1,\widehat{\vartheta}}$ lies in $\mathcal{X}^{1,++}$ with $\widehat{\vartheta} = \widehat{\lambda}/\widehat{Z}$. Moreover, $\widehat{Z}S$ is a local P-martingale by (the proof of) Proposition 4. A straightforward application of the product rule then shows that also $\widehat{Z}X^{1,\vartheta}$ is a local P-martingale, for every $X^{1,\vartheta} \in \mathcal{X}^{1,++}$, and so this product is also a P-supermartingale since it is nonnegative. Thus $1/\widehat{Z}$ satisfies the requirements for the numeraire portfolio and therefore agrees with X^{np} by uniqueness. $\square$

Of course, also Lemma 5 is not really new; the result can essentially already be found in [3, Corollary 4.10]. If one admits that the numeraire portfolio X^{np} coincides with the growth-optimal portfolio X^{go}, one can also quote [5, Corollary 7.4]. And finally, one could even use the description of X^{np} in [18, Theorem 3.15], because this becomes explicit when S is continuous.

For a complete and detailed connection between Sects. 2 and 3, the next result provides the last link in the chain. In the present formulation, it seems due to [12, Theorem 1.25]; the proof we give here is perhaps a little bit more compact.

Proposition 6 *Suppose S is an $\mathbb{R}^d$-valued continuous semimartingale. If the numeraire portfolio X^{np} exists, then S satisfies the structure condition (SC).*

Proof For every $X^{1,\vartheta} \in \mathcal{X}^{1,++}$, we can write $X^{1,\vartheta} = \mathcal{E}(\pi\cdot S)$ with $\pi := \vartheta/X^{1,\vartheta}$. Moreover, both ϑ and π are S-integrable and hence also in $L^2_{\mathrm{loc}}(M)$, since the processes $X^{1,\vartheta} > 0$, S and M are all continuous. Using the explicit expression $\mathcal{E}(\pi\cdot S) = \exp(\pi\cdot S - \frac{1}{2}\int \pi^{\mathrm{tr}} d\langle M\rangle \pi)$ for the stochastic exponential then gives for every $X^{1,\vartheta} \in \mathcal{X}^{1,++}$ that

$$
\begin{aligned}
\frac{X^{1,\vartheta}}{X^{\mathrm{np}}} &= \frac{\mathcal{E}(\pi\cdot S)}{\mathcal{E}(\pi^{\mathrm{np}}\cdot S)} \\
&= \exp\left((\pi - \pi^{\mathrm{np}})\cdot S - \frac{1}{2}\int \pi^{\mathrm{tr}} d\langle M\rangle \pi + \frac{1}{2}\int (\pi^{\mathrm{np}})^{\mathrm{tr}} d\langle M\rangle \pi^{\mathrm{np}}\right) \\
&= \exp\left((\pi - \pi^{\mathrm{np}})\cdot M - \frac{1}{2}\int (\pi - \pi^{\mathrm{np}})^{\mathrm{tr}} d\langle M\rangle (\pi - \pi^{\mathrm{np}})\right) \\
&\quad \times \exp\left(\int (\pi - \pi^{\mathrm{np}})^{\mathrm{tr}}\left(dA - d\langle M\rangle \pi^{\mathrm{np}}\right)\right),
\end{aligned}
$$

where the last equality is readily verified by multiplying out and collecting terms. But this means that

$$
\frac{X^{1,\vartheta}}{X^{\mathrm{np}}} = \mathcal{E}(L)\exp(B), \tag{1}
$$

where $L := (\pi - \pi^{\mathrm{np}})\cdot M$ (and hence also $\mathcal{E}(L)$) is a continuous local P-martingale and

$$
B := B(\pi) := \int (\pi - \pi^{\mathrm{np}})^{\mathrm{tr}}\left(dA - d\langle M\rangle \pi^{\mathrm{np}}\right)
$$

is a continuous adapted process of finite variation. Because X^{np} is the numeraire portfolio, the left-hand side of (1) is a P-supermartingale for every ϑ, and the right-hand side gives a multiplicative decomposition of that P-special semimartingale as the product of a local martingale and a process of finite variation; see Théorème 6.17 in [13]. But now the uniqueness of the multiplicative decomposition and the fact that $\mathcal{E}(L)\exp(B)$ is a P-supermartingale together imply that $B = B(\pi)$ must be a decreasing process, for every π (coming from a ϑ such that $X^{1,\vartheta}$ lies in $\mathcal{X}^{1,++}$). By a standard variational argument, this is only possible if $A = \int d\langle M\rangle\,\pi^{np}$, and because π^{np} is in $L^2_{loc}(M)$, we see that (SC) is satisfied with $\widehat{\lambda} = \pi^{np}$. $\qquad\square$

Putting everything together, we now obtain the main result of this section.

Theorem 7 *For an $\mathbb{R}^d$-valued continuous semimartingale S, the following are equivalent:*

1) *S satisfies NUPBR.*
2) *The numeraire portfolio X^{np} exists.*
3) *The growth-optimal portfolio X^{go} exists.*
4) *There exists an equivalent supermartingale deflator for $\mathcal{X}^{1,++}$: $\mathcal{Y} \neq \emptyset$.*
5) *There exists a strictly positive local P-martingale density for S: $\mathcal{D}^{++}_{loc}(S, P) \neq \emptyset$.*
6) *S satisfies the structure condition (SC).*
7) *S satisfies the weak structure condition (SC') and $\widehat{\lambda} \in L^2_{loc}(M)$.*
8) *S satisfies the weak structure condition (SC') and $\widehat{K}_T < \infty$ P-a.s.*
9) *S satisfies the weak structure condition (SC') and the minimal martingale density $\widehat{Z}$ exists in $\mathcal{D}^{++}_{loc}(S, P)$.*

In each of these cases, we then have $X^{np} = X^{go} = 1/\widehat{Z}$.

Proof The equivalence of 1)–4) is the statement of Theorem 3. The equivalence of 5)–9) comes from Proposition 4 and directly from the definitions. Lemma 5 shows that 6) implies 2), and Proposition 6 conversely shows that 2) implies 6). The final statement is due to Theorem 3 and Lemma 5. $\qquad\square$

We emphasise once again that all the individual results in this section are known. However, we have not seen anywhere so far the full list of equivalences compiled in Theorem 7, and so we hope that the result may be viewed as useful.

Remarks

1) Because our main interest here lies on the numeraire or growth-optimal portfolio, we have focussed exclusively on the (equivalent) condition NUPBR. There is in fact a whole zoo of absence-of-arbitrage conditions, and an extensive discussion and comparison of these in the framework of a continuous financial market can be found in Chap. 1 of [12]. That work also contains many more details as well as explicit examples and counterexamples.
2) We already know from Proposition 1 that there is a close connection between the log-utility-optimal portfolio and the numeraire portfolio. If S is continuous, this

turns out to be rather transparent. Indeed, if S satisfies (SC), then $1/\widehat{Z}$ maximises $E[\log X_T^{1,\vartheta}]$ over all $X^{1,\vartheta} \in \mathcal{X}^{1,++}$, and the maximal expected utility is

$$E\left[\log(1/\widehat{Z}_T)\right] = \frac{1}{2}E\left[\widehat{K}_T\right] \in [0, \infty].$$

So existence of the log-utility-optimal portfolio with finite maximal expected utility is equivalent to the structure condition (SC) plus the extra requirement that $E[\widehat{K}_T] < \infty$. For more details, we refer to part 1) of Theorem 3.5 in [1].

3) We have already seen in Sect. 2 that the condition NFLVR is equivalent to the combination of the conditions NUPBR and NA. The latter can be formulated as saying that whenever ϑ is admissible, i.e. a-admissible for some $a \geq 0$, $X_T^{0,\vartheta} \geq 0$ P-a.s. implies that $X_T^{0,\vartheta} = 0$ P-a.s. A slightly different condition is (NA$_+$) which stipulates that whenever ϑ is 0-admissible, $X_T^{0,\vartheta} \geq 0$ P-a.s. implies that $X^{0,\vartheta} \equiv 0$ P-a.s. We mention this condition because it is just a little weaker than NUPBR. In fact, if S is continuous, Theorem 3.5 of [27] shows that S satisfies (NA$_+$) if and only if S satisfies the weak structure condition (SC$'$) and the mean-variance tradeoff process $\widehat{K}$ does not jump to $+\infty$, i.e.

$$\inf\left\{t > 0 \,\middle|\, \int_t^{t+\delta} \widehat{\lambda}_u^{\text{tr}} \, d\langle M \rangle_u \widehat{\lambda}_u = +\infty \quad \text{for all } \delta \in (0, T - t]\right\} = \infty.$$

The second condition follows from, but does not imply, $\widehat{K}_T < \infty$ P-a.s., so that (NA$_+$) is a little weaker than (SC), or equivalently NUPBR.

4 Minimal Market Models

The notion of a minimal market model is due to E. Platen and has been introduced in a series of papers with various co-authors; see Chap. 13 of [24] for a recent textbook account. Our goal in this section is to link that concept to the notions introduced in Sects. 2 and 3 and to exhibit a fundamental probabilistic structure result for such models. The presentation here is strongly inspired by Chap. 5 of [12], but extends and simplifies the analysis and results given there. The latter Chap. 5 is in turn based on Chaps. 10 and 13 from [24], although the presentation is a bit different.

The key idea behind the formulation of a minimal market model is an asymptotic diversification result due to E. Platen. Theorem 3.6 of [23] states that under fairly weak assumptions, a sequence of well-diversified portfolios converges in a suitable sense to the growth-optimal portfolio. It is therefore natural to model a broadly based (hence diversified) *index* by the same structure as the growth-optimal portfolio, and to call such a model for an index a *minimal market model*. To study this, we therefore have to take a closer look at the probabilistic behaviour of the growth-optimal portfolio.

We begin by considering a continuous financial market model almost as in Sect. 3. More precisely, let $S = (S_t)_{t \geq 0}$ be a continuous $\mathbb{R}^d$-valued semimartingale on $(\Omega, \mathcal{F}, \mathbb{F}, P)$ with $\mathbb{F} = (\mathcal{F}_t)_{t \geq 0}$. We assume that S satisfies NUPBR on $[0, T]$ for every $T \in (0, \infty)$ so that we can use Theorem 7 for every fixed finite T. We thus obtain $S = S_0 + M + \int d\langle M\rangle\widehat{\lambda}$ with $\widehat{\lambda} \in L^2_{loc}(M)$, the minimal martingale density $\widehat{Z} = \mathcal{E}(-\widehat{\lambda}\cdot M)$ exists, and so does the growth-optimal portfolio X^{go}, which coincides with $1/\widehat{Z}$. All this is true on $[0, \infty)$ since it holds on every interval $[0, T]$ and we can simply paste things together.

The minimal market model for the index $I = (I_t)_{t\geq 0}$ is now defined by

$$I := X^{go} = 1/\widehat{Z} = 1/\mathcal{E}(-\widehat{\lambda}\cdot M). \tag{2}$$

In view of the remark after Proposition 4, this shows that the *minimal market model (MMM)* is directly connected to the *minimal martingale measure (MMM)*, or more precisely to the minimal martingale density $\widehat{Z}$. It also explains why we have deliberately avoided the use of the abbreviation MMM and gives a clear hint where the title of this paper comes from.

The next result is the key for understanding the probabilistic structure of the process I in (2). Recall the notation $\cdot$ for stochastic integrals.

Proposition 8 *Suppose $N = (N_t)_{t\geq 0}$ is a real-valued continuous local martingale null at 0, and $V = (V_t)_{t\geq 0}$ is defined by $V := 1/\mathcal{E}(-N) = \mathcal{E}(N + \langle N\rangle)$. Then:*

1) $\{V\cdot\langle N\rangle_\infty = +\infty\} = \{\langle N\rangle_\infty = +\infty\}$ *P-a.s.*
2) *If $\langle N\rangle_\infty = +\infty$ P-a.s., then*

$$V_t = C_{\varrho_t}, \quad t \geq 0,$$

where $C = (C_t)_{t\geq 0}$ is a squared Bessel process of dimension 4 (a $BESQ^4$, for short) and the time change $t \mapsto \varrho_t$ is explicitly given by the increasing process

$$\varrho_t = \frac{1}{4}\int_0^t \frac{1}{\mathcal{E}(-N)_s}\,d\langle N\rangle_s = \frac{1}{4}\int_0^t V_s\,d\langle N\rangle_s, \quad t \geq 0.$$

3) *The result in 2) is also valid without the assumption that $\langle N\rangle_\infty = +\infty$ P-a.s. if we are allowed to enlarge the underlying probability space in a suitable way.*

Proof The second expression for V follows directly from the explicit formula for the stochastic exponential $\mathcal{E}(-N)$.

1) This is fairly easy, but for completeness we give details. By Proposition V.1.8 of [25], the sets $\{N_\infty := \lim_{t\to\infty} N_t \text{ exists in } \mathbb{R}\}$ and $\{\langle N\rangle_\infty < \infty\}$ are equal with probability 1. So on $\{\langle N\rangle_\infty < \infty\}$, the process $\mathcal{E}(-N)_s = \exp(-N_s - \frac{1}{2}\langle N\rangle_s)$ converges to $\exp(-N_\infty - \frac{1}{2}\langle N\rangle_\infty) > 0$ P-a.s. which implies that $s \mapsto V_s = 1/\mathcal{E}(-N)_s$ remains bounded P-a.s. as $s \to \infty$. Hence

$$\big(V\cdot\langle N\rangle_\infty\big)(\omega) \leq \text{const.}\,(\omega)\langle N\rangle_\infty(\omega) < \infty \quad P\text{-a.s. on } \{\langle N\rangle_\infty < \infty\}.$$

On the other hand, $N \in \mathcal{M}^c_{0,loc}$ implies by Fatou's lemma that $s \mapsto \mathcal{E}(-N)_s$ is a nonnegative supermartingale and therefore converges P-a.s. to a finite limit (which might be 0) as $s \to \infty$. So $s \mapsto 1/V_s = \mathcal{E}(-N)_s$ is bounded P-a.s. and thus

$$\big(V\cdot\langle N\rangle_\infty\big)(\omega) \geq \frac{1}{\text{const.}\,(\omega)}\langle N\rangle_\infty(\omega) = +\infty \quad P\text{-a.s. on } \{\langle N\rangle_\infty = +\infty\}.$$

This proves the assertion.

2) Because $V = \mathcal{E}(N + \langle N \rangle)$ satisfies $dV = V\,dN + V\,d\langle N \rangle$, defining $L \in \mathcal{M}_{0,\text{loc}}^c$ by $dL = \frac{1}{2}\sqrt{V}\,dN$ yields $d\langle L \rangle = \frac{1}{4}V\,d\langle N \rangle$ and

$$dV = 2\sqrt{V}\,dL + 4\,d\langle L \rangle.$$

By 1), $\langle N \rangle_\infty = +\infty$ P-a.s. implies that $\langle L \rangle_\infty = \frac{1}{4}V \cdot \langle N \rangle_\infty = +\infty$ P-a.s., and so the Dambis–Dubins–Schwarz theorem (see Theorem V.1.6 in [25]) yields the existence of some Brownian motion $B = (B_t)_{t \geq 0}$ such that $L_t = B_{\langle L \rangle_t}$ for $t \geq 0$. Hence

$$dV_t = 2\sqrt{V_t}\,dB_{\langle L \rangle_t} + 4\,d\langle L \rangle_t,$$

and if $t \mapsto \tau_t$ denotes the inverse of $t \mapsto \langle L \rangle_t$, we see that $C_t := V_{\tau_t}$, $t \geq 0$, satisfies

$$dC_t = 2\sqrt{C_t}\,dB_t + 4\,dt,$$

so that C is a BESQ^4 process; see Chap. XI of [25]. Finally, since τ and $\langle L \rangle$ are inverse to each other, $V_t = C_{\langle L \rangle_t}$ and as claimed, the time change $t \mapsto \varrho_t$ is given by

$$\varrho_t = \langle L \rangle_t = \frac{1}{4}V \cdot \langle N \rangle_t, \quad t \geq 0.$$

3) If we may enlarge the probability space to guarantee the existence of an independent Brownian motion, we can still use the Dambis–Dubins–Schwarz theorem; see Theorem V.1.7 in [25]. Thus the same argument as for 2) still works. □

In view of Theorem 7, applying Proposition 8 with $N = \widehat{\lambda} \cdot M$ and noting that $\langle N \rangle = \widehat{K}$ now immediately gives the main result of this section.

Theorem 9 *Let $S = (S_t)_{t \geq 0}$ be an $\mathbb{R}^d$-valued continuous semimartingale and suppose that S satisfies NUPBR on $[0, T]$ for every $T \in (0, \infty)$ (or equivalently that the growth-optimal portfolio X^{go} exists for every finite time horizon T). Denote by $\widehat{Z} = \mathcal{E}(-\widehat{\lambda} \cdot M)$ the minimal martingale density for S and model the index $I = (I_t)_{t \geq 0}$ by $I := X^{go} = 1/\widehat{Z}$. Then $I_t = C_{\varrho_t}$, $t \geq 0$, is a time change of a squared Bessel process C of dimension 4, with the time change given by*

$$\varrho_t = \frac{1}{4}\int_0^t \frac{1}{\widehat{Z}_s}\,d\widehat{K}_s, \quad t \geq 0. \tag{3}$$

Theorem 9 is a generalisation of Proposition 5.8 in [12], where the same conclusion is obtained under the more restrictive assumption that S is given by a multidimensional Itô process model which is complete. But most of the key ideas for the above proof can already be seen in that result of [12]. Even considerably earlier, the same result as in [12] can be found in Sect. 3.1 of [22], although it is not stated as a theorem. The main contribution of Theorem 9 is to show that neither completeness nor the Itô process structure are needed.

Example 10 To illustrate the theory developed so far, we briefly consider the standard, but *incomplete multidimensional Itô process model* for S. Suppose discounted asset prices are given by the stochastic differential equations

$$\frac{dS_t^i}{S_t^i} = (\mu_t^i - r_t)\,dt + \sum_{k=1}^m \sigma_t^{ik}\,dW_t^k \quad \text{for } i = 1, \ldots, d \text{ and } 0 \le t \le T.$$

Here $W = (W^1, \ldots, W^m)^{\mathrm{tr}}$ is an $\mathbb{R}^m$-valued standard Brownian motion on $(\Omega, \mathcal{F}, P)$ with respect to $\mathbb{F}$; there is no assumption that $\mathbb{F}$ is generated by W, and we only suppose that $m \ge d$ so that we have at least as many sources of uncertainty as risky assets available for trade. The processes $r = (r_t)_{t\ge 0}$ (the *instantaneous short rate*), $\mu^i = (\mu_t^i)_{t\ge 0}$ (the *instantaneous drift rate* of asset i) for $i = 1, \ldots, d$ and $\sigma^{ik} = (\sigma_t^{ik})_{t\ge 0}$ for $i = 1, \ldots, d$ and $k = 1, \ldots, m$ (the *instantaneous volatilities*) are predictable (or even progressively measurable) and satisfy

$$\int_0^T |r_u|\,du + \sum_{i=1}^d \int_0^T |\mu_u^i|\,du + \sum_{i=1}^d \sum_{k=1}^m \int_0^T (\sigma_u^{ik})^2\,du < \infty$$

P-a.s. for each $T \in (0, \infty)$.

Moreover, to avoid redundant assets (locally in time), we assume that for each $t \ge 0$,

the $d \times m$-matrix σ_t has P-a.s. full rank d.

Then $\sigma_t \sigma_t^{\mathrm{tr}}$ is P-a.s. invertible and we can define the predictable (or progressively measurable) $\mathbb{R}^m$-valued process $\lambda = (\lambda_t)_{t\ge 0}$ by

$$\lambda_t := \sigma_t^{\mathrm{tr}} (\sigma_t \sigma_t^{\mathrm{tr}})^{-1} (\mu_t - r_t \mathbf{1}), \quad t \ge 0$$

with $\mathbf{1} = (1, \ldots, 1)^{\mathrm{tr}} \in \mathbb{R}^d$. Our final assumption is that $\lambda \in L^2_{\mathrm{loc}}(W)$ or equivalently that

$$\int_0^T |\lambda_u|^2\,du < \infty \quad P\text{-a.s. for each } T \in (0, \infty). \tag{4}$$

Sometimes λ (instead of $\widehat{\lambda}$ below) is called the *(instantaneous) market price of risk* for S.

It is straightforward to verify that the canonical decomposition of the continuous semimartingale S^i is given by

$$dM_t^i = S_t^i \sum_{k=1}^m \sigma_t^{ik}\,dW_t^k \quad \text{and} \quad dA_t^i = S_t^i (\mu_t^i - r_t)\,dt,$$

so that

$$d\langle M^i, M^j \rangle_t = S_t^i S_t^j \sum_{k=1}^m \sigma_t^{ik} \sigma_t^{jk}\,dt = S_t^i S_t^j (\sigma_t \sigma_t^{\mathrm{tr}})^{ij}\,dt.$$

The weak structure condition (SC$'$) is therefore satisfied with the $\mathbb{R}^d$-valued process $\widehat{\lambda} = (\widehat{\lambda}_t)_{t\ge 0}$ given by

$$\widehat{\lambda}_t^i = \frac{1}{S_t^i} \left((\sigma_t \sigma_t^{\mathrm{tr}})^{-1} (\mu_t - r_t \mathbf{1}) \right)^i \quad \text{for } i = 1, \ldots, d \text{ and } t \ge 0.$$

This gives $\int \widehat{\lambda}\,dM = \int \lambda\,dW$ and therefore the mean-variance tradeoff process as

$$\widehat{K}_t = \langle \widehat{\lambda} \cdot M \rangle_t = \langle \lambda \cdot W \rangle_t = \int_0^t |\lambda_u|^2\,du \quad \text{for } t \ge 0,$$

and so (4) immediately implies that S satisfies (SC) on $[0, T]$ for each $T \in (0, \infty)$. Therefore this model directly falls into the scope of Theorem 7 (for each fixed T) and of Theorem 9. In particular, we of course recover Proposition 5.8 of [12] or the result from Sect. 13.1 of [24] as a special case (for $m = d$, even without the stronger assumptions imposed there).

Remark To obtain a more concrete model just for the index I, Theorem 9 makes it very tempting to start with a BESQ^4 process C and choose some time change $t \mapsto \varrho_t$ to then define the index by

$$I_t := C_{\varrho_t}, \quad t \geq 0. \tag{5}$$

Depending on the choice of ϱ, this may provide a good fit to observed data and hence yield a plausible and useful model; see Sect. 13.2 of [24] on the stylised minimal market model. However, a word of caution seems indicated here. In fact, if we accept (as in this section) the basic modelling of the index I by the growth-optimal portfolio X^{go}, then the approach (5) raises the following *inverse problem*:

Given a time change $t \mapsto \varrho_t$, when does there exist an $\mathbb{R}^d$-valued continuous semimartingale $S = (S_t)_{t \geq 0}$ which satisfies the structure condition (SC) and whose growth-optimal portfolio is given by the process I defined by (5)?

We do not have an answer to this question, but we suspect that the problem is non-trivial. One first indication for this is the observation that the explicit form (3) of the time change in Theorem 9 implies that

$$\frac{d\widehat{K}_t}{d\varrho_t} = 4\widehat{Z}_t, \quad t \geq 0. \tag{6}$$

Since the right-hand side of (6) is a local martingale, the processes $\widehat{K}$ and ϱ cannot be chosen with an arbitrarily simple structure — for example they cannot both be deterministic.

Acknowledgements   MS gratefully acknowledges financial support by the National Centre of Competence in Research "Financial Valuation and Risk Management" (NCCR FINRISK), Project D1 (Mathematical Methods in Financial Risk Management). The NCCR FINRISK is a research instrument of the Swiss National Science Foundation. We also thank an anonymous referee for helpful suggestions and useful comments.

References

1. Amendinger, J., Imkeller, P., Schweizer, M.: Additional logarithmic utility of an insider. Stoch. Process. Appl. **75**, 263–286 (1998)
2. Ansel, J.P., Stricker, C.: Couverture des actifs contingents et prix maximum. Ann. Inst. Henri Poincaré Probab. Stat. **30**, 303–315 (1994)
3. Becherer, D.: The numeraire portfolio for unbounded semimartingales. Finance Stoch. **5**, 327–341 (2001)
4. Chatelain, M., Stricker, C.: On componentwise and vector stochastic integration. Math. Finance **4**, 57–65 (1994)

5. Christensen, M.M., Larsen, K.: No arbitrage and the growth optimal portfolio. Stoch. Anal. Appl. **25**, 255–280 (2007)
6. Delbaen, F., Schachermayer, W.: A general version of the fundamental theorem of asset pricing. Math. Ann. **300**, 463–520 (1994)
7. Delbaen, F., Schachermayer, W.: The existence of absolutely continuous local martingale measures. Ann. Appl. Probab. **5**, 926–945 (1995)
8. Delbaen, F., Schachermayer, W.: The fundamental theorem of asset pricing for unbounded stochastic processes. Math. Ann. **312**, 215–250 (1998)
9. Delbaen, F., Schachermayer, W.: The Mathematics of Arbitrage. Springer, Berlin (2006)
10. Föllmer, H., Schweizer, M.: Hedging of contingent claims under incomplete information. In: Davis, M.H.A., Elliott, R.J. (eds.) Applied Stochastic Analysis. *Stochastics Monographs*, vol. 5, pp. 389–414. Gordon and Breach, London (1991)
11. Goll, T., Kallsen, J.: A complete explicit solution to the log-optimal portfolio problem. Ann. Appl. Probab. **13**, 774–799 (2003)
12. Hulley, H.: Strict Local Martingales in Continuous Financial Market Models. PhD thesis, University of Technology, Sydney (2009)
13. Jacod, J.: Calcul Stochastique et Problèmes de Martingales. *Lecture Notes in Mathematics*, vol. 714. Springer, Berlin (1979)
14. Jacod, J.: Intégrales stochastiques par rapport à une semimartingale vectorielle et changements de filtration. In: Séminaire de Probabilités XIV. *Lecture Notes in Mathematics*, vol. 784, pp. 161–172. Springer, Berlin (1980)
15. Kabanov, Y.M.: On the FTAP of Kreps–Delbaen–Schachermayer. In: Kabanov, Y.M., Rozovskii, B.L., Shiryaev, A.N. (eds.) Statistics and Control of Stochastic Processes: The Liptser Festschrift, pp. 191–203. World Scientific, Singapore (1997)
16. Kabanov, Y.M., Stricker, C.: Remarks on the true no-arbitrage property. In: Séminaire de Probabilités XXXVIII. *Lecture Notes in Mathematics*, vol. 1857, pp. 186–194. Springer, Berlin (2005)
17. Karatzas, I., Fernholz, R.: Stochastic portfolio theory: An overview. In: Ciarlet, P.G. (ed.) Handbook of Numerical Analysis, vol. XV, pp. 89–167. North Holland, Amsterdam (2009)
18. Karatzas, I., Kardaras, C.: The numéraire portfolio in semimartingale financial models. Finance Stoch. **11**, 447–493 (2007)
19. Kardaras, C.: Market viability via absence of arbitrages of the first kind. Preprint, Boston University, September 2009, http://arxiv.org/abs/0904.1798v2 (2009)
20. Kramkov, D., Schachermayer, W.: The asymptotic elasticity of utility functions and optimal investment in incomplete markets. Ann. Appl. Probab. **9**, 904–950 (1999)
21. Loewenstein, M., Willard, G.A.: Local martingales, arbitrage, and viability. Free snacks and cheap thrills. Econ. Theory **16**, 135–161 (2000)
22. Platen, E.: Modeling the volatility and expected value of a diversified world index. Int. J. Theor. Appl. Finance **7**, 511–529 (2004)
23. Platen, E.: Diversified portfolios with jumps in a benchmark framework. Asia-Pacific Financ. Mark. **11**, 1–22 (2005)
24. Platen, E., Heath, D.: A Benchmark Approach to Quantitative Finance. Springer, Berlin (2006)
25. Revuz, D., Yor, M.: Continuous Martingales and Brownian Motion, 3rd edn. Springer, Berlin (2005)
26. Schweizer, M.: On the minimal martingale measure and the Föllmer–Schweizer decomposition. Stoch. Anal. Appl. **13**, 573–599 (1995)
27. Strasser, E.: Characterization of arbitrage-free markets. Ann. Appl. Probab. **15**, 116–124 (2005)

The Economic Plausibility of Strict Local Martingales in Financial Modelling

Hardy Hulley

Abstract The context for this article is a continuous financial market consisting of a risk-free savings account and a single non-dividend-paying risky security. We present two concrete models for this market, in which strict local martingales play decisive roles. The first admits an equivalent risk-neutral probability measure under which the discounted price of the risky security is a strict local martingale, while the second model does not even admit an equivalent risk-neutral probability measure, since the putative density process for such a measure is itself a strict local martingale. We highlight a number of apparent anomalies associated with both models that may offend the sensibilities of the classically-educated reader. However, we also demonstrate that these issues are easily resolved if one thinks economically about the models in the right way. In particular, we argue that there is nothing inherently objectionable about either model.

1 Introduction

The setting for this article is a continuous financial market consisting of a single non-dividend-paying risky security and a risk-free money-market account. We consider two models for the price of the risky security, both of which are affected by the presence of strict local martingales. The first model admits an equivalent risk-neutral probability measure under which the discounted price of the risky security is a strict local martingale. In particular, this model satisfies the strong no-arbitrage requirement of "no free lunch with vanishing risk" (NFLVR) (see [5]). The second model, on the other hand, does not even admit an equivalent risk-neutral probability measure, since the density process associated with the putative equivalent risk-neutral probability measure is a strict local martingale. Nevertheless, we shall see

For Eckhard Platen on the occasion of his 60th birthday, with gratitude.

H. Hulley (✉)
School of Finance and Economics, University of Technology, Sydney, P.O. Box 123, Broadway, NSW 2007, Australia
e-mail: hardy.hulley@uts.edu.au

C. Chiarella, A. Novikov (eds.), *Contemporary Quantitative Finance*,
DOI 10.1007/978-3-642-03479-4_4, © Springer-Verlag Berlin Heidelberg 2010

that this model does possess a numéraire portfolio, which ensures that it satisfies the weaker no-arbitrage requirement of "no unbounded profit with bounded risk" (NUPBR) (see [15]).

Following the terminology introduced by [9], we shall say that the first model is an example of a "stock price bubble," while the second model is an example of a "bond price bubble." Stock price bubbles have recently become a popular topic in the literature (see e.g. [3, 7, 9, 11, 12]). Although such models remain firmly within the ambit of the risk-neutral approach to contingent claim pricing, they nevertheless exhibit a number of anomalies that have occupied the research articles cited above. For example, the risk-neutral forward price of the risky security is strictly less than its spot price grossed-up at the risk-free interest rate. Furthermore, the risk-neutral prices of European puts and calls on the risky security no longer obey the put-call parity relationship.

By contrast, bond price bubbles have received comparatively little attention in the literature—possibly due to a misguided concern about their arbitrage properties. The only obvious exception is the so-called "benchmark approach" of [17], which develops a systematic approach to the pricing and hedging of contingent claims in financial markets that possess a numéraire portfolio, without necessarily admitting an equivalent risk-neutral probability measure. The resulting pricing methodology is referred to as "real-world pricing," since it does not rely on any equivalent change of probability measure. At first blush, the real-world prices of contingent claims for the bond price bubble also appear to exhibit a number of anomalies. For example, the real-world price of a zero-coupon bond turns out to be strictly less than the discounted value of its principal. It is also possible to generate a risk-free profit from no initial investment, by constructing a portfolio whose value is bounded from below.

In this article we analyze the pricing and hedging of contingent claims written on the risky security, for both the stock price bubble and the bond price bubble. Our main objective is to demonstrate that the anomalies associated with strict local martingales are actually not so strange after all. In particular, if one thinks economically about the strict local martingales in our two models, then the apparently strange behaviour of contingent claim prices begins to seem quite natural.

The remainder of the paper is structured as follows: Sect. 2 introduces the two models, both of which are based on a Bessel process of dimension three. Section 3 then derives pricing formulae and hedge ratios for a number of European contingent claims written on the risky security, for the case of the stock price bubble. By analyzing the behaviour of these pricing formulae and hedge ratios, we argue that the anomalies associated with the claim prices for the stock price bubble are easily resolved. Section 4 then performs a similar analysis, for the case of the bond price bubble.

2 An Overview of the Two Models

As stated in the introduction, the subject of this article is a financial market consisting of a single non-dividend-paying risky security and a risk-free money-market account. We are concerned with the following two models for the price S of the risky security:

$$dS_t = -S_t^2 \, d\beta_t \tag{1a}$$

and

$$dS_t = \frac{1}{S_t} \, dt + d\beta_t, \tag{1b}$$

for all $t \in \mathbb{R}_+$. Here β is a standard Brownian motion residing on a filtered probability space $(\Omega, \mathscr{F}, \mathfrak{F}, \mathsf{P})$, whose filtration $\mathfrak{F} = (\mathscr{F}_t)_{t \in \mathbb{R}_+}$ satisfies the usual conditions. We also adopt the standard convention of assuming that the risk-free interest rate is zero, so that the value of the money-market account is given by $B = 1$.

Observe that the process specified by (1a) is by inspection a local martingale. The first of the two models presented above thus clearly satisfies the NFLVR condition, since P is itself the (unique) equivalent risk-neutral probability measure. In particular, the density process for the equivalent risk-neutral probability measure for that model is given by $Z = 1$. However, there is more to it than that, since the process described by (1a) is in fact the inverse of a Bessel process of dimension three, and is therefore a well-known example of a strict local martingale (see e.g. [18, Ex. V.2.13, p. 194]). This establishes the first model as an example of a stock price bubble.

In the case of the second model, (1b) expresses the price of the risky security as a Bessel process of dimension three. It then follows that S is the numéraire portfolio for that model (in the sense of [2, Definition 4.1]), since $B/S = 1/S$ and $S/S = 1$ are both non-negative local martingales, and consequently also supermartingales, by Fatou's lemma. The model therefore satisfies the NUPBR condition, by virtue of [15, Theorem 4.12].[1] However, it does not satisfy the NFLVR condition, since the density process for the putative equivalent risk-neutral probability measure may be expressed in terms of the numéraire portfolio as follows: $Z = (S_0/B_0)(B/S) = S_0/S$, which is a strict local martingale.[2] This establishes the second model as an example of a bond price bubble.

We acknowledge that the models presented by (1a) and (1b) have already been studied in detail by [4] and [6], respectively. However, whereas the above-mentioned articles focus on questions of arbitrage, we wish to address the pricing and hedging of contingent claims. Our paper may therefore be regarded as a companion for those articles.

We begin our discussion of contingent claim pricing by introducing a finite maturity date $T > 0$ and a payoff function $h : (0, \infty) \to \mathbb{R}_+$, which satisfy the following three conditions:

---

[1] Actually, this is not completely true, since Theorem 4.12 of [15] requires the value of the numéraire portfolio to be a "semimartingale up to infinity." The Bessel process of dimension three does not meet this requirement, since its sample paths diverge in the limit as time goes to infinity. Nevertheless, this provides no practical obstacle, since our applications only involve a finite time-horizon.

[2] Strictly speaking, we are abusing terminology by calling Z a density process in this instance, since most of the literature adopts the convention that density processes are positive martingales. In the case of the second model, it may therefore be more appropriate to describe Z as a "predensity process" or a "candidate density process." However, in the interest of maintaining a uniform terminology for discussing both models, we shall continue to refer to Z simply as a density process, with the understanding that the reader is sufficiently flexible on this point.

(i) h is continuous and of polynomial growth;

(ii) $\mathsf{E}\big(Z_T h(S_T)\big) < \infty$; and

(iii) $\mathsf{P}\big(h(S_T) > 0\big) > 0$.

We now consider a European contingent claim on the risky security, with maturity T and payoff $h(S_T)$. We denote the value of this claim by the process $V^h = (V_t^h)_{t \in [0,T]}$, which is determined as follows:

$$V_t^h := \frac{1}{Z_t} \mathsf{E}\big(Z_T h(S_T) \,\big|\, \mathscr{F}_t\big), \tag{2}$$

for all $t \in [0, T]$. Observe that $V_t^h < \infty$, for all $t \in [0, T]$, by virtue of condition (ii) above, while condition (iii) ensures that $V_t^h > 0$, for all $t \in [0, T)$. Also, note that since $V_T^h = h(S_T)$, it easily follows that the process $(Z_t V_t^h)_{t \in [0,T]}$ is a martingale.

We shall refer to (2) as the "real-world" price of the claim, since it does not require any transformation of probability measure.[3] For the first model it is easily seen that (2) agrees with the risk-neutral price of the claim, since Z is a martingale. However, it is also clear that (2) remains well-defined for the second model. In that case there is no equivalent risk-neutral probability measure, since Z is a strict local martingale, and hence risk-neutral pricing is infeasible. In other words, real-world pricing offers a proper extension of the risk-neutral pricing concept.

An important observation for the two models under consideration is that

$$\frac{Z_T}{Z_t} = \begin{cases} 1 & \text{if the risky security price satisfies (1a),} \\ \frac{S_t}{S_T} & \text{if the risky security price satisfies (1b),} \end{cases}$$

for all $t \in [0, T]$. We may therefore exploit the Markov property of the price of the risky security, for both models, by writing $V_t^h = V^h(t, S_t)$, for all $t \in [0, T]$, where the "pricing function" $V^h : [0, T] \times (0, \infty) \to \mathbb{R}_+$ that appears in this expression may be computed as follows:

$$V^h(t, S) := \begin{cases} \mathsf{E}_{t,S}\big(h(S_T)\big) & \text{if the risky security price satisfies (1a),} \\ S\,\mathsf{E}_{t,S}\big(\frac{h(S_T)}{S_T}\big) & \text{if the risky security price satisfies (1b),} \end{cases} \tag{3}$$

for all $(t, S) \in [0, T] \times (0, \infty)$.[4] The following proposition will allow us to obtain convenient expressions for the pricing functions of a number of standard European contingent claims in Sects. 3 and 4:

---

[3] The reader is directed to Sect. 9.1 of [17] for a more detailed account of the real-world approach to contingent claim pricing. Note that real-world prices are expressed there in terms of the numéraire portfolio, whereas (2) expresses them in terms of the density process. These two formulations are equivalent, by dint of the fact that the density process is proportional to the numéraire-denominated money-market account.

[4] As usual, $\mathsf{E}_{t,S}$ denotes the expected value operator with respect to the probability measure $\mathsf{P}_{t,S}$, under which the price of the risky security at time $t \in \mathbb{R}_+$ is $S \in (0, \infty)$. Of course, we are abusing notation slightly by using S to denote a particular value for the price of the risky security, as well the price process for this asset.

Proposition 1 *A Bessel process ϱ of dimension three has the following (truncated) moments*:

$$E_x\left(1_{\{\varrho_t \leq z\}}\frac{1}{\varrho_t}\right) = \frac{1}{x}\left(\Phi\left(-\frac{x-z}{\sqrt{t}}\right) - \Phi\left(\frac{x+z}{\sqrt{t}}\right) + \Phi\left(\frac{x}{\sqrt{t}}\right) - \Phi\left(-\frac{x}{\sqrt{t}}\right)\right), \quad (4)$$

$$E_x\left(\frac{1}{\varrho_t}\right) = \frac{1}{x}\left(\Phi\left(\frac{x}{\sqrt{t}}\right) - \Phi\left(-\frac{x}{\sqrt{t}}\right)\right), \quad (5)$$

$$E_x\left(1_{\{\varrho_t > z\}}\frac{1}{\varrho_t}\right) = \frac{1}{x}\left(\Phi\left(\frac{x+z}{\sqrt{t}}\right) - \Phi\left(-\frac{x-z}{\sqrt{t}}\right)\right), \quad (6)$$

$$P_x(\varrho_t \leq z) = \Phi\left(\frac{x+z}{\sqrt{t}}\right) - \Phi\left(\frac{x-z}{\sqrt{t}}\right) + \frac{\sqrt{t}}{x}\left(\varphi\left(\frac{x+z}{\sqrt{t}}\right) - \varphi\left(\frac{x-z}{\sqrt{t}}\right)\right), \quad (7)$$

and

$$P_x(\varrho_t > z) = \Phi\left(-\frac{x+z}{\sqrt{t}}\right) + \Phi\left(\frac{x-z}{\sqrt{t}}\right) - \frac{\sqrt{t}}{x}\left(\varphi\left(\frac{x+z}{\sqrt{t}}\right) - \varphi\left(\frac{x-z}{\sqrt{t}}\right)\right), \quad (8)$$

for all $t > 0$, $x, y \in (0, \infty)$ and $z > 0$.[5] In these expressions Φ denotes the cumulative distribution function of a standard normal random variable, while φ is the associated density function.[6]

Proof To begin with, [18, Proposition VI.3.1, p. 251] provides the following expression for the transition density of ϱ:

$$q(t, x, y) := \frac{\partial}{\partial y}P_x(\varrho_t \leq y) = \frac{x}{t}\left(\frac{y}{x}\right)^{\frac{3}{2}}I_{\frac{1}{2}}\left(\frac{xy}{t}\right)\exp\left(-\frac{x^2 + y^2}{2t}\right),$$

---

[5]In this case E_x should be understood as the expected value operator with respect to the probability measure P_x, under which $\varrho_0 = x$, for all $x \in (0, \infty)$.

[6]In other words,

$$\Phi(z) := \int_{-\infty}^{z}\underbrace{\frac{1}{\sqrt{2\pi}}e^{-\zeta^2/2}}_{\varphi(\zeta)}\,d\zeta,$$

for all $z \in \mathbb{R}$. Also, recall that the mapping

$$\mathbb{R} \ni \zeta \mapsto \frac{1}{\sigma}\varphi\left(\frac{\zeta - \mu}{\sigma}\right)$$

is the probability density function of a normal random variable with mean $\mu \in \mathbb{R}$ and standard deviation $\sigma > 0$.

for all $t > 0$ and $x, y \in (0, \infty)$. The function $I_{1/2}$ that appears above is the modified Bessel function of the first kind with index one half (see e.g. [1, Chap. 9]). It satisfies the following identity:

$$I_{\frac{1}{2}}(z) = \sqrt{\frac{2}{\pi z}} \sinh z = \frac{1}{\sqrt{2\pi z}}(e^z - e^{-z}),$$

for all $z \in \mathbb{R}$, according to [16, Eq. 5.8.5, p. 112]. By combining these two expressions, we obtain

$$q(t, x, y) = \frac{y}{x}\left(\frac{1}{\sqrt{t}}\varphi\left(\frac{y-x}{\sqrt{t}}\right) - \frac{1}{\sqrt{t}}\varphi\left(\frac{y+x}{\sqrt{t}}\right)\right),$$

for all $t > 0$ and $x, y \in (0, \infty)$. Now, fix $t > 0$, $x \in (0, \infty)$ and $z > 0$, and observe that

$$\mathsf{E}_x\left(\mathbf{1}_{\{\varrho_t \le z\}}\frac{1}{\varrho_t}\right) = \int_0^z \frac{1}{y}q(t, x, y)\,dy$$

$$= \frac{1}{x}\left(\int_0^z \frac{1}{\sqrt{t}}\varphi\left(\frac{y-x}{\sqrt{t}}\right)dy - \int_0^z \frac{1}{\sqrt{t}}\varphi\left(\frac{y+x}{\sqrt{t}}\right)dy\right).$$

Since the final expression above involves the difference between the truncated zeroth moments of two normal random variables, we obtain (4) from [13, Table 1], for example. We may then derive (5) from (4), by an application of the monotone convergence theorem, while (6) follows directly from (4) and (5). Next, we see that

$$\mathsf{P}_x(\varrho_t \le z) = \int_0^z q(t, x, y)\,dy = \frac{1}{x}\left(\int_0^z \frac{y}{\sqrt{t}}\varphi\left(\frac{y-x}{\sqrt{t}}\right)dy - \int_0^z \frac{y}{\sqrt{t}}\varphi\left(\frac{y+x}{\sqrt{t}}\right)dy\right).$$

In this case we recognize that the final expression above involves the difference between the truncated first moments of two normal random variables. Once again, we may look these up in [13, Table 1], to get (7). Finally, (8) is obtained directly from (7). □

We turn our attention now to the question of how to hedge the contingent claim introduced above, for the two models under consideration. To begin with, throughout this article the phrase "viable portfolio" should be understood to mean a self-financing portfolio consisting of holdings in the risky security and the money-market account, whose value is strictly positive over $[0, T)$ and non-negative at time T.[7] Such a portfolio may be presented by a "trading strategy" $\pi = (\pi_t)_{t \in [0,T)}$, which specifies the fraction of its wealth invested in the risky security. Given an initial endowment $x > 0$ and a trading strategy $\pi = (\pi_t)_{t \in [0,T)}$ for a viable portfolio, the

---

[7]Strictly speaking, the last requirement is redundant, since all self-financing portfolios in the financial market introduced above have continuous sample paths.

associated "wealth process" $W^{x,\pi} = (W_t^{x,\pi})_{t \in [0,T]}$ is easily seen to satisfy one of the following two stochastic differential equations:

$$dW_t^{x,\pi} = -\pi_t S_t W_t^{x,\pi} d\beta_t \qquad (9a)$$

or

$$dW_t^{x,\pi} = \frac{\pi_t}{S_t^2} W_t^{x,\pi} dt + \frac{\pi_t}{S_t} W_t^{x,\pi} d\beta_t, \qquad (9b)$$

for all $t \in [0, T)$, depending on whether the price of the risky security is modelled by (1a) or by (1b).[8]

As an aside, observe that (9) suggests the following stochastic exponential representation for the wealth of a viable portfolio with initial endowment $x > 0$ and trading strategy $\pi = (\pi_t)_{t \in [0,T)}$:

$$W_t^{x,\pi} = x \mathscr{E} \left(\int_0^\cdot \pi_s \, dX_s \right)_t,$$

for all $t \in [0, T]$. The process X in the expression above is given by

$$dX_t = -S_t \, d\beta_t \qquad (10a)$$

or

$$dX_t = \frac{1}{S_t^2} dt + \frac{1}{S_t} d\beta_t, \qquad (10b)$$

for all $t \in \mathbb{R}_+$, depending on whether the price of the risky security is determined by (1a) or by (1b). The requirement that the portfolio wealth is strictly positive over $[0, T)$ then translates into the following concrete condition:

$$\int_0^t \pi_s^2 \, d\langle X \rangle_s < \infty,$$

for all $t \in [0, T)$. Moreover, we obtain the following correspondence between events:

$$\left\{ W_T^{x,\pi} = 0 \right\} = \left\{ \int_0^T \pi_s^2 \, d\langle X \rangle_s = \infty \right\},$$

by an application of the law of large numbers for local martingales (see e.g. [18, Ex. V.1.16, pp. 186–187]).

It will be useful for us to identify two notions of hedging, for the contingent claim introduced earlier. The first of these corresponds to the situation when the value of a viable portfolio precisely matches the payoff of the claim at its maturity, while the second describes the situation when the value of a viable portfolio precisely matches the real-world price of the claim at all times up to its maturity:

---

[8]Of course, π should belong to an appropriate class of processes in order for the stochastic integrals in (9) to be well-defined. For example, we could stipulate that all trading strategies should be progressively measurable.

Definition 2 Consider a viable portfolio, which is determined by an initial endowment $x > 0$ and a trading strategy $\pi = (\pi_t)_{t \in [0,T)}$.

(i) The portfolio is said to "hedge the claim" if and only if $W_T^{x,\pi} = h(S_T)$.
(ii) The portfolio is said to "replicate the claim" if and only if $W_t^{x,\pi} = V_t^h$, for all $t \in [0, T]$.

It is immediately evident from the definition above that any viable portfolio that replicates the claim must also hedge it. To make the relationship between hedging and replication more explicit, suppose that the initial endowment $x^{hed} > 0$ and trading strategy $\pi^{hed} = (\pi_t^{hed})_{t \in [0,T)}$ specify a viable portfolio that hedges the claim. Similarly, let the initial endowment $x^{rep} > 0$ and trading strategy $\pi^{rep} = (\pi_t^{rep})_{t \in [0,T)}$ specify a viable portfolio that replicates the claim. We begin by observing that the process $\left(Z_t W_t^{x^{rep},\pi^{rep}} \right)_{t \in [0,T]}$ is a martingale, since

$$Z_t W_t^{x^{rep},\pi^{rep}} = Z_t V_t^h = \mathsf{E}\left(Z_T V_T^h \mid \mathscr{F}_t \right) = \mathsf{E}\left(Z_T W_T^{x^{rep},\pi^{rep}} \mid \mathscr{F}_t \right),$$

for all $t \in [0, T]$, by virtue of the already established fact that $(Z_t V_t^h)_{t \in [0,T]}$ is a martingale. Next, we note that the process $\left(Z_t W_t^{x^{hed},\pi^{hed}} \right)_{t \in [0,T]}$ is a local martingale. In the case of the stock price bubble, this is an easy consequence of (9a) and the fact that $Z = 1$. In the case of the bond price bubble, the stochastic integration by parts formula yields

$$d\left(ZW^{x^{hed},\pi^{hed}} \right)_t = -\left(1 - \pi_t^{hed} \right) \frac{1}{S_t} \left(ZW^{x^{hed},\pi^{hed}} \right)_t d\beta_t,$$

for all $t \in [0, T)$, with the help (9b) and the fact that $Z = S_0/S$. We may therefore deduce that $\left(Z_t W_t^{x^{hed},\pi^{hed}} \right)_{t \in [0,T]}$ is in fact a supermartingale, for both models, by an application of Fatou's lemma. Finally, putting all of the above together, we obtain

$$Z_t W_t^{x^{rep},\pi^{rep}} = \mathsf{E}\left(Z_T W_T^{x^{rep},\pi^{rep}} \mid \mathscr{F}_t \right) = \mathsf{E}\left(Z_T W_T^{x^{hed},\pi^{hed}} \mid \mathscr{F}_t \right) \leq Z_t W_t^{x^{hed},\pi^{hed}},$$

for all $t \in [0, T]$. In particular, since $Z > 0$ for both the stock price bubble and the bond price bubble, we must have

$$W_t^{x^{rep},\pi^{rep}} \leq W_t^{x^{hed},\pi^{hed}}, \tag{11}$$

for all $t \in [0, T]$, irrespective of which model is chosen.

The importance of (11) lies in the justification it provides for the real-world pricing formula (2). In particular, since the value of the replicating portfolio—if it exists—corresponds with the real-world price of the claim, we see that the real-world price represents the value of the cheapest possible hedging portfolio. This raises the following two important questions: Does the replicating portfolio necessarily exist, and can we characterize it? These questions are addressed by the next proposition, for which the crucial ingredient is a converse for the Feynman-Kac Theorem, due to [10]:

Proposition 3 *Let $x^{rep} : (0, \infty) \to (0, \infty)$ and $\pi^{rep} : [0, T] \times (0, \infty) \to \mathbb{R}_+$ be defined as follows:*

$$x^{rep}(S) := \begin{cases} E_{0,S}(h(S_T)) & \text{if the risky security price satisfies (1a),} \\ SE_{0,S}\left(\frac{h(S_T)}{S_T}\right) & \text{if the risky security price satisfies (1b),} \end{cases} \tag{12}$$

for all $S \in (0, \infty)$, and

$$\pi^{rep}(t, S) := \frac{S}{V^h(t, S)} \frac{\partial V^h}{\partial S}(t, S), \tag{13}$$

for all $(t, S) \in [0, T) \times (0, \infty)$, respectively. In particular, (13) is well-defined, since the pricing function (3) is both strictly positive and differentiable. The replicating portfolio for the contingent claim under consideration is then determined by the initial endowment $x^{rep} = x^{rep}(S_0)$ and the trading strategy $\pi^{rep} = (\pi_t^{rep})_{t \in [0,T)}$, with $\pi_t^{rep} = \pi^{rep}(t, S_t)$, for all $t \in [0, T)$.

Proof We have already observed that by imposing condition (iii) on the payoff function, we guarantee that the pricing function is strictly positive, for both the stock price bubble and the bond price bubble. So to establish that the expression in (13) is well-defined, we need only demonstrate that the pricing function of the claim is differentiable. We shall now analyze the stock price bubble and the bond price bubble separately:

(i) Suppose the price of the risky security is modelled by (1a). It then follows from [10, Theorem 6.1] that the pricing function (3) is continuously differentiable with respect to the temporal variable and twice continuously differentiable with respect to the spatial variable, and that it satisfies the following partial differential equation:

$$\frac{\partial V^h}{\partial t}(t, S) + \frac{1}{2} S^4 \frac{\partial^2 V^h}{\partial S^2}(t, S) = 0, \tag{14}$$

for all $(t, S) \in [0, T) \times (0, \infty)$.[9] The differentiability of the pricing function ensures that the trading strategy (13) is well-defined, while Itô's formula combined with (14) yields

$$dV^h(t, S_t) = \left(\frac{\partial V^h}{\partial t}(t, S_t) + \frac{1}{2} S_t^4 \frac{\partial^2 V^h}{\partial S^2}(t, S_t) \right) dt - S_t^2 \frac{\partial V^h}{\partial S}(t, S_t) d\beta_t$$

$$= -S_t^2 \frac{\partial V^h}{\partial S}(t, S_t) d\beta_t$$

$$= -\pi^{rep}(t, S_t) S_t V^h(t, S_t) d\beta_t, \tag{15}$$

---

[9]Note that Theorem 6.1 of [10] requires a continuous payoff function of polynomial growth. This explains why we imposed condition (i) when we introduced the payoff function for the claim.

for all $t \in [0, T)$. Finally, since $x^{\text{rep}}(S_0) = V^h(0, S_0)$, we see that $W_t^{x^{\text{rep}}, \pi^{\text{rep}}} = V^h(t, S_t)$, for all $t \in [0, T]$, by comparing (15) with (9a).

(ii) Suppose the price of the risky security is modelled by (1b), and consider the function $\widehat{V}^h : [0, T] \times (0, \infty) \to \mathbb{R}_+$, given by

$$\widehat{V}^h(t, S) := \frac{V^h(t, S)}{S}, \tag{16}$$

for all $(t, S) \in [0, T] \times (0, \infty)$. It then follows from [10, Theorem 6.1] and (3) that the function defined by (16) is continuously differentiable with respect to the temporal variable and twice continuously differentiable with respect to the spatial variable, and that it satisfies the following partial differential equation:

$$\frac{\partial \widehat{V}^h}{\partial t}(t, S) + \frac{1}{S} \frac{\partial \widehat{V}^h}{\partial S}(t, S) + \frac{1}{2} \frac{\partial^2 \widehat{V}^h}{\partial S^2}(t, S) = 0, \tag{17}$$

for all $(t, S) \in [0, T) \times (0, \infty)$. In particular, the differentiability of this function implies that the pricing function is differentiable, from which it follows that the trading strategy (13) is well-defined. Next, by combing (17) with (16), we obtain the following partial differential equation for the pricing function itself:

$$\frac{\partial V^h}{\partial t}(t, S) + \frac{1}{S} \frac{\partial V^h}{\partial S}(t, S) + \frac{1}{2} \frac{\partial^2 V^h}{\partial S^2}(t, S) - \frac{1}{S} \frac{\partial V^h}{\partial S}(t, S) = 0, \tag{18}$$

for all $(t, S) \in [0, T) \times (0, \infty)$. By combining Itô's formula with (18), we then get

$$\begin{aligned}
dV^h(t, S_t) &= \left(\frac{\partial V^h}{\partial t}(t, S_t) + \frac{1}{S_t} \frac{\partial V^h}{\partial S}(t, S_t) + \frac{1}{2} \frac{\partial^2 V^h}{\partial S^2}(t, S_t) \right) dt \\
&\quad + \frac{\partial V^h}{\partial S}(t, S) \, d\beta_t \\
&= \frac{1}{S_t} \frac{\partial V^h}{\partial S}(t, S_t) \, dt + \frac{\partial V^h}{\partial S}(t, S) \, d\beta_t \\
&= \frac{\pi^{\text{rep}}(t, S_t)}{S_t^2} V^h(t, S_t) \, dt + \frac{\pi^{\text{rep}}(t, S_t)}{S_t} V^h(t, S_t) \, d\beta_t,
\end{aligned} \tag{19}$$

for all $t \in [0, T)$. Finally, since $x^{\text{rep}}(S_0) = V^h(0, S_0)$, we see that $W_t^{x^{\text{rep}}, \pi^{\text{rep}}} = V^h(t, S_t)$, for all $t \in [0, T]$, by comparing (19) with (9b). $\qquad \square$

Together, (11) and Proposition 3 establish the canonical nature of the real-world pricing formula: Given any payoff function satisfying conditions (i)–(iii) above, the real-world price of the corresponding claim provides the minimal cost of hedging it. Moreover, we see that the trading strategy (13) for the replicating portfolio may be expressed in terms of the "delta" of the claim. The latter is in turn completely determined by a function $\Delta^h : [0, T) \times (0, \infty) \to \mathbb{R}$, which is defined by

$$\Delta^h(t, S) := \frac{\partial V^h}{\partial S}(t, S),$$

for all $(t, S) \in [0, T) \times (0, \infty)$.

3 The Stock Price Bubble

In this section we focus on the situation when the price of the risky security is modelled by (1a). Since its price is then an inverted Bessel process of dimension three, we may use Proposition 1 to derive the pricing functions and deltas for a number of standard European claims on the risky security, all of which share a common maturity date $T > 0$ and strike $K > 0$ (when appropriate).

We begin by considering a zero-coupon bond with a face-value of one dollar.[10] As one would expect, this instrument is trivial in the case of the stock price bubble. In particular, its delta is uniformly zero, so that its replicating portfolio consists simply of an initial endowment—equal to (the discounted value of) its principal—invested in the money-market account:

Example 4 The pricing function $Z : [0, T) \times (0, \infty) \to (0, \infty)$ for a zero-coupon bond is given by

$$Z(t, S) := E_{t,S}(1) = 1, \tag{20}$$

for all $(t, S) \in [0, T) \times (0, \infty)$. The associated replicating portfolio is determined by the delta $\Delta^Z : [0, T) \times (0, \infty) \to \mathbb{R}$ of the contract, which is given by

$$\Delta^Z(t, S) := \frac{\partial Z}{\partial S}(t, S) = 0, \tag{21}$$

for all $(t, S) \in [0, T) \times (0, \infty)$.   □

Next, we examine a prepaid forward contract on the risky security.[11] Its pricing function and delta are presented below, and are plotted in Fig. 1.

Example 5 The pricing function $F : [0, T) \times (0, \infty) \to (0, \infty)$ for a prepaid forward on the risky security is given by

$$F(t, S) := E_{t,S}(S_T) = S\left(1 - 2\Phi\left(-\frac{1/S}{\sqrt{T-t}}\right)\right), \tag{22}$$

---

[10] Here a "dollar" should be interpreted as a generic unit of currency—we assume that the values of all instruments are denominated in "dollars."

[11] The principal difference between prepaid forwards and conventional forward contracts is that the latter are settled at maturity, while the purchaser of a prepaid forward pays an up-front premium for subsequent delivery of the underlying asset when the contract matures. In other words, a prepaid forward is simply a European call with a strike price of zero.

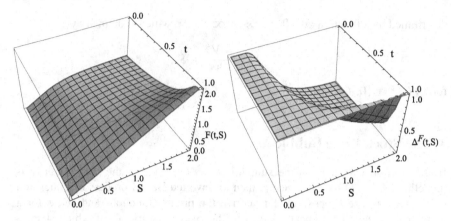

Fig. 1 The pricing function and the delta of a prepaid forward on the risky security ($T = 1$)

for all $(t, S) \in [0, T) \times (0, \infty)$, by an application of (5). The associated replicating portfolio is determined by the delta $\Delta^F : [0, T) \times (0, \infty) \to \mathbb{R}$ of the contract, which is given by

$$\Delta^F(t, S) := \frac{\partial F}{\partial S}(t, S) = 1 - 2\Phi\left(-\frac{1/S}{\sqrt{T-t}}\right) - \frac{2}{S\sqrt{T-t}}\varphi\left(\frac{1/S}{\sqrt{T-t}}\right), \quad (23)$$

for all $(t, S) \in [0, T) \times (0, \infty)$. $\qquad\square$

We immediately notice from Fig. 1 that the value of the prepaid forward contract is strictly less than the price of the risky security itself, at all times prior to maturity. In particular, we have

$$S - F(t, S) = \underbrace{2S\Phi\left(-\frac{1/S}{\sqrt{T-t}}\right)}_{\gamma(t,S)} > 0, \quad (24)$$

for all $(t, S) \in [0, T) \times (0, \infty)$. The process $(\gamma(t, S_t))_{t \in \mathbb{R}_+}$ introduced above corresponds with what [8] refer to as the "default" of a strict local martingale (in this case the price of the risky security).

Another interesting observation is that the price of the prepaid forward is bounded from above at any fixed time $t \in [0, T)$, since (22) yields

$$\lim_{S \uparrow \infty} F(t, S) = \frac{2}{\sqrt{2\pi(T-t)}}. \quad (25)$$

This explains why one cannot hope to exploit any arbitrage opportunity by simultaneously purchasing the prepaid forward and short-selling the underlying risky security—the value of the resulting portfolio is not bounded from below. It also explains why, for large values of the risky security price, the portfolio that replicates the prepaid forward is almost completely invested in the money-market account, as evidenced by the surface plot of its delta in Fig. 1. In particular, (25) ensures that the delta of the prepaid forward is zero in the limit, as the price of the risky security increases to infinity.

Let us now attempt to explain the intuition behind the fact that the prepaid forward is worth less than its underling risky security. To begin with, consider for a moment a model that does admit an equivalent risk-neutral probability measure, under which the (discounted) price of the risky security is in fact a martingale. In such a case the risk-neutral value of the prepaid forward would exactly match the market price of the risky security. This phenomenon is often interpreted to mean that risk-neutral valuation allows one to retrieve the price the risky security itself, by computing its discounted expected future value. By this line of reasoning the stock price bubble seems incoherent, since it appears that there are now two prices for the risky security: a market price and a model price.

To resolve the conundrum above, we should first point out that the prices of the underlying primary securities (i.e. the risky security and the money-market account) are completely exogenous in the modelling framework considered here—they are what they are, and they are not subject to any valuation principle. The fact that the risk-neutral value of the prepaid forward agrees with the market price of the risky security, in the case when the (discounted) price of the risky security is a martingale under the equivalent risk-neutral probability measure, is simply a curiosity induced by the properties of martingales. In truth, the risky security and the prepaid forward contract written on it are fundamentally different instruments. To make this point clear, observe that the prepaid forward may also be regarded as a European call on the risky security with a strike price of zero. By the same token, the risky security may be regarded as an American call on itself, also with a strike price of zero. Seen in this light, it is not strange that the value of the prepaid forward should be less than the price of the risky security—the difference is simply an early exercise premium! Moreover, in the case of the stock price bubble—when the price of the risky security is expected to decrease over time—this early exercise premium becomes significant. In this way we would argue that the inequality (24) is in fact quite natural.

The next instrument we consider is a European call on the risky security. Its pricing function and delta are presented below, and are plotted in Fig. 2.

Example 6 The pricing function $C : [0, T) \times (0, \infty) \to (0, \infty)$ for a European call on the risky security is given by

$$C(t, S) := E_{t,S}\big((S_T - K)^+\big) = E_{t,S}\big(1_{\{S_T > K\}} S_T\big) - K P_{t,S}(S_T > K)$$

$$= (S - K)\Phi\left(-\frac{1/S - 1/K}{\sqrt{T-t}}\right) + (S + K)\Phi\left(-\frac{1/S + 1/K}{\sqrt{T-t}}\right)$$

$$- 2S\Phi\left(-\frac{1/S}{\sqrt{T-t}}\right)$$

$$+ KS\sqrt{T-t}\left(\varphi\left(\frac{1/S - 1/K}{\sqrt{T-t}}\right) - \varphi\left(\frac{1/S + 1/K}{\sqrt{T-t}}\right)\right), \qquad (26)$$

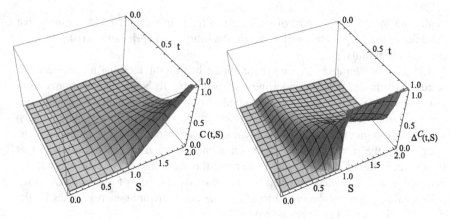

Fig. 2 The pricing function and the delta of a call on the risky security ($T = 1$ and $K = 1$)

for all $(t, S) \in [0, T) \times (0, \infty)$, by an application of (4) and (7). The associated replicating portfolio is determined by the delta $\Delta^C : [0, T) \times (0, \infty) \to \mathbb{R}$ of the call, which is given by

$$
\begin{aligned}
\Delta^C(t, S) := \frac{\partial C}{\partial S}(t, S) &= \Phi\left(-\frac{1/S - 1/K}{\sqrt{T-t}}\right) + \Phi\left(-\frac{1/S + 1/K}{\sqrt{T-t}}\right) \\
&+ \frac{S - K}{S^2\sqrt{T-t}}\varphi\left(\frac{1/S - 1/K}{\sqrt{T-t}}\right) + \frac{S + K}{S^2\sqrt{T-t}}\varphi\left(\frac{1/S + 1/K}{\sqrt{T-t}}\right) \\
&- 2\Phi\left(-\frac{1/S}{\sqrt{T-t}}\right) - \frac{2}{S\sqrt{T-t}}\varphi\left(\frac{1/S}{\sqrt{T-t}}\right) \\
&- \frac{S - K - KS^2(T-t)}{S^2\sqrt{T-t}}\varphi\left(\frac{1/S - 1/K}{\sqrt{T-t}}\right) \\
&- \frac{S + K + KS^2(T-t)}{S^2\sqrt{T-t}}\varphi\left(\frac{1/S + 1/K}{\sqrt{T-t}}\right),
\end{aligned}
\tag{27}
$$

for all $(t, S) \in [0, T) \times (0, \infty)$. $\square$

It is evident from Fig. 2 that the call option exhibits similar anomalies to the prepaid forward contract. In particular, for the price of the call we obtain the following limit from (26):

$$
\lim_{S \uparrow \infty} C(t, S) = \frac{2}{\sqrt{2\pi(T-t)}} - K\left(2\Phi\left(\frac{1/K}{\sqrt{T-t}}\right) - 1\right),
\tag{28}
$$

for all $t \in [0, T)$. Once again, this implies that the call price is bounded from above at any fixed time prior to maturity. In addition, it establishes that the pricing function for the call does not preserve the convexity of its payoff function, with respect to the

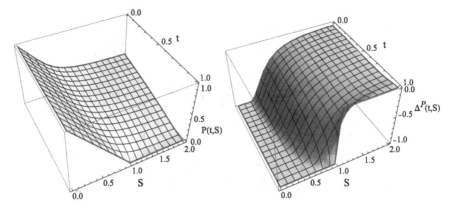

Fig. 3 The pricing function and the delta of a put on the risky security ($T = 1$ and $K = 1$)

price of the underlying risky security. This non-convexity is easily observed in the surface plot of the call delta in Fig. 2, where we see that the second derivative of the price of the call with respect to the price of the risky security is negative, for large enough prices of the risky security.

Another easy consequence of (28) is that the slope of the call price with respect to the price of the risky security must tend to zero asymptotically, at any fixed time before maturity, for large values of the risky security price. In other words, the call delta should decrease to zero, when the price of the risky security becomes large, which implies that the replicating portfolio for the call should ultimately become fully invested in the money-market account. Once again, this phenomenon is illustrated by Fig. 2.

The final claim considered in this section is a European put on the risky security. Expressions for its pricing function and delta are obtained below, and are plotted in Fig. 3.

Example 7 The pricing function $\mathsf{P} : [0, T) \times (0, \infty) \to (0, \infty)$ for a European put on the risky security is given by

$$\mathsf{P}(t, S) := \mathsf{E}_{t,S}\big((K - S_T)^+\big) = K\mathsf{P}_{t,S}(S_T \leq K) - \mathsf{E}_{t,S}\big(\mathbf{1}_{\{S_t \leq K\}} S_T\big)$$

$$= (S + K)\Phi\left(-\frac{1/S + 1/K}{\sqrt{T - t}}\right) - (S - K)\Phi\left(\frac{1/S - 1/K}{\sqrt{T - t}}\right)$$

$$+ K S\sqrt{T - t}\left(\varphi\left(\frac{1/S - 1/K}{\sqrt{T - t}}\right) - \varphi\left(\frac{1/S + 1/K}{\sqrt{T - t}}\right)\right), \quad (29)$$

for all $(t, S) \in [0, T) \times (0, \infty)$, by an application of (4) and (7). The associated replicating portfolio is determined by the delta $\Delta^{\mathsf{P}} : [0, T) \times (0, \infty) \to \mathbb{R}$ of the put, which is given by

$$
\begin{aligned}
\Delta^{P}(t, S) := \frac{\partial P}{\partial S}(t, S) = {} & \Phi\left(-\frac{1/S + 1/K}{\sqrt{T-t}}\right) - \Phi\left(\frac{1/S - 1/K}{\sqrt{T-t}}\right) \\
& + \frac{S+K}{S^2\sqrt{T-t}}\varphi\left(\frac{1/S + 1/K}{\sqrt{T-t}}\right) + \frac{S-K}{S^2\sqrt{T-t}}\varphi\left(\frac{1/S - 1/K}{\sqrt{T-t}}\right) \\
& - \frac{S - K - KS^2(T-t)}{S^2\sqrt{T-t}}\varphi\left(\frac{1/S - 1/K}{\sqrt{T-t}}\right) \\
& - \frac{S + K + KS^2(T-t)}{S^2\sqrt{T-t}}\varphi\left(\frac{1/S + 1/K}{\sqrt{T-t}}\right),
\end{aligned}
\tag{30}
$$

for all $(t, S) \in [0, T) \times (0, \infty)$.   □

It is immediately evident from Fig. 3 that the behaviour of the put is much more conventional than that of the call. In particular, we see that the pricing function for the put preserves the convexity of its payoff function, with respect the price of the risky security, at all times prior to maturity. Furthermore, the delta of the put indicates that the replicating strategy for the claim converges to a short position in one share of the risky security, if the price of this security is very low, but has almost no exposure to the risky security, if its price is high. Once again, this corresponds with the "normal" behaviour of European puts under the assumptions of the Black-Scholes model, for example.

Finally, we turn our attention to the question of put-call parity. It is already well-known that put-call parity fails for the stock price bubble, and we may confirm this explicitly, with the help of (26) and (29):

$$
C(t, S) - P(t, S) = S\left(2\Phi\left(\frac{1/S}{\sqrt{T-t}}\right) - 1\right) - K < S - K,
\tag{31}
$$

for all $(t, S) \in [0, T) \times (0, \infty)$. There are two ways of interpreting the above inequality: Either the model is pathological, or else the put-call parity relationship is misspecified. We shall argue for the latter interpretation. To wit, note that put-call parity asserts an equivalence between a portfolio containing a long European call and a short European put, on the one hand, and a long position in the risky security combined with a short position in the money-market account, on the other hand. Since the latter portfolio offers investors the possibility of taking advantage of favourable interim movements in the price of the risky security, while the portfolio of European options does not, the inequality (31) simply expresses a type of early exercise premium associated with holding the underlying securities instead of the European options. In the case of the stock price bubble, where the price of the risky security is expected to decline over time, the flexibility associated with holding the underlying securities becomes valuable.

A properly specified put-call relationship should only involve instruments that are European by nature, to avoid the type of phenomenon described above. We therefore propose the following as the correct formulation:

$$C(t, S) - P(t, S) = F(t, S) - KZ(t, S),　\qquad (32)$$

for all $(t, S) \in [0, T) \times (0, \infty)$. Note that this identity is easily verified for the stock price bubble, using the pricing functions presented in this section. Moreover, in Sect. 4 we shall see that (32) continues to hold for the case of the bond price bubble. It is therefore, in some sense, the fundamental statement of put-call parity.

4 The Bond Price Bubble

This section examines the bond price bubble, when the price of the risky security is modelled by (1b). Since its price is a Bessel process of dimension three, in this case, we may once again apply Proposition 1 to derive the pricing functions and deltas for a number of European claims on the risky security, all of which share a common maturity date $T > 0$ and strike $K > 0$ (when appropriate).

The first instrument we consider is a zero-coupon bond with a face-value of one dollar, which we see is no longer trivial. In fact, we make the rather startling discovery that such a claim is actually an equity derivative, in the setting of the bond price bubble. Its pricing function and delta are presented below, and are plotted in Fig. 4:

Example 8 The pricing function $Z : [0, T) \times (0, \infty) \to (0, \infty)$ for a zero-coupon bond is given by

$$Z(t, S) := S \mathbb{E}_{t,S}\left(\frac{1}{S_T}\right) = 1 - 2\Phi\left(-\frac{S}{\sqrt{T-t}}\right),　\qquad (33)$$

for all $(t, S) \in [0, T) \times (0, \infty)$, by an application of (5). The associated replicating portfolio is determined by the delta $\Delta^Z : [0, T) \times (0, \infty) \to \mathbb{R}$ of the zero-coupon bond, which is given by

$$\Delta^Z(t, S) := \frac{\partial Z}{\partial S}(t, S) = \frac{2}{\sqrt{T-t}}\varphi\left(\frac{S}{\sqrt{T-t}}\right),　\qquad (34)$$

for all $(t, S) \in [0, T) \times (0, \infty)$.

Figure 4 reveals that the price of the bond is only significantly less than its (discounted) principal—which would have been its price if an equivalent risk-neutral probability measure existed—when the price of the risky security is low. The reason for this can be found in (1b), which reveals that when the price of the risky security is small, its drift rate explodes, resulting in a strong repulsion from the origin. Under these circumstances the risky security is an extremely attractive investment, since it generates a positive return over a short period of time, with a high degree of certainty. By comparison, the zero-coupon bond is relatively unattractive under these conditions, unless it trades at a price substantially lower than its discounted face-value. This explains why the price of the zero-coupon bond vanishes as the price of the risky security approaches zero.

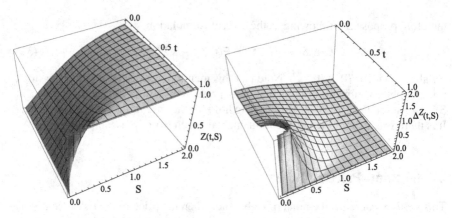

Fig. 4 The pricing function and the delta of a zero-coupon bond ($T = 1$)

It is also interesting to analyze the zero-coupon bond from the perspective of its replicating strategy. The surface plot of the delta of this instrument in Fig. 4 indicates that, at any fixed time before maturity, its replicating portfolio becomes progressively more heavily invested in the risky security as the price of that asset decreases. Once again, this simply takes advantage of the strong growth in the price of the risky security when it is low. In particular, (34) yields

$$\lim_{S \downarrow 0} \Delta^Z(t, S) = \frac{2}{\sqrt{2\pi(T - t)}},$$

for all $t \in [0, T)$. An interesting consequence of this is that when maturity is imminent and the price of the risky security is close to zero, the certain principal payment of the zero-coupon bond can be hedged by purchasing an arbitrarily large number of units of the risky security. In other words, over short periods the price of the risky security exhibits growth that is almost deterministic, if its initial value is close to zero. This can be exploited to produce a non-random payoff at very low cost.

We have already pointed out that the bond price bubble does not admit an equivalent risk-neutral probability measure, from which it follows that this model does not satisfy the NFLVR condition, by the results of [5]. As we shall now demonstrate, Example 8 allows us to establish the failure of NFLVR explicitly. To do so, consider a portfolio comprising a long position in the zero-coupon bond, which is funded by borrowing $Z(0, S_0)$ from the money-market account. The initial value of this portfolio is obviously zero, but its payoff at maturity is

$$Z(T, S_T) - Z(0, S_0) = 2\Phi\left(-\frac{S_0}{\sqrt{T}}\right) > 0,$$

by (33). Moreover, since $Z(t, S_t) > 0$, for all $t \in [0, T]$, it follows that the value of this portfolio is uniformly bounded from below by $-Z(0, S_0)$. In other words, the portfolio described above violates what [14] refer to as the "true no-arbitrage property," which is in fact a weaker condition than NFLVR.

In Sect. 3 we saw that a prepaid forward contract on the risky security is non-trivial, in the case of the stock price bubble. The opposite is true for the bond price bubble, as demonstrated below:

Example 9 The pricing function $F : [0, T) \times (0, \infty) \to (0, \infty)$ for a prepaid forward on the risky security is given by

$$F(t, S) := SE_{t,S}\left(\frac{S_T}{S_T}\right) = S, \tag{35}$$

for all $(t, S) \in [0, T) \times (0, \infty)$. The associated replicating portfolio is determined by the delta $\Delta^F : [0, T) \times (0, \infty) \to \mathbb{R}$ of the contract, which is given by

$$\Delta^F(t, S) := \frac{\partial F}{\partial S}(t, S) = 1, \tag{36}$$

for all $(t, S) \in [0, T) \times (0, \infty)$.

The next instrument we consider is a European call on the risky security. Its pricing function and delta are presented below, and are plotted in Fig. 5:

Example 10 The pricing function $C : [0, T) \times (0, \infty) \to (0, \infty)$ for a European call on the risky security is given by

$$C(t, S) := SE_{t,S}\left(\frac{(S_T - K)^+}{S_T}\right) = SP_{t,S}(S_T > K) - K SE_{t,S}\left(1_{\{S_T > K\}}\frac{1}{S_T}\right)$$

$$= (S - K)\Phi\left(\frac{S - K}{\sqrt{T - t}}\right) + (S + K)\Phi\left(-\frac{S + K}{\sqrt{T - t}}\right)$$

$$+ \sqrt{T - t}\left(\varphi\left(\frac{S - K}{\sqrt{T - t}}\right) - \varphi\left(\frac{S + K}{\sqrt{T - t}}\right)\right), \tag{37}$$

for all $(t, S) \in [0, T) \times (0, \infty)$, by an application of (6) and (8). The associated replicating portfolio is determined by the delta $\Delta^C : [0, T) \times (0, \infty) \to \mathbb{R}$ of the call, which is given by

$$\Delta^C(t, S) := \frac{\partial C}{\partial S}(t, S) = \Phi\left(\frac{S - K}{\sqrt{T - t}}\right) + \Phi\left(-\frac{S + K}{\sqrt{T - t}}\right), \tag{38}$$

for all $(t, S) \in [0, T) \times (0, \infty)$.

By inspecting Fig. 5 we see that the price of the call and its delta behave quite conventionally, when compared with European calls in the Black-Scholes model, for example. In particular, the call price preserves the convexity of its payoff function, with respect to the price of the underlying risky security, at all times prior to maturity. Also, contrary to what we observed for the case of the stock price bubble, we see that the call price is an unbounded function of the price of the risky security, at any fixed time prior to maturity.

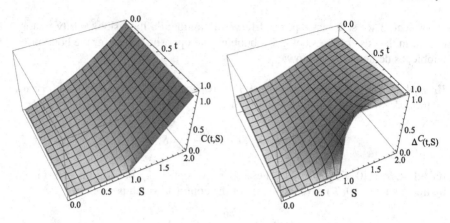

Fig. 5 The pricing function and the delta of a call on the risky security ($T = 1$ and $K = 1$)

The delta of the call behaves as one would expect, for large values of risky security price, by converging to one. In other words, the replicating portfolio for the call is almost completely invested in the risky security when the contract is deep in-the-money. However, the behaviour of the call delta is a little unusual when the price of the risky security is very low. In particular, we obtain the following from (38):

$$\lim_{S\downarrow 0} \Delta^C(t, S) = 2\Phi\left(-\frac{K}{\sqrt{T-t}}\right),$$

for all $t \in [0, T)$. In other words, at any fixed time before to maturity, the replicating portfolio for the call always holds at least a minimal number of units of the risky security. Once again, this phenomenon is explained by the strong growth of the price of the risky security when it is small. By comparison, under the assumptions of the Black-Scholes model, for example, the price of the risky security does not exhibit such growth behaviour near the origin, and the delta of a European call converges to zero as the price of the underlying asset becomes very small.

To complete this section, we finally consider a European put on the risky security. Its pricing function and delta are presented below, and are plotted in Fig. 6.

Example 11 The pricing function $P : [0, T) \times (0, \infty) \to (0, \infty)$ for a European put on the risky security is given by

$$P(t, S) := SE_{t,S}\left(\frac{(K - S_T)^+}{S_T}\right) = K SE_{t,S}\left(1_{\{S_T \leq K\}}\frac{1}{S_T}\right) - SP_{t,S}(S_T \leq K)$$

$$= (S - K)\Phi\left(\frac{S - K}{\sqrt{T-t}}\right) - (S + K)\Phi\left(\frac{S + K}{\sqrt{T-t}}\right) + 2K\Phi\left(\frac{S}{\sqrt{T-t}}\right)$$

$$+ \sqrt{T-t}\left(\varphi\left(\frac{S - K}{\sqrt{T-t}}\right) - \varphi\left(\frac{S + K}{\sqrt{T-t}}\right)\right),\tag{39}$$

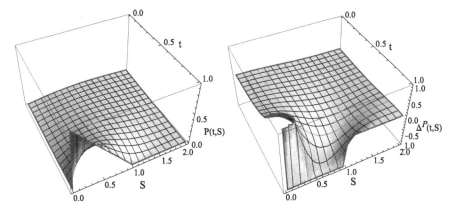

Fig. 6 The pricing function and the delta of a put on the risky security ($T = 1$ and $K = 1$)

for all $(t, S) \in [0, T) \times (0, \infty)$, by an application of (6) and (8). The associated replicating portfolio is determined by the delta $\Delta^P : [0, T) \times (0, \infty) \to \mathbb{R}$ of the put, which is given by

$$\Delta^P(t, S) := \frac{\partial P}{\partial S}(t, S)$$

$$= \Phi\left(\frac{S - K}{\sqrt{T - t}}\right) - \Phi\left(\frac{S + K}{\sqrt{T - t}}\right) + \frac{2K}{\sqrt{T - t}}\varphi\left(\frac{S}{\sqrt{T - t}}\right), \quad (40)$$

for all $(t, S) \in [0, T) \times (0, \infty)$.

Figure 6 reveals that the put option shares some of the strange features of the zero-coupon bond, discussed earlier in this section. In particular, we observe that the put price converges to zero as the price of the risky security decreases, for any fixed time before maturity. Once again, there is a sound economic explanation for this phenomenon, which takes into account the behaviour of the risky security price near zero. In detail, since the drift term in (1b) explodes near the origin, an almost certain profit can be earned in a short time by investing in the risky security when its price is very low. Due to its bounded payoff structure, the put therefore becomes unattractive relative to an investment in the risky security, when the price of the latter is very low.

Another feature of the put, which is clearly visible from the surface plot of its pricing function in Fig. 6, is the fact that its price does not preserve the convexity of its payoff, with respect to the underlying risky security price, at any fixed time prior to maturity. This non-convexity is also observable in the surface plot of the put delta, where we see that the second derivative of the put price with respect to the price of the risky security is negative, for small enough prices of the risky security.

In addition to the above, the expression (40) for the delta of the put reveals the following anomaly as well:

$$
\begin{aligned}
\lim_{S \downarrow 0} \Delta^{P}(t, S) &= 1 - 2\Phi\left(\frac{K}{\sqrt{T-t}}\right) + \frac{2K}{\sqrt{2\pi(T-t)}} \\
&= 1 - 2\left(\frac{1}{2} + \int_{0}^{\frac{K}{\sqrt{T-t}}} \varphi(\zeta)\,\mathrm{d}\zeta\right) + \frac{2K}{\sqrt{2\pi(T-t)}} \\
&= 2 \int_{0}^{\frac{K}{\sqrt{T-t}}} \big(\varphi(0) - \varphi(\zeta)\big)\,\mathrm{d}\zeta > 0,
\end{aligned}
$$

for all $t \in [0, T)$, by virtue of the fact that $\varphi(\zeta) \le \varphi(0) = \frac{1}{\sqrt{2\pi}}$, for all $\zeta \in \mathbb{R}$. This tells us that the replicating portfolio for the put option contains a long position in the risky security, if the price of the latter is very low. In other words, a long position in the risky security may be required to hedge a contract whose payoff is negatively related to the price of the risky security at maturity! In fact, as we see from the surface plot of the put delta in Fig. 6, when the price of the risky security is very low and maturity is imminent, then a very long position in the risky security may be required replicate the put. The explanation for this apparent contradiction is to be found, once more, in the strong growth of the risky security price near the origin.

Finally, it is easily verified that the pricing functions derived in this section are all related via the put-call parity relationship (32) proposed in Sect. 3. The fact that this identity holds for the stock price bubble and the bond price bubble (as well as for models where all local martingales are martingales) supports our claim that it is the proper formulation of put-call parity.

References

1. Abramowitz, M., Stegun, I.A. (eds.): Handbook of Mathematical Functions. Dover, New York (1972)
2. Becherer, D.: The numeraire portfolio for unbounded semimartingales. Finance Stoch. 5(3), 327–341 (2001)
3. Cox, A.M.G., Hobson, D.G.: Local martingales, bubbles and option prices. Finance Stoch. 9(4), 477–492 (2005)
4. Delbaen, F., Schachermayer, W.: Arbitrage and free lunch with bounded risk for unbounded continuous processes. Math. Finance 4(4), 343–348 (1994)
5. Delbaen, F., Schachermayer, W.: A general version of the fundamental theorem of asset pricing. Math. Ann. 300(3), 463–520 (1994)
6. Delbaen, F., Schachermayer, W.: Arbitrage possibilities in Bessel processes and their relations to local martingales. Probab. Theory Relat. Fields 102(3), 357–366 (1995)
7. Ekström, E., Tysk, J.: Bubbles, convexity and the Black-Scholes equation. Ann. Appl. Probab. 19(4), 1369–1384 (2009)
8. Elworthy, K.D., Li, X.M., Yor, M.: The importance of strictly local martingales; applications to radial Ornstein-Uhlenbeck processes. Probab. Theory Relat. Fields 115(3), 325–355 (1999)
9. Heston, S.L., Loewenstein, M., Willard, G.A.: Options and bubbles. Rev. Financ. Stud. 20(2), 359–389 (2007)

10. Janson, S., Tysk, J.: Feynman-Kac formulas for Black-Scholes-type operators. Bull. Lond. Math. Soc. **38**(2), 269–282 (2006)
11. Jarrow, R.A., Protter, P., Shimbo, K.: Asset price bubbles in complete markets. In: Fu, M.C., Jarrow, R.A., Yen, J.Y.J., Elliott, R.J. (eds.) Advances in Mathematical Finance, pp. 97–121. Birkhäuser, Boston (2007)
12. Jarrow, R.A., Protter, P., Shimbo, K.: Asset price bubbles in incomplete markets. Johnson School Research Paper Series No. 03-07, Cornell University (2007)
13. Jawitz, J.W.: Moments of truncated continuous univariate distributions. Adv. Water Resour. **27**(3), 269–281 (2004)
14. Kabanov, Y., Stricker, C.: Remarks on the true no-arbitrage property. In: Séminaire de Probabilités XXXVIII. *Lecture Notes in Mathematics*, vol. 1857, pp. 186–194. Springer, Berlin (2004)
15. Karatzas, I., Kardaras, C.: The numéraire portfolio in semimartingale financial models. Finance Stoch. **11**(4), 447–493 (2007)
16. Lebedev, N.N.: Special Functions and Their Applications. Dover, New York (1972)
17. Platen, E., Heath, D.: A Benchmark Approach to Quantitative Finance. Springer, Berlin (2006)
18. Revuz, D., Yor, M.: Continuous Martingales and Brownian Motion, 3rd edn. Springer, Berlin (1999)

A Remarkable σ-finite Measure Associated with Last Passage Times and Penalisation Problems

Joseph Najnudel and Ashkan Nikeghbali

Abstract In this paper, we give a global view of the results we have obtained in relation with a remarkable class of submartingales, called (Σ), and which are stated in (Najnudel and Nikeghbali, 0906.1782 (2009), 0910.4959 (2009), 0911.2571 (2009) and 0911.4365 (2009)). More precisely, we associate to a given submartingale in this class (Σ), defined on a filtered probability space $(\Omega, \mathscr{F}, \mathbb{P}, (\mathscr{F}_t)_{t \geq 0})$, satisfying some technical conditions, a σ-finite measure $\mathscr{Q}$ on $(\Omega, \mathscr{F})$, such that for all $t \geq 0$, and for all events $\Lambda_t \in \mathscr{F}_t$:

$$\mathscr{Q}[\Lambda_t, g \leq t] = \mathbb{E}_{\mathbb{P}}[\mathbb{1}_{\Lambda_t} X_t]$$

where g is the last hitting time of zero of the process X. This measure $\mathscr{Q}$ has already been defined in several particular cases, some of them are involved in the study of Brownian penalisation, and others are related with problems in mathematical finance. More precisely, the existence of $\mathscr{Q}$ in the general case solves a problem stated by D. Madan, B. Roynette and M. Yor, in a paper studying the link between Black-Scholes formula and last passage times of certain submartingales. Once the measure $\mathscr{Q}$ is constructed, we define a family of nonnegative martingales, corresponding to the local densities (with respect to $\mathbb{P}$) of the finite measures which are absolutely continuous with respect to $\mathscr{Q}$. We study in detail the relation between $\mathscr{Q}$ and this class of martingales, and we deduce a decomposition of any nonnegative martingale into three parts, corresponding to the decomposition of finite measures on $(\Omega, \mathscr{F})$ as the sum of three measures, such that the first one is absolutely continuous with respect to $\mathbb{P}$, the second one is absolutely continuous with respect to $\mathscr{Q}$ and the third one is singular with respect to $\mathbb{P}$ and $\mathscr{Q}$. This decomposition can be generalized to supermartingales. Moreover, if under $\mathbb{P}$, the process $(X_t)_{t \geq 0}$ is

J. Najnudel · A. Nikeghbali (✉)
Institut für Mathematik, Universität Zürich, Winterthurerstrasse 190, 8057 Zürich, Switzerland
e-mail: ashkan.nikeghbali@math.uzh.ch

J. Najnudel
e-mail: joseph.najnudel@math.uzh.ch

C. Chiarella, A. Novikov (eds.), *Contemporary Quantitative Finance*,
DOI 10.1007/978-3-642-03479-4_5, © Springer-Verlag Berlin Heidelberg 2010

a diffusion satisfying some technical conditions, one can state a penalisation result involving the measure $\mathcal{Q}$, and generalizing a theorem given in (Najnudel et al., A Global View of Brownian Penalisations, 2009). Now, in the construction of the measure $\mathcal{Q}$, we encounter the following problem: if $(\Omega, \mathcal{F}, (\mathcal{F}_t)_{t\geq 0}, \mathbb{P})$ is a filtered probability space satisfying the usual assumptions, then it is usually not possible to extend to $\mathcal{F}_\infty$ (the σ-algebra generated by $(\mathcal{F}_t)_{t\geq 0}$) a coherent family of probability measures $(\mathbb{Q}_t)$ indexed by $t \geq 0$, each of them being defined on $\mathcal{F}_t$. That is why we must not assume the usual assumptions in our case. On the other hand, the usual assumptions are crucial in order to obtain the existence of regular versions of paths (typically adapted and continuous or adapted and càdlàg versions) for most stochastic processes of interest, such as the local time of the standard Brownian motion, stochastic integrals, etc. In order to fix this problem, we introduce another augmentation of filtrations, intermediate between the right continuity and the usual conditions, and call it N-augmentation in this paper. This augmentation has also been considered by Bichteler (Stochastic integration and stochastic differential equations, 2002). Most of the important results of the theory of stochastic processes which are generally proved under the usual augmentation still hold under the N-augmentation; moreover this new augmentation allows the extension of a coherent family of probability measures whenever this is possible with the original filtration.

1 Notation

In this paper, $(\Omega, \mathcal{F}, (\mathcal{F}_t)_{t\geq 0}, \mathbb{P})$ will denote a filtered probability space. $\mathscr{C}(\mathbb{R}_+, \mathbb{R})$ is the space of continuous functions from $\mathbb{R}_+$ to $\mathbb{R}$. $\mathscr{D}(\mathbb{R}_+, \mathbb{R})$ is the space of càdlàg functions from $\mathbb{R}_+$ to $\mathbb{R}$. If Y is a random variable, we denote indifferently by $\mathbb{P}[Y]$ or by $\mathbb{E}_{\mathbb{P}}[Y]$ the expectation of Y with respect to $\mathbb{P}$.

2 Introduction

This paper reviews some recent results obtained by Cheridito, Nikeghbali and Platen [4] and by the authors of this paper [9–12] on the last zero of some remarkable stochastic processes and its relation with a universal σ-finite measure which has many remarkable properties and which seems to be in many places the key object to interpret in a unified way results which do not seem to be related at first sight. The last zero of càdlàg and adapted processes will play an essential role in our discussions: this fact is quite surprising since such a random time is not a stopping time and hence falls outside the domain of applications of the classical theorems in stochastic analysis.

The problem originally came from a paper of Madan, Roynette and Yor [8] on the pricing of European put options where they are able to represent the price of a European put option in terms of the probability distribution of some last passage time. More precisely, they prove that if $(M_t)_{t\geq 0}$ is a continuous nonnegative local martingale defined on a filtered probability space $(\Omega, \mathscr{F}, (\mathscr{F}_t)_{t \geq 0}, \mathbb{P})$ satisfying the usual assumptions, and such that $\lim_{t \to \infty} M_t = 0$, then

$$(K - M_t)^+ = K\mathbb{P}(g_K \leq t | \mathscr{F}_t) \tag{1}$$

where $K \geq 0$ is a constant and $g_K = \sup\{t \geq 0 : M_t = K\}$. Formula (1) tells that it is enough to know the terminal value of the submartingale $(K - M_t)^+$ and its last zero g_K to reconstruct it. Yet a nicer interpretation of (1) is suggested in [8]: there exists a measure $\mathscr{Q}$, a random time g, such that the submartingale $X_t = (K - M_t)^+$ satisfies

$$\mathscr{Q}\left[F_t \mathbb{1}_{g \leq t}\right] = \mathbb{E}\left[F_t X_t\right], \tag{2}$$

for any $t \geq 0$ and for any bounded $\mathscr{F}_t$-measurable random variable F_t. Indeed, it easily follows from (1) that, in this case, $\mathscr{Q} = K.\mathbb{P}$ and $g = g_K$. It is also clear that if a stochastic process X satisfies (2), then it is a submartingale. The problem of finding the class of submartingales which satisfy (2) is posed in [8]:

Problem 1 ([8]) For which nonnegative submartingales X can we find a σ-finite measure $\mathscr{Q}$ and the end of an optional set g such that

$$\mathscr{Q}\left[F_t \mathbb{1}_{g \leq t}\right] = \mathbb{E}\left[F_t X_t\right]? \tag{3}$$

It is also noticed in [8] that other instances of (2) have already been discovered: for example, in [2], Azéma and Yor proved that for any continuous and uniformly integrable martingale M, (3) holds for $X_t = |M_t|$, $\mathscr{Q} = |M_\infty|.\mathbb{P}$ and $g = \sup\{t \geq 0 : M_t = 0\}$, or equivalently

$$|M_t| = \mathbb{E}[|M_\infty| \mathbb{1}_{g \leq t} | \mathscr{F}_t].$$

Here again the measure $\mathscr{Q}$ is finite. Recently, another particular case where the measure $\mathscr{Q}$ is not finite was obtained by Najnudel, Roynette and Yor in their study of Brownian penalisation (see [13]). They prove the existence of the measure $\mathscr{Q}$ when $X_t = |Y_t|$ is the absolute value of the canonical process $(Y_t)_{t \geq 0}$ under the Wiener measure $\mathbb{W}$, on the space $\mathscr{C}(\mathbb{R}_+, \mathbb{R})$ equipped with the filtration generated by the canonical process. In this case, the measure $\mathscr{Q}$, which we shall denote hereafter $\mathscr{W}$ to remind we are working on the Wiener space, is not finite but σ-finite and is singular with respect to the Wiener measure: it satisfies $\mathscr{W}(g = \infty) = 0$, where $g = \sup\{t \geq 0 : X_t = 0\}$. However, a closer look at this last example reveals that Problem 1 may lead to some paradox. Indeed, the existence of the measure $\mathscr{Q}$ implies that the filtration $(\mathscr{F}_t)_{t \geq 0}$ should not satisfy the usual assumptions. Indeed, if this were the case, then for all $t \geq 0$, the event $\{g > t\}$ would have probability one (under $\mathbb{W}$) and then, would be in $\mathscr{F}_0$ and, a fortiori, in $\mathscr{F}_t$. If one assumes that $\mathscr{W}$ exists, this implies:

$$\mathscr{W}[g > t, g \leq t] = \mathbb{E}_\mathbb{W}[\mathbb{1}_{g > t} X_t],$$

and then

$$\mathbb{E}_{\mathbb{W}}[X_t] = 0,$$

which is absurd. But now, if one does not complete the original probability space, then it is possible to show (see Sect. 3) that there does not exist a continuous and adapted version of the local time for the Wiener process Y, which is one of the key processes in the study of Brownian penalisations! More generally, one cannot apply most of the useful results from the general theory of stochastic processes such as the existence of càdlàg versions for martingales, the Doob-Meyer decomposition, the début theorem, etc. Consequently, one has to provide conditions on the underlying filtered probability space $(\Omega, \mathscr{F}, (\mathscr{F}_t)_{t\geq 0}, \mathbb{P})$ under which not only (3) in Problem 1 can hold, but also under which the existence of regular versions (e.g. continuous and adapted version for the Brownian local time) of stochastic processes of interest do exist:

Problem 2 What are the natural conditions to impose to $(\Omega, \mathscr{F}, (\mathscr{F}_t)_{t\geq 0}, \mathbb{P})$ in order to have (3) and at the same time the existence of regular versions for stochastic processes of interest?

In fact we shall see that the existence of the measure $\mathscr{Q}$ is related to the problem of extension of a coherent family of probability measures: given a coherent family of probability measures $(\mathbb{Q}_t)_{t\geq 0}$ with $\mathbb{Q}_t$ defined on $\mathscr{F}_t$ (i.e.: the restriction of $\mathbb{Q}_t$ to $\mathscr{F}_s$ is $\mathbb{Q}_s$ for $t \geq s$), does there exist a probability measure $\mathbb{Q}_\infty$ defined on $\mathscr{F}_\infty = \bigvee_{t\geq 0} \mathscr{F}_t$ such that the restriction of $\mathbb{Q}_\infty$ to $\mathscr{F}_t$ is $\mathbb{Q}_t$? The solution to such extension problems is well-known and very well detailed in the book by Parthasarathy [15], however, it is surprising that such a fundamental problem (if one thinks about the widely used changes of probability measures which are only locally absolutely continuous with respect to a reference probability measure) has rarely been considered together with the existence of càdlàg versions for martingales, the Doob-Meyer decomposition, the début theorem, etc.

In Sect. 3, we shall propose an alternative augmentation of filtrations, intermediate between the right-continuous version and the usual augmentation, under which most of the properties generally proved under usual conditions (existence of càdlàg versions, existence of Doob-Meyer decomposition, etc.) are preserved, but for which it is still possible to extend compatible families of probability measures. We proposed this augmentation in [9] and discovered later than Bichteler has also proposed it in his book [3]. By using this augmentation, we are able to prove the existence of the measure $\mathscr{Q}$ under very general assumptions. The relevant class of submartingales is called (Σ), and the precise conditions under which we can show that $\mathscr{Q}$ exists are stated in Sect. 4. Just before this statement, we give a detailed solution of Problem 1 in the particular case where $\mathscr{Q}$ is absolutely continuous with respect to $\mathbb{P}$, with some applications to financial modeling.

The measure $\mathscr{Q}$ has some remarkable properties which we shall detail in Sect. 5. One of its most striking properties is that it allows a unified treatment of many problems of penalisation on the Wiener space which do not seem to be related at first sight. The framework which is generally used it the following (see the book [13] for more details and references): we consider $\mathbb{W}$, the Wiener measure on $\mathscr{C}(\mathbb{R}_+, \mathbb{R})$, endowed with its natural filtration $(\mathscr{F}_s)_{s \geq 0}$, $(\Gamma_t)_{t \geq 0}$ a family of nonnegative random variables on the same space, such that

$$0 < \mathbb{W}[\Gamma_t] < \infty,$$

and for $t \geq 0$, the probability measure

$$Q_t := \frac{\Gamma_t}{\mathbb{W}[\Gamma_t]} \cdot \mathbb{W}.$$

Under these assumptions, Roynette, Vallois and Yor have proven (see [16]) that for many examples of family of functionals $(\Gamma_t)_{t \geq 0}$, there exists a probability measure $\mathbb{Q}_\infty$ which can be considered as the weak limit of $(\mathbb{Q}_t)_{t \geq 0}$ when t goes to infinity, in the following sense: for all $s \geq 0$ and for all bounded, $\mathscr{F}_s$-measurable random variables F_s, one has

$$\mathbb{Q}_t[F_s] \xrightarrow[t \to \infty]{} \mathbb{Q}_\infty[F_s].$$

For example, the measure $\mathbb{Q}_\infty$ exists for the following families of functionals[1] $(\Gamma_t)_{t \geq 0}$:

- $\Gamma_t = \phi(L_t)$, where $(L_t)_{t \geq 0}$ is the local time at zero of the canonical process X, and ϕ is a nonnegative, integrable function from $\mathbb{R}_+$ to $\mathbb{R}_+$.
- $\Gamma_t = \phi(S_t)$, where S_t is the supremum of X on the interval $[0, t]$, and ϕ is, again, a nonnegative, integrable function from $\mathbb{R}_+$ to $\mathbb{R}_+$.
- $\Gamma_t = e^{\lambda L_t + \mu |X_t|}$, where $(L_t)_{t \geq 0}$ is, again, the local time at zero of X.

In [13], Najnudel, Roynette and Yor obtain a result which gives the existence of $\mathbb{Q}_\infty$ for a large class of families of functionals $(\Gamma_t)_{t \geq 0}$. The proof of this penalisation result involves, in an essential way, the σ-finite measure $\mathscr{W}$ on the space $\mathscr{C}(\mathbb{R}_+, \mathbb{R})$ described above. More precisely, they prove that for a relatively large class of functionals $(\Gamma_t)_{t \geq 0}$,

$$\mathbb{Q}_\infty = \frac{\Gamma_\infty}{\mathscr{W}[\Gamma_\infty]} \cdot \mathscr{W}, \tag{4}$$

where Γ_∞ is the limit of Γ_t when t goes to infinity, which is supposed to exist everywhere.

Problem 3 Can we use our general existence theorem on the measure $\mathscr{Q}$ in Sect. 4 to extend the general result on the Brownian penalisation problem and $\mathscr{W}$ to a larger class of stochastic processes?

---

[1] The discussion around Problem 2 shows that the results described below are not correct because a continuous and adapted version of the Brownian local time does not exist with the natural filtration. The conditions given in Sect. 3 will remedy this gap.

At end of Sect. 5, we shall see that the answer to Problem 3 is positive (which extends (4)), if under $\mathbb{P}$, the submartingale $(X_t)_{t \geq 0}$ is a diffusion satisfying some technical conditions. In particular, our result can be applied to suitable powers of Bessel processes if the dimension is in the interval $(0, 2)$ (which includes the case of the reflected Brownian motion). Unlike all our other results, for which we are able to get rid of the Markov and scaling properties that have been used so far in the Brownian studies, the Markov property plays here a crucial role.

3 A New Kind of Augmentation of Filtrations Consistent with the Problem of Extension of Measures

The discussion in this section follows closely our paper [9]; in particular the proofs which are not provided can be found there.

3.1 Understanding the Problem

In stochastic analysis, most of the interesting properties of continuous time random processes cannot be established if one does not assume that their trajectories satisfy some regularity conditions. For example, a nonnegative càdlàg martingale converges almost surely, but if the càdlàg assumption is removed, the result becomes false in general. Recall a very simple counter-example: on the filtered probability space $(\mathscr{C}(\mathbb{R}_+, \mathbb{R}), \mathscr{F}, (\mathscr{F}_t)_{t \geq 0}, \mathbb{W})$, where $\mathscr{F}_t = \sigma\{X_s, 0 \leq s \leq t\}$, $\mathscr{F} = \sigma\{X_s, s \geq 0\}$, $(X_s)_{s \geq 0}$ is the canonical process and $\mathbb{W}$ the Wiener measure, the martingale

$$\left(M_t := \mathbb{1}_{X_t=1}\right)_{t \geq 0},$$

which is a.s. equal to zero for each fixed $t \geq 0$, does not converge at infinity. That is the reason why one generally considers a càdlàg version of a martingale. However there are fundamental examples of stochastic processes for which such a version does not exist. Indeed let us define on the filtered probability space $(\mathscr{C}(\mathbb{R}_+, \mathbb{R}), \mathscr{F}, (\mathscr{F}_t)_{t \geq 0}, \mathbb{W})$ described above, the stochastic process $(\mathscr{L}_t)_{t \geq 0}$ as follows:

$$\mathscr{L}_t = \Phi\left(\liminf_{m \to \infty} \int_0^t f_m(X_s)ds\right),$$

where f_m denotes the density of a centered Gaussian variable with variance $1/m$ and Φ is the function from $\mathbb{R}_+ \cup \{\infty\}$ to $\mathbb{R}_+$ such that $\Phi(x) = x$ for $x < \infty$ and $\Phi(\infty) = 0$. The process $(\mathscr{L}_t)_{t \geq 0}$ is a version of the local time of the canonical process at level zero, which is defined everywhere and $(\mathscr{F}_t)_{t \geq 0}$-adapted. It is known that the process:

$$\left(M_t := |X_t| - \mathscr{L}_t\right)_{t \geq 0}$$

is an $(\mathscr{F}_t)_{t\geq 0}$-martingale. However, $(M_t)_{t\geq 0}$ does not admit a càdlàg version which is adapted. In other words, there exists no càdlàg, adapted version $(L_t)_{t\geq 0}$ for the local time at level zero of the canonical process! This property can be proved in the following way: let us consider an Ornstein-Uhlenbeck process $(U_t)_{t\geq 0}$, starting from zero, and let us define the process $(V_t)_{t\geq 0}$ by:

$$V_t = (1-t)U_{t/(1-t)}$$

for $t < 1$, and

$$V_t = 0$$

for $t \geq 1$. This process is a.s. continuous: we denote by $\mathbb{Q}$ its distribution. One can check the following properties:

- For all $t \in [0, 1)$, the restriction of $\mathbb{Q}$ to $\mathscr{F}_t$ is absolutely continuous with respect to the corresponding restriction of $\mathbb{W}$.
- Under $\mathbb{Q}$, $\mathscr{L}_t \to \infty$ a.s. when $t \to 1$, $t < 1$.

By the second property, the set $\{\mathscr{L}_t \underset{t\to 1, t<1}{\longrightarrow} \infty\}$ has probability one under $\mathbb{Q}$. Since it is negligible under $\mathbb{P}$, it is essential to suppose that it is not contained in $\mathscr{F}_0$, if we need to have the first property: the filtration must not be completed. The two properties above imply

$$\mathbb{Q}\left[L_{1-2^{-n}} \underset{n\to\infty}{\longrightarrow} \infty\right] \geq \mathbb{Q}\left[\mathscr{L}_{1-2^{-n}} \underset{n\to\infty}{\longrightarrow} \infty, \forall n \in \mathbb{N}, L_{1-2^{-n}} = \mathscr{L}_{1-2^{-n}}\right]$$

$$\geq 1 - \sum_{n\in\mathbb{N}} \mathbb{Q}\left[L_{1-2^{-n}} \neq \mathscr{L}_{1-2^{-n}}\right] = 1.$$

The last equality is due to the fact that for all $n \in \mathbb{N}$,

$$\mathbb{W}\left[L_{1-2^{-n}} \neq \mathscr{L}_{1-2^{-n}}\right] = 0,$$

and then

$$\mathbb{Q}\left[L_{1-2^{-n}} \neq \mathscr{L}_{1-2^{-n}}\right] = 0,$$

since $L_{1-2^{-n}}$ and $\mathscr{L}_{1-2^{-n}}$ are $\mathscr{F}_{1-2^{-n}}$-measurable and since the restriction of $\mathbb{Q}$ to this σ-algebra is absolutely continuous with respect to $\mathbb{W}$. We have thus proved that there exist some paths such that $L_{1-2^{-n}}$ tends to infinity with n, which contradicts the fact that $(L_t)_{t\geq 0}$ is càdlàg. From this we also deduce that in general there do not exist càdlàg versions for martingales. Similarly many other important results from stochastic analysis cannot be proved on the most general filtered probability space, e.g. the existence of the Doob-Meyer decomposition for submartingales and the début theorem (see for instance [5] and [6]). In order to avoid this technical problem, it is generally assumed that the filtered probability space on which the processes are constructed satisfies the usual conditions, i.e. the filtration is complete and right-continuous.

But now, if we wish to perform a change of probability measure (for example, by using the Girsanov theorem), this assumption reveals to be too restrictive. Let us illustrate this fact by a simple example. Let us consider the filtered probability space $(\mathscr{C}(\mathbb{R}_+, \mathbb{R}), \widetilde{\mathscr{F}}, (\widetilde{\mathscr{F}}_t)_{t \geq 0}, \widetilde{\mathbb{W}})$ obtained, from the Wiener space $(\mathscr{C}(\mathbb{R}_+, \mathbb{R}), \mathscr{F}, (\mathscr{F}_t)_{t \geq 0}, \mathbb{W})$ described above, by taking its usual augmentation, i.e.:

- $\widetilde{\mathscr{F}}$ is the σ-algebra generated by $\mathscr{F}$ and its negligible sets.
- For all $t \geq 0$, $\widetilde{\mathscr{F}}_t$ is σ-algebra generated by $\mathscr{F}_t$ and the negligible sets of $\mathscr{F}$.
- $\widetilde{\mathbb{W}}$ is the unique possible extension of $\mathbb{W}$ to the completed σ-algebra $\widetilde{\mathscr{F}}$.

Let us also consider the family of probability measures $(\mathbb{Q}_t)_{t \geq 0}$, such that $\mathbb{Q}_t$ is defined on $\widetilde{\mathscr{F}}_t$ by

$$\mathbb{Q}_t = e^{X_t - \frac{t}{2}} . \widetilde{\mathbb{W}}_{|\widetilde{\mathscr{F}}_t}.$$

This family of probability measures is coherent, i.e. for $0 \leq s \leq t$, the restriction of $\mathbb{Q}_t$ to $\widetilde{\mathscr{F}}_s$ is equal to $\mathbb{Q}_s$. However, unlike what one would expect, there does not exist a probability measure $\mathbb{Q}$ on $\widetilde{\mathscr{F}}$ such that its restriction to $\widetilde{\mathscr{F}}_s$ is equal to $\mathbb{Q}_s$ for all $s \geq 0$. Indeed, let us assume that $\mathbb{Q}$ exists. The event

$$A := \{\forall t \geq 0, X_t \geq -1\}$$

satisfies $\widetilde{\mathbb{W}}[A] = 0$, and then $A \in \widetilde{\mathscr{F}}_0$ by completeness, which implies that $\mathbb{Q}[A] = 0$. On the other hand, under $\mathbb{Q}_t$, for all $t \geq 0$, the process $(X_s)_{0 \leq s \leq t}$ is a Brownian motion with drift 1, and hence under $\mathbb{Q}$. One deduces that:

$$\mathbb{Q}[\forall s \in [0, t], X_s \geq -1] = \widetilde{\mathbb{W}}[\forall s \in [0, t], X_s \geq -s - 1]$$
$$\geq \widetilde{\mathbb{W}}[\forall s \geq 0, X_s \geq -s - 1].$$

Consequently, by letting t go to infinity, one obtains:

$$\mathbb{Q}[A] \geq \widetilde{\mathbb{W}}[\forall s \geq 0, X_s \geq -s - 1] > 0,$$

which is a contradiction. Therefore, the usual conditions are not suitable for the problem of extension of coherent probability measures. In fact one can observe that the argument above does not depend on the completeness of $\widetilde{\mathscr{F}}$, but only on the fact that $\widetilde{\mathscr{F}}_0$ contains all the sets in $\widetilde{\mathscr{F}}$ of probability zero. That is why it still remains available if we consider, with the notation above, the space $(\mathscr{C}(\mathbb{R}_+, \mathbb{R}), \mathscr{F}, (\mathscr{F}'_t)_{t \geq 0}, \mathbb{W})$, where for all $t \geq 0$, $\mathscr{F}'_t$ is the σ-algebra generated by $\mathscr{F}_t$ and the sets in $\mathscr{F}$ of probability zero.

3.2 The N-usual Augmentation

In order to make compatible the general results on the extension of probability measures problem and the existence of regular versions for stochastic processes, we propose an augmentation which is intermediate between right continuity and the usual augmentation, and we call it the *N-augmentation*. As already mentioned, Bichteler has introduced this augmentation before us in his book [3], and has called it the *natural augmentation*. But when we discovered this augmentation, we were not aware of Bichteler's work and this is reflected in the difference in our approaches. Hence the interested reader would benefit by looking at both [3] and [9].

Definition 1 ([9]) Let $(\Omega, \mathscr{F}, (\mathscr{F}_t)_{t \geq 0}, \mathbb{P})$ be a filtered probability space. A subset A of Ω is N-negligible with respect to the space $(\Omega, \mathscr{F}, (\mathscr{F}_t)_{t \geq 0}, \mathbb{P})$, iff there exists a sequence $(B_n)_{n \geq 0}$ of subsets of Ω, such that for all $n \geq 0$, $B_n \in \mathscr{F}_n$, $\mathbb{P}[B_n] = 0$, and

$$A \subset \bigcup_{n \geq 0} B_n.$$

Remark 1 The integers do not play a crucial rôle in Definition 1. If $(t_n)_{n \geq 0}$ is an unbounded sequence in $\mathbb{R}_+$, one can replace the condition $B_n \in \mathscr{F}_n$ by the condition $B_n \in \mathscr{F}_{t_n}$.

Let us now define a notion which is the analog of completeness for N-negligible sets. It is the main ingredient in the definition of what we shall call the N-usual conditions:

Definition 2 ([9]) A filtered probability space $(\Omega, \mathscr{F}, (\mathscr{F}_t)_{t \geq 0}, \mathbb{P})$, is N-complete if and only if all the N-negligible sets of this space are contained in $\mathscr{F}_0$. It satisfies the N-usual conditions if and only if it is N-complete and the filtration $(\mathscr{F}_t)_{t \geq 0}$ is right-continuous.

It is natural to ask if from a given filtered probability space, one can define in a canonical way a space which satisfies the N-usual conditions and which is as "close" as possible to the initial space. The answer to this question is positive in the following sense:

Proposition 1 ([9]) *Let* $(\Omega, \mathscr{F}, (\mathscr{F}_t)_{t \geq 0}, \mathbb{P})$ *be a filtered probability space, and* $\mathscr{N}$ *the family of its N-negligible sets. Let* $\widetilde{\mathscr{F}}$ *be the σ-algebra generated by* $\mathscr{N}$ *and* $\mathscr{F}$, *and for all* $t \geq 0$, $\widetilde{\mathscr{F}}_t$ *the σ-algebra generated by* $\mathscr{N}$ *and* $\mathscr{F}_{t+}$, *where*

$$\mathscr{F}_{t+} := \bigcap_{u > t} \mathscr{F}_t.$$

Then there exists a unique probability measure $\widetilde{\mathbb{P}}$ *on* $(\Omega, \widetilde{\mathscr{F}})$ *which coincides with* $\mathbb{P}$ *on* $\mathscr{F}$, *and the space* $(\Omega, \widetilde{\mathscr{F}}, (\widetilde{\mathscr{F}}_t)_{t \geq 0}, \widetilde{\mathbb{P}})$ *satisfies the N-usual conditions. Moreover, if* $(\Omega, \mathscr{F}', (\mathscr{F}'_t)_{t \geq 0}, \mathbb{P}')$ *is a filtered probability space satisfying the N-usual conditions, such that* $\mathscr{F}'$ *contains* $\mathscr{F}$, $\mathscr{F}'_t$ *contains* $\mathscr{F}_t$ *for all* $t \geq 0$, *and if* $\mathbb{P}'$ *is an extension of* $\mathbb{P}$, *then* $\mathscr{F}'$ *contains* $\widetilde{\mathscr{F}}$, $\mathscr{F}'_t$ *contains* $\widetilde{\mathscr{F}}_t$, *for all* $t \geq 0$ *and* $\mathbb{P}'$ *is an extension of* $\widetilde{\mathbb{P}}$. *In other words,* $(\Omega, \widetilde{\mathscr{F}}, (\widetilde{\mathscr{F}}_t)_{t \geq 0}, \widetilde{\mathbb{P}})$ *is the smallest extension of* $(\Omega, \mathscr{F}, (\mathscr{F}_t)_{t \geq 0}, \mathbb{P})$ *which satisfies the N-usual conditions: we call it the N-augmentation of* $(\Omega, \mathscr{F}, (\mathscr{F}_t)_{t \geq 0}, \mathbb{P})$.

Once the N-usual conditions are defined, it is natural to compare them with the usual conditions. One has the following result:

Proposition 2 *Let $(\Omega, \mathscr{F}, (\mathscr{F}_t)_{t \geq 0}, \mathbb{P})$ be a filtered probability space which satisfies the N-usual conditions. Then for all $t \geq 0$, the space $(\Omega, \mathscr{F}_t, (\mathscr{F}_s)_{0 \leq s \leq t}, \mathbb{P})$ satisfies the usual conditions.*

Proof The right-continuity of $(\mathscr{F}_s)_{0 \leq s \leq t}$ is obvious, let us prove the completeness. If A is a negligible set of $(\Omega, \mathscr{F}_t, \mathbb{P})$, there exists $B \in \mathscr{F}_t$, such that $A \subset B$ and $\mathbb{P}[B] = 0$. One deduces immediately that A is N-negligible with respect to $(\Omega, \mathscr{F}, (\mathscr{F}_t)_{t \geq 0}, \mathbb{P})$, and by N-completeness of this filtered probability space, $A \in \mathscr{F}_0$.   □

This relation between the usual conditions and the N-usual conditions is the main ingredient to prove that one can replace the usual conditions by the N-usual conditions in most of the classical results in stochastic calculus. For example, the following can be proved:

- If (X_t) is a martingale, then it admits a càdlàg modification, which is unique up to indistinguishability.
- If $(\Omega, \mathscr{F}, (\mathscr{F}_t)_{t \geq 0}, \mathbb{P})$ is a filtered probability space satisfying the N-usual conditions, and if $(X_t)_{t \geq 0}$ is an adapted process defined on this space such that there exists a càdlàg version (resp. continuous version) $(Y_t)_{t \geq 0}$ of $(X_t)_{t \geq 0}$, then there exists a càdlàg and adapted version (resp. continuous and adapted version) of $(X_t)_{t \geq 0}$, which is necessarily indistinguishable from $(Y_t)_{t \geq 0}$.
- Let $(\Omega, \mathscr{F}, (\mathscr{F}_t)_{t \geq 0}, \mathbb{P})$ be a filtered probability space satisfying the N-usual conditions, and let A be a progressive subset of $\mathbb{R}_+ \times \Omega$. Then the début of A, i.e. the random time $D(A)$ such that for all $\omega \in \Omega$:

$$D(A)(\omega) := \inf\{t \geq 0, (t, \omega) \in A\}$$

is an $(\mathscr{F}_t)_{t \geq 0}$-stopping time.
- Let $(X_t)_{t \geq 0}$ be a right-continuous submartingale defined on a filtered probability space $(\Omega, \mathscr{F}, (\mathscr{F}_t)_{t \geq 0}, \mathbb{P})$, satisfying the N-usual conditions. We suppose that $(X_t)_{t \geq 0}$ is of class (DL), i.e. for all $a \geq 0$, $(X_T)_{T \in \mathscr{T}_a}$ is uniformly integrable, where $\mathscr{T}_a$ is the family of the $(\mathscr{F}_t)_{t \geq 0}$-stopping times which are bounded by a (for example, every nonnegative submartingale is of class (DL)). Then, there exist a right-continuous $(\mathscr{F}_t)_{t \geq 0}$-martingale $(M_t)_{t \geq 0}$ and an increasing process $(A_t)_{t \geq 0}$ starting at zero, such that:

$$X_t = M_t + A_t$$

for all $t \geq 0$, and for every bounded, right-continuous martingale $(\xi_s)_{s \geq 0}$,

$$\mathbb{E}[\xi_t A_t] = \mathbb{E}\left[\int_{(0,t]} \xi_{s-} dA_s\right],$$

where ξ_{s-} is the left-limit of ξ at s, almost surely well-defined for all $s > 0$. The processes $(M_t)_{t\geq 0}$ and $(A_t)_{t\geq 0}$ are uniquely determined, up to indistinguishability. Moreover, they can be chosen to be continuous if $(X_t)_{t\geq 0}$ is a continuous process.
- The section theorem holds in a filtered probability space satisfying the N-usual assumptions and hence optional projections are well-defined.

3.3 Extension of Measures and the N-usual Augmentation

We first state a well-known Parthasarathy type condition for the probability measures extension problem.

Definition 3 Let $(\Omega, \mathscr{F}, (\mathscr{F}_t)_{t\geq 0})$ be a filtered measurable space, such that $\mathscr{F}$ is the σ-algebra generated by $\mathscr{F}_t, t \geq 0$: $\mathscr{F} = \bigvee_{t\geq 0} \mathscr{F}_t$. We shall say that the property (P) holds if and only if $(\mathscr{F}_t)_{t\geq 0}$ enjoys the following conditions:

- For all $t \geq 0$, $\mathscr{F}_t$ is generated by a countable number of sets;
- For all $t \geq 0$, there exist a Polish space Ω_t, and a surjective map π_t from Ω to Ω_t, such that $\mathscr{F}_t$ is the σ-algebra of the inverse images, by π_t, of Borel sets in Ω_t, and such that for all $B \in \mathscr{F}_t, \omega \in \Omega, \pi_t(\omega) \in \pi_t(B)$ implies $\omega \in B$;
- If $(\omega_n)_{n\geq 0}$ is a sequence of elements of Ω, such that for all $N \geq 0$,

$$\bigcap_{n=0}^{N} A_n(\omega_n) \neq \emptyset,$$

where $A_n(\omega_n)$ is the intersection of the sets in $\mathscr{F}_n$ containing ω_n, then:

$$\bigcap_{n=0}^{\infty} A_n(\omega_n) \neq \emptyset.$$

Given this technical definition, one can state the following result:

Proposition 3 ([9, 15]) *Let $(\Omega, \mathscr{F}, (\mathscr{F}_t)_{t\geq 0})$ be a filtered measurable space satisfying the property (P), and let, for $t \geq 0$, $\mathbb{Q}_t$ be a probability measure on $(\Omega, \mathscr{F}_t)$, such that for all $t \geq s \geq 0$, $\mathbb{Q}_s$ is the restriction of $\mathbb{Q}_t$ to $\mathscr{F}_s$. Then, there exists a unique measure $\mathbb{Q}$ on $(\Omega, \mathscr{F})$ such that for all $t \geq 0$, its restriction to $\mathscr{F}_t$ is equal to $\mathbb{Q}_t$.*

One can easily deduce the following corollary which is often used in practice:

Corollary 1 *Let Ω be $\mathscr{C}(\mathbb{R}_+, \mathbb{R}^d)$, the space of continuous functions from $\mathbb{R}_+$ to $\mathbb{R}^d$, or $\mathscr{D}(\mathbb{R}_+, \mathbb{R}^d)$, the space of càdlàg functions from $\mathbb{R}_+$ to $\mathbb{R}^d$ (for some $d \geq 1$). For $t \geq 0$, define $(\mathscr{F}_t)_{t\geq 0}$ as the natural filtration of the canonical process Y, and $\mathscr{F} = \bigvee_{t\geq 0} \mathscr{F}_t$. Then $(\Omega, \mathscr{F}, (\mathscr{F}_t)_{t\geq 0})$ satisfies property (P).*

Proof Let us prove this result for càdlàg functions (for continuous functions, the result is similar and proved in [17]). For all $t \geq 0$, $\mathscr{F}_t$ is generated by the variables Y_{rt}, for r, rational, in $[0, 1]$, hence, it is countably generated. For the second property, one can take for Ω_t, the set of càdlàg functions from $[0, t]$ to $\mathbb{R}^d$, and for π_t, the restriction to the interval $[0, t]$. The space Ω_t is Polish if one endows it with the Skorokhod metric. Moreover, its Borel σ-algebra is equal to the σ-algebra generated by the coordinates, a result from which one easily deduces the properties of π_t which need to be satisfied. The third property is easy to check: let us suppose that $(\omega_n)_{n \geq 0}$ is a sequence of elements of Ω, such that for all $N \geq 0$,

$$\bigcap_{n=0}^{N} A_n(\omega_n) \neq \emptyset,$$

where $A_n(\omega_n)$ is the intersection of the sets in $\mathscr{F}_n$ containing ω_n. Here, $A_n(\omega_n)$ is the set of functions ω' which coincide with ω_n on $[0, n]$. Moreover, for $n \leq n'$, integers, the intersection of $A_n(\omega_n)$ and $A_{n'}(\omega_{n'})$ is not empty, and then ω_n and $\omega_{n'}$ coincide on $[0, n]$. Therefore, there exists a càdlàg function ω which coincides with ω_n on $[0, n]$, for all n, which implies:

$$\bigcap_{n=0}^{\infty} A_n(\omega_n) \neq \emptyset.$$

$\square$

Remark 2 It is easily seen that the conditions of Proposition 3 are not satisfied by the space $\mathscr{C}([0, 1], \mathbb{R})$ endowed with the filtration $(\mathscr{F}_t)_{t \geq 0}$, where $\mathscr{F}_t$ is the σ-algebra generated by the canonical process up to time $t/(1+t)$. An explicit counterexample is provided in [7].

The next proposition shows how condition (P) combines with the N-usual augmentation:

Proposition 4 ([9]) *Let $(\Omega, \mathscr{F}, (\mathscr{F}_t)_{t \geq 0}, \mathbb{P})$ be the N-augmentation of a filtered probability space satisfying the property (P). Then if $(\mathbb{Q}_t)_{t \geq 0}$ is a coherent family of probability measures, $\mathbb{Q}_t$ defined on $\mathscr{F}_t$, and absolutely continuous with respect to the restriction of $\mathbb{P}$ to $\mathscr{F}_t$, there exists a unique probability measure $\mathbb{Q}$ on $\mathscr{F}$ which coincides with $\mathbb{Q}_t$ on $\mathscr{F}_t$, for all $t \geq 0$.*

4 A Universal σ-finite Measure $\mathscr{Q}$

4.1 The Class (Σ)

We now have all the ingredients to rigorously answer Problem 1 raised in the Introduction. For this we will first need to introduce a special class of local submartingales which was first introduced by Yor [18] and further studied by Nikeghbali [14] and Cheridito, Nikeghbali and Platen [4].

Definition 4 Let $(\Omega, \mathscr{F}, (\mathscr{F}_t)_{t \geq 0}, \mathbb{P})$ be a filtered probability space. A nonnegative (local) submartingale $(X_t)_{t \geq 0}$ is of class (Σ), if it can be decomposed as $X_t = N_t + A_t$ where $(N_t)_{t \geq 0}$ and $(A_t)_{t \geq 0}$ are $(\mathscr{F}_t)_{t \geq 0}$-adapted processes satisfying the following assumptions:

- $(N_t)_{t \geq 0}$ is a càdlàg (local) martingale.
- $(A_t)_{t \geq 0}$ is a continuous increasing process, with $A_0 = 0$.
- The measure (dA_t) is carried by the set $\{t \geq 0, X_t = 0\}$.

We shall say that $(X_t)_{t \geq 0}$ is of class (ΣD) if X is of class (Σ) and of class (D).

The class (Σ) contains many well-known examples of stochastic processes (see e.g. [14]) such as nonnegative local martingales, $|M_t|$, M_t^+, M_t^- if M is a continuous local martingale, the drawdown process $S_t - M_t$ where M is a local martingale with only negative jumps and $S_t = \sup_{u \leq t} M_u$, the relative drawdown process $1 - \frac{M_t}{S_t}$ if $M_0 \neq 0$, the age process of the standard Brownian motion W_t in the filtration of the zeros of the Brownian motion, namely $\sqrt{t - g_t}$, where $g_t = \sup\{u \leq t : W_u = 0\}$, etc. Moreover one notes that if X is of (Σ), then $X_t + M_t$ is also of class (Σ) for any strictly positive càdlàg local martingale M. Another key property of the class (Σ) is the following stability result which follows from an application of Ito's formula and a monotone class argument.

Proposition 5 ([14]) *Let $(X_t)_{t \geq 0}$ be of class (Σ) and let $f : \mathbb{R} \to \mathbb{R}$ be a locally bounded Borel function. Let us further assume that $(\Omega, \mathscr{F}, (\mathscr{F}_t)_{t \geq 0}, \mathbb{P})$ satisfies the usual assumptions or is N-complete. Denote $F(x) = \int_0^x f(y) dy$. Then the process $(f(A_t)X_t)_{t \geq 0}$ is again of class (Σ) with decomposition:*

$$f(A_t)X_t = f(0)X_0 + \int_0^t f(A_u) dN_u + F(A_t). \tag{5}$$

4.2 A Special Case Related to Financial Modeling

Now we state a theorem which gives sufficient conditions under which a process of the class (Σ) which converges to X_∞ a.s. satisfies (3). This result is an extension of a result by Azéma-Yor [2] and Azéma-Meyer-Yor [1]. Indeed, in the case when X is of class (ΣD), one could deduce it from part 1 of Theorem 8.1 in [1]. The proof is very simple in this case and we give it.

Theorem 1 ([4]) *Assume that $(\Omega, \mathscr{F}, (\mathscr{F}_t)_{t \geq 0}, \mathbb{P})$ satisfies the usual assumptions or is N-complete. Let $(X_t)_{t \geq 0}$ be a process of class (Σ) such that $\lim_{t \to \infty} X_t = X_\infty$ exists a.s. and is finite (in particular N_∞ and A_∞ exist and are a.s. finite). Let*

$$g := \sup\{t : X_t = 0\} \quad \text{with the convention } \sup \emptyset = 0.$$

1. *If $(X_t)_{t \geq 0}$ is of class (D), then*

$$X_T = \mathbb{E}[X_\infty 1_{\{g \leq T\}} \mid \mathscr{F}_T] \quad \text{for every stopping time } T. \tag{6}$$

2. *More generally, if there exists a strictly positive Borel function f such that $(f(A_t)X_t)_{t \geq 0}$ is of class (D), then (6) holds.*
3. *If $(N_t^+)_{t \geq 0}$ is of class (D), then (6) holds.*

Proof (1) For a given stopping time T, denote

$$d_T = \inf\{t > T : X_t = 0\} \quad \text{with the convention } \inf \emptyset = \infty.$$

One checks that d_T is a stopping time. Since $X_\infty 1_{\{g \leq T\}} = X_{d_T}$ and $A_T = A_{d_T}$, it follows from Doob's optional stopping theorem that

$$\mathbb{E}[X_\infty 1_{\{g \leq T\}} \mid \mathscr{F}_T] = \mathbb{E}[N_{d_T} + A_{d_T} \mid \mathscr{F}_T] = \mathbb{E}[N_{d_T} + A_T \mid \mathscr{F}_T] = N_T + A_T = X_T.$$

(2) Assume that there exists a strictly positive Borel function such that $(f(A_t)X_t)_{t \geq 0}$ is of class (D). This property is preserved if one replaces f by a smaller strictly positive Borel function, hence, one can suppose that f is locally bounded. Then $(f(A_t)X_t)_{t \geq 0}$ is of class (ΣD), and from part (1) of the theorem, we have:

$$f(A_T)X_T = \mathbb{E}[f(A_\infty)X_\infty 1_{\{g \leq T\}} \mid \mathscr{F}_T].$$

But on the set $\{g \leq T\}$, we have $A_\infty = A_T$, and consequently

$$f(A_T)X_T = f(A_T)\mathbb{E}[X_\infty 1_{\{g \leq T\}} \mid \mathscr{F}_T].$$

The result follows by dividing both sides by $f(A_T)$ which is strictly positive.

(3) Since $X \geq 0$ and since $(N_t^+)_{t \geq 0}$ is of class (D), we note that $(\exp(-A_t)X_t)_{t \geq 0}$ is of class (D) and the result follows from (2).   □

The above theorem is used in [4] for financial applications. For example, if $(M_t)_{t \geq 0}$ is a nonnegative local martingale, converging to 0 (which is a reasonable model for stock prices or for portfolios under the benchmark approach), then the drawdown process $DD_t = \max_{u \leq t} M_u - M_t$ is of class (Σ) as well as the relative drawdown process $rDD_t = 1 - \frac{M_t}{\max_{u \leq t} M_u}$. For example, if $(M_t)_{t \geq 0}$ is a strict nonnegative local martingale such as the inverse of the 3 dimensional Bessel process, then $(DD_t)_{t \geq 0}$ is of class (Σ), satisfies condition (3) of the above theorem but is not uniformly integrable. Note that Theorem 1 applies in this case, even if the local martingale part of the process $(DD_t)_{t \geq 0}$, which is equal to $(-M_t)_{t \geq 0}$, is not a true martingale.

4.3 The General Case

A general solution for Problem 1 is provided in [12]; the measure $\mathscr{W}$ is a special case of this theorem. For the general case, we need to be very careful: the theorem below would be wrong under the usual assumptions and it would also be wrong if the filtration $(\mathscr{F}_t)_{t\geq 0}$ does not allow the extension of coherent probability measures. The N-usual assumptions are here to ensure in particular that there exists a continuous and adapted version of the process A which is defined everywhere (think of A_t being for example the local time at the level 0 of the Wiener process). According to Sect. 3, typical probability spaces where the following theorem holds are $\mathscr{C}(\mathbb{R}_+, \mathbb{R})$ and $\mathscr{D}(\mathbb{R}_+, \mathbb{R})$.

Theorem 2 *Let $(X_t)_{t\geq 0}$ be a true submartingale of the class (Σ): its local martingale part $(N_t)_{t\geq 0}$ is a true martingale, and X_t is integrable for all $t \geq 0$. We suppose that $(X_t)_{t\geq 0}$ is defined on a filtered probability space $(\Omega, \mathscr{F}, \mathbb{P}, (\mathscr{F}_t)_{t\geq 0})$ satisfying the property (NP), in particular, this space satisfies the N-usual conditions and $\mathscr{F}$ is the σ-algebra generated by $\mathscr{F}_t$ for $t \geq 0$. Then, there exists a unique σ-finite measure $\mathscr{Q}$, defined on $(\Omega, \mathscr{F}, \mathbb{P})$, such that for $g := \sup\{t \geq 0, X_t = 0\}$:*

- $\mathscr{Q}[g = \infty] = 0$;
- *For all $t \geq 0$, and for all $\mathscr{F}_t$-measurable, bounded random variables Γ_t,*

$$\mathscr{Q}\left[\Gamma_t \mathbb{1}_{g \leq t}\right] = \mathbb{P}\left[\Gamma_t X_t\right].$$

In [12], the measure $\mathscr{Q}$ is explicitly constructed, in the following way (with a slightly different notation). Let f be a Borel, integrable, strictly positive and bounded function from $\mathbb{R}$ to $\mathbb{R}$, and let us define the function G by the formula:

$$G(x) = \int_x^\infty f(y)\, dy.$$

One can prove that the process

$$\left(M_t^f := G(A_t) - \mathbb{E}_{\mathbb{P}}[G(A_\infty)|\mathscr{F}_t] + f(A_t)X_t\right)_{t\geq 0}, \tag{7}$$

is a martingale with respect to $\mathbb{P}$ and the filtration $(\mathscr{F}_t)_{t\geq 0}$. Since $(\Omega, \mathscr{F}, \mathbb{P}, (\mathscr{F}_t)_{t\geq 0})$ satisfies the N-usual conditions and since $G(A_t) \geq G(A_\infty)$, one can suppose that this martingale is nonnegative and càdlàg, by choosing carefully the version of $\mathbb{E}_{\mathbb{P}}[G(A_\infty)|\mathscr{F}_t]$. In this case, since $(\Omega, \mathscr{F}, \mathbb{P}, (\mathscr{F}_t)_{t\geq 0})$ satisfies the property (NP), there exists a unique finite measure $\mathscr{M}^f$ such that for all $t \geq 0$, and for all bounded, $\mathscr{F}_t$-measurable functionals Γ_t:

$$\mathscr{M}^f[\Gamma_t] = \mathbb{E}_{\mathbb{P}}[\Gamma_t M_t^f].$$

Now, since f is strictly positive, one can define a σ-finite measure $\mathscr{Q}^f$ by:

$$\mathscr{Q}^f := \frac{1}{f(A_\infty)} \cdot \mathscr{M}^f.$$

It is proved is [12] that if the function G/f is uniformly bounded (this condition is, for example, satisfied for $f(x) = e^{-x}$), then $\mathscr{Q}^f$ satisfies the conditions defining

$\mathscr{Q}$ in Theorem 2, which implies the existence part of this result. The uniqueness part is proved just after in a very easy way: one remarkable consequence of it is the fact that $\mathscr{Q}^f$ does not depend on the choice of f. The measure $\mathscr{Q}$ has many other interesting properties, which will be detailed in Sect. 5.

5 Further Properties of $\mathscr{Q}$ and Some Remarkable Associated Martingales, and Penalisation Results

The properties of $\mathscr{Q}$ given in this section are stated and proved in [11], except the results concerning penalisation, which are shown in [10]. By the construction of $\mathscr{Q}$ described above, it is clear that if f is a Borel, integrable, strictly positive and bounded function from $\mathbb{R}$ to $\mathbb{R}$, and if, with the notation above, G/f is uniformly bounded, then

$$\mathscr{M}^f = f(A_\infty) . \mathscr{Q}. \tag{8}$$

Now, by using (7), it is possible to construct $(M_t^f)_{t \geq 0}$, and then $\mathscr{M}^f$, for all Borel, integrable and nonnegative functions f. Moreover, it is proved in [11] that (8) remains true for any function f satisfying these weaker assumptions. The relation between the functional $f(A_\infty)$ and the martingale $(M_t^f)_{t \geq 0}$ can be generalized as follows:

Proposition 6 *We suppose that the assumptions of Theorem 2 hold, and we take the same notation. Let F be a $\mathscr{Q}$-integrable, nonnegative functional defined on $(\Omega, \mathscr{F})$. Then, there exists a càdlàg $\mathbb{P}$-martingale $(M_t(F))_{t \geq 0}$ such that the measure $\mathscr{M}^F :=$ $F.\mathscr{Q}$ is the unique finite measure satisfying, for all $t \geq 0$, and for all bounded, $\mathscr{F}_t$-measurable functionals Γ_t:*

$$\mathscr{M}^F[\Gamma_t] = \mathbb{P}[\Gamma_t M_t(F)].$$

The martingale $(M_t(F))_{t \geq 0}$ is unique up to indistinguishability.

If f is Borel, integrable and nonnegative, then $f(A_\infty)$ is integrable with respect to $\mathscr{Q}$ and one has

$$M_t(f(A_\infty)) = M_t^f,$$

which gives an explicit expression for $M_t(f(A_\infty))$. This explicit form can be generalized to the martingale $(M_t(F))_{t \geq 0}$ for all nonnegative and $\mathscr{Q}$-integrable functionals F, if the submartingale $(X_t)_{t \geq 0}$ is uniformly integrable. More precisely, one has the following result:

Proposition 7 *Let us suppose that the assumptions of Theorem 2 are satisfied, and that the process $(X_t)_{t \geq 0}$ is uniformly integrable. Then, X_t tends a.s. to a limit X_∞ when t goes to infinity, and the measure $\mathscr{Q}$ is absolutely continuous with respect to $\mathbb{P}$, with density X_∞. Moreover, a nonnegative functional F is integrable with respect* to $\mathcal{Q}$ if and only if FX_∞ is integrable with respect to $\mathbb{P}$, in this case, $(M_t(F))_{t\geq 0}$ a the càdlàg version of the conditional expectation $(\mathbb{P}[FX_\infty|\mathscr{F}_t])_{t\geq 0}$. In particular, it is uniformly integrable, and it converges a.s. and in L^1 to FX_∞ when t goes to infinity.

A more interesting case is when we suppose that A_∞ is infinite, $\mathbb{P}$-almost surely. From now, until the end of this section, we always implicitly make this assumption (which is satisfied, in particular, if $(X_t)_{t\geq 0}$ is a reflected Brownian motion). The following result gives the asymptotic behaviour of $(X_t)_{t\geq 0}$ under $\mathcal{Q}$, when t goes to infinity, and the behaviour of $M_t(F)$, which is not the same under $\mathbb{P}$ and under $\mathcal{Q}$:

Proposition 8 *Let us suppose that the assumptions of Theorem 2 are satisfied, and that $A_\infty = \infty$, $\mathbb{P}$-almost surely. Then, $\mathcal{Q}$-almost everywhere, X_t tends to infinity when t goes to infinity, and for all nonnegative, $\mathcal{Q}$-integrable functionals F, the martingale $(M_t(F))_{t\geq 0}$ tends $\mathbb{P}$-almost surely to zero and $\mathcal{Q}$-almost everywhere to infinity. Moreover, one has:*

$$\frac{M_t(F)}{X_t} \underset{t\to\infty}{\longrightarrow} F,$$

$\mathcal{Q}$-almost everywhere.

By definition, the martingales of the form $(M_t(F))_{t\geq 0}$ are exactly the local densities of the finite measures which are absolutely continuous with respect to $\mathcal{Q}$. This situation is similar to the case of uniformly integrable, nonnegative martingales, which are local densities of finite measures, absolutely continuous with respect to $\mathbb{P}$. The following decomposition of nonnegative supermartingales, already proved in [13] in the case of the reflected Brownian motion, involves simultaneously these two kind of martingales:

Proposition 9 *Let us suppose that the assumptions of Theorem 2 are satisfied, and that $A_\infty = \infty$, $\mathbb{P}$-almost surely. Let Z be a nonnegative, càdlàg $\mathbb{P}$-supermartingale. We denote by Z_∞ the $\mathbb{P}$-almost sure limit of Z_t when t goes to infinity. Then, $\mathcal{Q}$-almost everywhere, the quotient Z_t/X_t is well-defined for t large enough and converges, when t goes to infinity, to a limit z_∞, integrable with respect to $\mathcal{Q}$, and $(Z_t)_{t\geq 0}$ decomposes as*

$$(Z_t = M_t(z_\infty) + \mathbb{P}[Z_\infty|\mathscr{F}_t] + \xi_t)_{t\geq 0},$$

where $(\mathbb{P}[Z_\infty|\mathscr{F}_t])_{t\geq 0}$ denotes a càdlàg version of the conditional expectation of Z_∞ with respect to $\mathscr{F}_t$, and $(\xi_t)_{t\geq 0}$ is a nonnegative, càdlàg $\mathbb{P}$-supermartingale, such that:

- *$Z_\infty \in L_+^1(\mathscr{F},\mathbb{P})$, hence $\mathbb{P}[Z_\infty|\mathscr{F}_t]$ converges $\mathbb{P}$-almost surely and in $L^1(\mathscr{F},\mathbb{P})$ towards Z_∞.*
- *$\frac{\mathbb{P}[Z_\infty|\mathscr{F}_t]+\xi_t}{X_t} \underset{t\to\infty}{\longrightarrow} 0$, $\mathcal{Q}$-almost everywhere.*

- $M_t(z_\infty) + \xi_t \xrightarrow[t \to \infty]{} 0$, $\mathbb{P}$-*almost surely.*

Moreover, the decomposition is unique in the following sense: let z'_∞ be a $\mathscr{Q}$-integrable, nonnegative functional, Z'_∞ a $\mathbb{P}$-integrable, nonnegative random variable, $(\xi'_t)_{t \geq 0}$ a càdlàg, nonnegative $\mathbb{P}$-supermartingale, and let us suppose that for all $t \geq 0$,

$$Z_t = M_t(z'_\infty) + \mathbb{P}[Z'_\infty | \mathscr{F}_t] + \xi'_t.$$

Under these assumptions, if for t going to infinity, ξ'_t tends $\mathbb{P}$-almost surely to zero and ξ'_t / X_t tends $\mathscr{Q}$-almost everywhere to zero, then $z'_\infty = z_\infty$, $\mathscr{Q}$-almost everywhere, $Z'_\infty = Z_\infty$, $\mathbb{P}$-almost surely, and ξ' is $\mathbb{P}$-indistinguishable with ξ.

This result implies the following characterisation of martingales of the form $(M_t(F))_{t \geq 0}$:

Corollary 2 *Let us suppose that the assumptions of Theorem 2 are satisfied, and that $A_\infty = \infty$, $\mathbb{P}$-almost surely. Then, a càdlàg, nonnegative $\mathbb{P}$-martingale $(Z_t)_{t \geq 0}$ is of the form $(M_t(F))_{t \geq 0}$ for a nonnegative, $\mathscr{Q}$-integrable functional F, if and only if:*

$$\mathbb{P}[Z_0] = \mathscr{Q}\left(\lim_{t \to \infty} \frac{Z_t}{X_t} \right). \tag{9}$$

Note that, by Proposition 9, the limit above necessarily exists $\mathscr{Q}$-almost everywhere.

If in Proposition 9, $(Z_t)_{t \geq 0}$ is a nonnegative martingale, the corresponding decomposition can be interpreted as a Radon-Nykodym decomposition of finite measures. Indeed, let us observe that since the space satisfies the property (NP), there exists a unique finite measure $\mathbb{Q}_Z$ on $(\Omega, \mathscr{F})$, such that for all $t \geq 0$, its restriction to $\mathscr{F}_t$ has density Z_t with respect to $\mathbb{P}$. If one writes the decomposition

$$Z_t = M_t(z_\infty) + \mathbb{P}[Z_\infty | \mathscr{F}_t] + \xi_t,$$

one deduces:

$$\mathbb{Q}_Z = z_\infty . \mathscr{Q} + Z_\infty . \mathbb{P} + \mathbb{Q}_\xi,$$

where the restriction of $\mathbb{Q}_\xi$ to $\mathscr{F}_t$ has density ξ_t with respect to $\mathbb{P}$. In [11], it is proved that $\mathbb{Q}_\xi$ is singular with respect to $\mathbb{P}$ and $\mathscr{Q}$, hence one has a decomposition of $\mathbb{Q}_Z$ into three parts:

- A part which is absolutely continuous with respect to $\mathbb{P}$.
- A part which is absolutely continuous with respect to $\mathscr{Q}$.
- A part which is singular with respect to $\mathbb{P}$ and $\mathscr{Q}$.

This decomposition is unique, as a consequence of the uniqueness of the Radon-Nykodym decomposition.

Another interesting problem is to find a relation between the measure $\mathscr{Q}$ and the last passage times of $(X_t)_{t \geq 0}$ at a given level, which can be different from zero. The following result proves this relation when $(X_t)_{t \geq 0}$ is continuous:

Proposition 10 *Let us suppose that the assumptions of Theorem 2 are satisfied, the submartingale $(X_t)_{t\geq 0}$ is continuous and $A_\infty = \infty$ almost surely under $\mathbb{P}$. For $a \geq 0$, let $g^{[a]}$ be the last hitting time of the interval $[0, a]$:*

$$g^{[a]} = \sup\{t \geq 0, X_t \leq a\}.$$

Then, the measure $\mathcal{Q}$ satisfies the following formula, available for any $t \geq 0$, and for all $\mathscr{F}_t$-measurable, bounded variables Γ_t:

$$\mathcal{Q}\left[\Gamma_t \, \mathbb{1}_{g^{[a]}\leq t}\right] = \mathbb{P}\left[\Gamma_t(X_t - a)_+\right]. \tag{10}$$

Moreover, $((X_t - a)_+)_{t\geq 0}$ is a submartingale of class (Σ) and the σ-finite measure obtained by applying Theorem 2 to it is equal to $\mathcal{Q}$.

The relation between the measure $\mathcal{Q}$ and the last hitting time of a given level a is one of the main ingredients of the proof of our penalisation result stated below. Since we are able to show this statement only when $(X_t)_{t\geq 0}$ is a diffusion satisfying some technical conditions, let us give another result, strongly related to penalisations, but available under more general assumptions:

Proposition 11 *Let us suppose that the assumptions of Theorem 2 are satisfied, and $A_\infty = \infty$, $\mathbb{P}$-almost surely. Let $(F_t)_{t\geq 0}$ be a càdlàg, adapted, nonnegative, nonincreasing and uniformly bounded process, such that for some $a > 0$, one has for all $t \geq 0$, $X_t = X_{g^{[a]}}$ on the set $\{t \geq g^{[a]}\}$. Then, if F_g is $\mathcal{Q}$-integrable, and if one defines F_∞ as the limit of F_t for t going to infinity (in particular, $F_\infty = F_{g^{[a]}}$ for $g^{[a]} < \infty$), one has:*

$$\mathbb{P}[F_t X_t] \underset{t\to\infty}{\longrightarrow} \mathcal{Q}[F_\infty].$$

In [11], there are another version of this result, with slightly different assumptions. In order to obtain a penalisation result, we need to estimate the expectation of F_t under $\mathbb{P}$, instead of the expectation of $F_t X_t$. This gives an extra factor, depending on t, in the asymptotics, which justifies some restriction on the process $(X_t)_{t\geq 0}$. Let us now describe the precise framework which is considered. We define Ω as the space of continuous functions from $\mathbb{R}_+$ to $\mathbb{R}_+$, $(\mathscr{F}_t^0)_{t\geq 0}$ as the natural filtration of Ω, and $\mathscr{F}^0$, as the σ-algebra generated by $(\mathscr{F}_t^0)_{t\geq 0}$. The probability $\mathbb{P}^0$, defined on $(\Omega, \mathscr{F}^0)$, satisfies the following condition: under $\mathbb{P}^0$, the canonical process is a recurrent diffusion in natural scale, starting from a fixed point $x_0 \geq 0$, with zero as an instantaneously reflecting barrier, and such that its speed measure is absolutely continuous with respect to Lebesgue measure on $\mathbb{R}_+$, with a density $m : \mathbb{R}_+^* \to \mathbb{R}_+^*$, continuous, and such that $m(x)$ is equivalent to cx^β when x goes to infinity, for some $c > 0$ and $\beta > -1$. Moreover, we suppose that there exists $C > 0$ such that for all $x > 0$, $m(x) \leq Cx^\beta$ if $\beta \leq 0$, and $m(x) \leq C(1 + x^\beta)$ if $\beta > 0$. Let us now define the filtered probability space $(\Omega, \mathscr{F}, (\mathscr{F}_t)_{t\geq 0}, \mathbb{P})$ as the N-augmentation of $(\Omega, \mathscr{F}^0, (\mathscr{F}_t^0)_{t\geq 0}, \mathbb{P}^0)$: $(\Omega, \mathscr{F}, (\mathscr{F}_t)_{t\geq 0}, \mathbb{P})$ satisfies property (NP) and under $\mathbb{P}$, the law of the canonical process is a diffusion with the same parameters as under

$\mathbb{P}^0$. This diffusion is in natural scale, and one can deduce from this fact and the assumptions above that the canonical process $(X_t)_{t\geq 0}$ is a submartingale of class (Σ). In particular, X_t is integrable for all $t \geq 0$. Moreover, the local time $(L_t)_{t\geq 0}$ of $(X_t)_{t\geq 0}$ at level zero, is its increasing process. One deduces that Theorem 2 applies, which defines the corresponding measure $\mathscr{Q}$, and we can check that we are in the situation where $L_\infty = \infty$, $\mathbb{P}$-almost surely. Under these assumptions, we can prove the following result:

Proposition 12 *Let $(F_t)_{t\geq 0}$ be a càdlàg, adapted, nonnegative, uniformly bounded and nonincreasing process. We assume that there exists $a > 0$ such that for all $t \geq 0$, $X_t = X_{g^{[a]}}$ on the set $\{t \geq g^{[a]}\}$, we define F_∞ as the limit of F_t for t going to infinity, and we suppose that F_∞ is integrable with respect to $\mathscr{Q}$. Then, there exists $D > 0$ such that for all $s \geq 0$ and for all events $\Lambda_s \in \mathscr{F}_s$:*

$$t^{1/(\beta+2)} \mathbb{P}[F_t \mathbb{1}_{\Lambda_s}] \underset{t\to\infty}{\longrightarrow} D \mathscr{Q}[F_\infty \mathbb{1}_{\Lambda_s}].$$

The following penalisation result is a straightforward consequence of Proposition 12:

Proposition 13 *Let $(F_t)_{t\geq 0}$ be a càdlàg, adapted, nonnegative, uniformly bounded and nonincreasing process. We assume that there exists $a > 0$ such that for all $t \geq 0$, $X_t = X_{g^{[a]}}$ on the set $\{t \geq g^{[a]}\}$, we define F_∞ as the limit of F_t for t going to infinity, and we suppose that $0 < \mathscr{Q}[F_\infty] < \infty$. Then, for all $t \geq 0$:*

$$0 < \mathbb{P}[F_t] < \infty,$$

and one can then define a probability measure $\mathbb{Q}_t$ on $(\Omega, \mathscr{F})$ by

$$\mathbb{Q}_t := \frac{F_t}{\mathbb{P}[F_t]}.\mathbb{P}.$$

Moreover, the probability measure:

$$\mathbb{Q}_\infty := \frac{F_\infty}{\mathscr{Q}[F_\infty]}.\mathscr{Q}$$

is the weak limit of $\mathbb{Q}_t$ in the sense of penalisations, i.e. for all $s \geq 0$, and for all events $\Lambda_s \in \mathscr{F}_s$,

$$\mathbb{Q}_t[\Lambda_s] \underset{t\to\infty}{\longrightarrow} \mathbb{Q}_\infty[\Lambda_s].$$

This penalisation result applies, in particular, to the power $2r$ of a Bessel process of dimension $2(1-r)$, for any $r \in (0,1)$ (in this case, the parameter β is equal to $(1/r)-2$). For $r=1$, $(X_t)_{t\geq 0}$ is a reflected Brownian motion, and our result is very similar to the penalisation theorem stated in [13] for the Brownian motion.

References

1. Azéma, J., Meyer, P.-A., Yor, M.: Martingales relatives. In: Séminaire de Probabilités, XXVI. *Lecture Notes in Math.*, vol. 1526, pp. 307–321. Springer, Berlin (1992)
2. Azéma, J., Yor, M.: Sur les zéros des martingales continues. In: Séminaire de Probabilités, XXVI. *Lecture Notes in Math.*, vol. 1526, pp. 248–306. Springer, Berlin (1992)
3. Bichteler, K.: Stochastic Integration and Stochastic Differential Equations. *Encyclopedia of Mathematics and its Applications.* Cambridge University Press, Cambridge (2002)
4. Cheridito, P., Nikeghbali, A., Platen, E.: Processes of the class sigma, last zero and draw-down processes (2009)
5. Dellacherie, C., Meyer, P.-A.: Probabilité et Potentiel, vol. I. Hermann, Paris (1976)
6. Dellacherie, C., Meyer, P.-A.: Probabilité et Potentiel, vol. II. Hermann, Paris (1980)
7. Föllmer, H., Imkeller, P.: Anticipation cancelled by a Girsanov transformation: A paradox on Wiener space. Ann. Inst. Henri Poincaré Probab. Stat. **29**(4), 569–586 (1993)
8. Madan, D., Roynette, B., Yor, M.: From Black-Scholes formula, to local times and last passage times for certain submartingales. Prépublication IECN, No. 14 (2008)
9. Najnudel, J., Nikeghbali, A.: A new kind of augmentation of filtrations. 0910.4959 (2009)
10. Najnudel, J., Nikeghbali, A.: On penalisation results related with a remarkable class of submartingales. 0911.4365 (2009)
11. Najnudel, J., Nikeghbali, A.: On some properties of a universal sigma-finite measure associated with a remarkable class of submartingales. 0911.2571 (2009)
12. Najnudel, J., Nikeghbali, A.: On some universal σ-finite measures and some extensions of Doob's optional stopping theorem. 0906.1782 (2009)
13. Najnudel, J., Roynette, B., Yor, M.: A Global View of Brownian Penalisations. *MSJ Memoirs* 19. Mathematical Society of Japan, Tokyo (2009)
14. Nikeghbali, A.: A class of remarkable submartingales. Stoch. Process. Appl. **116**(6), 917–938 (2006)
15. Parthasarathy, K.-R.: Probability Measures on Metric Spaces. Academic Press, New York (1967)
16. Roynette, B., Vallois, P., Yor, M.: Some penalisations of the Wiener measure. Jpn. J. Math. **1**(1), 263–290 (2006)
17. Stroock, D.-W., Varadhan, S.-R.-S.: Multidimensional Diffusion Processes. *Classics in Mathematics.* Reprint of the 1997 edition. Springer, Berlin (2006)
18. Yor, M.: Les inégalités de sous-martingales, comme conséquences de la relation de domination. Stochastics **3**(1), 1–15 (1979)

Pricing Without Equivalent Martingale Measures Under Complete and Incomplete Observation

Giorgia Galesso and Wolfgang J. Runggaldier

Abstract Traditional arbitrage pricing theory is based on martingale measures. Recent studies show that there are situations when there does not exist an equivalent martingale measure and so the question arises: what can one do with pricing and hedging in this situation? We mention here two approaches to this effect that have appeared in the literature, namely the "Fernholz-Karatzas" approach and Platen's "Benchmark approach" and discuss their relationships both in models where all relevant quantities are fully observable as well as in models where this is not the case and, furthermore, not all observables are also investment instruments.

1 Introduction

Traditional arbitrage pricing theory is based on martingale measures, whereby the density process as well as the prices expressed in units of a same numeraire are martingales. Recent studies such as [1, 2] show that some form of arbitrage may exist in real markets and then processes, which in traditional theory are martingales, turn out to be strictly local martingales. Without necessarily passing through absence of arbitrage, also the authors of [5] describe instances in which classical martingale processes are strict local martingales. Typically such a process is the density process and then no equivalent martingale measure exists. In such a situation the question arises: what can one do with pricing and hedging?

This issue has already been dealt with in the literature and, after recalling the basic features of two of these approaches, our first purpose in the paper is to show that, although based on different viewpoints, they lead to the same values for the prices in

W.J. Runggaldier (✉) · G. Galesso
Department of Pure and Applied Mathematics, University of Padova, 63, Via Trieste, I-35121
Padova, Italy
e-mail: runggal@math.unipd.it

G. Galesso
e-mail: giorgiaigroig@alice.it

C. Chiarella, A. Novikov (eds.), *Contemporary Quantitative Finance*,
DOI 10.1007/978-3-642-03479-4_6, © Springer-Verlag Berlin Heidelberg 2010

the case when all relevant quantities are fully observable. The second purpose is to extend the two approaches to situations when some of the underlying quantities of the market model are not fully observable and, furthermore, not all observables are also investment instruments and to compare the two approaches also in this latter case.

The first approach is based on work mainly done by E.R. Fernholz and I. Karatzas together with co-workers and of which an overview is given in [2]. We shall refer to it as "Fernholz-Karatzas (FK) approach". The basic idea here is taken from a classical approach in incomplete markets whereby the price of a derivative asset is defined as the smallest superreplicating initial capital. By utilizing the exponential (local) martingale given by the density process $Z(\cdot)$, it is shown that this price can be obtained by utilizing $\frac{Z(\cdot)}{B(\cdot)}$ as "deflator" ($B(t)$ is the money market account/local riskless asset). In this context it is also shown that, even without existence of a martingale measure, the market can be completed and a hedging strategy is explicitly exhibited.

The second approach is based on work done by E. Platen and coworkers and a comprehensive account can be found in the book [5]. Platen calls it the "Benchmark approach" and we shall refer to it under this same name with the acronym (PB). Here the methodology is based on utility indifference pricing thereby showing that the resulting price is independent of the chosen utility function as well as of the initial wealth and it does not require the claim to be replicable. By showing, in this paper, the equivalence of the two approaches in the case of complete observation we also show that, although by (PB) the claim does not have a priori to be replicable, it is so a posteriori and a replicating strategy is given by the one constructed in the (FK) approach.

Since the (FK) approach relies on completing the market, while (PB) is of the utility indifference pricing type that is typically designed for incomplete markets, it appears to be rather intuitive that (FK) has to be restricted to those situations where all observables are also investment instruments. In fact we show that, while in the complete observation case the two approaches are equivalent, there are some differences in the case of incomplete observation if not all observables are also investment instruments.

We consider a classical diffusion type model, although extensions to discontinuous market models may be envisaged. A major role in both approaches is plaid by the growth optimal portfolio (GOP) and we briefly summarize some of its basic features. We also discuss the relationship between the (GOP) strategy and growth rate under complete and incomplete observation. We always assume however that the value of the GOP is observable (possibly via a proxy).

We start in Sect. 2 by describing the model and deriving the GOP. Section 3 is a brief account on how some form of classical arbitrage may arise implying that then an equivalent martingale measure does not exist. It is based on [2], further instances may be found in [5]. In Sect. 4 we summarize the basic features of the two approaches, adding some detail to the proofs of the main results and showing their equivalence under full observation of all relevant quantities. In Sect. 5 we first describe the incomplete observation model that is based on [7] and where, furthermore, not all observables are also investment instruments and discuss some consequences for the GOP and utility maximization. Then we describe extensions of the two approaches to this model and discuss a possible equivalence.

2 The Model

We consider a market $\mathcal{M}$ with a locally riskless asset and a certain number n of risky assets with prices $B(t)$ and $S_i(t)$ respectively that, on a given filtered probability space $(\Omega, \mathcal{F}, \mathcal{F}(t), P)$, satisfy the following diffusion-type model

$$dB(t) = B(t)r(t)dt, \quad B(0) = 1,$$

$$dS_i(t) = S_i(t)\left(b_i(t)dt + \sum_{j=1}^{n}\sigma_{ij}(t)dw_j(t)\right), \quad S_i(0) = s_i > 0, \ i = 1, \ldots, n. \tag{1}$$

We assume the number of driving Wiener processes $w_j(t)$ to be equal to the number of risky assets and that the diffusion matrix $\sigma(t) = \{\sigma_{ij}(t)\}_{i,j=1,\ldots,n}$ is non singular for each $t \in [0, T]$ with $T < \infty$. Furthermore, the coefficients in (1) are progressively measurable and satisfy

$$\int_0^T |r(t)|dt + \sum_{i=1}^{n}\int_0^T \left(|b_i(t)| + \sum_{j=1}^{n}(\sigma_{ij}(t))^2\right)dt < \infty \text{ a.s. for any } T \in (0, \infty). \tag{2}$$

While for most of the results below the filtration $\{\mathcal{F}(t)\}$ can be fully general, for some of them we shall, without much loss of generality, assume that $\mathcal{F}(t) = \mathcal{F}^w(t)$.

From (1) we obtain for the risky assets' log prices

$$d\log S_i(t) = \gamma_i(t)dt + \sum_{j=1}^{n}\sigma_{ij}(t)dw_j(t), \quad i = 1, \ldots, n \tag{3}$$

with

$$\gamma_i(t) := b_i(t) - \frac{1}{2}a_{ii}(t), \quad a(t) = \sigma\sigma'(t), \ i = 1, \ldots, n. \tag{4}$$

Under controlled growth of $a(t)$ as e.g.

$$\lim_{T \to \infty}\left(\frac{\log\log T}{T^2}\int_0^T a_{ii}(t)dt\right) = 0, \quad \text{a.s. } i = 1, \ldots, n \tag{5}$$

one has

$$\lim_{T \to \infty}\frac{1}{T}\left(\log S_i(T) - \int_0^T \gamma_i(t)dt\right) = 0 \tag{6}$$

which justifies the definition of $\gamma_i(t)$ as *growth rate* of the i-th asset.

A *portfolio strategy* (investment rates) is a progressively measurable process $\pi(t) = (\pi_1(t), \ldots, \pi_n(t))$ with values in

$$\bigcup_{k \in \mathbb{N}} \{ (\pi_1, \ldots, \pi_n) \in \mathbb{R} \mid \pi_1 + \cdots + \pi_n \leq 1, \ \pi_1^2 + \cdots + \pi_n^2 \leq k^2 \}.$$

The ratio $1 - \sum_{i=1}^{n} \pi_i(t)$ is invested in the risk-free asset.

Let $V^{v,\pi}(t)$ be the *value* at t of a self financing portfolio with $V^{v,\pi}(0) = v$. From the self financing condition one then has

$$\frac{dV^{v,\pi}(t)}{V^{v,\pi}(t)} = \left(1 - \sum_{i=1}^{n} \pi_i(t)\right) \frac{dB(t)}{B(t)} + \sum_{i=1}^{n} \pi_i(t) \frac{dS_i(t)}{S_i(t)}$$

$$= \left(1 - \sum_{i=1}^{n} \pi_i(t)\right) r(t)dt + \sum_{i=1}^{n} \pi_i(t) \left(b_i(t)dt + \sum_{j=1}^{n} \sigma_{ij}(t)dw_j(t)\right).$$

$$(7)$$

Putting $\sigma_{\pi j}(t) := \sum_{i=1}^{n} \pi_i(t)\sigma_{ij}(t)$, we have, analogously to (3) and (4),

$$d \log V^{v,\pi}(t) = \gamma_\pi(t)dt + \sum_{j=1}^{n} \sigma_{\pi j}(t)dw_j(t), \quad i = 1, \ldots, n \qquad (8)$$

with the *portfolio growth rate*

$$\gamma_\pi(t) = r(t) + \sum_{j=1}^{n} \sigma_{\pi j}(t)\theta_j(t) - \frac{1}{2}\left(\sum_{j=1}^{n} \sigma_{\pi j}(t)\right)^2 \qquad (9)$$

where, in vector notation,

$$\theta(t) := (\sigma(t))^{-1}(b(t) - r(t)\underline{1}) \qquad (10)$$

is the "*market price of risk*". Under a controlled growth of $a(t)$ analogous to (5), also here one has

$$\lim_{T \to \infty} \frac{1}{T}\left(\log V^{v,\pi}(T) - \int_0^T \gamma_\pi(t)dt\right) = 0, \qquad (11)$$

justifying again the name "growth rate" and notice that $V^{v,\pi} = vV^{1,\pi} := vV^{\pi}$. This latter scaling property of V allows us to derive most of the results for the simpler V^π, they will then hold automatically also for $V^{v,\pi}$ for any $v > 0$.

Definition 1 Given an initial wealth v, the set of *admissible strategies* is $\mathscr{H}(v) := \bigcup_{T>0} \mathscr{H}(v; T)$ where

$$\mathscr{H}(v; T) := \{\pi(\cdot) \mid P\{V^{v,\pi}(t) \geq 0, \forall\, 0 \leq t \leq T\} = 1\}. \qquad (12)$$

A major role is plaid by the so-called **growth optimal portfolio (GOP)** that maximizes the growth rate among the admissible portfolios (*it maximizes also the expected log-utility*).

In vector form we have for the growth rate for a generic portfolio π

$$\gamma_\pi(t) = r(t) + \pi'(t)\sigma(t)\theta(t) - \frac{1}{2}\pi'(t)\sigma\sigma'(t)\pi(t)$$

which is a concave function of π so that, denoting by $\rho(t)$ the GOP strategy that maximizes this growth rate, one has

$$\rho(t) = (\sigma'(t))^{-1}\theta(t). \tag{13}$$

From (8) and (9) we have for any admissible π

$$\frac{dV^\pi(t)}{V^\pi(t)} = r(t)dt + \pi'(t)\sigma(t)\big(\theta(t)dt + dw(t)\big) \tag{14}$$

which implies for $\pi = \rho$

$$\frac{dV^\rho(t)}{V^\rho(t)} = \big(r(t) + \|\theta(t)\|^2\big)dt + \theta'(t)dw(t) \tag{15}$$

and so

$$V^\rho(t) = \exp\left[\int_0^t r(s)ds + \frac{1}{2}\int_0^t \|\theta(s)\|^2 ds + \int_0^t \theta'(s)dw(s)\right]. \tag{16}$$

Furthermore, the growth rate of the GOP is

$$\gamma_\rho(t) = r(t) + \frac{1}{2}\|\theta(t)\|^2. \tag{17}$$

3 Relative Arbitrage and Absence of Martingale Measures

Traditional arbitrage pricing theory is based on equivalent martingale measures. In particular, defining

$$Z(t) := \exp\left\{-\int_0^t \theta'(s)dw(s) - \frac{1}{2}\int_0^t \|\theta(s)\|^2 ds\right\}, \quad 0 \le t < \infty \tag{18}$$

with $\theta(t)$ as in (10), the following holds: if $E\{Z(T)\} = 1$ for any given $T \in (0, \infty)$, then $Z(t)$ is a mean one martingale and, defining a measure $Q \sim P$ with $\frac{dQ}{dP}|_{\mathscr{F}(T)} = Z(T)$, one has that $\hat{w}(t) = w(t) + \int_0^t \theta(s)ds$ is a Wiener process on $(\Omega, \mathscr{F}, \mathscr{F}(t), Q)$ and $\bar{S}_i(t) := B^{-1}(t)S_i(t)$ are $(\mathscr{F}(t), Q)$-martingales.

There are however situations, and below we shall mention some instances, where $E\{Z(T)\} < 1$ for some $T \in (0, \infty)$ and then there does not exist an equivalent martingale measure and the traditional pricing approach breaks down. In view of mentioning examples where $E\{Z(T)\} < 1$ let us give the following definition (see [2])

Definition 2 Given two portfolio strategies $\pi(\cdot)$ and $\rho(\cdot)$, we say that $\pi(\cdot)$ represents an arbitrage opportunity relative to $\rho(\cdot)$ on $[0, T]$ for $T > 0$ if, with $V^\pi(0) = V^\rho(0)$, one has

$$\mathbb{P}(V^\pi(T) \ge V^\rho(T)) = 1 \quad \text{and} \quad \mathbb{P}(V^\pi(T) > V^\rho(T)) > 0.$$

One can then show the following proposition (see again [2])

Proposition 1 *If model* (1) *satisfies* $\xi'(\sigma\sigma')(t)\xi \leq K\|\xi\|^2$, $\forall \xi \in \mathbb{R}^n$, $\forall t \in [0,\infty)$ *with* $K > 0$ *and for a* $T > 0$ *there exists an arbitrage relative to a given portfolio* $\rho(\cdot)$, *then* $Z(t)$ *is a strictly local martingale and* $E\{Z(T)\} < 1$.

As a corollary one can show (see always [2])

Corollary 2 *Under the assumptions of Proposition 1 and assuming* $\mathscr{F}(t) = \mathscr{F}^w(t)$, *there does not exist an equivalent martingale measure.*

In [2] it is shown how one can construct relative arbitrage also on arbitrary time horizons, which leads to one kind of examples where there does not exist an equivalent martingale measure. Other kinds of examples can be found in [5].

Since in the situations corresponding to the above examples the traditional pricing approach based on equivalent martingale measures breaks down, one may wonder what can be done in such situations for what concerns pricing and the rest of the paper is devoted to this issue.

4 Pricing Under Absence of an Equivalent Martingale Measure (the Case of Complete Observation)

We summarize here two approaches and show that, although based on different viewpoints, they lead to the same prices.

4.1 The "Fernholz-Karatzas Approach"

This subsection in based on [2]. The main idea here is taken from a classical approach in incomplete markets, whereby a reasonable way to assign a price to a derivative asset is to put it equal to the smallest initial amount needed to superreplicate a given claim with a self financing portfolio.

The methodology is based on showing that, even without existence of a martingale measure, the market can be completed and here the proof is constructive by exhibiting a hedging strategy.

One has the following theorem (see [2, 3])

Theorem 3 *Consider a given maturity* T *and a claim* $Y \in \mathscr{F}(T)$ *such that* $E\{Y \frac{Z(T)}{B(T)}\} \in (0,\infty)$ *with* $Z(t)$ *as in* (18). *For* $t \in [0,T]$, *let*

$$\mathscr{H}_t(v;T) := \left\{\pi(\cdot) \mid P\left\{V^{v,\pi}(s) \geq 0, \forall\, t \leq s \leq T\right\} = 1; V^{v,\pi}(t) = v \in \mathscr{F}(t)\right\}$$

namely the set of admissible strategies starting from a capital v at time t and put

$$\mathcal{U}^Y(t) := \operatorname{ess\,inf}\left\{v > 0 \mid \exists \pi \in \mathcal{H}_t(v; T) \text{ s.t. } V^{v,\pi}(T) \geq Y\right\} \tag{19}$$

which represents the minimal hedging price at time t. Let, furthermore,

$$Y(t) := \frac{B(t)}{Z(t)} E\left\{Y \frac{Z(T)}{B(T)} \mid \mathcal{F}(t)\right\} \tag{20}$$

with the expectation under the physical measure P. If $\mathcal{F}(t) = \mathcal{F}^w(t)$, then

$$Y(t) = \mathcal{U}^Y(t). \tag{21}$$

Remark 4 On the basis of this theorem we shall now take $Y(t)$ as price of the claim at time t. Notice that in formula (20) the process $\frac{Z(t)}{B(t)}$ acts as a "deflator". Notice furthermore that, if $Z(t)$ is a true martingale and $\mathcal{S}$ is the set of martingale measures then, since $\mathcal{U}^Y(t)$ is the minimal superhedging price,

$$Y(t) = \mathcal{U}^Y(t) = \sup_{Q \in \mathcal{S}} E^Q\left\{\frac{B(t)}{B(T)} Y \mid \mathcal{F}(t)\right\}$$

namely the here defined price coincides with the standard superhedging price.

Proof of Theorem 3 We proceed along several steps, whereby we shall also need the fact that, as will be shown in Sect. 4.3 below, $V^\rho(t) = \frac{B(t)}{Z(t)}$.

i) Notice first that from (14) and (15) we have for $\hat{V}^{v,\pi}(s) := \frac{V^{v,\pi}(s)}{V^\rho(s)} = \frac{Z(s)}{B(s)} V^{v,\pi}(s)$, with $s \in [t, T]$ and any $v > 0$, that (in vector form)

$$d\hat{V}^{v,\pi}(s) = \hat{V}^{v,\pi}(s)(\pi'(s)\sigma(s) - \theta'(s))dw(s) \tag{22}$$

so that $\hat{V}^{v,\pi}$ is a non-negative local martingale and therefore a supermartingale with $\hat{V}^{v,\pi}(t) = v$ (recall that $V^\rho(s)$ is the GOP with $V^\rho(t) = 1$). Since, as will be shown in iii) below, the set in the definition of $\mathcal{U}^Y(t)$ is not empty, then there exists $\pi \in \mathcal{H}_t(v; T)$ such that $V^{v,\pi}(T) \geq Y$ and so

$$Y(t) = \frac{B(t)}{Z(t)} E\left\{Y \frac{Z(T)}{B(T)} \mid \mathcal{F}(t)\right\} \leq \frac{B(t)}{Z(t)} E\left\{V^{v,\pi}(T) \frac{Z(T)}{B(T)} \mid \mathcal{F}(t)\right\}$$

$$\leq \frac{B(t)}{Z(t)} V^{v,\pi}(t) \frac{Z(t)}{B(t)} = v$$

which implies $Y(t) \leq \mathcal{U}^Y(t)$.

To complete the proof, in the next steps we show that $Y(t) \geq \mathcal{U}^Y(t)$ and we do it by explicitly constructing a hedging strategy.

ii) Under the assumptions of the theorem we can define the $(P, \mathcal{F}(t))$-martingale

$$M(s) := E\left\{Y \frac{Z(T)}{B(T)} \mid \mathcal{F}(s)\right\} \tag{23}$$

which, given that $\mathscr{F}(t) = \mathscr{F}^w(t)$, admits the following "martingale representation" for $s \in [t, T]$,

$$M(s) = M(t) + \int_t^s \psi'(u)dw(u) = Y(t)\frac{Z(t)}{B(t)} + \int_t^s \psi'(u)dw(u) \quad (24)$$

for a progressively measurable, a.s. square integrable process $\psi(t)$ and where the last equality follows from the definition of $Y(t)$ in (20).

iii) If we choose v such that

$$v = \hat{V}^{v,\pi}(t) = M(t) = E\left\{Y\frac{Z(T)}{B(T)} \mid \mathscr{F}(t)\right\}$$

and π such that

$$d\hat{V}^{v,\pi}(s) = dM(s) \quad \text{for} \ s \in [t, T] \quad (25)$$

then $\hat{V}^{v,\pi}(s) = M(s)$ for all $s \in [t, T]$ and we have

$$V^{v,\pi}(s) = V^\rho(s)M(s) = \frac{B(s)}{Z(s)}E\left\{Y\frac{Z(T)}{B(T)} \mid \mathscr{F}(s)\right\}. \quad (26)$$

This implies that

$$\begin{cases} v = V^{v,\pi}(t) = Y(t) \\ V^{v,\pi}(T) = Y \end{cases}$$

i.e. the strategy π satisfying (25) is an admissible hedging strategy and this also implies that the set in the definition of $\mathscr{U}^Y(t)$ is not empty and so $Y(t) \geq \mathscr{U}^Y(t)$.

iv) It remains to explicitly exhibit an admissible strategy π satisfying (25). It suffices to equalize the coefficients of $dw(t)$ in (22) and (24) namely, recalling that $\hat{V}^{v,\pi}(s) := \frac{V^{v,\pi}(s)}{V^\rho(s)} = \frac{Z(s)}{B(s)}V^{v,\pi}(s) = M(s)$, to impose the following

$$\frac{Z(s)}{B(s)}V^{v,\pi}(s)\left(\pi'(s)\sigma(s) - \theta'(s)\right) = \psi'(s)$$

which leads to

$$\pi(s) = \sigma^{-1}(s)\left(\frac{\psi(s)}{M(s)} + \theta(s)\right) \quad (27)$$

for $s \in [t, T]$ with $t \in [0, T]$.

$\square$

4.2 The "Benchmark Approach" of Platen

This subsection is based on [5]. Here the methodology builds on utility indifference pricing thereby showing that the resulting price is independent of the chosen utility function and of the initial wealth and it does not require the claim to be replicable.

To simplify the presentation, in what follows we shall denote with a "bar" the values expressed in units of the locally riskless asset, for example

$$\bar{V}^{v,\pi}(t) = B^{-1}(t)V^{v,\pi}(t).$$

We start with the following

Definition 3 An asset or portfolio value expressed in units of the GOP is called *benchmarked*. A price or portfolio process is *fair* if its benchmarked value is a $(\mathscr{F}_t, P)$-martingale.

This definition corresponds to the property of a discounted self financing portfolio value process to be a martingale under any martingale measure, provided there exists one. Notice also that it is reasonable to restrict oneself to fair portfolios because they are martingales and a martingale is the smallest among the supermartingales that reach a same value at a future (stopping) time.

Utility indifference pricing is based on expected utility maximization, which we recall next. Given a utility function $U(\cdot)$ and an initial capital $v > 0$, the *utility maximization problem* consists in determining an admissible strategy $\tilde{\pi}$ such that

$$E\{U(\bar{V}^{v,\tilde{\pi}}(T))\} \geq E\{U(\bar{V}^{v,\pi}(T))\} \quad \forall \pi \in \mathscr{H}(v; T). \tag{28}$$

Here we want the strategies also to generate a fair portfolio process.

A rather straightforward extension of the so-called "martingale approach" to utility maximization (see e.g. [8]), coupled with fairness of the corresponding portfolio process, leads to the following

Lemma 5 *Given a utility function $U(\cdot)$ and an initial wealth $v > 0$, the fair discounted terminal portfolio value solving (28) is given by*

$$\bar{V}^{v,\tilde{\pi}}(T) = \dot{U}^{-1}\left(\frac{\lambda}{\bar{V}^{v,\rho}(T)}\right) \tag{29}$$

where $\dot{U}$ denotes the derivative of U, $\bar{V}^{v,\rho}(T)$ is the discounted terminal value of the GOP and the (Lagrange) parameter λ is determined from the "budget equation"

$$E\left\{\frac{V^{v,\tilde{\pi}}(T)}{V^{v,\rho}(T)}\right\} = \frac{V^{v,\tilde{\pi}}(0)}{V^{v,\rho}(0)} = \frac{v}{v} = 1. \tag{30}$$

The strategy $\tilde{\pi}$ is any admissible strategy leading to the terminal portfolio value in (29).

Remark 6 In the standard martingale approach, where the existence of martingale measure is given, the budget equation results from the martingality of the discounted value process of a self financing portfolio. Not assuming here the existence of a martingale measure, we impose the condition of fairness instead.

Sketch of the proof The budget equation (30) is a consequence of the requirement of fairness. Given the budget constraint, by the Lagrange multiplier method the problem becomes that of maximizing

$$E\left\{U(\bar{V}^{v,\pi}(T)) - \lambda \frac{V^{v,\pi}(T)}{V^{v,\rho}(T)}\right\}. \tag{31}$$

The "martingale approach" now consists in maximizing

$$U(\bar{V}) - \lambda \frac{V}{V^{v,\rho}(T)} = U(\bar{V}) - \lambda \frac{\bar{V}}{\bar{V}^{v,\rho}(T)} \tag{32}$$

with respect to $\bar{V}$ that leads, for each "scenario" $\omega \in \Omega$, to the optimal terminal value $\bar{V}^{v,\tilde{\pi}}(T)$ given by (29). The strategy $\tilde{\pi}$ is then determined as the hedging strategy for the "claim" $\bar{V}^{v,\tilde{\pi}}(T)$. □

We come next to the actual *utility indifference pricing*. Given $U(\cdot)$ and the corresponding $\tilde{\pi}$, it is well known that the utility indifference price $P(t)$ at time t of a given claim Y is such that, defining

$$p_{\varepsilon,P(t)}^{v,\tilde{\pi}} := E\left\{U\left[(V(t) - \varepsilon P(t))\frac{\bar{V}^{v,\tilde{\pi}}(T)}{V(t)} + \varepsilon \bar{Y}\right] \mid \mathscr{F}(t)\right\}$$

where $V(t)$ is the investor's wealth at time t starting from v and following the strategy $\tilde{\pi}$, one has

$$\lim_{\varepsilon \to 0} \frac{p_{\varepsilon,P(t)}^{v,\tilde{\pi}} - p_{0,P(t)}^{v,\tilde{\pi}}}{\varepsilon} = 0 \quad \text{a.s.}$$

Using a Taylor expansion of $p_{\varepsilon,P(t)}^{v,\tilde{\pi}}$ around $p_{0,P(t)}^{v,\tilde{\pi}}$, it is easily seen that this leads to the following (see [5])

Proposition 7 *Given $U(\cdot)$ and the corresponding $\tilde{\pi}$, the utility indifference price is*

$$P(t) = \frac{E\left\{\dot{U}(\bar{V}^{v,\tilde{\pi}}(T))\bar{Y} \mid \mathscr{F}(t)\right\}}{E\left\{\dot{U}(\bar{V}^{v,\tilde{\pi}}(T))\frac{\bar{V}^{v,\tilde{\pi}}(T)}{V(t)} \mid \mathscr{F}(t)\right\}}, \tag{33}$$

where, again, $\dot{U}$ denotes the derivative of U.

We can now show the following (see [3, 5])

Theorem 8 *The utility indifference price is given by*

$$P(t) = V^{\rho}(t)E\left\{\frac{Y}{V^{\rho}(T)} \mid \mathscr{F}(t)\right\} \tag{34}$$

where $V^{\rho}(t) = V^{1,\rho}(t)$ is the portfolio value corresponding to the GOP strategy ρ.

Proof Substituting (29) into (33) results in the following chain of equalities, where we use the fairness property of the benchmarked portfolio values

$$P(t) = \frac{E\left\{\frac{Y}{V^{v,\rho}(T)} \mid \mathscr{F}(t)\right\}}{E\left\{\frac{V^{v,\tilde{\pi}}(T)}{V^{v,\rho}(T)} \frac{1}{V(t)} \mid \mathscr{F}(t)\right\}} = \frac{V^{v,\tilde{\pi}}(t) E\left\{\frac{Y}{V^{v,\rho}(T)} \mid \mathscr{F}(t)\right\}}{\frac{V^{v,\tilde{\pi}}(t)}{V^{v,\rho}(t)}}$$

$$= V^{v,\rho}(t) E\left\{\frac{Y}{V^{v,\rho}(T)} \mid \mathscr{F}(t)\right\} = v V^{\rho}(t) E\left\{\frac{Y}{v V^{\rho}(T)} \mid \mathscr{F}(t)\right\}$$

$$= V^{\rho}(t) E\left\{\frac{Y}{V^{\rho}(T)} \mid \mathscr{F}(t)\right\}$$

whereby the Lagrange parameter as well as the initial wealth v have canceled out. $\square$

Remark 9 Formula (34) corresponds to a standard arbitrage pricing formula, whereby one takes as numeraire the GOP and the resulting martingale measure is the physical measure P. It is thus an instance of the so-called "real world pricing". In line with Definition 3 the price $P(t)$ is a fair price. We stress once again that the pricing formula (34) is independent of the utility function $U(\cdot)$ and of the initial wealth $v > 0$ and does not require the existence of a martingale measure nor the claim to be a priori replicable. The claim will though turn out to be replicable a posteriori via the equivalence of the pricing approach of this Sect. 4.2 with that of the previous Sect. 4.1 that will be discussed next.

4.3 Equivalence of the Two Pricing Approaches

Recall that, according to the pricing approach in Sect. 4.1, the (superhedging) price $Y(t)$ at time $t \in [0, T]$ of a given claim Y with maturity T is given by (see (20))

$$Y(t) := \frac{B(t)}{Z(t)} E\left\{Y \frac{Z(T)}{B(T)} \mid \mathscr{F}(t)\right\}$$

and it corresponds to that of the standard martingale approach if $Z(t)$ is a true martingale. On the other hand, the (utility indifference) price $P(t)$ according to the approach in Sect. 4.2 is given by (see (34))

$$P(t) = V^{\rho}(t) E\left\{\frac{Y}{V^{\rho}(T)} \mid \mathscr{F}(t)\right\}.$$

We now show that the two pricing approaches lead to the same result by proving

Proposition 10 *One has*

$$Y(t) = P(t).$$

Proof It suffices to show that

$$V^{\rho}(t) = \frac{B(t)}{Z(t)}$$

which is equivalent to showing that the discounted GOP value is the inverse of the "density process" $Z(t)$ a fact that is generally known under standard assumptions. We show it here formally within our context. Recalling (16), namely

$$V^\rho(t) = \exp\left[\int_0^t r(s)ds + \frac{1}{2}\int_0^t \|\theta(s)\|^2 ds + \int_0^t \theta'(s)dw(s)\right]$$

and noticing that from (18) one has

$$\frac{Z(t)}{B(t)} = \frac{\exp\left[-\int_0^t \theta'(s)dw(s) - \frac{1}{2}\int_0^t \|\theta(s)\|^2 ds\right]}{\exp\left[\int_0^t r(s)ds\right]}$$

$$= \exp\left[-\int_0^t \theta'(s)dw(s) - \frac{1}{2}\int_0^t \|\theta(s)\|^2 ds - \int_0^t r(s)ds\right],$$

the result follows immediately.   □

Remark 11 Notice that none of the two approaches, that in Sect. 4.1 and that in 4.2, requires the existence of an equivalent martingale measure. From the first approach it is easily seen that, if $Z(t)$ is a martingale and therefore there exists an equivalent martingale measure, then the usual arbitrage (martingale-based) price coincides with that in Sect. 4.1 and, by the equivalence just shown, also with that in Sect. 4.2. The pricing formula in Sect. 4.2 and, by the equivalence, also that in Sect. 4.1, corresponds to that of a standard pricing approach by taking as numeraire the GOP, for which the physical measure is a martingale measure (this also justifies the fact that none of the two approaches required the existence of a martingale measure different from the physical measure P). The approach in Sect. 4.2 does not require the claim to be a priori replicable. The constructive approach of Sect. 4.1, which is based on a replication argument and exhibits explicitly a replication strategy, implies, via the equivalence, that also in the approach of Sect. 4.2 the claim is in fact replicable. We finally want to point out that the equivalence of the two approaches is implicit in some parts of [5], for example in Sect. 11.5 where it is emphasized that "for a nonreplicable payoff the real world pricing formula provides the minimal nonnegative price process" or also in later sections like 12.2 where Fig. 12.2.4 is intended to "visualize the fact that fair prices are the minimal prices that replicate a contingent claim" and for this the authors refer to Corollary 10.4.2 (always in [5]).

5 Pricing Under Absence of an Equivalent Martingale Measure (the Case of Incomplete Observation)

We now extend the results, summarized in the previous Sect. 4 for the case of complete observation on the market model, to a market model with incomplete observation. We first describe the model that we are considering.

5.1 The Model

The model is taken from [7]. Consider the model (1) with one riskless asset and n risky assets and assume that the observable quantities are the $n + 1$ prices of the primary assets, the empirical covariances of the risky assets' log-prices with that of the GOP (we assume that the values of the GOP or of a proxy thereof are observable) in addition to the short rate of interest $r(t)$. To simplify the notation, in what follows we shall consider all the prices expressed in units of the riskless asset (discounted values) and, whenever ambiguity may arise, denote their values with a bar. Let

$$C(t) = (C_1(t), \ldots, C_{2n}(t))$$

be the vector of observable (discounted) quantities (excluding the short rate). It is natural to assume that the underlying filtration $\mathscr{F}(t)$ measures $C(t)$. Assume then that there are also a certain number of unobservable quantities, denote them by

$$A(t) = (A_1(t), \ldots, A_m(t))$$

that we may think of as representing latent factors that affect the primary assets' price dynamics. We restrict the volatility matrix $\sigma(t)$ to depend only on the observable quantities and to stress this fact we shall write $\sigma(t) = \sigma(t, C(t))$ assuming furthermore that it is invertible for all $t \leq T < \infty$. Recalling the market price of risk in (10), here we then write it as

$$\theta(t, A(t)) = (\sigma(t, C(t)))^{-1} (b(t, A(t)) - r(t)\underline{1}) \tag{35}$$

to stress the dependence also on $A(t)$ (it depends also on $C(t)$ not only through $\sigma(t, C(t))$ but in general also through $b(\cdot)$ that is $\mathscr{F}(t)$-measurable).

In line with the above definitions we rewrite the dynamics of the discounted values of the risky assets as (see (1))

$$d\bar{S}_i(t) = \bar{S}_i(t) \left[\sum_{j=1}^{n} \sigma_{ij}(t, C(t))(\theta_j(t, A(t))dt + dw_j(t)) \right], \quad \bar{S}_i(0) = s_i > 0 \tag{36}$$

and the discounted value process $\bar{V}^{v,\pi}(t) = v\bar{V}^{1,\pi}(t) = v\bar{V}^{\pi}(t)$ of a self financing portfolio (that invests only in the risky assets) becomes (see (14))

$$\frac{d\bar{V}^{\pi}(t)}{\bar{V}^{\pi}(t)} = \pi'(t)\sigma(t, C(t))[\theta(t, A(t))dt + dw(t)]. \tag{37}$$

Finally, corresponding to (15)

$$\frac{d\bar{V}^{\rho}(t)}{\bar{V}^{\rho}(t)} = \|\theta(t, A(t))\|^2 + \theta'(t, A(t))dw(t). \tag{38}$$

Having assumed that also the *empirical log-covariances* are observable, we need a model for these quantities as well, and we shall represent them as noisy perturbations of the corresponding unobservable theoretical covariances.

Letting Δ denote the discrete time observation step and $t_\ell := \ell\Delta$ the grid points of the observations, for the i-th log-covariance we have

$$z_i(t) = \left[\ln(\bar{S}_i), \ln(\bar{V}^\rho)\right]_{\Delta,t}$$

$$= \sum_{l=1}^{\left[\frac{t}{\Delta}\right]} \left(\ln(\bar{S}_i(t_l)) - \ln(\bar{S}_i(t_{l-1}))\right)\left(\ln(\bar{V}^\rho(t_l)) - \ln(\bar{V}^\rho(t_{l-1}))\right) \qquad (39)$$

where $[a]$ denotes the largest integer less than or equal to a. Putting

$$D(t) = (D_1(t), \ldots, D_{2n+m}(t))' = (C_1(t), \ldots, C_{2n}(t), A_1(t), \ldots, A_m(t))' \qquad (40)$$

and

$$B_i(t, D(t)) := \sum_{j=1}^{n} \sigma_{ij}(t, C(t))\theta_j(t, A(t)) \qquad (41)$$

the i-th *theoretical covariance* is then

$$\left[\ln(\bar{S}_i), \ln(\bar{V}^\rho)\right]_t = \langle\ln(\bar{S}_i), \ln(\bar{V}^\rho)\rangle_t = \int_0^t B_i(s, D(s))ds. \qquad (42)$$

Since the theoretical covariances are the limit in probability of the empirical covariances, we may postulate for the latter the following model

$$dz_i(t) = B_i(t, D(t))dt + d\eta_i(t) \qquad (43)$$

where $\eta_i(t)$ are n independent Wiener processes, independent of the $w_j(t)$, and they represent the discrepancy between theoretical and empirical covariances.

5.2 Corresponding Complete Observation Model

Since the unobservable factors are inputs to the model, also solutions of problems for the model will depend on them. On the other hand prices and strategies should only depend on the available information that results from the observations. In the following two subsections we shall show that, for our pricing problem, we can equivalently consider a complete observation problem corresponding to the original one under incomplete observation and in this equivalent problem all inputs are given by quantities that are all observable either directly or indirectly. In this subsection we now derive this equivalent complete observation problem, sometimes also referred to as "separated problem".

Put, with some abuse of notation

$$C(t) = (\bar{S}_1(t), \ldots, \bar{S}_n(t), z_1(t), \ldots, z_n(t))' \qquad (44)$$

that now represents the totality of the observable quantities and consider the $2n$-dimensional Wiener process

$$W(t) = (w_1(t), \ldots, w_n(t), \eta_1(t), \ldots, \eta_n(t))'. \qquad (45)$$

Recalling (40) and (41) and putting

$$B(t, D(t)) = (B_1(t, D(t)), \ldots, B_n(t, D(t)))'$$

we obtain the following dynamics for the global $2n$-dimensional vector $C(t)$

$$dC(t) = \begin{pmatrix} diag\,(\bar{S}(t))\,B(t, D(t)) \\ B(t, D(t)) \end{pmatrix} dt + \begin{pmatrix} diag\,(\bar{S}(t))\,\sigma(t, C(t)) & 0 \\ 0 & I \end{pmatrix} dW(t)$$
$$:= F(t, D(t))dt + G(t, C(t))dW(t) \tag{46}$$

thereby implicitly defining the (vector) drift $F(t, D)$ and (matrix) volatility $G(t, C)$.

The information filtration is the one generated by the $k := 2n$ components of $C(t)$ and to stress the "dimension" k of the observations we write

$$\mathscr{F}^k(t) := \sigma\{C(s),\ s \le t\}. \tag{47}$$

As it is typically done in partial observation problems, we consider now instead of the unobservable quantities $A(t)$ their conditional distribution (*the so-called filter distribution*)

$$p\big(A(t) \mid \mathscr{F}^k(t)\big) = p(A(t) \mid \sigma\{C(s),\ s \le t\}). \tag{48}$$

We make the following

Assumption 12 *The filter for $A(t)$ given $\mathscr{F}^k(t)$ is finite-dimensional, namely the filter distribution $p(A(t) \mid \mathscr{F}^k(t))$ is parametrized by a finite number of parameters*

$$\xi(t) = \big(\xi_1(t), \ldots, \xi_q(t)\big)'.$$

More precisely, $p(A(t) \mid \mathscr{F}^k(t)) = p(A(t); \xi_1(t), \ldots, \xi_q(t))$

Remark 13 Well known finite dimensional filters are the *Kalman-Bucy filter* and the *Wonham filter* for the case when the unobserved quantities form a finite-state Markov chain. More generally, the Extended Kalman filter is a finite-dimensional filter that is applicable, modulo an approximation, to a variety of situations. In all the above cases the filter parameters form a Markov process driven by the *innovations* (for the definition of "innovations" see Remark 15 below). This, together with (46) motivates and also justifies the following additional assumption.

Assumption 14 *By analogy to (40) put*

$$\tilde{D}(t) = \big(C_1(t), \ldots, C_{2n}(t), \xi_1(t), \ldots, \xi_q(t)\big)' \tag{49}$$

that represents the extended $(2n + q)$-dimensional vector of "observables". We assume that the parameters $\xi(t)$ of the finite-dimensional filter satisfy

$$d\xi(t) = H(t, \tilde{D}(t))dt + K(t, \tilde{D}(t))dC(t). \tag{50}$$

We now apply Theorem 7.12 in [4] (see also Proposition 2.3 in [6]) according to which the $k = 2n$-dimensional process $C(t)$ satisfies, instead of (46) where the factor $D(t)$ in the drift is not fully observable, the following dynamics

$$dC(t) = \tilde{F}(t, \tilde{D}(t))dt + G(t, C(t))d\tilde{W}(t) \tag{51}$$

with

$$\tilde{F}(t, \tilde{D}(t)) = E\{F(t, D(t)) \mid \mathscr{F}^k(t)\}$$
$$= \int F(t, C(t), A)dp(A \mid \mathscr{F}^k(t)) = \int F(t, C(t), A)dp(A; \xi_1, \ldots, \xi_q) \tag{52}$$

and where $\tilde{W}(t)$ is a k-dimensional $\mathscr{F}^k$-Wiener process (with independent components) satisfying

$$d\tilde{W}(t) = (G(t, C(t))^{-1}[dC(t) - \tilde{F}(t, \tilde{D}(t))dt]. \tag{53}$$

Remark 15 Due to the representation in (53), the process $\tilde{W}(t)$ is also called "innovations process" and notice that by this same (53), combined with (50), the dynamics of the finite-dimensional filter parameters $\xi(t)$ can be expressed in terms of these innovations (see also the last part of Remark 13). Furthermore, from (51) and (53) and recalling the definition of $\mathscr{F}^k(t)$ in (47), one has that $\mathscr{F}^k(t) = \mathscr{F}^{\tilde{W}}(t)$, namely the observation filtration is equivalently generated by the innovations $\tilde{W}(t)$.

Remark 16 In what follows we shall use the symbol $\tilde{w}(t)$ to denote the n-dimensional subvector of the $k = 2n$ dimensional vector $\tilde{W}(t)$, which is formed by the first n components and which (see (51) and the definition of $G(t, C)$ in (46)) satisfies

$$d\tilde{w}(t) = (\sigma(t, C(t)))^{-1} \left[\left(diag\left(\bar{S}(t) \right) \right)^{-1} d\bar{S}(t) - \tilde{B}(t, \tilde{D}(t))dt \right] \tag{54}$$

where, by analogy to $\tilde{F}(t, \tilde{D}(t))$ in (52),

$$\tilde{B}(t, \tilde{D}(t)) = E\left\{ B(t, D(t)) \mid \mathscr{F}^k(t) \right\}$$
$$= \int B(t, C(t), A)dp(A \mid \mathscr{F}^k(t)) = \int B(t, C(t), A)dp(A; \xi_1, \ldots, \xi_q). \tag{55}$$

Combining (51) with (50), it is easily seen that, for appropriate coefficients $\alpha(\cdot)$ and $\beta(\cdot)$ one obtains for the extended $(2n + q)$-dimensional observation vector dynamics of the form

$$d\tilde{D}_i(t) = \alpha_i(t, \tilde{D}(t))dt + \sum_{j=1}^{2n} \beta_{ij}(t, \tilde{D}(t))d\tilde{W}_j(t) \tag{56}$$

We shall next derive particularized dynamics, in terms of the innovations, for the (discounted) prices and the GOP.

5.2.1 Price Dynamics

The (discounted) prices of the risky assets form (see (49) and (44)) the first n components of $\tilde{D}(t)$ and their dynamics are affected only by the first n components of $\tilde{W}(t)$, the vector of which we had denoted by $\tilde{w}(t)$. Recalling then (41) and (46) as well as (51), one obtains for the discounted asset prices

$$d\bar{S}_i(t) = \bar{S}_i(t)\left[\sum_{j=1}^{n}\sigma_{ij}(t, C(t))\left(\tilde{\theta}_j(t, \tilde{D}(t))dt + d\tilde{w}_j(t)\right)\right], \quad \bar{S}_i(0) = s_i > 0$$

(57)

where

$$\tilde{\theta}_j(t, \tilde{D}(t)) = E\left\{\theta_j(t, A(t)) \mid \mathscr{F}^k(t)\right\} = \int \theta_j(t, A)dp(A; \xi_1, \ldots, \xi_q). \quad (58)$$

(Recall from the comment after (35) that $\theta(t, A(t))$ hides a possible dependence also on $C(t)$.) Notice that joining (56) to (57), one obtains a Markovian system and in (57) all the coefficients depend on observable quantities.

5.2.2 GOP Dynamics

Given its importance in our analysis, we now want to obtain also for the GOP a dynamics analogous to (57). For this purpose notice first that, with price dynamics (57) instead of (36), the dynamics (37) for the discounted portfolio values become (we use $\bar{V}^{\pi,k}$ to stress the information level/observation dimension k)

$$\frac{d\bar{V}^{\pi,k}(t)}{\bar{V}^{\pi,k}(t)} = \pi'(t)\sigma(t, C(t))[\tilde{\theta}(t, \tilde{D}(t))dt + d\tilde{w}(t)] \quad (59)$$

where $\pi(t)$ has now to be adapted to $\mathscr{F}^k(t)$. This leads to the following adaptation of the definition of the set of admissible strategies, namely (see Definition 1)

Definition 4 Given an initial wealth v, the set of *admissible strategies* under partial observation of level k is $\mathscr{H}^k(v) := \bigcup_{T>0}\mathscr{H}^k(v; T)$ where

$$\mathscr{H}^k(v; T) := \left\{\pi(\cdot) \in \mathscr{F}^k(\cdot) \mid P\left\{V^{v,\pi,k}(t) \geq 0, \forall\, 0 \leq t \leq T\right\} = 1\right\}. \quad (60)$$

By analogy with the case of complete observations (see (9) and notice that now we use discounted values), the growth rate of a portfolio π is here

$$\gamma_\pi^k(t) = \pi'(t)\sigma(t, C(t))\tilde{\theta}(t, \tilde{D}(t)) - \frac{1}{2}\pi'(t)\sigma(t, C(t))\sigma'(t, C(t))\pi(t) \quad (61)$$

and it leads to the following GOP strategy

$$\rho^k(t) = \left(\sigma'(t, C(t))\right)^{-1}\tilde{\theta}(t, \tilde{D}(t)). \quad (62)$$

Substituting the expression for $\rho^k(t)$ into (59) we immediately obtain the

Proposition 17 *The discounted value of the GOP (with unit initial wealth) satisfies*

$$\frac{d\bar{V}^{\rho,k}(t)}{\bar{V}^{\rho,k}(t)} = \|\tilde{\theta}(t, \tilde{D}(t))\|^2 dt + \tilde{\theta}'(t, \tilde{D}(t)) d\tilde{w}(t) \tag{63}$$

and the growth rate of the GOP is given by

$$\gamma_\rho^k(t) = \frac{1}{2}\|\tilde{\theta}(t, \tilde{D}(t))\|^2. \tag{64}$$

5.2.3 GOP Strategy and Growth Rate Under Complete and Incomplete Observation

Comparing the expressions for the GOP strategy and growth rate under complete observations (see (13) and (17)) with those under incomplete observations in (62) and (64) (notice that the growth rate under incomplete observation was derived for discounted values so that in (64) the additive term $r(t) \in \mathscr{F}^k(t)$ is missing) we obtain

Corollary 18 *We have*

$$\text{i)} \quad \rho^k(t) = E\left\{\rho(t) \mid \mathscr{F}^k(t)\right\},$$

$$\text{ii)} \quad \gamma_\rho^k(t) \leq E\left\{\gamma_\rho(t) \mid \mathscr{F}^k(t)\right\}.$$

Proof i) is immediate considering the definition of $\tilde{\theta}(t, \tilde{D})$ in (58), while for ii) we use Jensen's inequality and the fact that, under full observation, we have simply written $\theta_i(t)$ for what under incomplete observations is $\theta(t, A(t))$. We thus obtain (for the definition of $\tilde{\theta}_i(t, \tilde{D}(t))$ see (58))

$$E\left\{\gamma_\rho^k(t)\right\} = \frac{1}{2}\sum_{i=1}^{n} E\left\{(\tilde{\theta}_i(t, \tilde{D}(t)))^2\right\} = \frac{1}{2}\sum_{i=1}^{n} E\left\{(E\{\theta_i(t, A(t)) \mid \mathscr{F}^k(t)\})^2\right\}$$

$$\leq \frac{1}{2}\sum_{i=1}^{n} E\left\{E\{\theta_i^2(t, A(t)) \mid \mathscr{F}^k(t)\}\right\}$$

$$= E\left\{E\left\{\frac{1}{2}\sum_{i=1}^{n}\theta_i^2(t, A(t)) \mid \mathscr{F}^k(t)\right\}\right\}$$

$$= E\{E\{\gamma_\rho(t) \mid \mathscr{F}^k(t)\}\} = E\{\gamma_\rho(t)\}. \qquad \square$$

5.2.4 Expected Utility Maximization Under Complete and Incomplete Observation

Here we furthermore point out that, whenever one has to solve a problem for a partial observation model where the strategies have to be adapted to the filtration representing the available information, then the original problem can equivalently be solved by solving it for the corresponding complete observation model. As an example, which will be relevant for Platen's benchmark approach, we show that expected utility for the original incomplete observation problem is equivalent to the same problem for the corresponding complete observation model.

The original problem can, for an arbitrary initial capital $v > 0$, be formulated as (see Definition 4 for the class of admissible strategies and (37) for the portfolio dynamics)

$$\begin{cases} \max_{\pi \in \mathcal{H}^k(v;T)} E\{U(\bar{V}^{v,\pi}(T))\} \\ \frac{d\bar{V}^{v,\pi(t)}}{\bar{V}^{v,\pi(t)}} = \pi'(t)\sigma(t, C(t))\,[\theta(t, A(t))dt + dw(t)]. \end{cases}$$

For the corresponding complete observation model the problem becomes (see (59) for the portfolio dynamics)

$$\begin{cases} \max_{\pi \in \mathcal{H}^k(v;T)} E\{U(\bar{V}^{v,\pi,k}(T))\} \\ \frac{d\bar{V}^{v,\pi,k}(t)}{\bar{V}^{v,\pi,k}(t)} = \pi'(t)\sigma(t, C(t))[\tilde{\theta}(t, \tilde{D}(t))dt + d\tilde{w}(t)]. \end{cases}$$

We show the following

Lemma 19

$$\max_{\pi \in \mathcal{H}^k(v;T)} E\{U(\bar{V}^{v,\pi}(T))\} = \max_{\pi \in \mathcal{H}^k(v;T)} E\{U(\bar{V}^{v,\pi,k}(T))\}.$$

Proof From (53) and (46) we obtain

$$d\tilde{W}(t) = (G(t, C(t)))^{-1}[F(t, D(t)) - \tilde{F}(t, \tilde{D}(t))]dt + dW(t).$$

Since $V^{v,\pi}(t)$ and $\bar{V}^{v,\pi,k}(t)$ are driven by $w(t)$ and $\tilde{w}(t)$ respectively, recalling the definitions of $F(\cdot)$, $G(\cdot)$ and $\tilde{F}(\cdot)$ in (46) and (52) as well as the fact that the first n components of $C(t)$ are (see (44)) the discounted prices $\bar{S}_i(t)$, we have

$$d\tilde{w}(t) = (\sigma(t, C(t)))^{-1}(diag(\bar{S}(t)))^{-1}\Big\{diag(\bar{S}(t))\sigma(t, C(t))\theta(t, A(t))$$

$$- E\big[diag(\bar{S}(t))\sigma(t, C(t))\theta(t, A(t)) \mid \mathscr{F}^k(t)\big]\Big\}dt + dw(t)$$

$$= [\theta(t, A(t)) - E\{\theta(t, A(t)) \mid \mathscr{F}^k(t)\}]dt + dw(t)$$

$$= [\theta(t, A(t)) - \tilde{\theta}(t, \tilde{D}(t))]dt + dw(t).$$

It follows that for any $\pi \in \mathcal{H}^k(v; T)$

$$E\{U(\bar{V}^{v,\pi}(T))\} = E\Big\{U\Big(\exp\Big[v + \int_0^T \pi'(t)\sigma(t, C(t))\Big[\theta(t, A(t))dt + dw(t)$$

$$- \frac{1}{2}\sigma'(t, C(t))\pi(t)dt\Big]\Big]\Big)\Big\}$$

$$
\begin{aligned}
&= E\left\{ U\left(\exp\left[\tilde{v} + \int_0^T \pi'(t)\sigma(t, C(t))\left[\tilde{\theta}(t, \tilde{D}(t))dt + d\tilde{w}(t) \right. \right. \right. \right. \\
&\qquad\qquad \left. \left. \left. \left. - \frac{1}{2}\sigma'(t, C(t))\pi(t)dt \right] \right] \right) \right\} \\
&= E\{ U(\bar{V}^{v,\pi,k}(T)) \}. \qquad\qquad\qquad\qquad\qquad\qquad \square
\end{aligned}
$$

5.3 Pricing According to the "Fernholz-Karatzas Approach"

As we had seen in the complete observation case, in this approach the price corresponds to the minimal superreplicating initial capital and thus involves implicitly hedging strategies that are $\mathscr{F}^k$-adapted. In line with Sect. 5.2.4 we then use the same methodology as in Sect. 4.1 for the complete observation case, but applied here to the complete observation model that corresponds to the original incomplete observation model and was derived in the previous Sect. 5.2.

When trying to carry over Theorem 3 to the present situation we face the problem that, although in addition to the primary assets also the empirical log-covariances are observable, the investment instruments are only the primary assets. We present now two possible variants of extensions of Theorem 3.

5.3.1 First Variant

In this variant we completely ignore the empirical covariances and consider $\bar{S}(t)$ as the only observables, namely $C(t)$ is the n-dimensional vector given by $C(t) = \bar{S}(t)$ and satisfying the subsystem of (46) consisting of the first n rows. The filtration $\mathscr{F}^k$ remains as defined in (47) and the filter has the same form as in (48) and we also keep Assumption 12. Furthermore, analogously to (49) we now put

$$
\tilde{D}(t) = \left(\bar{S}_1(t), \dots, \bar{S}_n(t), \xi_1(t), \dots, \xi_q(t) \right)'. \tag{65}
$$

It is also easily seen that in the given situation $\mathscr{F}^k(t) = \mathscr{F}^{\tilde{w}}(t)$ with $\tilde{w}(t)$ as in (54). Replacing then $\theta(t)$ in (22) by $\tilde{\theta}(t, \tilde{D}(t))$ from (58), and using, instead of $Z(t)$ in (18), the process

$$
Z^k(t) := \exp\left\{ -\int_0^t \tilde{\theta}'(s, \tilde{D}(s))d\tilde{w}(s) - \frac{1}{2}\int_0^t \|\tilde{\theta}(s, \tilde{D}(s))\|^2 ds \right\}, \quad 0 \le t < \infty \tag{66}
$$

as "Radon-Nikodym process", it can be easily verified that the following extension of Theorem 3 holds

Theorem 20 *Consider a given a maturity T and a claim $Y^k \in \mathscr{F}^k(T)$ such that $E\{Y^k \frac{Z^k(T)}{B(T)}\} \in (0, \infty)$ with $Z^k(t)$ as in (66). For $t \in [0, T]$ and $s \in [t, T]$ denote by*

$V^{v,\pi,k}(s)$ *the value in* $s \geq t$ *of a portfolio starting from* $V^{v,\pi,k}(t) = v$ *with* $\pi(\cdot) \in \mathscr{F}^k(\cdot)$. *Let*

$$\mathscr{H}_t^k(v; T) := \left\{ \pi(\cdot) \in \mathscr{F}^k(\cdot) \mid P\{V^{v,\pi,k}(s) \geq 0, \forall t \leq s \leq T\} = 1; \right.$$

$$\left. V^{v,\pi,k}(t) = v \in \mathscr{F}^k(t) \right\}$$

namely the set of admissible strategies starting from a capital v *at time* t *and put*

$$\mathscr{U}^{Y^k}(t) := \operatorname{ess\,inf}\left\{ v > 0 \mid \exists \pi \in \mathscr{H}_t^k(v; T) \ s.t. \ V^{v,\pi,k}(T) \geq Y^k \right\} \tag{67}$$

which represents the minimal hedging price at time t. *Let, furthermore,*

$$Y^k(t) := \frac{B(t)}{Z^k(t)} E\left\{ Y^k \frac{Z^k(T)}{B(T)} \mid \mathscr{F}^k(t) \right\} \tag{68}$$

with $Z^k(t)$ *as in* (66) *and with the expectation under the physical measure* P. *If* $\mathscr{F}^k(t) = \mathscr{F}^{\tilde{w}}(t)$, *where* $\tilde{w}(t)$ *is the n-dimensional Wiener process according to Remark* 16, *then*

$$Y^k(t) = \mathscr{U}^{Y^k}(t). \tag{69}$$

Proof The proof follows step by step that of Theorem 3 (with the appropriate adaptation of symbols) whereby $M(s)$ in (23) is replaced by (recall that $\mathscr{F}^k(t) = \mathscr{F}^{\tilde{w}}(t)$)

$$M(s) := E\left\{ Y^k \frac{Z^k(T)}{B(T)} \mid \mathscr{F}^k(s) \right\} = Y^k(t) \frac{Z^k(t)}{B(t)} + \int_t^s \psi'(u) d\tilde{w}(u) \tag{70}$$

and $\pi(s)$ turns out to be

$$\pi(s) = \sigma^{-1}(s, C(s)) \left(\frac{\psi(s)}{M(s)} + \tilde{\theta}(s, \tilde{D}(s)) \right) \tag{71}$$

for $s \in [t, T]$ with $t \in [0, T]$. $\square$

5.3.2 Second Variant

This is a more theoretical variant, where one imagines the possibility of investing also in the log-covariances (in general in all quantities that are supposed to be observable). In this case $C(t)$ and $\tilde{D}(t)$ remain as in Sect. 5.2 where $C(t)$ satisfies (46) with $W(t)$ as in (45) (recall also that this implies that the $diag\,(\bar{S}(t))\,\sigma(t, C(t))$ of the first variant now becomes $G(t, C(t))$). Furthermore (see Remark 15), this time $\mathscr{F}^k(t) = \mathscr{F}^{\tilde{W}}(t)$ with $\tilde{W}(t)$ as in (53). The extension of Theorem 3 for the present variant now reads exactly as Theorem 20. In the proof of the Theorem for this second variant the changes with respect to the first variant concern, besides $C(t)$, the process $\tilde{\theta}(t, \tilde{D}(t))$ that now has to be of dimension $k = 2n$. The other changes concern $Z^k(t)$ in (66) that now becomes

$$Z^k(t) := \exp\left\{ -\int_0^t \tilde{\theta}'(s, \tilde{D}(s)) d\tilde{W}(s) - \frac{1}{2} \int_0^t \|\tilde{\theta}(s, \tilde{D}(s))\|^2 ds \right\}, \quad 0 \leq t < \infty \tag{72}$$

with $\tilde{W}(t)$ as in (53). Furthermore, $M(t)$ here becomes

$$M(s) := E\left\{ Y^k \frac{Z^k(T)}{B(T)} \mid \mathscr{F}^k(s) \right\} = Y^k(t)\frac{Z^k(t)}{B(t)} + \int_t^s \psi'(u)d\tilde{W}(u) \qquad (73)$$

with $\psi(t)$ a $2n$-dimensional process and the hedging strategy is

$$\pi(t) := \left(G^{-1}(t, C(t)) \right)' \left[\frac{\psi(t)}{M(t)} + \tilde{\theta}(t, \tilde{D}(t)) \right]. \qquad (74)$$

5.4 Pricing According Platen's "Benchmark Approach"

As we had seen in the complete observation case, Platen's approach is of the type of utility indifference pricing. In line with Sect. 5.2.4 we then use the same methodology as in Sect. 4.2 for the complete observation case, but applied here to the complete observation model that corresponds to the original incomplete observation model and was derived in Sect. 5.2. Contrary to what happened for the Fernholz-Karatzas approach in Sect. 5.3, where we had to consider two somewhat restricted variants, here it can be seen that we can follow the exact same steps as in the proof of Theorem 8 for the complete observation case (with the appropriate adaptation of symbols). We thus conclude that the following formula holds for the price $P^k(t)$ under incomplete observation with information level/observation dimension k

$$P^k(t) = V^{\rho,k}(t)E\left\{ \frac{Y^k}{V^{\rho,k}(T)} \mid \mathscr{F}^k(t) \right\} \qquad (75)$$

where the discounted version of $V^{\rho,k}(t)$ satisfies (63).

5.5 Comments on the Equivalence of the Two Approaches in the Case of Incomplete Observations

In Sect. 4.3 we have shown that, in case of complete observations, the Fernholz-Karatzas (F.-K.) and Platen's benchmark (PB) approach lead to the same derivative prices. In the case of incomplete observations the formulas that we have derived in Sects. 5.3 and 5.4 correspond to those in Sects. 4.1 and 4.2 respectively and so the equivalence should also hold for incomplete observations. However, while PB can be applied also if not all observables can be used as investment instruments (in the case of this paper the empirical covariances), FK had to be restricted to situations where all observables are also investment instruments. This corresponds to intuition if one considers that FK relies on completing the market, while PB is of the utility indifference pricing type that is typically designed for incomplete markets. This difference did not come up in the complete observation case because there all investment instruments are automatically observable.

Acknowledgements   Part of the contribution by the second author was obtained while he was visiting professor 2009 for the chair Quantitative Finance and Insurance at the LMU University in Munich funded by LMU Excellent. Hospitality and financial support are gratefully acknowledged as are the useful suggestions by the participants to a series of seminars on the topic of the paper.

References

1. Fernholz, E.R.: Stochastic Portfolio Theory. Springer, New York (2002)
2. Fernholz, E.R., Karatzas, I.: Stochastic portfolio theory: an overview. In: Bensoussan, A., Zhang, Q. (eds.) Mathematical Modeling and Numerical Methods in Finance. *Handbook of Numerical Analysis*, vol. XV, pp. 89–168. North-Holland, Amsterdam (2008)
3. Galesso, G.: Arbitraggio e diversificazione in informazione completa e parziale. Thesis, Univesity of Padova (2008)
4. Liptser, R., Shiryaev, A.: Statistics of Random Processes: I. General Theory. Springer, Berlin (1977)
5. Platen, E., Heath, D.: A Benchmark Approach to Quantitative Finance. *Springer Finance*, Springer, Berlin (2006)
6. Platen, E., Runggaldier, W.J.: A benchmark approach to filtering in finance. Asia-Pac. Financ. Markets **11**, 79–105 (2005)
7. Platen, E., Runggaldier, W.J.: A benchmark approach to portfolio optimization under partial information. Asia-Pac. Financ. Markets **14**, 25–43 (2007)
8. Schachermayer, W.: Utility maximization in incomplete markets. In: Frittelli, M., Runggaldier, W. (eds.) Stochastic Methods in Finance. *Lecture Notes in Mathematics*, vol. 1856, pp. 255–293. Springer, Berlin (2004)

Existence and Non-uniqueness of Solutions for BSDE

Xiaobo Bao, Freddy Delbaen, and Ying Hu

Abstract We study BSDE where the driver is pathwise quadratically bounded. The associated utility function is always a solution but even in the class of Markovian solutions uniqueness is not guaranteed. we relate the problem to problems for quasi-linear parabolic PDE.

1 Notation

We will use standard notation. Time will be continuous and the time interval $[0, T]$ is bounded. The probability space $(\Omega, \mathscr{F}_T, \mathbb{P})$ is big enough to carry a d-dimensional Brownian Motion W. The filtration $\mathscr{F} = (\mathscr{F}_t)_{t \in [0,T]}$ is generated by W and $\mathscr{F}_0$ contains all the negligible sets of $\mathscr{F}_T$. It is well known that it satisfies the usual assumptions, i.e. $\mathscr{F}$ is also right continuous.

We will study Backward Stochastic Differential Equations (BSDE) of the type

$$dY_t = g(t, Z_t)\, dt - Z_t\, dW_t,$$

where the terminal condition $Y_T = \xi \in L^\infty$ is bounded and where the solution Y is supposed to be bounded. We know that for some drivers g, see below for precise conditions, there is a relation with utility theory. In [3] the general form of a Fatou utility function u_0 such that for all $\xi \in L^\infty, u_0(\xi) \leq \mathbb{E}_\mathbb{P}[\xi]$, was found. It is given by the expression:

$$u_0(\xi) = \inf \mathbb{E}_\mathbb{Q}\left[\xi + \int_0^T f_u(q_u)\, du\right].$$

The function f has the following properties

F. Delbaen (✉) · X. Bao
Department of Mathematics, ETH Zurich, Zurich, Switzerland
e-mail: delbaen@math.ethz.ch

Y. Hu
Département de Mathématiques, Université de Rennes-1, Rennes, France

C. Chiarella, A. Novikov (eds.), *Contemporary Quantitative Finance*,
DOI 10.1007/978-3-642-03479-4_7, © Springer-Verlag Berlin Heidelberg 2010

1. $f : [0, T] \times \Omega \times \mathbb{R}^d \to \overline{\mathbb{R}_+}$,
2. for each $q \in \mathbb{R}^d$, the partial function on $[0, T] \times \Omega$ is predictable,
3. for each (u, ω) the partial function is lsc and convex on $\mathbb{R}^d$,
4. the function f is measurable for $\mathscr{B} \times \mathscr{P}$, where $\mathscr{B}$ is the Borel σ-algebra on $\mathbb{R}^d$ and $\mathscr{P}$ is the predictable σ-algebra on $[0, T] \times \Omega$,
5. $f(u, \omega, 0) = 0$ for each (u, ω).

The infimum is taken over all probability measures $\mathbb{Q}$ that – on the sigma-algebra $\mathscr{F}_T$ – are equivalent to $\mathbb{P}$. The density process $(\frac{d\mathbb{Q}}{d\mathbb{P}})_t$ is then given by a stochastic exponential $\mathscr{E}(q \cdot W)_t = \exp(\int_0^t q_u \, dW_u - \frac{1}{2} \int_0^t |q_u|^2 \, du)$, where q is a d-dimensional predictable process that – because $\mathbb{Q} \sim \mathbb{P}$ – satisfies $\int_0^T |q_u|^2 \, du < \infty$ a.s. Each time we use a measure $\mathbb{Q}$, the process q will have this meaning. One can show that the penalty function, $c_0(\mathbb{Q}) = \mathbb{E}_{\mathbb{Q}}[\int_0^T f_u(q_u) \, du]$, can be extended to a function defined for all $\mathbb{Q} \ll \mathbb{P}$ and that it has the same expression, provided the integrals are taken with some liberal interpretation. For details we refer to [3].

There is only one time consistent extension of the utility function u_0 to a utility process. It is given by

$$u_t(\xi) = \text{ess.inf}_{\mathbb{Q} \sim \mathbb{P}} \, \mathbb{E}\left[\xi + \int_t^T f_u(q_u) \, du \mid \mathscr{F}_t\right].$$

It is shown, see [3] and [2], that there is a càdlàg version for the process u. Of course this is the version we will use.

The Legendre transform of $f(t, \omega, .)$ is denoted by $g(t, \omega, z)$ and it has similar properties as f. We make the standing assumption that g takes finite values. A statement equivalent to: for each (t, ω): $\lim_{|q| \to +\infty} \frac{f(t, \omega, q)}{|q|} = \infty$. This means that f has sufficient growth. It rules out the case that f is e.g. 0 on subspaces. We leave it as an exercise for the reader to show that also g is predictable in (t, ω). The following theorem gives an inequality between the solutions of BSDE and the utility process.

Theorem 1 *Suppose that the continuous process Y is bounded and that it satisfies*

$$\begin{cases} dY_t = g_t(Z_t) \, dt - Z_t \, dW_t \\ Y_T = \xi \end{cases}.$$

We then have $Y \le u(\xi)$.

Proof Let $\mathbb{Q} \sim \mathbb{P}$ be a measure such that $c_0(\mathbb{Q}) < \infty$, then

$$\mathbb{E}_{\mathbb{Q}}\left[\xi + \int_t^T f_u(q_u) \, du \mid \mathscr{F}_t\right]$$
$$= Y_t + \mathbb{E}_{\mathbb{Q}}\left[\int_t^T (g_u(Z_u) + f_u(q_u) - q_u Z_u) \, du \mid \mathscr{F}_t\right]$$
$$\ge Y_t.$$

Taking the infimum over all such measures $\mathbb{Q}$ gives $u_t(\xi) \ge Y_t$. □

The key to applications is the following "verification" theorem.

Theorem 2 *Suppose $\xi \in L^\infty$ and suppose that there exists an equivalent measure $\mathbb{Q} \sim \mathbb{P}$ such that*

$$u_0(\xi) = \mathbb{E}_\mathbb{Q}\left[\xi + \int_0^T f_u(q_u)\,du\right].$$

Then there exists a process Z such that for all $t \leq T$:

$$\xi = u_t(\xi) + \int_t^T g_u(Z_u)\,du - \int_t^T Z_u\,dW_u.$$

The process defined by $Y_0 = u_0(\xi)$ and $dY_t = g_t(Z_t)\,dt - Z_t\,dW_t$ is bounded and $Y_T = \xi$. The backward stochastic differential equation

$$\begin{cases} dY_t = g_t(Z_t)\,dt - Z_t\,dW_t \\ Y_T = \xi \end{cases}$$

has a bounded solution.

Proof Let $\mathbb{Q}$ be such that $u_0(\xi) = \mathbb{E}_\mathbb{Q}[\xi + \int_0^T f_u(q_u)\,du]$. The density process of $\mathbb{Q}$ is given by the stochastic exponential $\mathscr{E}(q \cdot W)$. We also have that

$$u_t(\xi) = \mathbb{E}_\mathbb{Q}\left[\xi + \int_t^T f_u(q_u)\,du \mid \mathscr{F}_t\right].$$

It follows that $u_t(\xi) + \int_0^t f_u(q_u)\,du$ is a $\mathbb{Q}$-martingale. But for other measure $\mathbb{Q}'$, in particular for $\mathbb{P}$, the process $u_t(\xi) + \int_0^t f_u(q'_u)\,du$ is a $\mathbb{Q}'$-submartingale. The Doob-Meyer decomposition – under $\mathbb{P}$ – of the $\mathbb{P}$-submartingale, $u(\xi)$, can be written as $du_t(\xi) = dA_t - Z_t dW_t$, where the process $Z \cdot W$ is a BMO-martingale, see below for a proof of this well known fact. The decomposition under $\mathbb{Q}$ therefore becomes $du_t(\xi) = dA_t - q_t Z_t dt - Z_t dW_t^\mathbb{Q}$. The martingale property shows that $dA_t = (q_t Z_t - f_t(q_t))dt$. But $\mathbb{Q}$ is a minimising measure and hence for all other density processes $\mathscr{E}(k \cdot W)$ we must have $dA_t \geq (k_t Z_t - f_t(k_t))dt$, which shows that $dA_t = g_t(Z_t)\,dt$. Combining these results yields $du_t(\xi) = g_t(Z_t)dt - Z_t dW_t$ as required. The other statements are trivial consequences. $\square$

Remark 1 We do not claim uniqueness of the solution. In fact we will see later that uniqueness is not always valid.

By Ekeland's variational principle, see [5] for the appropriate version of the Bishop-Phelps theorem, the set

$$\left\{\xi \mid \text{there is } \mathbb{Q} \ll \mathbb{P} \text{ such that } u_0(\xi) = \mathbb{E}_\mathbb{Q}\left[\xi + \int_0^T f_u(q_u)\,du\right]\right\}$$

is norm dense in L^∞. But there is no guarantee that a minimising measure is equivalent to $\mathbb{P}$. This is the topic of Sect. 3.

2 A Convergence Result

We now present some results on the decomposition of semimartingales. Let us introduce the following space

$$
\mathbf{S}^{BMO} = \left\{ X \left|
\begin{array}{l}
X \text{ is a continuous semi-martingale, } X_0 = 0 \\
X = A + M, A_0 = 0, \text{ is the Doob-Meyer decomposition} \\
A \text{ is of finite total variation} \\
M \text{ is a continuous } BMO\text{-martingale} \\
\text{there is } C \text{ such that for all } t : \mathbb{E}\left[\int_t^\infty |dA_u| \,|\, \mathscr{F}_t\right] \leq C < \infty
\end{array}
\right.\right\}.
$$

The space $\mathbf{S}$ can be given a norm

$$
\|X\| = \sup_t \left\| \mathbb{E}\left[\int_t^\infty |dA_u| \,|\, \mathscr{F}_t\right] \right\|_\infty + \|M\|_{BMO_2}.
$$

The space is a Banach space and using the martingale convergence theorem, one can show that both A and M converge when $t \to \infty$. An equivalent norm is

$$
\sup\{ \|\mathbb{E}[\,|(\theta \cdot X)_\infty - (\theta \cdot X)_t| \,|\, \mathscr{F}_t]\|_\infty \},
$$

where the sup is taken over all t and all predictable θ with $|\theta| \leq 1$. This is not trivial but can be proved as follows. For each $\varepsilon > 0$ there is a predictable process θ, $|\theta| \leq 1$ such that $|\theta \cdot A| \leq \varepsilon$. This allows to give an estimate for the $\|M\|_{BMO_1}$ norm. Once you know that M is BMO, one can choose θ_u such that $\theta \, dA_u = |dA_u|$. Then you get the bound for $\mathbb{E}[\int_t^\infty |dA_u| \,|\, \mathscr{F}_t]$. From here the rest is trivial. The mapping

$$
\mathbf{S}^{BMO} \to BMO; X \to M,
$$

is of course continuous. For applications in utility theory we need an estimate based on the supremum of the process. But even for deterministic processes there is a big difference between the quantity $\|X\|$ and expressions such as $\sup_t |X_t|$. Nevertheless the *martingale part* can be estimated using the supremum of the process.

Lemma 1 *Let X be a continuous bounded submartingale bounded by a constant c. Suppose that X has the continuous Doob-Meyer decomposition $X = A + M$, then the martingale part M is in BMO, more precisely $\|M\|_{BMO_2} \leq 2c$ and $\|X\|_{\mathbf{S}^{BMO}} \leq 4c$. Consequently A_∞ has exponential moments of some order.*

Proof The process A satisfies $\mathbb{E}[\int_t^\infty dA_u \,|\, \mathscr{F}_t] = \mathbb{E}[X_\infty - X_t \,|\, \mathscr{F}_t] \leq 2c$. This already shows that $X \in \mathbf{S}^{BMO}$. However this only yields a bound for the BMO_1 norm and the bound is not the best. We prefer to give a direct proof. First observe that the process $Y = (c - X)^2$ is given by the differential equation

$$
dY_t = 2(c - X_t)dA_t + d\langle M, M\rangle_t + 2(c - X_t)dM_t.
$$

Since the coefficient $c - X_t$ is nonnegative, the process Y is also a bounded sub-martingale and taking conditional expectations gives

$$\mathbb{E}\left[\int_t^\infty d\langle M, M\rangle_u \mid \mathscr{F}_t\right] \leq \mathbb{E}[Y_\infty - Y_t \mid \mathscr{F}_t] \leq \mathbb{E}[Y_\infty \mid \mathscr{F}_t].$$

Consequently

$$\mathbb{E}\left[\int_t^\infty d\langle M, M\rangle_u \mid \mathscr{F}_t\right] \leq 4c^2.$$

Hence $\|M\|_{BMO} \leq 2c$. Because M is BMO it has – by the John-Nirenberg inequality, see [4] – some exponential moments. Because X is bounded also A_∞ must have exponential moments. $\qquad\square$

Lemma 2 *For $X \in \mathbf{S}^{BMO}$ and X bounded, we have*

$$\|M\|_{BMO_2}^2 \leq 8\|\sup_t |X_t|\|_\infty^2 + 8\|\sup_t |X_t|\|_\infty \|X\|.$$

Proof Let $c = \|\sup_t |X_t|\|_\infty$. The following inequalities are now straightforward. If needed one can use a localisation argument by stopping the processes when they reach a level.

$$\begin{aligned}
\mathbb{E}[(M_\infty - M_t)^2 \mid \mathscr{F}_t] &\leq 2\mathbb{E}[(X_\infty - X_t)^2 \mid \mathscr{F}_t] + 2\mathbb{E}[(A_\infty - A_t)^2 \mid \mathscr{F}_t] \\
&\leq 8c^2 + 4\mathbb{E}\left[\int_t^\infty (A_u - A_t)dA_u \mid \mathscr{F}_t\right] \\
&\leq 8c^2 + 4\mathbb{E}\left[\int_t^\infty (A_u - A_t)dX_u \mid \mathscr{F}_t\right] \\
&\leq 8c^2 + 4\mathbb{E}\left[\int_t^\infty (X_\infty - X_u)dA_u \mid \mathscr{F}_t\right] \\
&\leq 8c^2 + 8c\mathbb{E}\left[\int_t^\infty |dA_u| \mid \mathscr{F}_t\right].
\end{aligned}$$

This shows the bound on $\|M\|_{BMO_2}$. $\qquad\square$

Corollary 1 *If X^n is a sequence in $\mathbf{S}^{BMO}$, $\sup_n \|X^n\| < \infty$, if $\|\sup_t |X_t^n - X_t|\|_\infty \to 0$, then $X \in \mathbf{S}^{BMO}$ and the martingale parts of the Doob-Meyer decompositions satisfy $M^n \to M$ in BMO.*

Proof Only the statement that $X \in \mathbf{S}^{BMO}$ needs to be shown. By the inequality we get that $\lim_{n,m\to\infty}(M^n - M^m) \to 0$ in BMO. So the Cauchy sequence converges to some $M \in BMO$. The finite variation part then also converges to a process A in the sense that $\sup_t |A_t^n - A_t|$ tends to zero at least in probability. That $\mathbb{E}[\int_t^\infty |dA_u| \mid \mathscr{F}_t] \leq \sup_n \|X^n\|$ is straightforward. So $X \in \mathbf{S}^{BMO}$, $\|X\| \leq \sup_n \|X^n\|$. $\qquad\square$

Remark 2 Of course we did not make a statement on norm convergence in the space S^{BMO}. As already observed above, even for deterministic bounded variation processes, the convergence in sup-norm is not the same as the convergence in variation norm.

Corollary 2 *If* $(X^n)_{n \geq 1}$ *is a uniformly bounded sequence of submartingales* $|X_t^n| \leq c$, *if* $\| \sup_t |X_t - X_t^n| \|_\infty$ *tends to* 0, *then the Doob-Meyer decompositions* $X^n = A^n + M^n$, $X = A + M$ *satisfy* M^n *tends to* M *in BMO.*

Proof The previous corollary shows that $\|X^n\| \leq C$ (where C only depends on c). The convergence of the martingale parts now follows. □

Remark 3 If in the corollary we replace the condition $\| \sup_t |X_t - X_t^n| \|_\infty$ tends to 0 by the weaker condition $\sup_t |X_t - X_t^n| \to 0$ a.s., then the conclusion is wrong, even for sequences of uniformly bounded martingales. Indeed there is a sequence of uniformly bounded martingales M^n such that $\sup_t |M_t^n - M_t|$ tends to 0 a.s., but the convergence in BMO is false. This has to do with the fact that the map $\mathcal{H}^1 \to L^1$ is not weakly compact. We do not give details.

3 Pathwise Subquadratic Drivers

Theorem 3 *Suppose that for every* $\omega \in \Omega$ *there is a constant* $K < \infty$ *depending on* ω *such that for all* t: $g_t(z) \leq K(1 + |z|^2)$. *Suppose that* $\mathbb{Q}$ *is a minimising measure for* ξ, *then necessarily* $\mathbb{Q} \sim \mathbb{P}$.

We need the following lemma:

Lemma 3 *If* $\phi: \mathbb{R}^d \to \mathbb{R}_+$ *is a convex function such that* $\phi(z) \leq K(1 + |z|^2)$ *for all* $z \in \mathbb{R}^d$. *Then every* $q \in \partial_z \phi$ (q *an element of the subgradient of* ϕ *in the point* z) *satisfies* $|q| \leq 5K(1 + |z|)$.

Proof Take $z \in \mathbb{R}^d$ and $q \in \partial_z \phi$. Because of convexity we have for all $w \in \mathbb{R}^d$: $\phi(z + w) - \phi(z) \geq q \cdot w$. If $|z| \geq 1$ we apply this inequality for all w with $|w| = |z|$ and we get $K(1 + 4|z|^2) \geq |q||z|$. This implies that $|q| \leq 4K(1 + |z|)$ for $|z| \geq 1$. For $|z| \leq 1$ we take the inequality for all $|w| = 1$ and we get $5K \geq |q|$. The two cases can be written as $|q| \leq 5K(1 + |z|)$. □

Proof We will need the Lebesgue measure m on $[0, T]$. Let $du_t(\xi) = dA_t - Z_t dW_t$ be the Doob-Meyer decomposition of the process $u(\xi)$. Let the density process D of $\mathbb{Q}$ be given by the stochastic exponential $\mathcal{E}(q \cdot W)$. Let $\tau = \inf\{t \mid D_t = 0\}$. Since $0 < \tau$ is predictable we have that there is a sequence $\tau_n < \tau$ such that $\tau_n \uparrow \tau$. Of course $\mathbb{Q} \sim \mathbb{P}$ on $\mathcal{F}_{\tau_n}$. Under $\mathbb{Q}$ the process $u_t(\xi) + \int_0^t f_s(q_s)$ is a martingale and hence $\mathbb{Q} \times m$ a.s.: $g_t(Z_t) + f_t(q_t) - q_t Z_t = 0$ on $[\![0, \tau[\![$. Hence on $[\![0, \tau[\![$, we have that q_t is a an element of the subgradient of g_t at Z_t. This is true for the measure $\mathbb{Q} \times m$ and hence also for $\mathbb{P} \times m$ (because on $\mathscr{F}_{\tau_n}$ both measures are equivalent). Hence on $[\![0, \tau[\![$, we have $|q_t| \leq C(1 + |Z_t|)$. But this shows that $\int_0^\tau |q_t|^2 dt \leq C + C \int_0^\tau |Z_t|^2 dt$ for some constant C. So we have $\mathbb{P}$ a.s. that $\int_0^\tau |Z_t|^2 dt \leq \int_0^T |Z_t|^2 dt < \infty$ and hence also $\int_0^\tau |q_t|^2 dt < \infty$ $\mathbb{P}$ a.s. Since $\{D_T = 0\} = \{\int_0^\tau |q_t|^2 dt = \infty\}$, we have $D_T > 0$ a.s., meaning $\mathbb{Q} \sim \mathbb{P}$. □

Theorem 4 *Suppose that for every $\omega \in \Omega$ there is a constant K depending on ω such that for all t: $g_t(z) \leq K(1 + |z|^2)$. For every $\xi \in L^\infty$ the process $u(\xi)$ is a bounded solution of the BSDE:*

$$\begin{cases} dY_t = g_t(Z_t) dt - Z_t dW_t \\ Y_T = \xi \end{cases}.$$

Proof By the Bishop-Phelps theorem there is a sequence $\xi_n \to \xi$ (in L^∞ norm) such that for ξ_n there is a minimising measure $\mathbb{Q}_n$. By the preceding theorem this measure is equivalent to $\mathbb{P}$. By the "verification" theorem we can write $du_t(\xi_n) = g_t(Z_t^n) dt - Z_t^n dW_t$. Since $\sup_t |u_t(\xi_n) - u_t(\xi)| \leq \|\xi_n - \xi\|$ we can apply the results of the previous section and hence we get that $Z^n \cdot W \to Z \cdot W$ in BMO. This implies $\int_0^T |Z_t^n - Z_t|^2 dt \to 0$ in probability. The hypothesis on g then implies that in probability: $\int_0^s g_t(Z_t^n) dt \to \int_0^s g_t(Z_t) dt$ for every $s \leq T$. We get $u_s(\xi) = u_0(\xi) + \int_0^s g_t(Z_t) dt - \int_0^s Z_t dW_t$. □

4 The Relation with Minimal Elements

For $\xi \in L^\infty$ we know that the process $u(\xi)$ is a $\mathbb{P}$-submartingale and hence has a Doob-Meyer decomposition $du_t(\xi) = dA_t - Z_t dW_t$. Because for $\mathbb{Q} \sim \mathbb{P}$ with $c_0(\mathbb{Q}) < \infty$, the process $u_t(\xi) + \int_0^t f_u(q_u) du$ is a $\mathbb{Q}$-submartingale we must have that $dA_t \geq g_t(Z_t) dt$. So we can write $du_t(\xi) = g_t(Z_t) dt + dC_t - Z_t dW_t$, where C is a nondecreasing process, with $C_0 = 0$. The discontinuity points of C must be the same as the discontinuity points of $u(\xi)$. So we get that the jumps of C must be bounded by the jumps of $u(\xi)$ and hence smaller than $2\|\xi\|$. It follows that the process C is locally bounded. We already saw that the existence of a minimising measure $\mathbb{Q} \sim \mathbb{P}$ implies that $C_T = 0$. There is also a relation with minimal elements, see also [1]. These are defined as follows

Definition 1 An element $\xi \in L^\infty$ is called minimal if for each $\eta \in L_+^\infty$, $\mathbb{P}[\eta > 0] > 0$ we have $u_0(\xi - \eta) < u_0(\xi)$.

Proposition 1 *If ξ is minimal then for each stopping time σ, the element $u_\sigma(\xi)$ is minimal.*

Proof Take $\eta \leq u_\sigma(\xi)$ such that $u_0(\eta) = u_0(u_\sigma(\xi))$ and $\mathbb{P}[\eta < u_\sigma(\xi)] > 0$. Obviously we have $u_\sigma(\eta) \leq u_\sigma(\xi)$. We first show that $u_\sigma(\eta) < u_\sigma(\xi)$ on a set of positive measure. Indeed if, $u_\sigma(\eta) = u_\sigma(\xi)$, then

$$u_\sigma(\xi) \geq \mathbb{E}_\mathbb{P}[\eta \mid \mathscr{F}_\sigma] \text{ since } \eta \leq u_\sigma(\xi)$$
$$\geq u_\sigma(\eta) = u_\sigma(\xi),$$

hence $u_\sigma(\xi) = \mathbb{E}_\mathbb{P}[\eta \mid \mathscr{F}_\sigma]$. Integrating with respect to $\mathbb{P}$ gives $\int u_\sigma(\xi)\,d\mathbb{P} = \int \eta\,d\mathbb{P}$, a contradiction to $\mathbb{P}[\eta < u_\sigma(\xi)] > 0$. This shows $\mathbb{P}[u_\sigma(\eta) < u_\sigma(\xi)] > 0$. We then get

$$u_0(\xi - (u_\sigma(\xi) - u_\sigma(\eta))) = u_0(u_\sigma(\xi - (u_\sigma(\xi) - u_\sigma(\eta))))$$
$$= u_0(u_\sigma(\eta) = u_0(\eta) = u_0(u_\sigma(\xi))$$
$$= u_0(\xi),$$

showing that ξ was not minimal, a contradiction.   □

Theorem 5 *If ξ is minimal, then $C_T = 0$ and hence the process $u(\xi)$ is a bounded solution of the BSDE:*

$$\begin{cases} dY_t = g_t(Z_t)\,dt - Z_t\,dW_t \\ Y_T = \xi \end{cases}.$$

Proof If $C_T \neq 0$ there is a stopping time σ, such that C_σ is bounded and $\mathbb{P}[C_\sigma > 0] > 0$. Indeed take $\varepsilon > 0$ such that $\mathbb{P}[C_T > \varepsilon] > 0$. Take now $\sigma = \inf\{t \mid C_t \geq \varepsilon\}$. Since the jumps of C are bounded we have that C_σ is bounded. We can now write

$$u_\sigma(\xi) - C_\sigma = u_0(\xi) + \int_0^\sigma g_u(Z_u)\,du - \int_0^\sigma Z_u\,dW_u.$$

This shows that we have a bounded solution of the BSDE with endpoint $u_\sigma(\xi) - C_\sigma$ and starting point $u_0(\xi)$. The proposition then gives $u_0(u_\sigma(\xi) - C_\sigma) \geq u_0(\xi)$. But we certainly have $u_0(u_\sigma(\xi) - C_\sigma) \leq u_0(u_\sigma(\xi)) = u_0(\xi)$. This shows $u_0(u_\sigma(\xi) - C_\sigma) = u_0(\xi)$ and hence $u_\sigma(\xi)$ cannot be minimal, a contradiction to the previous proposition. We conclude that $C_T = 0$.   □

Remark 4 The converse is not true. We can show that there is a utility function u and a random variable ξ such that $u(\xi)$ satisfies the associated BSDE but ξ is not minimal, see [1]. This shows that the characterisation of those ξ for which the BSDE has a solution is not an easy problem. But we can show the following:

Theorem 6 *Suppose that for every $\omega \in \Omega$ there is a constant $K < \infty$ depending on ω such that for all t: $g_t(z) \leq K(1 + |z|^2)$. Every $\xi \in L^\infty$ is then minimal.*

Proof The proof uses that the process $u(\xi)$ is a solution of the BSDE. Suppose that $u_0(\xi) = 0$ and $du_t(\xi) = g_t(Z_t)\,dt - Z_t\,dW_t$. For each $\omega \in \Omega$ there is a constant $K(\omega)$ such that $g_t(z) \leq K(1 + |z|^2)$. We can of course suppose that K is measurable for $\mathscr{F}_T$. To find a minimising measure we can try a selection q of $\partial_z g$ but of course there is no guarantee that the stochastic exponential $\mathscr{E}(q \cdot W)$ is uniformly integrable. Nevertheless this is the idea behind the proof. To show that ξ is minimal we take $A \in \mathscr{F}_T$, $\delta > 0$ and we must show that $u_0(\xi - \delta \mathbf{1}_A) < 0$. Take N big enough so that $\mathbb{P}[A \cap \{K \leq N\}] > 0$. Define the stopping time σ as

$$\sigma = \inf\{t \mid |q_t| > 5N(1 + |Z_t|\} \wedge T\}.$$

By the lemma above $\{\sigma < T\} \subset \{K > N\}$ and therefore $\mathbb{P}[A \cap \{\sigma = T\}] > 0$. We now show that the stochastic exponential $\mathscr{E}(q \cdot W)$ stopped at σ is uniformly integrable. Since $u_\sigma(\xi) = \int_0^\sigma g_t(Z_t) \, dt - \int_0^\sigma Z_t \, dW_t$, we have $Z \cdot W$ is in BMO. But

$$|q| \mathbf{1}_{[0,\sigma[} \leq 5N(1 + |Z|)$$

and hence $q \cdot W$ is also in BMO. Therefore $\mathscr{E}(q \mathbf{1}_{[0,\sigma[} \cdot W) = \mathscr{E}(q \cdot W)^\sigma$ is uniformly integrable, see e.g. [4]. The measure $d\mathbb{Q} = \mathscr{E}(q \cdot W)_\sigma \, d\mathbb{P}$ satisfies $0 = u_0(u_\sigma(\xi)) = \mathbb{E}_\mathbb{Q}[u_\sigma(\xi) + \int_0^\sigma f_t(q_t) \, dt]$. We also have that $\mathbb{Q} \sim \mathbb{P}$. The following inequalities are now obvious

$$\begin{aligned}
u_0(\xi - \delta \mathbf{1}_A) &= u_0(u_\sigma(\xi - \delta \mathbf{1}_A)) \\
&\leq u_0\big(u_\sigma(\xi - \delta \mathbf{1}_{A \cap \{\sigma = T\}})\big) \\
&\leq u_0\big(u_\sigma(\xi) - \delta \mathbf{1}_{A \cap \{\sigma = T\}}\big) \\
&\leq \mathbb{E}_\mathbb{Q}\left[u_\sigma(\xi) + \int_0^\sigma f_t(q_t) \, dt\right] - \delta \mathbb{Q}[A \cap \{\sigma = T\}] \\
&< 0.
\end{aligned}$$

$\square$

5 An Example

Here we take a coherent utility given by a non-weakly compact set $\mathscr{S}$, i.e.

$$u_0(\xi) = \inf_{\mathbb{Q} \in \mathscr{S}} \mathbb{E}_\mathbb{Q}.$$

The set will be taken in such a way that all its elements are equivalent to $\mathbb{P}$. We will make use of the BES^3 process which provides a good example of a strict local martingale. The notation is standard. We take $d = 1$ and $dR_t = \frac{dt}{R_t} + dW_t$, $L_t = 1/R_t$, $R_0 = L_0 = 1$, $dL_t = -L_t^2 \, dW_t$. The process L is a strict local martingale since $\mathbb{E}_\mathbb{P}[L_t] < 1$ for all $t > 0$. Then we take

$$\mathscr{S} = \left\{\mathbb{Q} \mid \frac{d\mathbb{Q}}{d\mathbb{P}} = \mathscr{E}(q \cdot W), |q| \leq \frac{L^2}{1 + L} = \frac{1}{R(R+1)}\right\}.$$

This set is not weakly compact. This can be seen as follows. If we take $q = \frac{-L_t^2}{1 + L_t}$, then the solution of $dV_t = q_t V_t \, dW_t$, $V_0 = 1$ is precisely $V = \frac{1+L}{2}$, which is a strict local martingale. The stopped processes V^{σ_n} where $\sigma_n = \inf\{t \mid V_t \geq n\} \wedge T$, are bounded and define elements of $\mathscr{S}$. Since $V_T^{\sigma_n}$ converges to V_T and since

$\mathbb{E}[V_T] < 1$, the set $\mathscr{S}$ is not uniformly integrable. The function f is given by $f_t(q) = 0$ if $|q| \leq \frac{L_t^2}{1+L_t}$ and equals $+\infty$ elsewhere. The conjugate is then

$$g_t(z) = \frac{L_t^2}{1 + L_t}|z| = \frac{1}{R_t(1 + R_t)}|z|,$$

which certainly satisfies the hypothesis of the theorems in Sects. 2 and 3. So we have that for $\xi \in L^\infty$, $u_t(\xi)$ is a solution of the BSDE

$$dY_t = g_t(Z_t)\,dt - Z_t\,dW_t, \quad Y_T = \xi, \quad Y \text{ bounded.}$$

We claim that even for functions of the form $\xi = \phi(R_T)$, there is no uniqueness, even in the class of processes which are functions of t and R_t. An easy application of Itô's formula gives that $\phi(R_t) = \frac{1}{1+L_t} = \frac{R_t}{R_t+1}$ is a bounded solution of the BSDE with terminal condition $\xi = \frac{1}{1+L_T}$. We will show that for T big enough (and maybe with some effort for every $T > 0$), $u_0(\xi) > 1/2$. In fact we will show that $\lim_{T\to\infty} u_0(\frac{1}{1+L_T}) = 1$, something to be expected since $L_T \to 0$ as $T \to +\infty$. However since $\mathscr{S}$ is not weakly compact such interchanges of limits and integrals is more delicate. The Markov property of R implies that also $u_t(\xi)$ is a function that only depends on t and R_t. We will see what this means for the related PDE.

Theorem 7 $\lim_{T\to\infty} u_0(\frac{1}{1+L_T}) = 1$.

Proof The proof is surprisingly difficult and maybe there is an easier way. We divide the proof in several steps.

Lemma 4 $\sup_{\mathbb{Q}\in\mathscr{S}} \mathbb{E}_\mathbb{Q}[R_T] \leq \sqrt{1 + 5T}$.

Proof We calculate $\mathbb{E}_\mathbb{Q}[R_T^2]$. Under $\mathbb{Q}$ the BESQ^3-process follows the equation

$$d(R^2)_t = 2R_t\,dR_t + dt = 3dt + 2R_t q_t\,dt + 2R_t\,dW_t^\mathbb{Q}.$$

Since $|q_t| \leq \frac{1}{R_t(R_t+1)}$, we get that

$$\mathbb{E}_\mathbb{Q}[R_T^2] \leq 1 + 3T + \mathbb{E}_\mathbb{Q}\left[\int_0^T \frac{2}{1 + R_t}\,dt\right] \leq 1 + 5T. \qquad \square$$

Lemma 5 $\sup_{\mathbb{Q}\in\mathscr{S}} \mathbb{E}_\mathbb{Q}\left[\frac{1}{1+R_T}\right] \leq \frac{1}{T}(\sqrt{1+5T} - 1) \to 0$ as $T \to \infty$.

Proof This is tricky. Under $\mathbb{Q}$ the process $\frac{1}{1+R_t}$ is a $\mathbb{Q}$-supermartingale as Itô's lemma shows. Also

$$dR_t = \left(\frac{1}{R_t} + q_t\right)dt + dW_t^\mathbb{Q} \text{ with } \left(\frac{1}{R_t} + q_t\right) \geq \frac{1}{1 + R_t}.$$

So we get

$$\mathbb{E}_{\mathbb{Q}}[R_T] \geq 1 + \mathbb{E}_{\mathbb{Q}}\left[\int_0^T \frac{1}{1+R_t}\, dt\right] \geq 1 + T\,\mathbb{E}_{\mathbb{Q}}\left[\frac{1}{1+R_T}\right].$$

Together with the previous lemma, we get the desired inequality.   □

We can now complete the proof of the theorem.

$$\inf_{Q \in \mathscr{S}} \mathbb{E}_{\mathbb{Q}}\left[\frac{R_T}{1+R_T}\right] = 1 - \sup_{Q \in \mathscr{S}} \mathbb{E}_{\mathbb{Q}}\left[\frac{1}{1+R_T}\right]$$

$$\geq 1 - \frac{1}{T}\left(\sqrt{1+5T} - 1\right)$$

$$\to 1 \text{ as } T \to \infty.$$ □

If we summarise what we have calculated so far, we get that for T big enough: $u_0(\xi) > 1/2$ and hence we have two Markovian solutions of the equation

$$dY_t = g_t(Z_t)\, dt - Z_t\, dW_t, \quad Y_T = \xi, \quad Y \text{ bounded}.$$

One solution is the process $\frac{1}{1+L_t} = \frac{R_t}{1+R_t}$, the other solution is $u(t, R_t) = u_t\left(\frac{R_T}{1+R_T}\right)$. If we write the PDE related to this problem we get

$$\partial_t u + \frac{1}{2}\partial_{xx} u + \frac{1}{x}\partial_x u = \frac{1}{x(x+1)}|\partial_x u|,$$

leading to

$$\partial_t u + \frac{1}{2}\partial_{xx} u + \left(\frac{1}{x} - \frac{1}{x(x+1)}\operatorname{sign}(\partial_x u)\right)\partial_x u = 0.$$

The domain where we look for the solution is $[0, T] \times (0, \infty)$ and the boundary condition is $u(T, x) = \frac{x}{1+x}$. Of course we did not discuss the regularity for the utility function, but PDE theory either shows that it is regular or we just see it as a viscosity solution. We do not go into details. For a terminal value of the form $\frac{R_T}{1+R_T}$ we can get rid of the sign-function. Indeed the non-confluence result for SDE's shows that the process R is increasing in its starting value. As a result we get that conditional on $R_t = x$, the endpoint R_T is increasing in x. The function $\frac{x}{1+x}$ itself is increasing and hence we get by monotonicity of u that $\partial_x u \geq 0$. The PDE then simplifies to

$$\partial_t u + \frac{1}{2}\partial_{xx} u + \frac{1}{x+1}\partial_x u = 0 \text{ on } [0, T] \times (0, \infty).$$

This seems a nice linear equation and we found two (hence infinitely many) bounded solutions of it. The reason can be explained using stochastic theory. The above equation is the equation that gives martingales for a diffusion of the form

$$dV_t = \frac{1}{V_t + 1}\, dt + dW_t, \quad V_0 = x > 0.$$

The solution is explicitly known as $V_t = R_t - 1$, where R is a BES^3-process starting at $x + 1$. The process V can reach the value 0 and hence the stochastic approach to solve the boundary value problem requires that either we give conditions for the final value on $(-1, +\infty)$ or we give a condition when crossing the level 0, i.e. we also need a boundary value on the set $(0, T] \times \{0\}$. In any case the linear PDE cannot have a unique solution.

Remark 5 Continuing the function $\frac{x}{x+1}, x \geq 0$ for $-1 < x < 0$ allows to produce a very big set of solutions. We do not pursue this rather delicate issues.

Acknowledgements This research was sponsored by a grant of Credit Suisse. The text only reflects the opinion of the authors. We also thank Mete Soner for discussions on PDE.

References

1. Bao, X.: Backward stochastic differential equations with super-quadratic growth. Ph.D. thesis, ETH-Zurich (2009)
2. Bion-Nadal, J.: Time consistent dynamic risk processes. Stoch. Process. Appl. **119**(2), 633–654 (2009)
3. Delbaen, F., Peng, S., Rosazza-Gianin, E.: Representation of the penalty term of dynamic concave utilities. Finance Stoch., accepted (2009)
4. Kazamaki, N.: Continuous Exponential Martingales and BMO. *Lecture Notes in Mathematics*, vol. 1579. Springer, Berlin (1994)
5. Phelps, R.R.: Convex Functions, Monotone Operators and Differentiability. *Lecture Notes in Mathematics*, vol. 1364, 2nd edn. Springer, Berlin (1993)

Comparison Theorems for Finite State Backward Stochastic Differential Equations

Samuel N. Cohen and Robert J. Elliott

Abstract Most previous contributions on BSDEs, and the related theories of non linear expectation and dynamic risk measures, have been in the framework of continuous time diffusions or jump diffusions. Using solutions of BSDEs on spaces related to finite state, continuous time Markov Chains, we discuss a theory of non-linear expectations in the spirit of Peng (math/0501415 (2005)). We prove basic properties of these expectations, and show their applications to dynamic risk measures on such spaces. In particular, we prove comparison theorems for scalar and vector valued solutions to BSDEs, and discuss arbitrage and risk measures in the scalar case.

1 Introduction

How to model and modify the future is a central human activity. In particular, a re-occurring concern is how to quantify and mitigate risk. Introducing negatives of quantities involved, this is related to investigating the maximisation of some expected future reward, taking into account the randomness involved.

Risk measures such as VaR (Value at Risk), and CVaR (Conditional Value at Risk), have well known deficiencies: in the case of VaR they can discourage diversification, and neither VaR nor CVaR is guaranteed to make consistent decisions through time. A topic of current interest is how to define practical measures of risk which are time consistent. Such measures of risk are of great importance, as they allow us to model and understand risk-averse behaviour, while avoiding paradoxes

S.N. Cohen · R.J. Elliott
School of Mathematical Sciences, University of Adelaide, Adelaide 5005, Australia

S.N. Cohen
e-mail: samuel.cohen@adelaide.edu.au

R.J. Elliott (✉)
Haskayne School of Business, University of Calgary, Calgary T2N 1N4, Canada
e-mail: relliott@ucalgary.ca

C. Chiarella, A. Novikov (eds.), *Contemporary Quantitative Finance*,
DOI 10.1007/978-3-642-03479-4_8, © Springer-Verlag Berlin Heidelberg 2010

and inconsistencies arising from decisions being made on different dates, when different information is available.

We now know that such measures can be defined as solutions of Backward Stochastic Differential Equations (see [8, 18]). In this paper and related recent work we consider backward equations on a finite state space, where the noise derives from a Markov chain. Related non linear expectations and dynamic risk measures can be defined once a comparison theorem is established.

The theory of Backward Stochastic Differential Equations (BSDEs) is applicable to many problems in Mathematical finance and elsewhere. As discussed in [11], BSDEs encompass the classical pricing theory of Black, Scholes, Merton and others, as well as many examples of pricing in incomplete markets. In [8, 18, 22] and others, BSDEs have been used to generate 'nonlinear expectations', that is, operators which behave in many ways like the classical expectations operator, but are not necessarily linear. Similarly, [10] developed a theory of recursive utilities in continuous time, while [3, 9, 23] and others have explored the use of BSDEs in the theory of *dynamic* risk measures, in the sense of [1, 15]. A key result in these applications is the 'comparison theorem', first given in [21], which establishes certain monotonicity properties of the operators considered.

The above contributions deal with the classical BSDE driven by a standard Wiener process. Other work has considered allowing jumps, primarily in the context of a Lévy process, for example [2, 4, 12, 20, 24]. A significant issue here is that the original comparison theorem no longer applies – see [2] for a counterexample – and so further restrictions must be placed on the terms of the equations to ensure the desired properties hold. These conditions do not easily extend to other settings, and apply only to the case where the equations considered are scalar.

In this paper a situation is considered where randomness is generated by a continuous time, finite state Markov chain. A corresponding discrete time theory is covered in our paper [7]. Both scalar and vector comparison theorems are stated in this framework. Following [22] we then develop a theory of nonlinear expectations. Full proofs are given in the paper [6].

2 A Mathematical Setting

Consider a continuous time, finite state Markov chain $X = \{X_t, t \in [0, T]\}$. Without loss of generality, we identify the states of this process with the standard unit vectors e_i in $\mathbb{R}^N$, where N is the number of states of the chain.

Consider stochastic processes defined on the filtered probability space $(\Omega, \mathscr{F}, \{\mathscr{F}_t\}, \mathbb{P})$, where $\{\mathscr{F}_t\}$ is the completed natural filtration generated by the σ-fields $\mathscr{F}_t = \sigma(\{X_u, u \le t\} \cup \{F \in \mathscr{F}_T : \mathbb{P}(F) = 0\})$, and $\mathscr{F} = \mathscr{F}_T$. As X is a right-continuous jump process with distinct jumps, this filtration is right-continuous. If A_t denotes the rate matrix for X at time t, then this chain has a semimartingale representation

$$X_t = X_0 + \int_{]0,t]} A_u X_{u-} du + M_t \tag{1}$$

where M_t is a martingale. (See Appendix B of [14].) As in [5], we know that the predictable quadratic covariation matrix of this martingale, $\langle M, M \rangle_t$, has the representation

$$\langle M, M \rangle_t = \int_{]0,t]} (\mathrm{diag}(A_u X_{u-}) - A_u \,\mathrm{diag}(X_{u-}) - \mathrm{diag}(X_{u-})A_u^*) du. \quad (2)$$

As in our earlier paper, [5], consider pairs (Y, Z), $Z_t \in \mathbb{R}^{K \times N}$ adapted and left continuous, $Y_t \in \mathbb{R}^K$ adapted and càdlàg, which are solutions to equations of the form

$$Y_t - \int_{]t,T]} F(\omega, u, Y_{u-}, Z_u) du + \int_{]t,T]} Z_u dM_u = Q. \quad (3)$$

Here Q is an $\mathscr{F}_T$ measurable, $\mathbb{R}^K$ valued, $\mathbb{P}$-square integrable random variable and F is a map $\Omega \times [0, T] \times \mathbb{R}^K \times \mathbb{R}^{K \times N} \to \mathbb{R}^K$ which is progressively measurable.

Let ψ_t be the nonnegative definite matrix

$$\psi_t := \mathrm{diag}(A_t X_{t-}) - A_t \,\mathrm{diag}(X_{t-}) - \mathrm{diag}(X_{t-})A_t^* \quad (4)$$

and

$$\|Z\|_{X_{t-}}^2 := \mathrm{Tr}(Z\psi_t Z^*). \quad (5)$$

Then $\| \cdot \|_{X_{t-}}$ defines a (stochastic) seminorm, with the property that

$$\mathrm{Tr}(Z_t d\langle M, M \rangle_t Z_t^*) = \|Z_t\|_{X_{t-}}^2 dt. \quad (6)$$

We have shown the following result in [5].

Theorem 1 *For $Q \in L^2(\mathscr{F}_T)$, if F is Lipschitz continuous, in the sense that there exists $c \geq 0$ such that, for any Z^1, Z^2, Y^1, Y^2 square integrable and of appropriate dimension,*

$$\|F(t, Y_{u-}^1, Z_t^1) - F(u, Y_{t-}^2, Z_t^2)\| \leq c(\|Z_t^1 - Z_t^2\|_{X_{t-}} + \|Y_{t-}^1 - Y_{t-}^2\|)$$

$dt \times \mathbb{P}$-a.s., then there exists a solution (Y, Z) to (3), such that

$$E\left[\int_{]0,T]} \|Y_t\|^2 du\right] < +\infty$$

$$E\left[\int_{]0,T]} \|Z_t\|_{X_{t-}}^2 du\right] < +\infty,$$

and this solution is the unique such solution, up to indistinguishability for Y and equality $d\langle M, M \rangle_t \times \mathbb{P}$-a.s. for Z.

We shall describe properties and applications of these solutions.

Note that, where appropriate, we shall denote by $\leq, \geq$, etc... an inequality holding simultaneously on all components of a vector quantity.

3 Some Lemmas

We restrict ourselves to deterministic T; however, it is clear that, as observed in [6], given appropriate modifications, the results stated could easily be modified to remain valid when T is an essentially bounded stopping time. We now state, without proof, some Lemmas whose proofs are given in [6].

We shall henceforth assume that F is $\mathbb{P}$-a.s. left continuous in t, is Lipschitz continuous as in Theorem 1 and satisfies

$$E\left[\int_{]0,T]} \|F(\omega, t, Y_{t-}, Z_t)\|^2 dt\right] < +\infty$$

for all Y, Z bounded as in Theorem 1. Such a driver will be called *standard*. If also $Q \in L^2(\mathscr{F}_T)$, then the pair (F, Q) will be called standard.

We shall assume that the rate matrix A of our chain is left-continuous, and there is an $0 < \varepsilon_r < 1$ such that it satisfies

$$e_i^* A_t e_j \in [\varepsilon_r, 1/\varepsilon_r] \cup \{0\} \tag{7}$$

dt-a.s., for all i and j, $i \neq j$. This assumption is satisfied if the chain X is time-homogenous, and essentially states that we shall not consider chains with positive transition rates unbounded or arbitrarily close to zero.

Throughout this section, $\mathbf{1}$ will denote a column vector of appropriate dimension with all components equal to one.

We first note that, from the basic properties of stochastic integrals, (see, for example, [19, p. 28]), the following isometry holds.

Lemma 1 *For any predictable (matrix) process Z (of appropriate dimension), any $s < t$,*

$$E\left[\left\|\int_{]s,t]} Z_u dM_u\right\|^2\right] = E\left[\int_{]s,t]} \|Z_u\|^2_{X_{u-}} du\right]$$

Lemma 2 *For (F, Q) standard, if Y is the solution to (3) given by Theorem 1, then Y satisfies*

$$\sup_{t \in [0,T]} E[\|Y_t\|^2] < +\infty$$

Lemma 3 *For all t, the matrix ψ_t is symmetric and bounded, and has all row and column sums equal to zero. The matrix given by its Moore-Penrose inverse, ψ_t^+, is also bounded and has all row and column sums equal to zero.*

Lemma 4 *At each time t, for all j such that $e_j^* A_t X_{t-} \neq 0$,*

$$\psi_t^+ \psi_t (e_j - X_{t-}) = \psi_t \psi_t^+ (e_j - X_{t-}) = e_j - X_{t-}.$$

Furthermore, the vectors $(e_j - X_{t-})$ form a basis for the range of the projections $\psi_t^+ \psi_t$ and $\psi_t \psi_t^+$.

Lemma 5 *For any standard driver F, for any Y and Z, up to indistinguishability*

$$F(\omega, t, Y_{t-}, Z_t) = F(\omega, t, Y_{t-}, Z_t \psi_t \psi_t^+)$$

and

$$\int_{]0,t]} Z_u d M_u = \int_{]0,t]} Z_u \psi_u \psi_u^+ d M_u.$$

Therefore, it is no loss of generality to assume that $Z = Z \psi \psi^+$.

Lemma 6 $\|Z_t\|_{X_{t-}} = 0$ *only if* $Z_t \psi_t \psi_t^+ = 0$. *Hence without loss of generality,* $\|Z_t\|_{X_{t-}} = 0$ *if and only if* $Z_t = 0$.

Lemma 7 *For all $t \in [0, T]$*

$$\psi_t X_{t-} = -A_t X_{t-} = -\sum_j (e_j^* A_t X_{t-})(e_j - X_{t-}). \tag{8}$$

Lemma 8 *For processes Z solving (3), without loss of generality, the $\|\cdot\|_{X_{t-}}$ norm has two equivalent forms*:

$$\|Z_t\|_{X_{t-}}^2 = \mathrm{Tr}(Z_t \psi_t Z_t^*)$$
$$= \sum_{i,j} (e_j^* A_t X_{t-})[e_i^* Z_t (e_j - X_{t-})]^2.$$

Lemma 9 *Consider a process $Z \in \mathbb{R}^{1 \times N}$ solving a standard, scalar, $(K = 1)$, BSDE of the form of (3). Suppose that for a given t, Z_t is such that $\|Z_t\|_{X_{t-}} \neq 0$, and, for some ε*

$$0 < \varepsilon < \varepsilon_r^{3/2} N^{-3/2}$$

with ε_r as in (7), we know

$$-\varepsilon \|Z_t\|_{X_{t-}} \leq (e_j^* A_t X_{t-}) Z_t (e_j - X_{t-})$$

for all j. Then

$$Z_t \psi_t X_{t-} \leq -\varepsilon \|Z_t\|_{X_{t-}}.$$

4 Linear BSDEs

Theorem 2 (Linear BSDEs) *Let (α, β, γ) be a $du \times \mathbb{P}$-a.s. bounded $(\mathbb{R}^{K \times N}, \mathbb{R}^{K \times K}, \mathbb{R}^K)$ valued predictable process, ϕ a predictable $\mathbb{R}^K$ valued process with $E[\int_{]0,T]} \|\phi_t\|^2 dt] < +\infty$, T a deterministic terminal time and Q a square-integrable $\mathscr{F}_T$ measurable $\mathbb{R}^K$ valued random variable. Then the linear BSDE given by*

$$Y_t - \int_{]t,T]} [\phi_u + \beta_u Y_{u-} + \alpha_u Z_u^* \gamma_u] du + \int_{]t,T]} Z_u d M_u = Q \tag{9}$$

has a unique square integrable solution (Y, Z), (up to appropriate sets of measure zero). Furthermore, if for all $s \in [t, T]$

$$I + \alpha_s \psi_s^+ (e_j - X_{s-}) \gamma_s^* \tag{10}$$

is invertible for all j such that $e_j^* A_s X_{s-} > 0$, except possibly on some evanescent set, then Y is given by the explicit formula

$$Y_t = E\left[\Gamma_t^T Q + \int_{]t,T]} \Gamma_t^u \phi_u du \,\middle|\, \mathcal{F}_t \right] \tag{11}$$

up to indistinguishability. Here Γ_t^s is the adjoint process defined for $t \leq s \leq T$ by the forward linear SDE

$$\Gamma_t^s = I + \int_{]t,s]} \Gamma_t^{u-} [\beta_u du + \alpha_u \psi_u^+ dM_u \gamma_u^*]. \tag{12}$$

Before proving this, we establish the following results:

Lemma 10 *The adjoint process defined by (12) will satisfy, for $t \leq r \leq s$,*

$$\Gamma_t^r \Gamma_r^s = \Gamma_t^s.$$

Proof Assume without loss of generality $t < s < T$. Write

$$dV_u = \beta_u du + \alpha_u \psi_u^+ dM_u \gamma_u^*.$$

Then Γ is defined by the forward SDE

$$\Gamma_t^s = I + \int_{]t,s]} \Gamma_t^{u-} dV_u.$$

If H is $\mathcal{F}_t$ measurable in $\mathbb{R}^{N \times N}$, then

$$L = H\Gamma$$

is the unique solution to

$$L_t^s = H + \int_{]t,s]} L_t^{u-} dV_u.$$

Suppose $t \leq r$ and define

$$D_t^s = \begin{cases} \Gamma_t^s & \text{for } t \leq s \leq r \\ \Gamma_t^r \Gamma_r^s & \text{for } t \leq r \leq s. \end{cases}$$

Then

$$D_t^s = I + \int_{]t,s]} D_t^{u-} dV_u$$

for $t \leq s \leq r$, and

$$D_t^s = \Gamma_t^r + \int_{]r,s]} D_t^{u-} dV_u$$

$$= I + \int_{]t,s]} D_t^{u-} dV_u$$

for $t \le r \le s$. By uniqueness, for $t \le r \le s$, this implies $D_t^s = \Gamma_t^s = \Gamma_t^r \Gamma_r^s$ as desired.   □

In this matrix framework, product integrals are useful, see [17]. In particular, solutions of first-order matrix differential equations can be expressed as product integrals.

Lemma 11 *Consider a deterministic first-order differential equation of the form*

$$G_s^t = I + \int_{]s,t]} G_s^{u^-} H_u \, du. \tag{13}$$

Here G and H are $K \times K$ matrix-valued functions, H is bounded and Lebesgue integrable, and $s < t$. Then (13) has an invertible solution, which can be expressed as a product integral, denoted by

$$G_s^t = \Pi_{]s,t]}(I + H_u \, du).$$

If H_t has nonnegative entries off the main diagonal for dt-almost all t, then this solution has all entries nonnegative.

Proof As H is bounded and Lebesgue integrable, we can appeal to the theory of the product integral. Using this,

$$G_s^t = \Pi_{]s,t]}(I + H_u \, du) = \lim_{n \to \infty} \prod_{i=0}^{n-1} \left\{ I + \int_{](s_n)_i,(s_n)_{i+1}]} H_u \, du \right\}, \tag{14}$$

where, for each n, (s_n) is an arbitrary partition of $]s, t]$ into n parts,

$$s = (s_n)_0 < (s_n)_1 < (s_n)_2 < \cdots < (s_n)_n = t,$$

converging, as $n \to \infty$, in the sense that

$$\lim_{n \to \infty} \max_i |(s_n)_i - (s_n)_{i+1}| = 0,$$

$\prod$ indicates products taken sequentially on the right, and the limit is taken in the matrix norm $\mathrm{Tr}(G_s^t (G_s^t)^*)$. This is called the product integral of H, exists by [17, Thm 1], and solves the integral equation (13) by [17, Thm 5].

G_s^t has an inverse, given by

$$(G_s^t)^{-1} = \lim_{n \to \infty} \prod_{i=n-1}^{0} \left\{ I - \int_{](s_n)_i,(s_n)_{i+1}]} H_u \, du \right\}, \tag{15}$$

as noted in [16, p. 134].

Finally, if H has nonnegative entries off the main diagonal dt-a.e., as H is bounded, for sufficiently large n, $\int_{](s_n)_i,(s_n)_{i+1}]} H_u \, du$ has all diagonal entries greater than -1 for all i. Therefore,

$$I + \int_{](s_n)_i,(s_n)_{i+1}]} H_u \, du$$

is nonnegative for all i. Consequently, the product in (14) must be nonnegative. As the set of matrices with nonnegative components is closed, the limit $\Pi_{]s,t]}(I + H_u du)$ will also have all components nonnegative. $\square$

Lemma 12 *Let t be a time at which X jumps, that is, $\Delta X_t \neq 0$. Then*

$$\mathbb{P}(\Delta X_t = e_j - X_{t-} \text{ for some } j \text{ with } e_j^* A_t X_{t-} > 0) = 1.$$

Proof This Lemma is quite intuitive – the jump times of a Markov chain are at points where the rate of jumping is nonzero. A formal proof is given in [6]. $\square$

Lemma 13 *Suppose that for all $s \in]t, T]$,*

$$I + \alpha_s \psi_s^+ (e_j - X_{s-}) \gamma_s^*$$

is invertible for all j such that $e_j^* A_s X_{s-} > 0$, except possibly on some evanescent set. Then the adjoint process Γ_t^s defined by (12) is invertible (except possibly on this evanescent set).

Proof By the definition of M in (1), we can rewrite (12) as

$$\Gamma_t^s = I + \int_{]t,s]} \Gamma_t^u [\beta_u - \alpha_u \psi_u^+ A_u X_{u-} \gamma_u^*] du + \sum_{t < u \leq s} \Gamma_t^{(u-)} \alpha_u \psi_u^+ \Delta X_u \gamma_u^*. \quad (16)$$

When $\Delta X_u = 0$, (16) is of the form of a classical, deterministic, linear matrix differential equation, with nonsingular solution $\Gamma_{s_0}^s$ given by the analogue of (14). Therefore, if two consecutive jump times are s_0, s_1, for $s \in [s_0, s_1[$ we have

$$\Gamma_t^s = \Gamma_t^{s_0} \Pi_{]s_0,s]} \{ I + [\beta_u - \alpha_u \psi_u^+ A_u X_{u-} \gamma_u^*] du \} = \Gamma_t^{s_0} \Gamma_{s_0}^s$$

by Lemma 11. If $\Gamma_t^{s_0}$ is invertible, then

$$(\Gamma_t^s)^{-1} = (\Gamma_{s_0}^s)^{-1} (\Gamma_t^{s_0})^{-1} \quad (17)$$

exists.

At each jump time, (16) implies we have

$$\Delta \Gamma_t^s = \Gamma_t^{(s-)} \alpha_s \psi_s^+ \Delta X_s \gamma_s^*.$$

Hence

$$\Gamma_t^s = \Gamma_t^{(s-)} (I + \alpha_s \psi_s^+ \Delta X_s \gamma_s^*).$$

By, Lemma 12, at each jump time s,

$$\mathbb{P}(\Delta X_s = e_j - X_{s-} \text{ for some } j \text{ with } e_j^* A_s X_{s-} > 0) = 1.$$

By assumption, $I + \alpha_s \psi_s^+ (e_j - X_{s-}) \gamma_s^*$ is invertible for all such j (up to evanescence), and

$$(\Gamma_t^s)^{-1} = (I + \alpha_s \psi_s^+ \Delta X_s \gamma_s^*)^{-1} (\Gamma_t^{(s-)})^{-1}. \quad (18)$$

The process X almost surely has finitely many jumps in $[0, T]$. Through a process of induction using the starting value $\Gamma_t^t = I$ and equations (17) and (18), we can conclude Γ_t^s is invertible (up to evanescence). $\square$

Theorem 3 *Suppose* $\beta_u - \alpha_u \psi_u^+ A_u X_{u-} \gamma_u^*$ *has all nonnegative components off the main diagonal* $\mathbb{P} \times du$-*a.s., for* $u \in]t, T]$, *and, except possibly on some evanescent subset of* $\Omega \times]t, T]$,

$$I + \alpha_u \psi_u^+ (e_j - X_{u-}) \gamma_u^*$$

has nonnegative components for all j *such that* $e_j^* A_u X_{u-} > 0$. *Then the adjoint process* Γ_t^s *has all entries nonnegative for all* $s \in]t, T]$, *up to evanescence.*

Proof As above, if two consecutive jump times are s_0, s_1, for $s \in [s_0, s_1[$ we have from (16),

$$\Gamma_t^s = \Gamma_t^{s_0} \Pi_{]s_0, s]} \{ I + [\beta_u - \alpha_u \psi_u^+ A_u X_{u-} \gamma_u^*] du \}.$$

By assumption, $[\beta_u - \alpha_u \psi_u^+ A_u X_{u-} \gamma_u^*] du$ has nonnegative components off the main diagonal, and therefore by Lemma 11, $\Gamma_{s_0}^s$ has nonnegative components. The product of matrices with nonnegative components has nonnegative components, so if $\Gamma_t^{s_0}$ has nonnegative components, Γ_t^s has nonnegative components.

At each jump time, we have from (16) that $\Delta \Gamma_t^s = \Gamma_t^{(s-)} \alpha_s \psi_s^+ \Delta X_s \gamma_s^*$. This implies

$$\Gamma_t^s = \Gamma_t^{(s-)} (I + \alpha_s \psi_s^+ \Delta X_s \gamma_s^*)$$

and the term in parentheses has all nonnegative components by our assumption and the argument of Lemma 13 regarding the values of ΔX_s.

The process X almost surely has finitely many jumps in $[0, T]$, and therefore through a process of induction using the starting value $\Gamma_t^t = I$ we conclude that Γ_t^s has nonnegative entries up to evanescence. $\qquad \square$

We can then deduce

Corollary 1 *The conditions of Theorem 3 are necessary for* Γ_t^s *to have all entries nonnegative for all* s *and* t.

Proof See [6]. $\qquad \square$

A useful result when applying this Lemma is the following.

Lemma 14 *For a nonnegative column vector* $x \in \mathbb{R}^K$ *and any basis vector* $e_i \in \mathbb{R}^K$, *the matrix* $I + x e_i^*$ *has nonnegative components and is invertible.*

Lemma 15 *Under the conditions of Theorem 2,*

$$\sup_t E[\|\Gamma_0^t\|^2] < +\infty$$

and

$$\sup_t E\left[\left\| \Gamma_0^t Y_t + \int_{]0,t]} \Gamma_0^{u-} \phi_u du \right\|^2 \right] < +\infty.$$

Proof See [6].   □

We now proceed with the proof of Theorem 2.

Proof (Theorem 2) It is clear that (9) is of the form of (3), where

$$F(\omega, t, Y_{t-}, Z_t) = \phi_t + \beta_t Y_{t-} + \alpha_t Z_t \gamma_t^*$$

is a Lipschitz continuous, square integrable driver for the equation. The uniqueness of the solution follows from Theorem 1. We now search for the closed-form solution.
   The first required result is that

$$\Gamma_0^t Y_t + \int_{]0,t]} \Gamma_0^u \phi_u du$$

is a uniformly integrable martingale. To see this, observe that in this context all integrals are Stieltjes integrals, and hence

$$\Gamma_0^t Y_t = \Gamma_0^0 Y_0 + \int_{]0,t]} [\Gamma_0^{u-} dY_u + d\Gamma_0^u Y_{u-}] + \sum_{0<u\leq t} \Delta\Gamma_0^u \Delta Y_u$$

$$= \Gamma_0^0 Y_0 - \int_{]0,t]} \Gamma_0^{u-} [\phi_u + \beta_u Y_{u-} + \alpha_u Z_u^* \gamma_u] du + \int_{]0,t]} \Gamma_0^{u-} Z_u dM_u$$

$$+ \int_{]0,t]} \Gamma_t^{u-} \beta_u Y_{u-} du + \int_{]0,t]} \Gamma_t^{u-} \alpha_u \psi_u^+ dM_u \gamma_u^* Y_{u-}$$

$$+ \sum_{0<u\leq t} \Gamma_0^{u-} \alpha_u \psi_u^+ \Delta M_u \gamma_u^* Z_u \Delta M_u.$$

Then, as a 1×1 matrix is its own transpose,

$$\Gamma_0^{u-} \alpha_u \psi_u^+ \Delta M_u \gamma_u^* Z_u \Delta M_u = \Gamma_0^{u-} \alpha_u \psi_u^+ \Delta M_u \Delta M_u^* Z_u^* \gamma_u.$$

   The quantity $\Delta M_u \Delta M_u^*$ is equal to $d[M, M]_u$, the measure induced by the optional quadratic covariation matrix of M. $d\langle M, M\rangle_u$ is then the dual predictable projection of $d[M, M]_u$ (see [13]). Therefore,

$$\Gamma_0^t Y_t + \int_{]0,t]} \Gamma_0^{u-} \phi_u du = L_t - \int_{]0,t]} \Gamma_0^{u-} \alpha_u Z_u^* \gamma_u du$$

$$+ \int_{]0,t]} \Gamma_0^{u-} \alpha_u \psi_u^+ d\langle M, M\rangle_u Z_u^* \gamma_u$$

for some local martingale L. It follows that

$$\Gamma_0^t Y_t + \int_{]0,t]} \Gamma_0^{u-} \phi_u du = L_t + \int_{]0,t]} \Gamma_0^{u-} [\alpha_u \psi_u^+ \psi_u - \alpha_u] Z_u^* \gamma_u du.$$

and from Lemmas 3 and 5, $\psi_u^+ \psi_u Z^* = Z^*$ without loss of generality, so the latter of these terms is zero, and, therefore, the left side is a local martingale.

By Lemma 15, $\Gamma_0^t Y_t + \int_{]0,t]} \Gamma_0^u \phi_u du$ is square integrable, and hence is a uniformly integrable martingale [19, p. 12]. Hence it is indistinguishable from the conditional expectation of its terminal value [19, p. 11],

$$\Gamma_0^t Y_t + \int_{]0,t]} \Gamma_0^{u-} \phi_u du = E\left[\Gamma_0^T Q + \int_{]0,T]} \Gamma_0^u \phi_u du \Big| \mathscr{F}_t\right].$$

Through standard calculations and the use of Lemmas 10 and 13, we can conclude that, up to indistinguishability,

$$Y_t = E\left[\Gamma_t^T Q + \int_{]t,T]} \Gamma_t^u \phi_u du \Big| \mathscr{F}_t\right]. \qquad \square$$

Corollary 2 *If Q and ϕ are nonnegative, and the assumptions of Theorem 3 are satisfied, then Y is nonnegative. If, in addition $Y_0 = 0$, then, $Y = 0$ up to indistinguishability (and hence $Q = 0$ $\mathbb{P}$-a.s.) and $\phi_t = 0$ $\mathbb{P} \times dt$-a.s.*

Proof This follows from Theorem 2 and the nonnegativity result of Theorem 3. The strict comparison is trivial, given the invertibility of Γ_s^t for all s and t, and the fact that Y is càdlàg. $\qquad \square$

5 A Scalar Comparison Theorem

Comparison theorems play a key role when non-linear expectations are defined using BSDEs. We shall now establish a comparison theorem relating the solutions of two BSDEs in the scalar case.

Remark 1 In the scalar case, when $K = 1$, the assumptions of Theorem 3 simplify considerably. The assumption of nonnegativity of

$$\beta_t - \alpha_t \psi_t^+ A_t X_{t-} \gamma_t^*$$

off the main diagonal becomes trivial (as there are no off diagonal terms) and the assumption that

$$I + \alpha_s \psi_s^+ (e_j - X_{s-}) \gamma_s^*$$

is $\mathbb{P}$-a.s. invertible and nonnegative for all j such that $e_j^* A_s X_{s-} > 0$ can be simplified to

$$\alpha_s \psi_s^+ (e_j - X_{s-}) > -1, \tag{19}$$

as $\gamma = 1$, without loss of generality.

Theorem 4 (Scalar Comparison Theorem) *Suppose we have two standard, scalar ($K = 1$) BSDEs corresponding to coefficients and terminal values (F^1, Q^1) and (F^2, Q^2). Let (Y^1, Z^1) and (Y^2, Z^2) be the associated solutions. We suppose the following conditions hold:*

(i) $Q^1 \geq Q^2$ $\mathbb{P}$-a.s.

(ii) $dt \times \mathbb{P}$-a.s.,

$$F^1(\omega, t, Y_{t-}^2, Z_t^2) \geq F^2(\omega, t, Y_{t-}^2, Z_t^2).$$

(iii) There exists an $\varepsilon > 0$ such that $\mathbb{P}$-a.s., for all $t \in [0, T]$, if Z_t^1, Z_t^2 are such that

$$(e_j^* A_t X_{t-})[Z_t^1 - Z_t^2](e_j - X_{t-}) \geq -\varepsilon \|Z_t^1 - Z_t^2\|_{X_{t-}}$$

for all e_j, then

$$F^1(\omega, t, Y_{t-}^2, Z_t^1) - F^1(\omega, t, Y_{t-}^2, Z_t^2) \geq -[Z_t^1 - Z_t^2]A_t X_{t-},$$

with equality only if $\|Z_t^1 - Z_t^2\|_{X_{t-}} = 0$.

It is then true that $Y^1 \geq Y^2$ $\mathbb{P}$-a.s. Moreover, this comparison is strict, that is, if on some $A \in \mathscr{F}_t$ we have $Y_t^1 = Y_t^2$, then $Q^1 = Q^2$ $\mathbb{P}$-a.s. on A, $F^1(\omega, s, Y_s^2, Z_s^2) = F^2(\omega, s, Y_s^2, Z_s^2)$ $ds \times \mathbb{P}$-a.s. on $[t, T] \times A$ and Y^1 is indistinguishable from Y^2 on $[t, T] \times A$.

Remark 2 Note that Assumption (iii) need only hold for Z^1 and Z^2; there may well be other processes Z for which it fails. In practise, as Z^1 and Z^2 are typically a priori unknown, it may be more convenient to assume that this Assumption holds for all Z^1 and Z^2.

An intuitive interpretation of this assumption is: Consider the difference between the SDEs with starting value Y_{t-}^2, trend $F(\omega, t, Y_t, Z_t^i)$, and hedging processes Z_t^1 and Z_t^2. If the only sizeable jumps that can occur in this difference are positive, then the overall trend through time, excluding jumps, should be negative.

Proof (Theorem 4) We can write

$$Y_t^1 - Y_t^2 - \int_{]t,T]} [F^1(\omega, u, Y_{u-}^1, Z_u^1) - F^2(\omega, u, Y_{u-}^2, Z_u^2)]du$$

$$+ \int_{]t,T]} [Z_u^1 - Z_u^2]dM_u = Q_1 - Q_2. \tag{20}$$

Taking (20), the equation satisfied by $\delta Y := Y^1 - Y^2$ and $\delta Z := Z^1 - Z^2$, we shall form an equivalent linear BSDE for δY and apply Corollary 2 to prove the desired result. For notational simplicity, we shall omit the ω, t arguments of F^1, F^2 as implicit. We also define $0/0 := 0$ wherever needed.

Without loss of generality, we can assume

$$\varepsilon < \varepsilon_r^{3/2} N^{-3/2},$$

where ε_r is as in (7).

We consider three cases.

1. If $F^1(Y_{t-}^2, Z_t^1) - F^1(Y_{t-}^2, Z_t^2) \geq 0$, then let

$$\phi_t = F^1(Y_{t-}^2, Z_t^1) - F^2(Y_{t-}^2, Z_t^2),$$

$$\beta_t = \frac{F^1(Y_{t-}^1, Z_t^1) - F^1(Y_{t-}^2, Z_t^1)}{\delta Y_{t-}},$$

$$\alpha_t = 0,$$

$$\gamma_t = 1.$$

Note that our Assumption (*ii*) and the fact $F^1(Y_{t-}^2, Z_t^1) - F^1(Y_{t-}^2, Z_t^2) \geq 0$ implies that $\phi_t \geq 0$.

2. If $F^1(Y_{t-}^2, Z_t^1) - F^1(Y_{t-}^2, Z_t^2) < 0$ and there is a j such that

$$(e_j^* A_t X_{t-})[Z_t^1 - Z_t^2](e_j - X_{t-}) < -\varepsilon \|Z_t^1 - Z_t^2\|_{X_{t-}},$$

then let

$$\phi_t = F^1(Y_{t-}^2, Z_t^2) - F^2(Y_{t-}^2, Z_t^2),$$

$$\beta_t = \frac{F^1(Y_{t-}^1, Z_t^1) - F^1(Y_{t-}^2, Z_t^1)}{\delta Y_{t-}},$$

$$\alpha_t = \frac{F^1(Y_{t-}^2, Z_t^1) - F^1(Y_{t-}^2, Z_t^2)}{[Z_t^1 - Z_t^2]\psi_t e_j} \cdot e_j^* \psi_t,$$

$$\gamma_t = 1.$$

3. If $F^1(Y_{t-}^2, Z_t^1) - F^1(Y_{t-}^2, Z_t^2) < 0$ and

$$(e_j^* A_t X_{t-})[Z_t^1 - Z_t^2](e_j - X_{t-}) \geq -\varepsilon \|Z_t^1 - Z_t^2\|_{X_{t-}} \tag{21}$$

for all j, then let

$$\phi_t = F^1(Y_{t-}^2, Z_t^2) - F^2(Y_{t-}^2, Z_t^2),$$

$$\beta_t = \frac{F^1(Y_{t-}^1, Z_t^1) - F^1(Y_{t-}^2, Z_t^1)}{\delta Y_{t-}},$$

$$\alpha_t = \frac{F^1(Y_{t-}^2, Z_t^1) - F^1(Y_{t-}^2, Z_t^2)}{[Z_t^1 - Z_t^2]\psi_t X_{t-}} \cdot X_{t-}^* \psi_t,$$

$$\gamma_t = 1.$$

In all three cases, it is clear that

$$F^1(Y_{t-}^1, Z_t^1) - F^2(Y_{t-}^2, Z_t^2) = \phi_t + \beta_t(\delta Y_{t-}) + \alpha_t(\delta Z_t)^* \gamma_t,$$

and so the linear BSDE with these values of ϕ_t, β_t, α_t and γ_t is equivalent to (20).

Furthermore, $E[\int_{]0,T]} \|\phi_t\|^2 dt] < +\infty$ as F is standard, and in each case, β_t, α_t and γ_t are $dt \times \mathbb{P}$-a.s. bounded. This is trivial for γ_t, and follows directly from Lipschitz continuity for β_t in all cases. We need to show this holds for α_t.

In Case 1, $\alpha_t = 0$. In Case 2, we know that $\psi_t e_j = (e_j^* A_t X_{t-})[e_j - X_{t-}]$ by the definition of ψ in (4). By assumption, the absolute value of $[Z_t^1 - Z_t^2]\psi_t e_j$ is then at least $\varepsilon \|Z_t^1 - Z_t^2\|_{X_{t-}}$, and therefore α_t is bounded by Lipschitz continuity.

In Case 3, by Lemma 6 and the fact $F^1(Y_{t-}^2, Z_t^1) \neq F^1(Y_{t-}^2, Z_t^2)$, we know $\|Z_t^1 - Z_t^2\|_{X_{t-}} \neq 0$, and so we are in precisely the situation considered in Lemma 9.

Therefore $[Z_t^1 - Z_t^2]\psi_t X_{t-} < -\varepsilon\|Z_t^1 - Z_t^2\|_{X_{t-}}$, and so Lipschitz continuity implies the boundedness of α_t in Case 3.

We have assumed that F^1 and F^2 are standard, and hence $\mathbb{P}$-a.s. left continuous in t, as are Y_{t-}^i and Z_t^i for $i = 1, 2$ by construction, we see that the variable C_t indicating which of Cases 1, 2 and 3 is in force at any time t is predictable. Furthermore, the processes α, β and ϕ defined in each case are also predictable. We then use C_t to piece together the various definitions of α β and ϕ to give a single predictable linear BSDE, with the same driver values as $F^1(Y_{t-}^1, Z_t^1) - F^2(Y_{t-}^2, Z_t^2) \, dt \times \mathbb{P}$-a.s. This linear BSDE satisfies all the requirements for Theorem 2.

We now appeal to Remark 1 to determine that the only requirement to show nonnegativity of δY is that, for all t, $\alpha_t \psi_t^+(e_k - X_{t-}) > -1$ for all k with $e_k^* A_{t-} X_{t-} > 0$.

In Case 1 this is clear. In Case 2, we can write

$$\alpha_t \psi_t^+(e_k - X_{t-}) = \frac{F^1(Y_{t-}^2, Z_t^1) - F^1(Y_{t-}^2, Z_t^2)}{e_j^* \psi_t [Z_t^1 - Z_t^2]^*} e_j^* \psi_t \psi_t^+(e_k - X_{t-})$$

and, by Lemma 4, $e_j^* \psi_t \psi^+(e_k - X_{t-}) = e_j^*(e_k - X_{t-})$, which is zero unless $k = j$. If $k = j$ we see that

$$\alpha_t \psi_t^+(e_k - X_{t-}) = \frac{F^1(Y_{t-}^2, Z_t^1) - F^1(Y_{t-}^2, Z_t^2)}{e_j^* \psi_t [Z_t^1 - Z_t^2]^*} > 0$$

by construction.

In Case 3, we know from Lemma 9 that $[Z_t^1 - Z_t^2]\psi_t X_{t-}$ is negative. For any k satisfying Lemma 4, this implies

$$\frac{X_{t-}^* \psi_t \psi_t^+(e_k - X_{t-})}{X_{t-}^* \psi_t [Z_t^1 - Z_t^2]^*} = \frac{X_{t-}^*(e_k - X_{t-})}{X_{t-}^* \psi_t [Z_t^1 - Z_t^2]^*}$$

is positive. We then use Assumption (iii) of the theorem, along with Lemma 7, to show that,

$$\alpha_t \psi^+(e_k - X_{t-})$$
$$= \frac{F^1(\omega, t, Y_{t-}^2, Z_t^1) - F^1(\omega, t, Y_{t-}^2, Z_t^2)}{[Z_t^1 - Z_t^2]\psi_t X_{t-}} X_{t-}^*(e_k - X_{t-})$$
$$> -\frac{[Z_t^1 - Z_t^2]A_t X_{t-}}{[Z_t^1 - Z_t^2]\psi_t X_{t-}} X_{t-}^*(e_k - X_{t-})$$
$$= -\frac{[Z_t^1 - Z_t^2]\psi_t X_{t-}}{[Z_t^1 - Z_t^2]\psi_t X_{t-}}$$
$$= -1,$$

as desired.

Therefore, we have shown that $\delta Y = Y^1 - Y^2$, the difference of our processes, satisfies the requirements of Remark 1. That is, the assumptions of Corollary 2 are satisfied by this process and δY is therefore nonnegative $\mathbb{P}$-a.s. The rest of the theorem also follows by Corollary 2. □

Remark 3 In general Assumption (*iii*) cannot be omitted, and is closely related to various geometric interpretations of no arbitrage. The fact that it is possible to create such Z^1 and Z^2, (due to the linear redundancy in M), indicates a significant difference between the Markov chain theory considered here and that based on Brownian motion considered elsewhere.

Examples can be given that show the comparison theorem does not hold when any of the assumptions of Theorem 4 are violated. See [6].

6 General Comparison Theorems

We now wish to extend Theorem 4 to the vector case. This is nontrivial, as the simplifications of Remark 1 are not possible, and we must satisfy the more difficult conditions of Corollary 2 directly. We also do not have the useful result of Lemma 9. We present here, without proof, some alternative generalisations. (Proofs are given in [6].)

Theorem 5 (Vector Comparison Theorem 1) *Suppose we have two standard BSDE parameters* (F^1, Q^1) *and* (F^2, Q^2). *Let* (Y^1, Z^1) *and* (Y^2, Z^2) *be the associated solutions.*

Recall the dimensions of each of the terms in our BSDEs: $F(\omega, t, Y_{t-}, Z_t)$, Y_{t-}, $Q \in \mathbb{R}^K$ *and* $Z_t \in \mathbb{R}^{K \times N}$, *where* N *is the number of states of the Markov chain* X.

We suppose the following conditions hold:

(i) $Q^1 \geq Q^2$ $\mathbb{P}$-*a.s.*

(ii) $dt \times \mathbb{P}$-*a.s.*,

$$F^1(\omega, t, Y^2_{t-}, Z^2_t) \geq F^2(\omega, t, Y^2_{t-}, Z^2_t).$$

(iii) *There exists an* $\varepsilon > 0$ *such that* $\mathbb{P}$-*a.s., for all* $t \in [0, T]$, *for any basis vector* $e_k \in \mathbb{R}^K$, *if* Z^1_t, Z^2_t *are such that*

$$(e_j^* A_t X_{t-}) e_k^* [Z^1_t - Z^2_t](e_j - X_{t-}) \geq -\varepsilon \|e_k^*[Z^1_t - Z^2_t]\| x_{t-}$$

for all e_j, *then*

$$e_k^*[F^1(\omega, t, Y^2_{t-}, Z^1_t) - F^1(\omega, t, Y^2_{t-}, Z^2_t)] \geq -e_k^*[Z^1_t - Z^2_t]A_t X_{t-}$$

with equality only if $\|e_k^*[Z^1_t - Z^2_t]\| x_{t-} = 0$.

(iv) *For each* i, $e_i^* F^1$ *can be written as a function*

$$e_i^* F^1(\omega, t, Y_{t-}, Z_t) = F_i^1(\omega, t, e_i^* Y_{t-}, e_i^* Z_t),$$

that is, the ith *component of* F^1 *is depends only on the* ith *component of* Y *and the* ith *row of* Z *(and* ω *and* t).

It is then true that $Y^1 \geq Y^2$ $\mathbb{P}$-*a.s. Moreover, this comparison is strict, that is, if on some* $A \in \mathscr{F}_t$ *we have* $Y^1_t = Y^2_t$, *then* $Q^1 = Q^2$ $\mathbb{P}$-*a.s. on* A, $F^1(\omega, s, Y^2_s, Z^2_s) = F^2(\omega, s, Y^2_s, Z^2_s) \, ds \times \mathbb{P}$-*a.s. on* $[t, T] \times A$ *and* Y^1 *is indistinguishable from* Y^2 *on* $[t, T] \times A$.

Corollary 3 *For any pair of BSDEs satisfying Assumption (iv) of Theorem 5, if Assumptions (i)–(iii) are satisfied by any component of the terminal condition and driver, then the comparison theorem will hold on that component.*

It is clear that this theorem does not provide for much more behaviour than the scalar case, as it is considering each component separately from the others. Removing this requirement is nontrivial, but can be done. To do so using the machinery of the solutions to linear BSDEs, we must make the following, in some sense stronger, assumption.

Theorem 6 (Vector Comparison Theorem 2) *Suppose we have two standard BSDE parameters (F^1, Q^1) and (F^2, Q^2). Let (Y^1, Z^1) and (Y^2, Z^2) be the associated solutions, where the dimensions of F, Q, Y and Z are as in Theorem 5. We suppose the following conditions hold:*

(i) $Q^1 \geq Q^2$ $\mathbb{P}$-a.s.
(ii) $dt \times \mathbb{P}$-a.s.,

$$F^1(\omega, t, Y_{t-}^2, Z_t^2) \geq F^2(\omega, t, Y_{t-}^2, Z_t^2).$$

(iii) *There exists an $\varepsilon > 0$ such that, $\mathbb{P}$-a.s., for all $t \in [0, T]$, for any basis vector $e_j \in \mathbb{R}^K$, if*

$$(e_k^* A_t X_{t-}) e_j^* [Z_t^1 - Z_t^2](e_k - X_{t-}) \geq -\varepsilon \| Z_t^1 - Z_t^2 \| X_{t-}$$

for all $e_k \neq X_{t-}$, then

$$e_j^* [F^1(\omega, t, Y_{t-}^2, Z_t^1) - F^1(\omega, t, Y_{t-}^2, Z_t^2)] \geq 0.$$

(iv) *For each i, $e_i^* F^1$ can be written as a function*

$$e_i^* F^1(\omega, t, Y_{t-}, Z_t) = F_i^1(\omega, t, e_i^* Y_{t-}, Z_t),$$

that is, the ith component of F^1 depends only on the ith component of Y (and on ω, t and Z_t).

It is then true that $Y^1 \geq Y^2$ $\mathbb{P}$-a.s. Moreover, this comparison is strict, that is, if on some $A \in \mathscr{F}_t$ we have $Y_t^1 = Y_t^2$, then $Q^1 = Q^2$ $\mathbb{P}$-a.s. on A, $F^1(\omega, s, Y_s^2, Z_s^2) = F^2(\omega, s, Y_s^2, Z_s^2) ds \times \mathbb{P}$-a.s. on $[t, T] \times A$ and Y^1 is indistinguishable from Y^2 on $[t, T] \times A$.

Remark 4 The key differences between Theorem 6 and Theorem 5 are that in Theorem 6, a weaker assumption is placed on the behaviour of the driver in relation to interactions between the rows of the Z matrix, but a stronger assumption is placed on the behaviour of a component of F when no jump is significantly negative in that component.

Assumption (iii) in Theorem 6 is a stronger assumption than for Theorem 5. In the scalar case, which is the foundation of Theorem 5, we used Lemma 9 to show that each row $e_i^* [Z_t^1 - Z_t^2] \psi_t X_{t-}$ is negative, and our assumption was equivalent to

$$e_k^* [F^1(\omega, t, Y_{t-}^2, Z_t^1) - F^1(\omega, t, Y_{t-}^2, Z_t^2)] \geq e_k^* [Z_t^1 - Z_t^2] \psi_t X_{t-},$$

which is clearly satisfied when

$$e_k^*[F^1(\omega, t, Y_{t-}^2, Z_t^1) - F^1(\omega, t, Y_{t-}^2, Z_t^2)] \geq 0.$$

The assumption that the ith component of F can depend only on the ith component of Y may be overly restrictive. Because of this, we have the following alternative generalisation, where we instead assume that F does not depend on Z.

Theorem 7 (Vector Comparison Theorem 3) *Suppose we have two standard BSDE parameters (F^1, Q^1) and (F^2, Q^2). Let (Y^1, Z^1) and (Y^2, Z^2) be the associated solutions. We suppose the following conditions hold:*

(i) $Q^1 \geq Q^2$ $\mathbb{P}$-a.s.
(ii) $dt \times \mathbb{P}$-a.s.,
$$F^1(\omega, t, Y_{t-}^2, Z_t^2) \geq F^2(\omega, t, Y_{t-}^2, Z_t^2).$$
(iii) *There exists an $\varepsilon > 0$ such that, $dt \times \mathbb{P}$-a.s., for each i, if*
$$e_i^* F^1(\omega, t, Y_{t-}^1, Z_t) < e_i^* F^1(\omega, t, Y_{t-}^2, Z_t)$$
then either
$$|e_i^*[Y_{t-}^1 - Y_{t-}^2]| > \varepsilon \|Y_{t-}^1 - Y_{t-}^2\|$$
or there is a j with
$$e_j^*[Y_{t-}^1 - Y_{t-}^2] < -\varepsilon \|Y_{t-}^1 - Y_{t-}^2\|.$$
(iv) F^1 *does not depend on Z.*

It is then true that $Y^1 \geq Y^2$ $\mathbb{P}$-a.s. Moreover, this comparison is strict, that is, if on some $A \in \mathscr{F}_t$ we have $Y_t^1 = Y_t^2$, then $Q^1 = Q^2$ $\mathbb{P}$-a.s. on A, $F^1(\omega, s, Y_s^2, Z_s^2) = F^2(\omega, s, Y_s^2, Z_s^2)$ $ds \times \mathbb{P}$-a.s. on $[t, T] \times A$ and Y^1 is indistinguishable from Y^2 on $[t, T] \times A$.

This theorem is counterintuitive, and when examined closely would appear to create a contradiction. The only resolution to this is the following corollary.

Corollary 4 *For a pair of BSDEs satisfying Theorem 7, the strict comparison must hold componentwise. That is, if for some t and some e_i, on some $A \in \mathscr{F}_t$ we have $e_i^* Y_t^1 = e_i^* Y_t^2$, then $e_i^* Q^1 = e_i^* Q^2$ $\mathbb{P}$-a.s. on A, $e_i^* F^1(\omega, s, Y_s^2, Z_s^2) = e_i^* F^2(\omega, s, Y_s^2, Z_s^2)$ $ds \times \mathbb{P}$-a.s. on $[t, T] \times A$ and $e_i^* Y^1$ is indistinguishable from $e_i^* Y^2$ on $[t, T] \times A$.*

7 F-Expectations

One interpretation of the solution to a BSDE is as a type of generalised expectation. In particular, for a fixed standard driver F and $Q \in \mathbb{R}^K$ an $\mathscr{F}_t$ measurable, square integrable random variable, we can define the conditional F-evaluation to be

$$\mathscr{E}_{s,t}^F(Q) = Y_s \tag{22}$$

for $s \leq t$, where Y_s is the solution to

$$Y_s - \int_{]s,t]} F(\omega, u, Y_{u-}, Z_u)du + \int_{]s,t]} Z_u dM_u = Q.$$

Definition 1 Following [22], we shall call a system of operators

$$\mathscr{E}_{s,t} : L^2(\mathscr{F}_t) \to L^2(\mathscr{F}_s), 0 \leq s \leq t \leq T$$

an $\mathscr{F}_t$-consistent **nonlinear evaluation** for $\{\mathscr{Q}_{s,t} \subset L^2(\mathscr{F}_t)|0 \leq s \leq t \leq T\}$ defined on $[0, T]$ if it satisfies the following properties.

1. For $Q, Q' \in \mathscr{Q}_{s,t}$,

$$\mathscr{E}_{s,t}(Q) \geq \mathscr{E}_{s,t}(Q')$$

   $\mathbb{P}$-a.s. componentwise whenever $Q \geq Q'$ $\mathbb{P}$-a.s. componentwise, with equality iff $Q = Q'$ $\mathbb{P}$-a.s.
2. $\mathscr{E}_{t,t}(Q) = Q$ $\mathbb{P}$-a.s.
3. $\mathscr{E}_{r,s}(\mathscr{E}_{s,t}(Q)) = \mathscr{E}_{r,t}(Q)$ $\mathbb{P}$-a.s. for any $r \leq s \leq t$.
4. For any $A \in \mathscr{F}_s$, $I_A \mathscr{E}_{s,t}(Q) = I_A \mathscr{E}_{s,t}(I_A Q)$ $\mathbb{P}$-a.s.

We consider here a significant generalisation of [22], as our evaluations can all be vector valued, provided they are square integrable. This generalisation allows the use of these evaluations in multi-objective problems, where a scalar evaluation process may be insufficient.

We also only require these properties to hold on some subset $\mathscr{Q}_{s,t}$ of the set of square integrable terminal conditions, a distinction which shall become important when we consider arbitrage and finite market modelling. Note however that $\mathscr{E}_{s,t}$ is defined over the whole of $L^2(\mathscr{F}_t)$, but it is only on $\mathscr{Q}_s^t$ that Property 1 holds. Furthermore, we add the restriction that if $Q \geq Q'$ $\mathbb{P}$-a.s., $Q, Q' \in \mathscr{Q}_t$, then $\mathscr{E}_{s,t}(Q) = \mathscr{E}_{s,t}(Q')$ $\mathbb{P}$-a.s. iff $Q = Q'$ $\mathbb{P}$-a.s.

Remark 5 If $\mathscr{E}_{s,t}$ is an $\mathscr{F}_t$-consistent nonlinear evaluation for $\mathscr{Q}_{s,t}$, it is also an $\mathscr{F}_t$-consistent nonlinear evaluation for any subset of $\mathscr{Q}_{s,t}$.

Definition 2 We often wish for the sets $\{Q_{s,t}\}$ to be stable through time. That is, $\mathscr{E}$ will be called **dynamically monotone** for $\{\mathscr{Q}_{s,t}\}$ if and only if, for all $r \leq s \leq t$,

   (i) $\mathscr{E}$ is an $\mathscr{F}_t$ consistent nonlinear evaluation for $\{\mathscr{Q}_{s,t}\}$,
   (ii) $\mathscr{Q}_{r,t} \subseteq \mathscr{Q}_{s,t}$, ($\mathscr{Q}_{s,t}$ is **nondecreasing** in s),
   (iii) $\mathscr{E}_{s,t}(Q) \in \mathscr{Q}_{r,s}$ for all $Q \in \mathscr{Q}_{r,t}$.

Remark 6 In an economic context, where $\mathscr{E}_{s,t}$ represents the price at s of an asset to be sold at time t, dynamic monotonicity corresponds, in some sense, to the statements

   (ii) "An asset with an arbitrage free price, when bought at a time r and sold at time t, should also have an arbitrage free price when bought at any time s following r, (with s prior to t)," and

(iii) "An asset with an arbitrage free price, when the asset is bought at a time r and sold at time t, should be arbitrage free when sold at any time s prior to t, (with s following r)."

Theorem 8 *Fix a driver F. Consider a collection of sets $\{\mathcal{Q}_{s,t} \subset L^2(\mathcal{F}_t)\}$ with $\mathcal{Q}_{r,t} \subseteq \mathcal{Q}_{s,t}$ for all $r \leq s \leq t$. Suppose that, for any $Q^1, Q^2 \in \mathcal{Q}_{s,t}$, at least one of Theorems 4, 5, 6 and 7 holds on $[s, t]$, with $F^1 = F^2 = F$, whenever $Q^1 \geq Q^2$ $\mathbb{P}$-a.s. Then $\mathcal{E}_{s,t}^F$ defined in (22) is an $\mathcal{F}_t$-consistent nonlinear evaluation for $\{\mathcal{Q}_{s,t}\}$.*

Proof

1. The statement $\mathcal{E}_{s,t}^F(Q_1) \geq \mathcal{E}_{s,t}^F(Q_2)$ $\mathbb{P}$-a.s. whenever $Q_1 \geq Q_2$ $\mathbb{P}$-a.s. is simply the main result of each comparison theorem, one of which holds by assumption. The strict comparison then establishes the second statement.
2. The fact $\mathcal{E}_{t,t}^F(Q) = Q$, $\mathbb{P}$-a.s. for any $\mathcal{F}_t$ measurable Q is trivial, as we have defined $\mathcal{E}_{t,t}^F(Q)$ by the solution to a BSDE, which reaches its terminal value Q at time t by construction.
3. For any $r \leq s \leq t$, let Y denote the solution to the relevant BSDE. Then we have

$$Q = Y_r - \int_{]r,t]} F(\omega, u, Y_{u-}, Z_u)du + \int_{]r,t]} Z_u dM_u$$

which implies

$$Y_s = Y_r - \int_{]r,s]} F(\omega, u, Y_{u-}, Z_u)du + \int_{]r,s]} Z_u dM_u.$$

Hence Y_r is also the time r value of a solution to the BSDE with terminal time s and value Y_s. Hence

$$\mathcal{E}_{r,s}^F(\mathcal{E}_{s,t}^F(Q)) = \mathcal{E}_{r,t}^F(Q)$$

$\mathbb{P}$-a.s. as desired.
4. We wish to show that for $A \in \mathcal{F}_s$, $I_A \mathcal{E}_{s,t}(Q) = I_A \mathcal{E}_{s,t}(I_A Q)$ $\mathbb{P}$-a.s. Write $\mathcal{E}_{s,t}(Q)$ as the solution to the BSDE with terminal value Q, and $\mathcal{E}_{s,t}(I_A Q)$ as the solution to the BSDE with terminal value $I_A Q$. Premultiplying these BSDEs by I_A gives two BSDEs, both with terminal value $I_A Q$, driver $\tilde{F}(\omega, t, Y_t, Z_t) = I_A F(\omega, t, Y_t, Z_t) = I_A F(\omega, t, I_A Y_t, I_A Z_t)$, and solutions $I_A \mathcal{E}_{s,t}(Q)$ and $I_A \mathcal{E}_{s,t}(I_A Q)$ respectively. From Theorem 1, the solution to this BSDE is unique, hence $I_A \mathcal{E}_{s,t}(Q) = I_A \mathcal{E}_{s,t}(I_A Q)$ up to indistinguishability. $\qquad\square$

Definition 3 Let $\{\mathcal{Q}_{s,t} \subset L^2(\mathcal{F}_t)\}$ be a family of sets which are nondecreasing in s. For a BSDE driver F^1, suppose that, for all $s \leq t \leq T$, Assumption (*iii*), (and Assumption (*iv*), when applicable), of at least one of Theorems 4, 5, 6 and 7 hold on $]s, t]$ for all $Q^1, Q^2 \in \mathcal{Q}_{s,t}$, (whether or not Assumption (*i*) holds). Then F^1 is said to be a **balanced** driver on $\{\mathcal{Q}_{s,t}\}$.

Remark 7 The logic of this name is due to the geometry of the problem, as, in some sense, F^1 here 'balances' the outcomes with zero hedging within the range of outcomes with hedging.

Lemma 16 *Let F be a balanced driver on* $\{\mathcal{Q}_{s,t}\}$, *where* $\mathcal{Q}_{s,t}$ *is nondecreasing in* s. *Then* $\mathcal{E}^F$ *is an* $\mathcal{F}_t$ *consistent nonlinear evaluation on* $\{\mathcal{Q}_{s,t}\}$.

Proof The requirements for F to be balanced are stronger than those needed for Theorem 8, and so the result follows. □

Definition 4 Again following [22], we shall call a system of operators

$$\mathcal{E}(.|\mathcal{F}_t) : L^2(\mathcal{F}_T) \to L^2(\mathcal{F}_t), 0 \le t \le T$$

an $\mathcal{F}_t$-consistent **nonlinear expectation** for $\{\mathcal{Q}_t \subset L^2(\mathcal{F}_T)\}$ defined on $[0, T]$ if it satisfies the following properties.

1. For $Q, Q' \in \mathcal{Q}_t$,

$$\mathcal{E}(Q|\mathcal{F}_t) \ge \mathcal{E}(Q'|\mathcal{F}_t)$$

   $\mathbb{P}$-a.s. componentwise whenever $Q \ge Q'$ $\mathbb{P}$-a.s. componentwise, with equality iff $Q = Q'$ $\mathbb{P}$-a.s.
2. $\mathcal{E}(Q|\mathcal{F}_t) = Q$ $\mathbb{P}$-a.s. for any $\mathcal{F}_t$ measurable Q.
3. $\mathcal{E}(\mathcal{E}(Q|\mathcal{F}_t)|\mathcal{F}_s) = \mathcal{E}(Q|\mathcal{F}_s)$ $\mathbb{P}$-a.s. for any $s \le t$.
4. For any $A \in \mathcal{F}_t$, $I_A \mathcal{E}(Q|\mathcal{F}_t) = \mathcal{E}(I_A Q|\mathcal{F}_t)$ $\mathbb{P}$-a.s.

Remark 8 It is clear that any nonlinear expectation is also a nonlinear evaluation, with $\mathcal{E}_{s,t}(\cdot) = \mathcal{E}(\cdot|\mathcal{F}_s)$ for all $s \le t$, and hence the concept of dynamic monotonicity extends to this new setting. On the other hand, as the terminal time is irrelevant in this context, one can simply specify the sets $\mathcal{Q}_t = \mathcal{Q}_t^T$, as the expectation refers without loss of generality to the terminal values at time T. Therefore, we say $\mathcal{E}(\cdot|\mathcal{F}_t)$ is **dynamically monotone** for $\{\mathcal{Q}_t\}$ if, for all $s \le t$,

   (i) $\mathcal{E}$ is an $\mathcal{F}_t$ consistent nonlinear expectation for $\{\mathcal{Q}_t\}$,
   (ii) $\mathcal{Q}_s \subseteq \mathcal{Q}_t$, ($\mathcal{Q}_t$ is **nondecreasing** in t),
   (iii) $\mathcal{E}(Q|\mathcal{F}_t) \in \mathcal{Q}_s$ for all $Q \in \mathcal{Q}_s$.

Theorem 9 *Fix a driver F such that* $F(\omega, t, Y_{t-}, 0) = 0$ $dt \times \mathbb{P}$-a.s. *Consider a family of sets* $\{\mathcal{Q}_t \subset L^2(\mathcal{F}_T)\}$ *such that for any* $Q, Q' \in \mathcal{Q}_t$ *with* $Q \ge Q'$, *at least one of Theorems 4, 5, 6 and 7 holds on* $]t, T]$ *with* $F^1 = F^2 = F$. *The functional* $\mathcal{E}^F(.|\mathcal{F}_t)$ *defined for each t by*

$$\mathcal{E}^F(Q|\mathcal{F}_t) = Y_t, \tag{23}$$

where Y_t *is the solution to*

$$Y_t - \int_{]t,T]} F(\omega, u, Y_{u-}, Z_u) du + \int_{]t,T]} Z_u dM_u = Q,$$

is an $\mathcal{F}_t$-consistent nonlinear expectation for $\{\mathcal{Q}_t\}$.

Proof Properties 1 and 3 follow exactly as in the proof of Theorem 8.

Property 2 follows because we know that for $t < T$, if Q is $\mathscr{F}_t$ measurable then the solution of

$$Y_t - \int_{]t,T]} F(\omega, u, Y_{u-}, Z_u) du + \int_{]t,T]} Z_u dM_u = Q$$

has $Z_u = 0$ $d\langle M, M \rangle_u \times \mathbb{P}$-a.s. (Simply take an $\mathscr{F}_t$ conditional expectation as done earlier.) Hence for $t \leq u \leq T$, $F(\omega, u, Y_{u-}, Z_u) = 0$, and therefore $Y_t = Q$, $\mathbb{P}$-a.s., as desired.

For Property 3, we know that $I_A \mathscr{E}(Q|\mathscr{F}_t)$ is the solution to a BSDE with terminal value $I_A Q$ and driver $I_A F(\omega, t, Y, Z)$. As $F(\omega, t, Y_{t-}, 0) = 0$, $I_A F(\omega, t, Y, Z) = F(\omega, t, Y, I_A Z)$, $\mathbb{P}$-a.s. We also know $\mathscr{E}(I_A Q|\mathscr{F}_t)$ is the solution to a BSDE with driver F, and taking an $\mathscr{F}_t$ conditional expectation shows that the solution Z process for $\mathscr{E}(I_A Q|\mathscr{F}_t)$ satisfies $Z = I_A Z$. Hence the two quantities solve the same BSDE, and so by the uniqueness of Theorem 1 must be equal. $\qquad\square$

8 Applications to Risk Measures

One use of BSDEs which has developed recently is as a framework for developing dynamically consistent, convex or coherent risk measures. This can be seen in [3], and more generally in [23]. We here present the key results in the context of BSDEs driven by Markov Chains.

As noted in [22], the theory of nonlinear expectations provides an ideal setting for the discussion of risk measures in the sense of [1] and others. We shall consider dynamic risk measures ϱ_t^F of the form

$$\varrho_t^F(Q) := -\mathscr{E}_{t,T}^F(Q)$$

where $\mathscr{E}^F$ is a nonlinear F-evaluation for some $\{\mathscr{Q}_{s,T} \subset L^2(\mathscr{F}_T)\}$. As before, our approach allows a considerable generalisation of earlier results, as the quantities considered can be vector valued.

An alternative specification, used in [3] and [23], is to let $\varrho_t^F(Q) = \mathscr{E}_{t,T}^F(-Q)$, the solution to a BSDE with terminal condition $-Q$. Under this alternative specification, the following results remain valid, however the result of Theorem 13 holds for ϱ with F convex rather than concave.

We seek to determine properties of ϱ_t^F, or equivalently of $\mathscr{E}_{t,T}^F$, which are particularly relevant in the context of risk measures.

The first four properties below follow directly from the properties of nonlinear evaluations, recalling the proofs of Theorems 8 and 9, with $\mathscr{Q}_t = \emptyset$ where appropriate.

Lemma 17 (Terminal Equality) *For any F, ϱ_t^F satisfies the terminal condition $\varrho_T^F(Q) = -Q$.*

Proof Equivalent to Property 2 of nonlinear evaluations. $\qquad\square$

Theorem 10 (Dynamic Consistency) *For any* F, ϱ_t^F *satisfies the recursivity and dynamic consistency requirements*

$$\varrho_s^F(Q) = \varrho_s^F(-\varrho_t^F(Q))$$

and

$$\varrho_t^F(Q) = \varrho_t^F(Q') \Rightarrow \varrho_s^F(Q) = \varrho_s^F(Q')$$

for times $s \leq t \leq T$, *all equalities being* $\mathbb{P}$-*a.s.*

Proof The first statement is equivalent to Property 3 of nonlinear evaluations. The second follows by uniqueness of solutions to BSDEs with given terminal conditions, as in Theorem 1. $\square$

Lemma 18 (Constants) *Let* Q *be* $\mathbb{P}$-*a.s. equal to a* $\mathscr{F}_t$ *measurable terminal condition. Then if* F *is a driver with normalisation* $F(\omega, t, Q, 0) = 0$ $dt \times \mathbb{P}$-*a.s.*, $\varrho_t^F(Q) = -Q$ *for all* t.

Proof Equivalent to Property 4 of nonlinear expectations. $\square$

Theorem 11 (Monotonicity) *Let* F *be a balanced driver for* $\{\mathcal{Q}_{t,T}\}$. *Then for any* $Q^1, Q^2 \in \mathcal{Q}_{t,T}$ *with* $Q^1 \geq Q^2$ $\mathbb{P}$-*a.s., we have* $\varrho_t^F(Q^1) \leq \varrho_t^F(Q^2)$, *with equality if and only if* $Q^1 = Q^2$, $\mathbb{P}$-*a.s.*

Proof Equivalent to Property 1 of nonlinear evaluations. $\square$

Lemma 19 (Translation invariance) *Let* F *be a driver for a BSDE with normalisation* $F(\omega, u, Y_{u-}, 0) = 0$ $\mathbb{P} \times du$-*a.s. on* $]t, T]$. *Suppose* F *does not depend on* Y_{s-}. *Then for any* $q \in L^2(\mathscr{F}_t)$,

$$\varrho_t^F(Q + q) = \varrho_t^F(Q) - q$$
$$\mathscr{E}_{t,T}^F(Q + q) = \mathscr{E}_{t,T}^F(Q) + q$$

Proof As F does not depend on Y_{u-}, we know that

$$Y_t - \int_{]t,T]} F(\omega, u, Y_{u-}, Z_u) du + \int_{]t,T]} Z_u dM_u = Q$$

implies

$$[Y_t + q] - \int_{]t,T]} F(\omega, u, [Y_{u-} + q], Z_u) du + \int_{]t,T]} Z_u dM_u = [Q + q].$$

The result follows. $\square$

Theorem 12 (Positive Homogeneity) *Let* F *be a positively homogeneous driver, that is, for any* $\lambda > 0$, *if* (Y, Z) *is the solution corresponding to some* Q, $\mathbb{P} \times du$-*a.s. on* $]t, T]$,

$$F(\omega, u, \lambda Y_{u-}, \lambda Z_u) = \lambda F(\omega, u, Y_{u-}, Z_u).$$

Then for all such Q, $\varrho_t^F(\lambda Q) = \lambda \varrho_t^F(Q)$ and $\mathscr{E}_{t,T}^F(\lambda Q) = \lambda \mathscr{E}_{t,T}^F(Q)$.

Proof Simply multiply the BSDE with terminal condition Q through by λ.   □

Theorem 13 (Convexity/Concavity) *Suppose $\mathscr{Q}_{t,T}$ is a convex set. Let F be a concave balanced driver for $\mathscr{Q}_{t,T}$, that is, for any $\lambda \in [0,1]$, any (Y^1, Z^1), (Y^2, Z^2) the solutions corresponding to $Q^1, Q^2 \in \mathscr{Q}_{t,T}$, $\mathbb{P} \times du$-a.s. on $]t, T]$,*

$$F(\omega, u, \lambda Y_{u-}^1 + (1-\lambda)Y_{u-}^2, \lambda Z_u^1 + (1-\lambda)Z_u^2)$$
$$\geq \lambda F(\omega, u, Y_{u-}^1, Z_u) + (1-\lambda)F(\omega, u, Y_{u-}^2, Z_u^2).$$

Then for any $\lambda \in [0,1]$ and any $Q^1, Q^2 \in \mathscr{Q}_{t,T}$,

$$\varrho_t^F(\lambda Q^1 + (1-\lambda)Q^2) \leq \lambda \varrho_t^F(Q^1) + (1-\lambda)\varrho_t^F(Q^2),$$
$$\mathscr{E}_{t,T}^F(\lambda Q^1 + (1-\lambda)Q^2) \geq \lambda \mathscr{E}_{t,T}^F(Q^1) + (1-\lambda)\mathscr{E}_{t,T}^F(Q^2).$$

Proof Taking a convex combination of the BSDEs with terminal conditions Q^1 and Q^2 gives the equation

$$\lambda Y_t^1 + (1-\lambda)Y_t^2 - \int_{]t,T]} [\lambda F(\omega, u, Y_{u-}^1, Z_u) + (1-\lambda)F(\omega, u, Y_{u-}^2, Z_u^2)]du$$
$$+ \int_{]t,T]} [\lambda Z_u^1 + (1-\lambda)Z_u^2]dM_u = \lambda Q^1 + (1-\lambda)Q^2,$$

which is a BSDE with terminal condition $\lambda Q^1 - (1-\lambda)Q^2$ and driver

$$\tilde{F} = \lambda F(\omega, u, Y_{u-}^1, Z_u) + (1-\lambda)F(\omega, u, Y_{u-}^2, Z_u^2).$$

We next consider the BSDE with terminal condition $\lambda Q^1 + (1-\lambda)Q^2$ and driver F. Denote the solution to this by Y^λ. We can compare these BSDEs using the relevant comparison theorem – the first assumption of the theorem is trivial, the second is satisfied by the convexity of F, and the remaining because F is balanced on $\mathscr{Q}_{t,T}$. Hence, our solutions satisfy

$$Y^\lambda \geq \lambda Y^1 + (1-\lambda)Y^2$$

with equality if and only if the terminal conditions are equal with conditional probability one. The inequality for $\mathscr{E}_{t,T}^F$ follows. The inequality for ϱ_t^F can then be established by noting that as $\mathscr{E}_{t,T}^F(Q)$ is concave in Q, it follows that $\varrho_t^F(Q) = -\mathscr{E}_{t,T}^F(Q)$ is convex in Q.   □

References

1. Artzner, P., Delbaen, F., Eber, J.M., Heath, D.: Coherent measures of risk. Math. Finance **9**(3), 203–228 (1999)
2. Barles, G., Buckdahn, R., Pardoux, E.: Backward stochastic differential equations and integral-partial differential equations. Stoch. Stoch. Rep. **60**(1), 57–83 (1997)

3. Barrieu, P., El Karoui, N.: Optimal derivatives design under dynamic risk measures. In: Yin, G., Zhang, Q. (eds.) Mathematics of Finance. *Contemporary Mathematics*, vol. 351, pp. 13–26. AMS, Providence (2004)
4. Becherer, D.: Bounded solutions to backward SDE's with jumps for utility optimization and indifference hedging. Ann. Appl. Probab. **16**(4), 2027–2054 (2006)
5. Cohen, S.N., Elliott, R.J.: Solutions of backward stochastic differential equations on Markov chains. Commun. Stoch. Anal. **2**(2), 251–262 (2008)
6. Cohen, S.N., Elliott, R.J.: Comparisons for backward stochastic differential equations on Markov chains and related no-arbitrage conditions. Ann. Appl. Probab. **20**(1), 267–311 (2010)
7. Cohen, S.N., Elliott, R.J.: A general theory of infinite state backward stochastic difference equations. Stoch. Process. Appl. **120**(4), 442–466 (2010)
8. Coquet, F., Hu, Y., Memin, J., Peng, S.: Filtration consistent nonlinear expectations and related g-expectations. Probab. Theory Relat. Fields **123**(1), 1–27 (2002)
9. Delbaen, F., Peng, S., Rosazza Gianin, E.: Representation of the penalty term of dynamic concave utilities. Working Paper. Available at 0802.1121 (2008)
10. Duffie, D., Epstein, L.G.: Asset pricing with stochastic differential utility. Rev. Financ. Stud. **5**(3), 411–436 (1992)
11. El Karoui, N., Peng, S., Quenez, M.: Backward stochastic differential equations in finance. Math. Finance **7**(1), 1–71 (1997)
12. El Otmani, M.: Backward stochastic differential equations associated with Lévy processes and partial integro-differential equations. Commun. Stoch. Anal. **2**(2), 277–288 (2008)
13. Elliott, R.J.: Stochastic Calculus and its Applications. Springer, Berlin (1982)
14. Elliott, R.J., Aggoun, L., Moore, J.: Hidden Markov Models: Estimation and Control. Springer, Berlin (1994)
15. Föllmer, H., Schied, A.: Stochastic Finance: An Introduction in Discrete Time. *Studies in Mathematics*, vol. 27. de Gruyter, Berlin (2002)
16. Gantmacher, F.: Matrix Theory, vol. 2. Chelsea, New York (1960)
17. Gill, R.D., Johansen, S.: A survey of product-integration with a view toward application in survival analysis. Ann. Stat. **18**(4), 1501–1555 (1990)
18. Hu, Y., Ma, J., Peng, S., Yao, S.: Representation theorems for quadratic $\mathscr{F}$-consistent nonlinear expectations. Stoch. Process. Appl. **118**, 1518–1551 (2008)
19. Jacod, J., Shiryaev, A.N.: Limit Theorems for Stochastic Processes. *Grundlehren der mathematischen Wissenschaften*, vol. 288. Springer, Berlin (2003)
20. Lin, Q.Q.: Nonlinear Doob-Meyer decomposition with jumps. Acta Math. Sin. Engl. Ser. **19**(1), 69–78 (2003)
21. Peng, S.: A generalized dynamic programming principle and Hamilton-Jacobi-Bellman equation. Stoch. Stoch. Rep. **38**, 119–134 (1992)
22. Peng, S.: Dynamically consistent nonlinear evaluations and expectations. Preprint No. 2004-1, Institute of Mathematics, Shandong University. Available at math/0501415 (2005)
23. Rosazza Gianin, E.: Risk measures via g-expectations. Insur. Math. Econ. **39**, 19–34 (2006)
24. Royer, M.: Backward stochastic differential equations with jumps and related non-linear expectations. Stoch. Process. Appl. **116**(10), 1358–1376 (2006)

Results on Numerics for FBSDE with Drivers of Quadratic Growth

**Peter Imkeller, Gonçalo Dos Reis,
and Jianing Zhang**

Abstract We consider the problem of numerical approximation for forward-backward stochastic differential equations with drivers of quadratic growth (qgFBSDE). To illustrate the significance of qgFBSDE, we discuss a problem of cross hedging of an insurance related financial derivative using correlated assets. For the convergence of numerical approximation schemes for such systems of stochastic equations, path regularity of the solution processes is instrumental. We present a method based on the truncation of the driver, and explicitly exhibit error estimates as functions of the truncation height. We discuss a reduction method to FBSDE with globally Lipschitz continuous drivers, by using the Cole-Hopf exponential transformation. We finally illustrate our numerical approximation methods by giving simulations for prices and optimal hedges of simple insurance derivatives.

1 Introduction

Owing to their central significance in optimization problems for instance in stochastic finance and insurance, backward stochastic differential equations (BSDE), one of the most efficient tools of stochastic control theory, have been receiving much attention in the last 15 years. A particularly important class, BSDE with drivers of quadratic growth, for example, arise in the context of utility optimization problems on incomplete markets with exponential utility functions, or alternatively in

P. Imkeller (✉) · J. Zhang
Department of Mathematics, Humboldt University Berlin, Unter den Linden 6, 10099 Berlin,
Germany
e-mail: imkeller@math.hu-berlin.de

J. Zhang
e-mail: zhangj@math.hu-berlin.de

G. Dos Reis
CMAP, École Polytechnique, Route de Saclay, 91128 Palaiseau Cedex, France
e-mail: dosreis@cmap.polytechnique.fr

C. Chiarella, A. Novikov (eds.), *Contemporary Quantitative Finance*,
DOI 10.1007/978-3-642-03479-4_9, © Springer-Verlag Berlin Heidelberg 2010

questions related to risk minimization for the entropic risk measure. BSDE provide the genuinely stochastic approach of control problems which find their analytical expression in the Hamilton-Jacobi-Bellman formalism. BSDE with drivers of this type keep being a source of intensive research.

As Monte-Carlo methods to simulate random processes, numerical schemes for BSDE provide a robust method for simulating and approximating solutions of control problems. Much has been done in recent years to create schemes for BSDE with Lipschitz continuous drivers (see [4] or [9] and references therein). The numerical approximation of BSDE with drivers of quadratic growth (qgBSDE) or systems of forward-backward stochastic equations with drivers of this kind (qgFBSDE) turned out to be more complicated. Only recently, in [7], one of the main obstacles was overcome. Following [4] in the setting of Lipschitz drivers, the strategy to prove convergence of a numerical approximation combines two ingredients: regularity of the trajectories of the control component of a solution pair of the BSDE in the L^2-sense, a tool first investigated in the framework of globally Lipschitz BSDE by [19], and a convenient a priori estimate for the solution. See [4, 6, 11] or [3] for numerical schemes of BSDE with globally Lipschitz continuous drivers, and an implementation of these ideas. The main difficulty treated in [7] consisted of establishing path regularity for the control component of the solution pair of the qgBSDE. For this purpose, the control component, known to be represented by the Malliavin trace of the other component, had to be thoroughly investigated in a subtle and complex study of Malliavin derivatives of solutions of BSDE. This study extends a thorough investigation of smoothness of systems of FBSDE by methods based on Malliavin's calculus and BMO martingales independently conducted in [1] and [5]. The knowledge of path regularity obtained this way is implemented in a second step of the approach in [7]. The quadratic growth part of the driver is truncated to create a sequence of approximating BSDE with Lipschitz continuous drivers. Path regularity is exploited to explicitly capture the convergence rate for the solutions of the truncated BSDE as a function of the truncation height. The error estimate for the truncation, which is of high polynomial order, combines with the ones for the numerical approximation in any existent numerical scheme for BSDE with Lipschitz continuous drivers, to control the convergence of a numerical scheme for qgBSDE. It allows in particular to establish the convergence order for the approximation of the control component in the solution process.

An elegant way to avoid the difficulties related to drivers of quadratic growth, and to fall back into the setting of globally Lipschitz ones, consists of using a coordinate transform well known in related PDE theory under the name "exponential Cole-Hopf transformation". The transformation eliminates the quadratic growth of the driver in the control component at the cost of producing a transformed driver of a new BSDE which in general lacks global Lipschitz continuity in the other component. This difficulty can be avoided by some structure hypotheses on the driver. Once this is done, the transformed BSDE enjoys global Lipschitz continuity properties. Therefore the problem of numerical approximation can be tackled in the framework of transformed coordinates by schemes well known in the Lipschitz setting. As stated before, this again requires path regularity results in the L^2-sense for the control component of the solution pair of the transformed BSDE. For globally Lipschitz continuous drivers [19] provides path regularity under simple and weak additional assumptions such as $\frac{1}{2}$-Hölder continuity of the driver in the time variable. The smoothness of the Cole-Hopf transformation allows passing back to the original coordinates without losing path regularity. In summary, if one accepts the additional structural assumptions on the driver, the exponential transformation approach provides numerical approximation schemes for qgBSDE under weaker smoothness conditions for the driver.

In this paper we aim to give a survey of these two approaches to obtain numerical results for qgBSDE. Doing this, we always keep an eye on one of the most important applications of qgBSDE, which consists of providing a genuinely probabilistic approach to utility optimization problems for exponential utility, or equivalently risk minimization problems with respect to the entropic risk measure, that lead to explicit descriptions of prices and hedges. We motivate qgBSDE by reviewing a simple exponential utility optimization problem resulting from a method to determine the utility indifference price of an insurance related asset in a typical incomplete market situation, following [1] and [10]. The setting of the problem allows in particular the calculation of the driver of quadratic growth of the associated BSDE. After discussing the problem of numerical approximations, in this case by applying the method related to the exponential transform, we are finally able to illustrate our findings by giving some numerical simulations obtained with the resulting scheme.

The paper is organized as follows. In Sect. 2 we fix the notation used for treating problems about qgFBSDE and recall some basic results. Section 3 is devoted to presenting utility optimization problems used for pricing and hedging derivatives on non-tradable underlyings using correlated assets in a utility indifference approach. In Sect. 4.2 we review smoothness results for the solutions processes of qgFBSDE, and apply them to show L^2-regularity of the control component of the solution process of a qgFBSDE. In Sect. 5 we discuss the truncation method for the quadratic terms of the driver to derive a numerical approximation scheme for qgFBSDE. Section 6 is reserved for a discussion of the applicability of the exponential transform in the qgBSDE setting. In Sect. 7 we return to the motivating pricing and hedging problem and use it as a platform for illustrating our results by numerical simulations.

2 Preliminaries

Fix $T \in \mathbb{R}_+ = [0, \infty)$. We work on the canonical Wiener space $(\Omega, \mathscr{F}, \mathbb{P})$ on which a d-dimensional Wiener process $W = (W^1, \ldots, W^d)$ restricted to the time interval $[0, T]$ is defined. We denote by $\mathscr{F} = (\mathscr{F}_t)_{t \in [0,T]}$ its natural filtration enlarged in the usual way by the $\mathbb{P}$-zero sets.

Let $p \geq 2$, $m, d \in \mathbb{N}$, $\mathbb{Q}$ be a probability measure on $(\Omega, \mathscr{F})$. We use the symbol $\mathbb{E}^{\mathbb{Q}}$ for the expectation with respect to $\mathbb{Q}$, and omit the superscript for the canonical measure $\mathbb{P}$. To denote the stochastic integral process of an adapted process Z with respect to the Wiener process on $[0, T]$, we write $Z * W = \int_0^{\cdot} Z_s dW_s$.

For vectors $x = (x^1, \ldots, x^m)$ in Euclidean space $\mathbb{R}^m$ we denote $|x| = (\sum_{i=1}^m (x^i)^2)^{\frac{1}{2}}$. In our analysis the following normed vector spaces will play a role. We denote by

- $L^p(\mathbb{R}^m; \mathbb{Q})$ the space of $\mathscr{F}_T$-measurable random variables $X : \Omega \mapsto \mathbb{R}^m$, normed by $\|X\|_{L^p} = \mathbb{E}^{\mathbb{Q}}[|X|^p]^{\frac{1}{p}}$; L^{∞} the space of bounded random variables;
- $\mathscr{S}^p(\mathbb{R}^m)$ the space of all measurable processes $(Y_t)_{t \in [0,T]}$ with values in $\mathbb{R}^m$ normed by $\|Y\|_{\mathscr{S}^p} = \mathbb{E}[(\sup_{t \in [0,T]} |Y_t|)^p]^{\frac{1}{p}}$; $\mathscr{S}^{\infty}(\mathbb{R}^m)$ the space of bounded measurable processes;
- $\mathscr{H}^p(\mathbb{R}^m, \mathbb{Q})$ the space of all progressively measurable processes $(Z_t)_{t \in [0,T]}$ with values in $\mathbb{R}^m$ normed by $\|Z\|_{\mathscr{H}^p} = \mathbb{E}^{\mathbb{Q}}[(\int_0^T |Z_s|^2 ds)^{p/2}]^{\frac{1}{p}}$;
- $BMO(\mathscr{F}, \mathbb{Q})$ or $BMO_2(\mathscr{F}, \mathbb{Q})$ the space of square integrable $\mathscr{F}$-martingales Φ with $\Phi_0 = 0$ and we set

$$\|\Phi\|_{BMO(\mathscr{F}, \mathbb{Q})}^2 = \sup_{\tau} \left\| \mathbb{E}^{\mathbb{Q}}[\langle \Phi \rangle_T - \langle \Phi \rangle_{\tau} | \mathscr{F}_{\tau}] \right\|_{\infty} < \infty,$$

  where the supremum is taken over all stopping times $\tau \in [0, T]$. More details on this space can be found in Appendix 1. In case $\mathbb{Q}$ resp. $\mathscr{F}$ is clear from the context, we may omit the arguments $\mathbb{Q}$ or $\mathscr{F}$ and simply write $BMO(\mathbb{Q})$ resp. $BMO(\mathscr{F})$ etc;
- $\mathbb{D}^{k,p}(\mathbb{R}^d)$ and $\mathbb{L}_{k,d}(\mathbb{R}^d)$ the spaces of Malliavin differentiable random variables and processes, see Appendix 2.

In case there is no ambiguity about m or $\mathbb{Q}$, we may omit the reference to $\mathbb{R}^m$ or $\mathbb{Q}$ and simply write $\mathscr{S}^{\infty}$ or $\mathscr{H}^p$ etc.

We investigate systems of forward diffusions coupled with backward stochastic differential equations with quadratic growth in the control variable (qgFBSDE for short), i.e. given $x \in \mathbb{R}^m$, $t \in [0, T]$, and four continuous measurable functions b, σ, g and f we analyze systems of the form

$$X_t^x = x + \int_0^t b(s, X_s^x) ds + \int_0^t \sigma(s, X_s^x) dW_s, \tag{1}$$

$$Y_t^x = g(X_T^x) + \int_t^T f(s, X_s^x, Y_s^x, Z_s^x) ds - \int_t^T Z_s^x dW_s. \tag{2}$$

In case there is no ambiguity about the initial state x of the forward system, we may and do suppress the superscript x and just write X, Y, Z for the solution components. For the coefficients of this system we make the following assumptions:

(H0) There exists a positive constant K such that $b, \sigma_i : [0, T] \times \mathbb{R}^m \to \mathbb{R}^m, 1 \leq i \leq d$, are uniformly Lipschitz continuous with Lipschitz constant K, and $b(\cdot, 0)$ and $\sigma_i(\cdot, 0)$, $1 \leq i \leq d$, are bounded by K.

There exists a constant $M \in \mathbb{R}_+$ such that $g : \mathbb{R}^m \to \mathbb{R}$ is absolutely bounded by M, $f : [0, T] \times \mathbb{R}^m \times \mathbb{R} \times \mathbb{R}^d \to \mathbb{R}$ is measurable and continuous in (x, y, z) and for $(t, x) \in [0, T] \times \mathbb{R}^m$, $y, y' \in \mathbb{R}$ and $z, z' \in \mathbb{R}^d$ we have

$$|f(t, x, y, z)| \leq M(1 + |y| + |z|^2),$$

$$|f(t, x, y, z) - f(t, x, y', z')| \leq M\{|y - y'| + (1 + |z| + |z'|)|z - z'|\}.$$

The theory of SDE is well established. Since we wish to focus on the backward equation component of our system we emphasize that the relevant results for SDE are summarized in Appendix 3.

Theorem 1 (Properties of qgFBSDE) *Under* (H0), *the system* (1), (2) *has a unique solution* $(X, Y, Z) \in \mathscr{S}^2 \times \mathscr{S}^\infty \times \mathscr{H}^2$. *The respective norms of Y and Z can be dominated from above by constants depending only on T and M as given by assumption* (H0). *Furthermore*

$$Z * W = \int_0^{\cdot} Z_s \, dW_s \in BMO(\mathbb{P}) \text{ and hence for all } p \geq 2 \text{ one has } Z \in \mathscr{H}^p.$$

It is possible to go beyond the bounded terminal condition hypothesis by imposing the existence of all its exponential moments instead. In this case $Z * W$ is no longer in *BMO*. As we shall see in Sect. 4, the *BMO* property of $Z * W$ plays a crucial role in all of our smoothness results for systems of FBSDE. It combines with the inverse Hölder inequality for the exponentials generated by *BMO* martingales to control moments of functionals of the solutions of FBSDE. Smoothness of solutions is instrumental for instance in estimates for numerical approximations of solutions.

3 Pricing and Hedging with Correlated Assets

The pivotal task of mathematical finance is to provide solid foundations for the valuation of contingent claims. In recent years, markets have displayed an increasing need for financial instruments pegged to non-tradable underlyings such as temperature and energy indices or toxic matter emission rates. In the same manner as liquidly traded underlyings, securities on non-tradable underlyings are used to measure, control and manage risks, as well as to speculate and take advantage of market imperfections. Since non-tradability produces residual risks which are innate and inaccessible to hedging, institutional investors look for tradable assets which are correlated to the non-tradable ones. In incomplete markets, one established pricing paradigm is the utility maximization principle. Upon choosing a risk preference, investors evaluate contingent claims by replicating according to an investment strategy that yields the most favorable utility value. Interplays and connections between the pricing of contingent claims on non-tradable underlyings and the theory of qgFBSDE were studied, among others, by [2, 12, 17], and recently by [10]. Based on this setup, we consider the problem of numerically evaluating contingent claims based on non-tradable underlyings. This will be done by intervention of the exponential transformation of qgBSDE, to be introduced in Sect. 6. It allows to work under weaker assumptions than the numerical schemes for qgFBSDE based on the results reviewed in Sect. 4, and will allow some illustrative numerical simulations in Sect. 7.

The following toy market setup can be found in Sect. 4 of [10]. Assume $d = 2$, so that $W = (W^1, W^2)$ is our basic two-dimensional Brownian motion. We use them to define a third Brownian motion W^3 correlated to W^1 with respect to a correlation coefficient $\rho \in [-1, 1]$ according to

$$W_s^3 := \int_0^s \rho dW_u^1 + \int_0^s \sqrt{1 - \rho^2} dW_u^2, \quad 0 \le s \le T.$$

Contingent claims are assumed to be tied to a one-dimensional non-tradable index that is subject to

$$dR_t = \mu(t, R_t)dt + \sigma(t, R_t)dW_t^1, \quad R_0 = r_0 > 0, \tag{3}$$

where $\mu, \sigma : [0, T] \times \mathbb{R} \to \mathbb{R}$ are deterministic measurable and uniformly Lipschitz continuous functions, uniformly of (at most) linear growth in their state variable. The securities market is governed by a risk free bank account yielding zero interest and one correlated risky asset whose dynamics (with respect to the zero interest bank account *numéraire*) are governed by

$$\frac{dS_s}{S_s} = \alpha(s, R_s)ds + \beta(s, R_s)dW_s^3, \quad S_0 = s_0 > 0. \tag{4}$$

In compliance with [2], we assume that $\alpha, \beta : [0, T] \times \mathbb{R} \to \mathbb{R}$ are bounded and measurable functions, and furthermore $\beta^2(t, r) \ge \varepsilon > 0$ holds uniformly for some fixed $\varepsilon > 0$. Next, we set

$$\theta(s, r) := \frac{\alpha(s, r)}{\beta(s, r)}, \quad (s, r) \in [0, T] \times \mathbb{R},$$

and note that the conditions on α and β imply that θ is uniformly bounded.

An admissible investment strategy is defined to be a real-valued, measurable predictable process λ such that $\int_0^T \lambda_u^2 du < \infty$ holds $\mathbb{P}$-almost surely and such that the family

$$\left\{ e^{-\eta \int_0^\tau \lambda_u \frac{dS_u}{S_u}} : \tau \text{ stopping time with values in } [0, T] \right\} \tag{5}$$

is uniformly integrable. The set of all admissible investment strategies is denoted by $\mathscr{A}$. In the following, let $t \in [0, T]$ denote a fixed time. Then the set of all admissible investment strategies living on the time interval $[t, T]$ is defined analogously and we denote it by $\mathscr{A}_t$. Let v_t denote the investor's initial endowment at time t, that is, v_t is an $\mathscr{F}_t$-measurable bounded random variable. The gain of the investor at time $s \in [t, T]$, denoted by G_s, is subject to trading according to investing λ into the risky asset, and therefore given by

$$dG_s^\lambda = \lambda_s \frac{dS_s}{S_s}, \quad G_t^\lambda = 0.$$

We focus on European style contingent claims, i.e. payoff profiles resuming the form $F(R_T)$ where we assume, in accordance with [2], that $F : \mathbb{R} \to \mathbb{R}$ is measurable and bounded. Moreover the investor's risk assessment presumes that her utility preference is reflected by the exponential utility function, so given a nonzero constant risk attitude parameter η, the investor's utility function is

$$U(x) = -e^{-\eta x}, \quad x \in \mathbb{R}.$$

The evolution of the investor's portfolio over the time interval $[t, T]$ consists of her initial endowment v_t, her gains (or losses) via her investment into the risky asset under an investment strategy λ and holding one share of the contingent claim $F(R_T)$. Her objective is to find an investment strategy such that her time-t utility is maximized, i.e. her maximization problem is given by

$$V_t^F(v_t) := \sup \left\{ \mathbb{E} \left[U(v_t + G_T^\lambda + F(R_T)) \big| \mathscr{F}_t \right] : \lambda \in \mathscr{A}_t \right\}$$

$$= \exp\{-\eta v_t\} \sup \left\{ \mathbb{E} \left[U(G_T^\lambda + F(R_T)) \big| \mathscr{F}_t \right] : \lambda \in \mathscr{A}_t \right\}. \tag{6}$$

For the sake of notational convenience, we write

$$V_t^F := V_t^F(0) = \sup \left\{ \mathbb{E} \left[U(G_T^\lambda + F(R_T)) \big| \mathscr{F}_t \right] : \lambda \in \mathscr{A}_t \right\}. \tag{7}$$

Now pricing $F(R_T)$ within the utility maximization paradigm is based on the identity

$$V_t^0(v_t) = V_t^F(v_t - p_t),$$

where $V_t^0(v_t)$ denotes the time-t utility with initial endowment v_t and with $F = 0$ (see also Sect. 2 of [2] and Sect. 3 of [10]). According to this identity, the investor is indifferent about a portfolio with initial endowment v_t without receiving one quantity of the contingent claim $F(R_T)$ and a portfolio with initial endowment $v_t - p_t$, now receiving one quantity of the contingent claim in addition. Hence p_t is interpreted as the time-t indifference price of the contingent claim $F(R_T)$. By the equality $V_t^F(v_t) = \exp\{-\eta v_t\} V_t^F$, it follows that

$$p_t = \frac{1}{\eta} \log \frac{V_t^0}{V_t^F}, \tag{8}$$

which means that the indifference price does not depend on the initial endowment v_t. Since the time-t indifference price (8) is fully characterized by V_t^0 and V_t^F, the focus now lies in the investigation of (6). In fact, [2] and [10] have already pointed out that (7) yields a characterization by means of a qgFBSDE. In accordance with [10], let us denote by $(\mathscr{G}_u)_{0 \le u \le T}$ the filtration generated by W^1, completed by $\mathbb{P}$-null sets. [10]'s main ideas for rephrasing (6) in terms of a qgBSDE are summarized in the following

Lemma 1 *The qgFBSDE*

$$Y_s = F(R_T) + \int_s^T f(u, R_u, Z_u)du - \int_s^T Z_u dW_u^1, \quad s \in [0, T], \tag{9}$$

$$f(u, r, z) = \frac{\theta^2(u, r)}{2\eta} - z\rho\theta(u, r) - \frac{\eta}{2}(1 - \rho^2)z^2, \tag{10}$$

has a unique solution $(Y, Z) \in \mathscr{S}^\infty \times \mathscr{H}^2$ *such that* $V_t^F = -e^{-\eta Y_t}$ *holds* $\mathbb{P}$-*almost surely.*

Proof Since $\theta(\cdot, r)$ is uniformly bounded and $\mathscr{G}$-predictable, the driver of (9) satisfies the conditions of [14]; thus (9) admits a unique solution $(Y, Z) \in \mathscr{S}^\infty \times \mathscr{H}^2$. Moreover, [16] have shown that $Z * W^1$ is both a BMO($\mathscr{F}$)- and BMO($\mathscr{G}$)-martingale. See also [2]. To prove the identity $V_t^F = -e^{-\eta Y_t}$, we notice that

$$e^{-\eta(G_T^\lambda + Y_T)} = e^{-\eta G_t^\lambda} e^{-\eta Y_t} e^{-\eta(Y_T - Y_t)} e^{-\eta(G_T^\lambda - G_t^\lambda)}$$
$$= e^{-\eta Y_t} e^{-\eta(Y_T - Y_t)} e^{-\eta(G_T^\lambda - G_t^\lambda)},$$

because $G_t^\lambda = 0$. We then have

$$\exp\{-\eta(Y_T - Y_t)\} \exp\{-\eta(G_T^\lambda - G_t^\lambda)\}$$
$$= \exp\left\{-\eta\left(\int_t^T Z_u dW_u^1 + \int_t^T \lambda_u \beta(u, R_u) dW_u^3\right)\right.$$
$$\left. + \int_t^T [\lambda_u \alpha(u, R_u) - f(u, R_u, Z_u)] du\right\}.$$

Denoting $\mathscr{E}_t^s(M) = \mathscr{E}(M)_s / \mathscr{E}(M)_t$ for $t \le s \le T$ where $\mathscr{E}(M)_s$ is the stochastic exponential of a given semi-martingale M, we introduce

$$K_u := \frac{1}{2}\left(\eta(\rho Z_u + \beta(u, R_u)\lambda_u) - \theta_u\right)^2, \quad t \le u \le T.$$

Then a simple calculation yields

$$\exp\{-\eta(Y_T - Y_t)\} \exp\{-\eta(G_T^\lambda - G_t^\lambda)\}$$
$$= \mathscr{E}_t^T\left(\int -\eta Z dW^1 - \int \eta\lambda\beta(\cdot, R) dW^3\right) \exp\left\{\int_t^T K_u du\right\}.$$

Since $\lambda\beta(\cdot, R) * W^3$ is a BMO-martingale, we can condition with respect to the σ-algebra $\mathscr{F}_t$ and get

$$\mathbb{E}\left[e^{-\eta(G_T^\lambda + F(R_T))} \mid \mathscr{F}_t\right] \ge e^{-\eta Y_t}. \tag{11}$$

By (5) and a localization argument, this inequality holds for every $\lambda \in \mathscr{A}_t$, and therefore we have $V_t^F \le -e^{-\eta Y_t}$. To prove equality, note that the inequality (11) becomes an equality for $\tilde{\lambda}_u = -\frac{\rho}{\beta(u, R_u)} Z_u + \frac{\theta(u, R_u)}{\eta\beta(u, R_u)}$; this in conjunction with the observation that

$$\exp\left\{-\eta \int_t^T \tilde{\lambda}_u \frac{dS_u}{S_u}\right\} = \exp\{-\eta G_T^{\tilde{\lambda}}\} = \exp\{-\eta(G_T^{\tilde{\lambda}} - G_t^{\tilde{\lambda}})\}$$
$$= \mathscr{E}_t^T\left(\int -\eta Z dW^1 - \int \eta\tilde{\lambda}\beta(\cdot, R) dW^3\right) \times \exp\{\eta(Y_T - Y_t)\}$$

is the product of a bounded process and true $\mathscr{F}$-martingale yields that condition (5) is satisfied. Hence $\tilde{\lambda} \in \mathscr{A}_t$ and we have shown $V_t^F = -e^{-\eta Y_t}$. $\square$

The proof of the previous Lemma 1 yields the following

Corollary 1 *The investment strategy*

$$\tilde{\lambda}_s := -\frac{\rho}{\beta(s, R_s)} Z_s + \frac{\theta(s, R_s)}{\eta \beta(s, R_s)}, \quad t \leq s \leq T, \tag{12}$$

where Z is the control component of the solution to (9), *belongs to $\mathscr{A}_t$ and satisfies*

$$\mathbb{E}\big[U(v_t + G^{\tilde{\lambda}_T} + F(R_T)\big|\mathscr{F}_t\big] = \sup\{\mathbb{E}[U(v_t + G^{\lambda}_T + F(R_T)\big|\mathscr{F}_t\big] : \lambda \in \mathscr{A}_t\} = V^F_t(v_t).$$

Example 1 [Put option on kerosene, compare with Example 1.2 from [2]] Facing recent considerable declines in world oil prices, companies producing kerosene wish to partially cover their risk of such a depreciation. European put options are an established financial instrument to comply with this demand of risk covering. Since kerosene is not traded in a liquid market, derivative contracts on this underlying must be arranged on an over-the-counter basis. Knowing that the price of heating oil is highly correlated with the price of kerosene, the pricing and hedging of a European put option on kerosene can be done by a dynamic investment in (the liquid market of) heating oil. A numerical treatment of this pricing problem will be displayed in Sect. 7.

4 Smoothness and Path Regularity Results

The principal aim of this paper is to survey some recent results on the numerical approximation of prices and hedging strategies of financial derivatives such as the liability $F(R_T)$ in the setting of the previous section. As we saw, this leads us directly to qgFBSDE. In the subsequent sections we shall discuss an approach based on a truncation of the driver's quadratic part in the control variable. It will be crucial to give an estimate for the error committed by truncating. Our error estimate will be based on smoothness results for the control component Z^x of solutions of the BSDE part of our system. Smoothness is understood both in the sense of regular sensitivity to initial states x of the forward component, as well as in the sense of the stochastic calculus of variations. Since the control component of the solution of a BSDE is related to the Malliavin trace of the other component, we will be led to look at variational derivatives of the first order.

Our first result concerns the smoothness of the map $[0, T] \times \mathbb{R}^m \ni (t, x) \mapsto Z^x_t$, especially its differentiability in x. The second result refers to the variational differentiability of (Y^x, Z^x) in the sense of Malliavin's calculus. We shall work under the following hypothesis, where we denote the gradient by the common symbol ∇, and by ∇_u if we wish to emphasize the variable u with respect to which the derivative is taken.

(H1) Assume that (H0) holds. For any $0 \leq t \leq T$ the functions $b(t, \cdot), \sigma_i(t, \cdot)$, $1 \leq i \leq d$, are continuously differentiable with bounded derivatives in the spatial variable. There exists a positive constant c such that

$$y^T \sigma(t, x) \sigma^T(t, x) y \geq c|y|^2, \quad x, y \in \mathbb{R}^m, \ t \in [0, T]. \tag{13}$$

f is continuously partially differentiable in (x, y, z) and there exists $M \in \mathbb{R}_+$ such that for $(t, x, y, z) \in [0, T] \times \mathbb{R}^m \times \mathbb{R} \times \mathbb{R}^d$

$$|\nabla_x f(t, x, y, z)| \leq M(1 + |y| + |z|^2),$$
$$|\nabla_y f(t, x, y, z)| \leq M,$$
$$|\nabla_z f(t, x, y, z)| \leq M(1 + |z|).$$

$g : \mathbb{R}^m \to \mathbb{R}$ is a continuously differentiable function satisfying $|\nabla g| \leq M$.

4.1 Smoothness Results

The following differentiability results are extensions of Theorems proved in [1] and [5]. For further details, comments and complete proofs we refer to the mentioned works or to [7].

Theorem 2 (Classical differentiability) *Suppose that* (H1) *holds. Then for all* $p \geq 2$ *the solution process* $\Theta^x = (X^x, Y^x, Z^x)$ *of the qgFBSDE* (1), (2) *with initial vector* $x \in \mathbb{R}^m$ *for the forward component belongs to* $\mathscr{S}^p \times \mathscr{S}^p \times \mathscr{H}^p$. *The application* $\mathbb{R}^m \ni x \mapsto (X^x, Y^x, Z^x) \in \mathscr{S}^p(\mathbb{R}^m) \times \mathscr{S}^p(\mathbb{R}) \times \mathscr{H}^p(\mathbb{R}^d)$ *is differentiable. The derivatives of* $x \mapsto X^x$ *satisfy* (21) *while the derivatives of the map* $x \mapsto (Y^x, Z^x)$ *satisfy the linear BSDE*

$$\nabla Y_t^x = \nabla g(X_T^x)\nabla X_T^x - \int_t^T \nabla Z_s^x \mathrm{d}W_s + \int_t^T \langle \nabla f(s, \Theta_s^x), \nabla \Theta_s^x \rangle \mathrm{d}s, \quad t \in [0, T].$$

Theorem 3 (Malliavin differentiability) *Suppose that* (H1) *holds. Then the solution process* (X, Y, Z) *of FBSDE* (1), (2) *has the following properties. For* $x \in \mathbb{R}^m$,

- X^x *satisfies* (22) *and for any* $0 \leq t \leq T$, $x \in \mathbb{R}^m$ *we have* $(Y^x, Z^x) \in \mathbb{L}_{1,2} \times (\mathbb{L}_{1,2})^d$. X^x *fulfills the statement of Theorem* 11, *and a version of* $(D_u Y_t^x, D_u Z_t^x)_{0 \leq u, t \leq T}$ *satisfies*

$$D_u Y_t^x = 0, \qquad D_u Z_t^x = 0, \quad t < u \leq T,$$

$$D_u Y_t^x = \nabla g(X_T^x) D_u X_T^x + \int_t^T \langle \nabla f(s, \Theta_s^x), D_u \Theta_s^x \rangle \mathrm{d}s - \int_t^T D_u Z_s^x \mathrm{d}W_s, \quad t \in [u, T].$$

Moreover, $(D_t Y_t^x)_{0 \leq t \leq T}$ *defined by the above equation is a version of* $(Z_t^x)_{0 \leq t \leq T}$.
- *The following representation holds for any* $0 \leq u \leq t \leq T$ *and* $x \in \mathbb{R}^m$

$$D_u Y_t^x = \nabla_x Y_t^x (\nabla_x X_u^x)^{-1} \sigma(u, X_u^x), \quad a.s.,$$

$$Z_t = \nabla_x Y_t^x (\nabla_x X_t^x)^{-1} \sigma(s, X_t^x), \quad a.s..$$

4.2 Regularity and Bounds for the Solution Process

A careful analysis of DY in both its variables under the smoothness assumptions on the coefficients of our system formulated earlier reveals the following continuity properties for the control process Z.

Theorem 4 (Time continuity and bounds) *Assume* (H1). *Then the control process Z of the qgFBSDE* (1), (2) *has a continuous version on* $[0, T]$. *Furthermore for all $p \geq 2$ it satisfies*

$$\|Z\|_{\mathscr{S}^p} < \infty. \tag{14}$$

Theorem 5 (Regularity) *Under* (H1) *the solution process* (X, Y, Z) *of the qgFB-SDE* (1), (2) *satisfies for all $p \geq 2$*

i) *there exists a constant $C_p > 0$ such that for $0 \leq s \leq t \leq T$ we have*

$$\mathbb{E}\left[\sup_{s \leq u \leq t} |Y_u - Y_s|^p \right] \leq C_p |t - s|^{\frac{p}{2}};$$

ii) *there exists a constant $C_p > 0$ such that for any partition $\pi = \{t_0, \ldots, t_N\}$ with $0 = t_0 < \cdots < t_N = T$ of $[0, T]$ with mesh size $|\pi|$*

$$\sum_{i=0}^{N-1} \mathbb{E}\left[\left(\int_{t_i}^{t_{i+1}} |Z_t - Z_{t_i}|^2 dt \right)^{\frac{p}{2}} \right] \leq C_p |\pi|^{\frac{p}{2}}.$$

Now let $h = T/N$, $\pi^N = \{t_i = ih : i = 0, \ldots, N\}$ be an equidistant partition of $[0, T]$ with $N + 1$ points and constant mesh size h. Let Z be the control component in the solution of the qgFBSDE (1), (2) under (H1) and define the family of random variables

$$\bar{Z}_{t_i}^{\pi^N} = \frac{1}{h} \mathbb{E}\left[\int_{t_i}^{t_{i+1}} Z_s ds \,\Big|\, \mathscr{F}_{t_i} \right], \quad t_i \in \pi^N \setminus \{t_N\}. \tag{15}$$

For $0 \leq i \leq N - 1$ the random variable $\bar{Z}_{t_i}^{\pi^N}$ is the best $\mathscr{F}_{t_i}$-measurable approximation of Z in $\mathscr{H}^2([t_i, t_{i+1}])$, i.e.

$$\mathbb{E}\left[\int_{t_i}^{t_{i+1}} |Z_s - \bar{Z}_{t_i}^{\pi^N}|^2 ds \right] = \inf_{\Lambda} \mathbb{E}\left[\int_{t_i}^{t_{i+1}} |Z_s - \Lambda|^2 ds \right],$$

where Λ is allowed to vary in the space of all square integrable $\mathscr{F}_{t_i}$-measurable random variables. By constant interpolation we define $\bar{Z}_t^{\pi^N} = \bar{Z}_{t_i}^{\pi^N}$ for $t \in [t_i, t_{i+1}[$, $0 \leq i \leq N - 1$. It is easy to see that $(\bar{Z}_t^{\pi^N})_{t \in [0, T]}$ converges to $(Z_t)_{t \in [0, T]}$ in $\mathscr{H}^2[0, T]$ as h vanishes. Since Z is adapted there exists a family of adapted processes Z^{π^N} indexed by our equidistant partitions such that $Z_t^{\pi^N} = Z_{t_i}$ for $t \in [t_i, t_{i+1})$ and that Z^{π^N} converges to Z in $\mathscr{H}^2$ as h tends to zero. Since $\bar{Z}^{\pi^N}$ is the best $\mathscr{H}^2$-approximation of Z, we obtain

$$\|Z - \bar{Z}^{\pi^N}\|_{\mathscr{H}^2} \leq \|Z - Z^{\pi^N}\|_{\mathscr{H}^2} \to 0, \quad \text{as } h \to 0.$$

The following Corollary of Theorem 5 extends Theorem 3.4.3 in [19] (see Theorem 12) to the setting of qgFBSDE.

Corollary 2 (L^2-regularity of Z) *Under* (H1) *and for the sequence of equidistant partitions* $(\pi^N)_{N\in\mathbb{N}}$ *of* $[0, T]$ *with mesh size* $h = \frac{T}{N}$, *we have*

$$\max_{0\leq i\leq N-1}\left\{\sup_{t\in[t_i,t_{i+1})}\mathbb{E}\big[|Y_t - Y_{t_i}|^2\big]\right\} + \sum_{i=0}^{N-1}\mathbb{E}\left[\int_{t_i}^{t_{i+1}}|Z_s - \bar{Z}_{t_i}^{\pi^N}|^2 ds\right] \leq Ch,$$

where C *is a positive constant independent of* N.

Remark 1 The above corollary still holds if (H1) is weakened. More precisely, the corollary's statement remains valid if one replaces in (H1) the sentence

"$g : \mathbb{R}^m \to \mathbb{R}$ is a continuously differentiable function satisfying $|\nabla g| \leq M$."

by

"$g : \mathbb{R}^m \to \mathbb{R}$ is uniformly Lipschitz continuous in all its variables."

The proof requires a regularization argument.

5 A Truncation Procedure

To the best of our knowledge so far none of the usual discretization schemes for FBSDE has been shown to converge in the case of systems of FBSDE considered in this paper, the driver of which is of quadratic growth in the control variable. The regularity results derived in the preceding section have the potential to play a crucial role in numerical approximation schemes for qgFBSDE. We shall now give arguments to substantiate this claim. In fact, the regularity of the control component of the solution processes of our BSDE will lead to precise estimates for the error committed in truncating the quadratic growth part of the driver. We will next explain how this truncation is done in our setting.

We start by introducing a sequence of real valued functions $(\tilde{h}_n)_{n\in\mathbb{N}}$ that truncate the identity on the real line. For $n \in \mathbb{N}$ the map $\tilde{h}_n$ is continuously differentiable and satisfies

- $\tilde{h}_n \to$ id locally uniformly, $|\tilde{h}_n| \leq |\text{id}|$ and $|\tilde{h}_n| \leq n + 1$; moreover

$$\tilde{h}_n(x) = \begin{cases} (n+1), & x > n+2, \\ x, & |x| \leq n, \\ -(n+1), & x < -(n+2); \end{cases}$$

- the derivative of $\tilde{h}_n$ is absolutely bounded by 1 and converges to 1 locally uniformly.

We remark that such a sequence of functions exists. The above requirements are for instance consistent with

$$\tilde{h}_n(x) = \begin{cases} \left(-n^2 + 2nx - x(x-4)\right)/4, & x \in [n, n+2], \\ \left(n^2 + 2nx + x(x+4)\right)/4, & x \in [-(n+2), -n]. \end{cases}$$

We then define $h_n : \mathbb{R}^d \to \mathbb{R}^d$ by $z \mapsto h_n(z) = (\tilde{h}_n(z_1), \dots, \tilde{h}_n(z_d))$, $n \in \mathbb{N}$. The sequence $(h_n)_{n \in \mathbb{N}}$ is chosen to be continuously differentiable because the properties stated in Theorem 4 need to hold for the solution processes of the family of FBSDE that the truncation sequence generates by modifying the driver according to the following definition.

Recalling the driver f of BSDE (2), for $n \in \mathbb{N}$ we define $f_n(t, x, y, z) := f(t, x, y, h_n(z))$, $(t, x, y, z) \in [0, T] \times \mathbb{R}^m \times \mathbb{R} \times \mathbb{R}^d$. With this driver and (1) we obtain a family of truncated BSDE by

$$Y_t^n = g(X_T) + \int_t^T f_n\big(s, X_s, Y_s^n, Z_s^n\big)ds - \int_t^T Z_s^n dW_s, \quad t \in [0, T], n \in \mathbb{N}. \quad (16)$$

The following Theorem proves that the truncation error leads to a polynomial deviation of the corresponding solution processes in their natural norms, formulated for polynomial order 12.

Theorem 6 *Assume that* (H1) *is satisfied. Fix* $n \in \mathbb{N}$ *and let* X *be the solution of* (1). *Let* (Y, Z) *and* $(Y^n, Z^n)_{n \in \mathbb{N}}$ *be the solution pairs of* (2) *and* (16) *respectively. Then for all* $p \geq 2$ *there exists a positive constant* C_p *such that for all* $n \in \mathbb{N}$

$$\mathbb{E}\left[\sup_{t \in [0,T]} |Y_t^n - Y_t|^p \right] + \mathbb{E}\left[\left(\int_0^T |Z_s^n - Z_s|^2 ds \right)^{\frac{p}{2}} \right] \leq C_p \frac{1}{n^{12}}.$$

The proof of Theorem 6 roughly involves estimating the probability that Z^n exceeds the threshold n as a function of $n \in \mathbb{N}$ through Markov's inequality. The application of Markov's inequality is possible thanks to (14).

6 The Exponential Transformation Method

In the preceding sections we exhibited the significance of path regularity for the solution of systems of qgFBSDE, in particular the control component, for their numerical approximation. In this section we shall discuss an alternative route to path regularity of solutions in a particular situation that allows for weaker conditions than in the preceding sections. We will use the exponential transform known in PDE theory as the Cole-Hopf transformation. This mapping takes the exponential of the component Y of a solution pair as the new first component of a solution pair of a modified BSDE. It makes a quadratic term in the control variable of the form $z \mapsto \gamma |z|^2$ vanish in the driver of the new system. The price one has to pay for this approach is a possibly missing global Lipschitz condition in the variable y for the modified driver. It is therefore not clear if the new BSDE is amenable to the usual numerical discretization techniques. We give sufficient conditions for the transformed driver to satisfy a global Lipschitz condition. In this simpler setting our techniques allow an easier access to smoothness results for the solutions of the transformed BSDE. The Cole-Hopf transformation being one-to-one, it is clear that regularity results carry over to the original qgFBSDE.

Under (H0), we consider the transformation $P = e^{\gamma Y}$ and $Q = \gamma P Z$. It transforms our qgBSDE (2) with driver f into the new BSDE

$$P_t = e^{\gamma g(X_T)} + \int_t^T \left[\gamma P_s f\left(s, X_s, \frac{\log P_s}{\gamma}, \frac{Q_s}{\gamma P_s} \right) - \frac{1}{2} \frac{|Q|_s^2}{P_s} \right] ds$$

$$- \int_t^T Q_s dW_s, \quad t \in [0, T]. \tag{17}$$

Combining (17) with SDE (1), we see that for any $p \geq 2$ a unique solution $(X, P, Q) \in \mathscr{S}^p \times \mathscr{S}^\infty \times \mathscr{H}^p$ of (1) and (17) exists. The properties of this triple follow from the properties of the solution (X, Y, Z) of the original qgFBSDE (1) and (2). For clarity, we remark that since Y is bounded, P is also bounded and bounded away from 0. The latter property allows us to deduce from the BMO martingale property of $Z * W$ the BMO martingale property of $Q * W$. For the rest of this section we denote by $\mathscr{K}$ a compact subset of $(\delta, +\infty)$ for some constant $\delta \in \mathbb{R}_+$ in which P takes its values.

The form of the driver in (17) indicates that after transforming drivers of the form of the following hypothesis, we have good chances to deal with a Lipschitz continuous one.

(H0*) Assume that (H0) holds. For $\gamma \in \mathbb{R}$ let $f : [0, T] \times \mathbb{R}^m \times \mathbb{R} \times \mathbb{R}^d \to \mathbb{R}$ be of the form

$$f(t, x, y, z) = l(t, x, y) + a(t, z) + \frac{\gamma}{2} |z|^2,$$

where l and a are measurable, l is uniformly Lipschitz continuous in x and y, a is uniformly Lipschitz continuous and homogeneous in z, i.e. for $c \in \mathbb{R}$, $(s, z) \in [0, T] \times \mathbb{R}^d$ we have $a(s, cz) = ca(s, z)$; l and a continuous in t.

Assumption (H0*) allows us to simplify the BSDE obtained from the exponential transformation to

$$P_t = e^{\gamma g(X_T)} + \int_t^T F(s, X_s, P_s, Q_s) ds - \int_t^T Q_s dW_s, \quad t \in [0, T], \tag{18}$$

where the driver is defined by

$$F : [0, T] \times \mathbb{R}^m \times \mathscr{K} \times \mathbb{R}^d \to \mathbb{R},$$

$$(s, x, p, q) \mapsto \gamma p \, l\left(s, x, \frac{\log p}{\gamma} \right) + \gamma p \, a\left(s, \frac{q}{\gamma p} \right). \tag{19}$$

Thanks to the homogeneity assumption on a our driver simplifies further. Indeed, we have for $(s, x, p, q) \in [0, T] \times \mathbb{R}^m \times \mathbb{R} \times \mathbb{R}^d$

$$F(s, x, p, q) = \gamma p \, l \left(s, x, \frac{\log p}{\gamma} \right) + a(s, q). \tag{20}$$

The terminal condition of the transformed BSDE still keeps the properties it had in the original setting. Indeed, boundedness of g is inherited by $\exp(\gamma g)$. Furthermore, if g is uniformly Lipschitz, then clearly by boundedness of g, the function $e^{\gamma g}$ is uniformly Lipschitz as well.

Let us next discuss the properties of the driver (19) in the transformed BSDE. We recall that since l and a are Lipschitz continuous, there is a constant $C > 0$ such that for all $(s, x, p, q) \in [0, T] \times \mathbb{R}^m \times \mathscr{K} \times \mathbb{R}^d$ we have

$$|F(s, x, p, q)| \leq \left| \gamma p \, l \left(s, x, \frac{\log p}{\gamma} \right) + a(s, q) \right|$$
$$\leq C|p|(1 + |x| + |\log p| + |q|) \leq C(1 + |x| + |p| + |q|).$$

This means that F is of linear growth in x, p and q.

To verify Lipschitz continuity properties of F in its variables x, p and q, by (20) and the Lipschitz continuity assumptions on a, it remains to verify that

$$(x, p) \mapsto \gamma p \, l \left(s, x, \frac{\log p}{\gamma} \right)$$

is Lipschitz continuous in x and p, with a Lipschitz constant independent of $s \in [0, T]$. As for x, this is an immediate consequence of the Lipschitz continuity of l in x. For p we have to recall that p is restricted to a compact set $\mathscr{K} \subset \mathbb{R}_+$ not containing 0, to be able to appeal to the Lipschitz continuity of l in y. This shows that F is globally Lipschitz continuous in its variables x, p and q.

We may summarize these observations in the following Theorem.

Theorem 7 *Let $f : [0, T] \times \mathbb{R}^m \times \mathbb{R} \times \mathbb{R}^d \to \mathbb{R}$ be a measurable function, continuous on $\mathbb{R}^m \times \mathbb{R} \times \mathbb{R}^d$, and satisfying (H0*). Then F as defined by (19) is a uniformly Lipschitz continuous function in the spatial variables.*

Theorem 7 opens another route to tackle convergence of numerical schemes via path regularity of the control component of a solution pair of a qgFBSDE system whose driver satisfies (H0*). Look at the new BSDE after applying the Cole-Hopf transform. Since it possesses a Lipschitz continuous driver, path regularity for the control component Q of the transformed BSDE will follow from Zhang's path regularity result stated in (12) provided the driver is $\frac{1}{2}$-Hölder continuous in time. Of course, by the smoothness of the Cole-Hopf transform, the control component Z of the original BSDE will inherit path regularity from Q. This way we circumvent the more stringent assumption (H1) which was made in Sect. 4.

In what follows the triples (X, Y, Z) and (X, P, Q) will always refer to the solution of qgFBSDE (1), (2) and FBSDE (1), (18) respectively.

Theorem 8 *Let* (H0*) *hold. Assume that* $[0, T] \times \mathbb{R}^m \times \mathcal{K} \times \mathbb{R}^d \ni (s, x, p, q) \mapsto F(s, x, p, q) \in \mathbb{R}$, *the driver of BSDE* (18), *is uniformly Lipschitz in* x, p *and* q *and is* $\frac{1}{2}$*-Hölder continuous in* s. *Suppose further that the map* $g : \mathbb{R}^d \to \mathbb{R}$, *as indicated in* (H0), *is globally Lipschitz continuous with Lipschitz constant* K. *Let* (X, Y, Z) *be the solution of qgFBSDE* (1), (2), *and* $\varepsilon > 0$ *be given. There exists a positive constant* C *such that for any partition* $\pi = \{t_0, \dots, t_N\}$ *with* $0 = t_0, T = t_N, t_0 < \cdots < t_N$ *of the interval* $[0, T]$, *with mesh size* $|\pi|$ *we have*

$$\max_{0 \leq i \leq N-1} \left\{ \sup_{t \in [t_i, t_{i+1})} \mathbb{E}\left[|Y_t - Y_{t_i}|^2 \right] \right\} \leq C|\pi| \quad \text{and}$$

$$\sum_{i=0}^{N-1} \mathbb{E}\left[\int_{t_i}^{t_{i+1}} |Z_s - \bar{Z}_{t_i}^\pi|^2 \mathrm{d}s \right] \leq C|\pi|^{1-\varepsilon}.$$

Moreover, if the functions b *and* σ *are continuously differentiable in* $x \in \mathbb{R}^m$ *then* $t \mapsto Z_t$ *is a.s. continuous in* $[0, T]$.

Proof Throughout this proof C will always denote a positive constant the value of which may change from line to line. Let (X, P, Q) be the solution of (1) and (18), where P takes its values in $\mathcal{K}$ and $Q * W$ is a BMO martingale. Applying Theorem 12 yields a positive constant C such that for any partition $\pi = \{t_0, \dots, t_N\}$ of $[0, T]$ with mesh size $|\pi|$

$$\max_{0 \leq i \leq N-1} \left\{ \sup_{t \in [t_i, t_{i+1})} \mathbb{E}\left[|P_t - P_{t_i}|^2 \right] \right\} + \sum_{i=0}^{N-1} \mathbb{E}\left[\int_{t_i}^{t_{i+1}} |Q_s - \bar{Q}_{t_i}^\pi|^2 \mathrm{d}s \right] \leq C|\pi|.$$

Since P takes its values in the compact set $\mathcal{K} \subset \mathbb{R}_+$ not containing 0 there exists a constant C such that for any $0 \leq i \leq N - 1, t \in [t_i, t_{i+1})$

$$|Y_t - Y_{t_i}| = C|\log P_t - \log P_{t_i}| \leq C|P_t - P_{t_i}|.$$

Using the two above inequalities we have

$$\max_{0 \leq i \leq N-1} \left\{ \sup_{t \in [t_i, t_{i+1})} \mathbb{E}\left[|Y_t - Y_{t_i}|^2 \right] \right\} \leq C \max_{0 \leq i \leq N-1} \left\{ \sup_{t \in [t_i, t_{i+1})} \mathbb{E}\left[|P_t - P_{t_i}|^2 \right] \right\} \leq C|\pi|.$$

This proves the first inequality. For the second one, note that by definition for $0 \leq i \leq N - 1, t \in [t_i, t_{i+1})$

$$|Z_t - \bar{Z}_{t_i}| \leq |Z_t - Z_{t_i}| \leq \frac{1}{\gamma} \left\{ \left| \frac{Q_t}{P_t} - \frac{Q_t}{P_{t_i}} \right| + \left| \frac{Q_t}{P_{t_i}} - \frac{Q_{t_i}}{P_{t_i}} \right| \right\}$$

$$\leq \frac{1}{\gamma} \left\{ |Q_t| \left| \frac{1}{P_t} - \frac{1}{P_{t_i}} \right| + \frac{1}{|P_{t_i}|} |Q_t - Q_{t_i}| \right\}$$

$$\leq C\{ |Q_t| \, |P_t - P_{t_i}| + |Q_t - Q_{t_i}| \}.$$

We therefore have for $0 \leq i \leq N - 1$

$$\mathbb{E}\left[\int_{t_i}^{t_{i+1}} |Z_s - \bar{Z}_{t_i}^{\pi}|^2 ds\right] \leq \mathbb{E}\left[\int_{t_i}^{t_{i+1}} |Z_s - Z_{t_i}|^2 ds\right]$$

$$\leq 2C\left\{\mathbb{E}\left[\sup_{t \in [t_i, t_{i+1})} |P_t - P_{t_i}|^2 \int_{t_i}^{t_{i+1}} |Q_s|^2 ds\right]\right.$$

$$\left. + \mathbb{E}\left[\int_{t_i}^{t_{i+1}} |Q_t - Q_{t_i}|^2 ds\right]\right\}.$$

Since $Q \in \mathscr{H}^p$ for all $p \geq 2$, for any two real numbers $\alpha, \beta \in (1, \infty)$ satisfying $1/\alpha + 1/\beta = 1$ we may continue using Hölder's inequality on the right hand side of the inequality just obtained, and then Theorem 12 to the term containing P. This yields the following inequality valid for any $0 \leq i \leq N - 1$ with a constant C not depending on i

$$\mathbb{E}\left[\int_{t_i}^{t_{i+1}} |Z_s - \bar{Z}_{t_i}^{\pi}|^2 ds\right] \leq C\left\{\mathbb{E}\left[\sup_{t \in [t_i, t_{i+1})} |P_t - P_{t_i}|^{2\alpha}\right]^{\frac{1}{\alpha}} \mathbb{E}\left[\left(\int_{t_i}^{t_{i+1}} |Q_s|^2 ds\right)^{\beta}\right]^{\frac{1}{\beta}} + |\pi|\right\}$$

$$\leq C\left\{\mathbb{E}\left[\sup_{t \in [t_i, t_{i+1})} |P_t - P_{t_i}|^2\right]^{\frac{1}{\alpha}} + |\pi|\right\} \leq C\{|\pi|^{\frac{1}{\alpha}} + |\pi|\}.$$

Now choose $\alpha = \frac{1}{1-\varepsilon}$, to complete the claimed estimate.

To prove that Z admits a.s. a continuous version, it is enough to remark that the Theorem's assumptions imply the conditions of Corollary 5.6 in [15]. The referred result yields that Q is a.s. continuous on $[0, T]$. Since P is continuous and bounded away from zero we conclude from the equation $\gamma P Z = Q$ that Z is a.s. continuous as well. $\qquad \square$

7 Back to the Pricing Problem

We now come back to the numerical valuation of the put option on kerosene as depicted in Example 1. Notations in the following are adopted from Sect. 3. Assume that the put option expires at $T = 1$. Let R and S denote the dynamics for the financial value of kerosene and heating oil respectively. In particular we assume both dynamics to be lognormally distributed according to

$$dR_t = \mu(t, R_t)dt + \sigma(t, R_t)dW_t^1 = 0.12\, R_t\, dt + 0.41\, R_t\, dW_t^1,$$

$$\frac{dS_t}{S_t} = \alpha(t, R_t)dt + \beta(t, R_t)dW_t^3 = 0.1\, dt + 0.35\, dW_t^3,$$

and we assume the spot price for heating oil to be $s_0 = 173$ money units (e.g. US Dollar, Euro), see also equations (3) and (4). Risk aversion is set at the level of

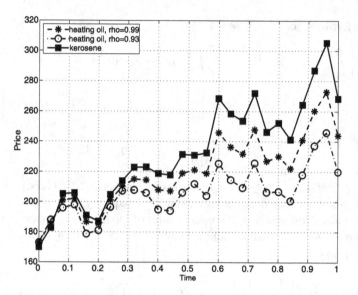

Fig. 1 Price paths of the nontradable asset kerosene and the correlated asset heating oil at different correlation levels. The spot of kerosene was set to $r_0 = 170$

$\eta = 0.3$. Figure 1 displays sample paths of the kerosene price with a spot price of $r_0 = 170$ and heating oil price at different correlation levels using the explicit solution formula for the geometric Brownian motion. We see that the higher the correlation, the better the approximation of the kerosene by heating oil becomes. We have seen that the valuation of the put option via utility maximization yields the pricing formula (8) which in conjunction with Lemma 1 becomes the difference of two solutions of a qgBSDE with the generator (10)

$$p_t = Y_t^F - Y_t^0, \quad 0 \le t \le T,$$

where $F(x) = (K - x)^+$ for some strike $K > 0$. For the numerical simulation of the qgFBSDE Y^F and Y^0, we apply the exponential transformation to both BSDE (see Sect. 6) and then employ the algorithm by [3] with $N = 100$ equidistant time points, 70000 paths and a regression basis consisting of five monomials and the payoff function of the put option. The Picard iteration stops as soon as the difference of two subsequent time zero values is less than 10^{-5}. Simulations reveal that 12 to 13 iterations are needed for solving one exponentially transformed qgFBSDE. Figures 2(a) and 2(b) depict the time zero price p_0 of the put option at different strike and kerosene spot levels. The lower the correlation, the lower the price becomes. This is clear because lower correlations between heating oil and kerosene lead to higher non-hedgeable residual risk which diminishes the risk covering effect of the contingent claim and thus also its value. Figures 3(a) and 3(b) depict sample paths of the dynamics for the price p_t and the optimal investment strategy π_t for an at the money put with strike $K = 180$ and kerosene spot $r_0 = 170$. The plots depict price and monetary investment for every fourth time point of the discretization. The

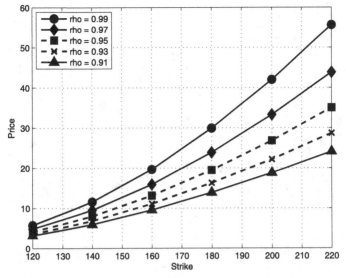

(a) Put option price in terms varying strikes at a fixed kerosene spot $r_0 = 170$

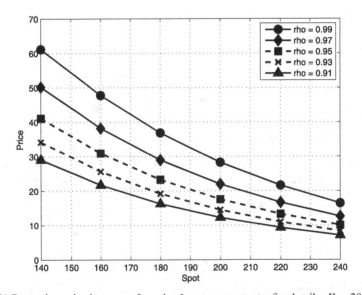

(b) Put option price in terms of varying kerosene spots at a fixed strike $K = 200$

Fig. 2 Values of the put option in terms of kerosene spot and strike for varying correlations. High correlations lead to high the prices for the contingent claim

price process and the dynamics of the optimal investment strategy are intertwined: high fluctuations of the price process result in high fluctuations of the investment strategy and vice versa. In general we observe that replication on high correlation levels tends to entail greater market activity because kerosene price risks can then be

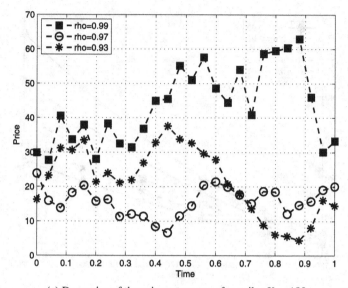

(a) Dynamics of the price process p_t for strike $K = 180$

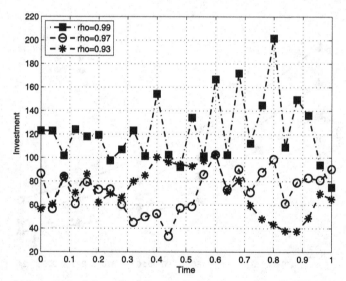

(b) Dynamics of the optimal investment strategy π_t for strike $K = 180$

Fig. 3 Paths of the price p_t and the optimal investment strategy π_t for varying correlation levels. In general high correlations entail greater market activity

well hedged by market transactions that move closely along the dynamics of heating oil. In contrast, replication on lower correlation levels leads to a higher amount of residual risk which is inaccessible for hedging and thus lower market activity is needed.

Acknowledgements Gonçalo Dos Reis would like to thank both Romuald Elie and Emmanuel Gobet for the helpful discussions. Jianing Zhang acknowledges financial support by IRTG 1339 SMCP.

Appendix 1: Some Results on BMO Martingales

BMO martingales play a key role for a priori estimates needed in our sensitivity analysis of solutions of BSDE. For details about this theory we refer the reader to [13].

Let Φ be a $BMO(\mathscr{F}, \mathbb{Q})$ martingale with $\Phi_0 = 0$. Φ being square integrable, the martingale representation theorem yields a square integrable process ϕ such that $\Phi_t = \int_0^t \phi_s dW_s, t \in [0, T]$. Hence the $BMO(\mathscr{F}, \mathbb{Q})$ norm of Φ can be alternatively expressed as

$$\sup_{\tau \; \mathscr{F}\text{-stopping time in } [0,T]} \mathbb{E}^{\mathbb{Q}}\left[\int_\tau^T \phi_s^2 ds | \mathscr{F}_\tau\right] < \infty.$$

Lemma 2 (Properties of BMO martingales) *Let Φ be a BMO martingale. Then we have:*

1) *The stochastic exponential $\mathscr{E}(\Phi)$ is uniformly integrable.*
2) *There exists a number $r > 1$ such that $\mathscr{E}(\Phi_T) \in L^r$. This property follows from the* Reverse Hölder inequality. *The maximal r with this property can be expressed explicitly in terms of the BMO norm of Φ.*
3) *If $\Phi = \int_0^\cdot \phi_s ds$ has BMO norm C, then for $p \geq 1$ the following estimate holds*

$$\mathbb{E}\left[\left(\int_0^T |\phi_s|^2 ds\right)^p\right] \leq 2p!(4C^2)^p.$$

Hence BMO $\subset \mathscr{H}^p$ for all $p \geq 1$.

Appendix 2: Basics of Malliavin's Calculus

We briefly introduce the main notation of the stochastic calculus of variations also known as Malliavin's calculus. For more details, we refer the reader to [18]. Let $\mathscr{S}$ be the space of random variables of the form

$$\xi = F\left(\left(\int_0^T h_s^{1,i} dW_s^1\right)_{1 \leq i \leq n}, \ldots, \left(\int_0^T h_s^{d,i} dW_s^d\right)_{1 \leq i \leq n}\right),$$

where $F \in C_b^\infty(\mathbb{R}^{n \times d})$, $h^1, \ldots, h^n \in L^2([0, T]; \mathbb{R}^d)$, $n \in \mathbb{N}$. To simplify notation, assume that all h^j are written as row vectors. For $\xi \in \mathscr{S}$, we define $D = (D^1, \ldots, D^d) : \mathscr{S} \to L^2(\Omega \times [0, T])^d$ by

$$D_\theta^i \xi = \sum_{j=1}^n \frac{\partial F}{\partial x_{i,j}}\left(\int_0^T h_t^1 dW_t, \ldots, \int_0^T h_t^n dW_t\right) h_\theta^{i,j}, \quad 0 \leq \theta \leq T, \; 1 \leq i \leq d,$$

and for $k \in \mathbb{N}$ its k-fold iteration by

$$D^{(k)} = (D^{i_1} \dots D^{i_k})_{1 \leq i_1, \dots, i_k \leq d}.$$

For $k \in \mathbb{N}$, $p \geq 1$ let $\mathbb{D}^{k,p}$ be the closure of $\mathscr{S}$ with respect to the norm

$$\|\xi\|_{k,p}^p = \mathbb{E}\left[\|\xi\|_{L^p}^p + \sum_{i=1}^{k} \||D^{(k)}\xi\||_{(\mathscr{H}^p)^i}^p\right].$$

$D^{(k)}$ is a closed linear operator on the space $\mathbb{D}^{k,p}$. Observe that if $\xi \in \mathbb{D}^{1,2}$ is $\mathscr{F}_t$-measurable then $D_\theta \xi = 0$ for $\theta \in (t, T]$. Further denote $\mathbb{D}^{k,\infty} = \cap_{p>1} \mathbb{D}^{k,p}$.

We also need Malliavin's calculus for smooth stochastic processes with values in $\mathbb{R}^m$. For $k \in \mathbb{N}$, $p \geq 1$, denote by $\mathbb{L}_{k,p}(\mathbb{R}^m)$ the set of $\mathbb{R}^m$-valued progressively measurable processes $u = (u^1, \dots, u^m)$ on $[0, T] \times \Omega$ such that

   i) For Lebesgue-a.a. $t \in [0, T]$, $u(t, \cdot) \in (\mathbb{D}^{k,p})^m$;
  ii) $[0, T] \times \Omega \ni (t, \omega) \mapsto D^{(k)} u(t, \omega) \in (L^2([0, T]^{1+k}))^{d \times n}$ admits a progressively measurable version;
 iii) $\|u\|_{k,p}^p = \|u\|_{\mathscr{H}^p}^p + \sum_{i=1}^{k} \|D^i u\|_{(\mathscr{H}^p)^{1+i}}^p < \infty$.

Note that Jensen's inequality gives for all $p \geq 2$

$$\mathbb{E}\left[\left(\int_0^T \int_0^T |D_u X_t|^2 du\, dt\right)^{\frac{p}{2}}\right] \leq T^{p/2-1} \int_0^T \|D_u X\|_{\mathscr{H}^p}^p du.$$

Appendix 3: Some Results on SDE

We recall results on SDE known from the literature that are relevant for this work. We state our assumptions in the multidimensional setting. However, for ease of notation we present some formulas in the one dimensional case.

Theorem 9 (Moment estimates for SDE) *Assume that* (H0) *holds. Then* (1) *has a unique solution* $X \in \mathscr{S}^2$ *and the following moment estimates hold: for any* $p \geq 2$ *there exists a constant* $C > 0$, *depending only on* T, K *and* p *such that for any* $x \in \mathbb{R}^m$, $s, t \in [0, T]$

$$\mathbb{E}\left[\sup_{0 \leq t \leq T} |X_t|^p\right] \leq C\mathbb{E}\left[|x|^p + \int_0^T (|b(t,0)|^p + |\sigma(t,0)|^p)dt\right],$$

$$\mathbb{E}\left[\sup_{s \leq u \leq t} |X_u - X_s|^p\right] \leq C\mathbb{E}\left[|x|^p + \sup_{0 \leq t \leq T} \{|b(t,0)|^p + |\sigma(t,0)|^p\}\right]|t - s|^{p/2}.$$

Furthermore, given two different initial conditions $x, x' \in \mathbb{R}^m$, *we have*

$$\mathbb{E}\left[\sup_{0 \leq t \leq T} |X_t^x - X_t^{x'}|^p\right] \leq C|x - x'|^p.$$

Theorem 10 (Classical differentiability) *Assume* (H1) *holds. Then the solution process* X *of* (1) *as a function of the initial condition* $x \in \mathbb{R}^m$ *is differentiable and satisfies for* $t \in [0, T]$

$$\nabla X_t = I_m + \int_0^t \nabla b(X_s) \nabla X_s ds + \int_0^t \nabla \sigma(X_s) \nabla X_s dW_s, \tag{21}$$

where I_m *denotes the* $m \times m$ *unit matrix. Moreover,* ∇X_t *as an* $m \times m$-*matrix is invertible for any* $t \in [0, T]$. *Its inverse* $(\nabla X_t)^{-1}$ *satisfies an SDE and for any* $p \geq 2$ *there are positive constants* C_p *and* c_p *such that*

$$\|\nabla X\|_{\mathscr{S}^p} + \|(\nabla X)^{-1}\|_{\mathscr{S}^p} \leq C_p$$

and

$$\mathbb{E}\left[\sup_{s \leq u \leq t} |(\nabla X_u) - (\nabla X_s)|^p + \sup_{s \leq u \leq t} |(\nabla X_u)^{-1} - (\nabla X_s)^{-1}|^p \right] \leq c_p |t - s|^{p/2}.$$

Theorem 11 (Malliavin Differentiability) *Under* (H1), $X \in \mathbb{L}_{1,2}$ *and its Malliavin derivative admits a version* $(u, t) \mapsto D_u X_t$ *satisfying for* $0 \leq u \leq t \leq T$ *the SDE*

$$D_u X_t = \sigma(X_u) + \int_u^t \nabla b(X_s) D_u X_s ds + \int_u^t \nabla \sigma(X_s) D_u X_s dW_s. \tag{22}$$

Moreover, for any $p \geq 2$ *there is a constant* $C_p > 0$ *such that for* $x \in \mathbb{R}^m$ *and* $0 \leq v \leq u \leq t \leq s \leq T$

$$\|D_u X\|_{\mathscr{S}^p}^p \leq C_p(1 + |x|^p),$$

$$\mathbb{E}[|D_u X_t - D_u X_s|^p] \leq C_p(1 + |x|^p)|t - s|^{\frac{p}{2}},$$

$$\|D_u X - D_v X\|_{\mathscr{S}^p}^p \leq C_p(1 + |x|^p)|u - v|^{\frac{p}{2}}.$$

By Theorem 10, we have the representation

$$D_u X_t = \nabla X_t (\nabla X_u)^{-1} \sigma(X_u) \mathbf{1}_{[0,u]}(t), \quad \text{for all } u, t \in [0, T].$$

Appendix 4: Path Regularity for Lipschitz FBSDE

We state a version of the L^2-regularity result for FBSDE satisfying a global Lipschitz condition. The result which was seen to be closely related to the convergence of numerical schemes for systems of FBSDE is due to [19]. For our FBSDE system (1), (2) we assume that b, σ, f, g are deterministic measurable functions that are Lipschitz continuous with respect to the spatial variables and $\frac{1}{2}$-Hölder continuous with respect to time. Furthermore we assume that σ satisfies (13). Then from [8] one easily obtains existence and uniqueness of a solution triple (X, Y, Z) of FBSDE (1), (2) belonging to $\mathscr{S}^2 \times \mathscr{S}^2 \times \mathscr{H}^2$. For a partition π of $[0, T]$ define the process $\bar{Z}^\pi$ as in (15). Then the following result holds.

Theorem 12 (Path regularity result of [19]) *Let* $(X, Y, Z) \in \mathscr{S}^2 \times \mathscr{S}^2 \times \mathscr{H}^2$ *be the solution of FBSDE* (1), (2) *in the setting described above. Then there exists* $C \in \mathbb{R}_+$ *such that for any partition* $\pi = \{t_0, \ldots, t_N\}$ *of the time interval* $[0, T]$ *with mesh size* $|\pi|$ *we have*

$$
\max_{0 \le i \le N-1} \left\{ \sup_{t \in [t_i, t_{i+1})} \mathbb{E}\left[|Y_t - Y_{t_i}|^2 \right] \right\}
$$

$$
+ \sum_{i=0}^{N-1} \mathbb{E}\left[\int_{t_i}^{t_{i+1}} |Z_s - \bar{Z}_{t_i}^\pi|^2 ds \right] + \sum_{i=0}^{N-1} \mathbb{E}\left[\int_{t_i}^{t_{i+1}} |Z_s - Z_{t_i}|^2 ds \right] \le C|\pi|.
$$

References

1. Ankirchner, S., Imkeller, P., dos Reis, G.: Classical and variational differentiability of BSDEs with quadratic growth. Electron. J. Probab. **12**(53), 1418–1453 (2007)
2. Ankirchner, S., Imkeller, P., dos Reis, G.: Pricing and hedging of derivatives based on non-tradable underlyings. Math. Finance **20**(2), 289–312 (2010)
3. Bender, C., Denk, R.: A forward simulation of backward SDEs. Stochastic Process. Appl. **117**(12), 1793–1812 (2007)
4. Bouchard, B., Touzi, N.: Discrete-time approximation and Monte-Carlo simulation of backward stochastic differential equations. Stochastic Process. Appl. **111**(2), 175–206 (2004)
5. Briand, P., Confortola, F.: BSDEs with stochastic Lipschitz condition and quadratic PDEs in Hilbert spaces. Stochastic Process. Appl. **118**(5), 818–838 (2008)
6. Delarue, F., Menozzi, S.: A forward-backward stochastic algorithm for quasi-linear PDEs. Ann. Appl. Probab. **16**(1), 140–184 (2006)
7. dos Reis, G.: On some properties of solutions of quadratic growth BSDE and applications in finance and insurance. PhD thesis, Humboldt University (2009)
8. El Karoui, N., Peng, S., Quenez, M.C.: Backward stochastic differential equations in finance. Math. Finance **7**(1), 1–71 (1997)
9. Elie, R.: Contrôle stochastique et méthodes numériques en finance mathématique. PhD thesis, Université Paris-Dauphine (2006)
10. Frei, C.: 2009 Convergence results for the indifference value in a Brownian setting with variable correlation. Preprint, available at www.cmapx.polytechnique.fr/~frei
11. Gobet, E., Lemor, J.P., Warin, X.: A regression-based Monte Carlo method to solve backward stochastic differential equations. Ann. Appl. Probab. **15**(3), 2172–2202 (2005)
12. Imkeller, P., Réveillac, A., Richter, A.: Differentiability of quadratic BSDE generated by continuous martingales and hedging in incomplete markets. arXiv:0907.0941v1 (2009)
13. Kazamaki, N.: Continuous Exponential Martingales and BMO. *Lecture Notes in Mathematics*, vol. 1579, p. 91. Springer, Berlin (1994)
14. Kobylanski, M.: Backward stochastic differential equations and partial differential equations with quadratic growth. Ann. Probab. **28**(2), 558–602 (2000)
15. Ma, J., Zhang, J.: Path regularity for solutions of backward stochastic differential equations. Probab. Theory Relat. Fields **122**(2), 163–190 (2002)
16. Mania, M., Schweizer, M.: Dynamic exponential utility indifference valuation. Ann. Appl. Probab. **15**(3), 2113–2143 (2005)
17. Morlais, M.A.: Quadratic BSDEs driven by a continuous martingale and applications to the utility maximization problem. Finance Stoch. **13**(1), 121–150 (2009)
18. Nualart, D.: The Malliavin Calculus and Related Topics, 2nd edn. *Probability and its Applications (New York)*, p. 382. Springer, Berlin (2006)
19. Zhang, J.: Some fine properties of BSDE. PhD thesis, Purdue University (2001)

Variance Swap Portfolio Theory

Dilip B. Madan

Abstract Optimal portfolios of variance swaps are constructed taking account of both autocorrelation and cross asset dependencies. Market prices of variance swaps are extracted from option surface calibrations. The methods developed permit simulation of cash flows to arbitrary portfolios of variance swaps. The optimal design maximizes the index of acceptability introduced in (Cherny and Madan, Review of Financial Studies, 2009). Full nonlinear optimization is contrasted with Simultaneous Perturbation Stochastic Approximation (*SPSA*). Preliminary out of sample results favor the use of *SPSA*.

1 Introduction

Financial markets now allow one to invest directly in the realized variances of stocks over a prespecified interval time. A natural question that arises is the construction or design of optimal portfolios of realized variance. This is the analog of the classical portfolio theory for stock investment, beginning with [5], except we now hold portfolios of realized variance and seek an optimal position. Our approach to this question is broadly similar to the traditional analysis offered by [5], requiring one to describe the distributional properties of various portfolios with the selected portfolio being one that optimizes a criterion measuring the quality of the distribution accessed. The context of trading variance swaps or realized squared returns over a prespecified future period, however, introduces a variety of complexities in the task of describing the distribution of cash flows accessed by potential portfolios.

The first among these is the recognition that squared returns are highly autocorrelated and are subject to considerable levels of clustering. Consequently one must account for such autocorrelations in describing future outcomes. Additionally, volatility movements are also correlated across assets and have distributions could

D.B. Madan (✉)
Robert H. Smith School of Business, University of Maryland, College Park, MD 20742, USA
e-mail: dmadan@rhsmith.umd.edu

C. Chiarella, A. Novikov (eds.), *Contemporary Quantitative Finance*,
DOI 10.1007/978-3-642-03479-4_10, © Springer-Verlag Berlin Heidelberg 2010

possess both skewness and excess kurtosis. The design of optimal portfolios should therefore reflect these aspects of the problem and a simulation of potential portfolio cash flows should preferably employ non-Gaussian yet correlated outcomes across assets and time, coupled with optimization criteria that are sensitive to skewness. The issues to be addressed are therefore somewhat more involved than those faced in studying stock portfolios. Ultimately we wish to be able to simulate the joint evolution of squared log returns on a number of stocks for a fixed number of days representing the maturity of the variance swap contracts to be traded. One may then take long and short positions in these cash flows to generate the outcomes for a candidate portfolio of variance swaps. In the process we need to ensure that the simulated squared returns remain positive.

We begin by directly modeling time series data for daily squared log returns. To account for autocorrelation in these data series it is natural to consider a regression model for daily squared log returns with their lagged values as explanatory variables. However, it is difficult to keep a linear regression model positive in a forward simulation. One could run then run the regressions in the logarithm of daily squared log returns, but placing a probability model on the residuals of such a transformation makes returns a double exponential and one is likely to end up with very unreasonable return outcomes. With a view to overcoming this difficulty we consider transformations other than the exponential that map the real line to the positive half line but have a linear rather than exponential behaviour for large arguments. We call this transformation a Hardy-Littlewood Gauss transform as it is based on using a Gaussian density in a transformation suggested by Hardy and Littlewood. The regression is run on data obtained by applying the inverse Hardy Littlewood Gauss transform to daily squared log returns.

After accounting for autocorrelation we model the marginal distributions of regression residuals, expecting them to display both some skewness and excess kurtosis. A simple and fairly robust distribution with these properties is given by the variance gamma, *VG* model. For the modeling of dependence we follow [4] and use the marginal distributions to transform the residuals to Gaussian random variates that we then correlate. Our model for the residuals is then a nonlinear transform of correlated multivariate normals. We may then simulate forward the regression model and obtain candidate observations on squared log returns by applying the Hardy Littlewood Gauss transform to the output of the simulated regression. These cash flows may be averaged and annualized to get the floating or random leg of the individual variance swaps. The price of the variance swap as estimated from a calibration of the individual option surface yields the fixed leg of the variance swap and subtracting this value from the floating leg one obtains readings on individual variance swap outcomes.

Portfolio cash flows can then be easily constructed and it remains to describe the criterion for selection of the optimal portfolio. For this purpose we follow recent advances in performance measurement introduced in [2] and maximize an index of acceptability. Such indices are infinite for arbitrages and hence if there is a sample space arbitrage the optimization of such a criterion will find the arbitrage. Like the Sharpe ratio or Gain Loss ratio, the criterion is scale invariant and the search for the portfolio design may be restricted to the compact set defined by the unit sphere. We employ the suggested criterion of maximizing the *MINMAXVAR* acceptability index as this criterion is particularly sensitive towards the avoidance of large losses, and is in this regard conservative.

In constructing the optimal portfolio one is optimizing on a simulated space of outcomes and a nonlinear optimizer may overly leverage aspects of the sample space that are not relevant to real out of sample performance. For this reason we employ both a general nonlinear optimizer and stochastic approximation using simultaneous perturbation for gradient estimation to economize on function evaluations in high dimensions.

The complete portfolio design strategy is illustrated on a sample of the top 50 stocks of the Standard and Poor Index along with a position in the index variance. We report on trading one month variance swaps at nonoverlapping one month intervals. The model is estimated, simulated, and the optimal portfolio is constructed at each month end using prior data on an adaptive basis. Sample empirical results for each section are summarized in that section. A final section reports on the actual out of sample performance of each portfolio constructed at month end.

The outline of the rest of the paper is as follows. Section 1 describes the autocorrelation strategy and the use of the Hardy Littlewood Gauss transform. The joint law for the regression residuals are formulated in Sect. 2. Section 3 presents the model used to price the fixed leg of the variance swap contracts. Section 4 presents the optimization criterion, the problem constraints, and the two optimization methods employed. Results are provided in Sect. 5 and Sect. 6 concludes.

2 Autocorrelation and the Hardy Littlewood Gauss Transform

We take data on the prices of the top 50 stocks on the S&P500 index as at November 11 2008 along with the time series of the index as well. Let $S_{i,t}$ denote the price of asset i at market close on day t for $i = 1, \ldots, 51$ with asset 51 being the index level. The daily realized variance for asset i on day t is defined by

$$v_{i,t} = \left(\ln \left(\frac{S_{i,t}}{S_{i,t-1}} \right) \right)^2$$

and the cash flow on a one month variance swap entered into on day t is given on day $t + 21$ by

$$Notional \times \left(\frac{252}{21} \sum_{n=t+1}^{t+21} v_{i,n} - k_{i,t}^2 \right)$$

where $k_{i,t}$ is the variance swap quote for asset i at market close on day t and is an annualized volatility. An analysis of variance swap portfolios requires a simultaneous generation of daily squared log returns for the coming months on all 51 assets.

As noted in the introduction, squared log returns are highly autocorrelated and the simulation of a linear model estimated on data for $v_{i,t}$ will not stay positive in the forecast period. We therefore wish to transform the data on $v_{i,t}$ from the positive half line to the real line and a popular transformation with this property is the logarithm. However, if we model $v_{i,t}$ as an exponential then returns are a double exponential and this introduces unreasonable return outcomes for any reasonable probability model placed on the logarithm of $v_{i,t}$. We therefore consider other transformations from the real line to the positive half line that do not grow as fast as an exponential.

There are a wide class of such transformations $g(x)$ that behave linearly for positive and large x values. Let $f(x)$ be any symmetric density on the real line with $f(-x) = f(x)$ and finite expectation of $|x|$. A candidate for the transformation $g(x)$ is given by

$$g(x) = \frac{\int_x^\infty u f(u) du}{\int_x^\infty f(u) du}. \tag{1}$$

The function g is easily seen to be positive, and tending to 0 as x tends to negative infinity, while for large x, $g(x)$ behaves like x. We refer to (1) as a Hardy Littlewood transform and we shall use the Hardy Littlewood Gauss transform when we take $f(x)$ to be the standard normal density. Applying the inverse of the Hardy Littlewood Gauss transform to data on $v_{i,t}$ we generate real valued data

$$x_{i,t} = g^{-1}(v_{i,t}).$$

We account for autocorrelation in squared returns by employing a linear model on the data $x_{i,t}$. In our specific application we take account of five lags and estimate the linear model

$$x_{i,t} = a_i + \sum_{j=1}^5 b_j x_{i,t-j} + u_{i,t}.$$

The linear model was estimated on six months of daily data prior to the date of portfolio construction. Forecasts of $x_{i,t}$ may be transformed to positive forecasts of squared daily log returns on applying the Hardy Littlewood Gauss transform to these forecasts.

3 NonGaussian Dependence Model

In addition to autocorrelation, we need to account for the possibility of skewness, excess kurtosis and correlation across assets. We observe that the residuals from the autocorrelation regressions do possess significant levels of skewness and excess kurtosis and it would be risky to overly understate extreme outcomes in modeling squared return cash flow outcomes. With these considerations in mind we estimate the variance gamma model on standardized residuals by maximum likelihood. This model can capture nonzero departures in skewness and excess kurtosis in standardized data. Given that we have non Gaussian marginal distributions for our regression residuals, it is not helpful to merely correlate these residuals as one has no joint probability law associated with these correlation computations from which one may draw on possible candidate outcomes.

Here we follow [4] and transform our standardized residuals to standard Gaussian variates by using the estimated VG distribution functions to obtain

$$z_{it} = N^{-1}\left(F_i^{VG}(u_{i,t})\right)$$

where $N(x)$ is the standard normal distribution function. We then correlate the data on z_{it} to get a residual correlation matrix C. One may then simulate outcomes from a multivariate Gaussian with correlation C to construct non Gaussian correlated residuals by

$$u_{i,t,s} = (F_i^{VG})^{-1}(N(z_{i,t,s})).$$

These residuals are fed into the regression model to generate $x_{i,t,s}$ a candidate occurrence for asset i at time t on paths s. Applying the Hardy Littlewood Gauss transform provides outcomes $v_{i,t,s}$. We employed 10000 paths and generated 21 forward squared returns on each of 51 assets after estimating 51 variance gamma models on the residuals $u_{i,t}$ of a regression run on six months of historical data.

The floating leg of the variance swap contract for unit notional can then be constructed on each of 10000 paths as

$$V_{is} = \frac{252}{21} \sum_{n=t+1}^{t+21} v_{i,n,s}.$$

The cash flow requires the construction of the fixed leg on each asset. This is accomplished in the next section.

4 Variance Swap Fixed Leg Pricing

For each of the 51 assets, the 50 stocks and the $S\&P500$ index, on each portfolio construction date, we have to obtain the price of the variance swap contract for the maturity of the variance contract being considered. For this purpose we extract prices of out-of-the-money options in the maturity range of one month to one year and follow [1] to summarize this surface using the four parameters of the Sato process based on the VG and termed $VGSSD$ in [1]. Once this model is estimated on the option surface one has an analytical expression for the risk distribution of the logarithm stock at each maturity. The price of the variance swap contract may be estimated by computing the expectation of the log price relative treated as a futures price process by employing a zero interest rate and dividend yield. Specifically let

$$\phi_M(u, t) = E^Q\left[\exp\left(iu\ln(M(t))\right)\right]$$

where the stock price process $S(t)$ for an interest rate of r and a dividend yield q is given by

$$S(t) = S(0)e^{(r-q)t}M(t).$$

The characteristic function for $M(t)$ is easily extracted from that for $S(t)$. The price of the variance swap contract, k, is then computed by

$$k^2 t = 2i \frac{\partial \phi_M(u, t)}{\partial u}|_{u=0}.$$

For each portfolio construction date we estimated the *VGSSD* model for all 51 underliers and then transformed the parameters into a variance swap price. We have tested this procedure applied to SPX options with one month variance swap quotes then being compared with the *VIX* index. The correlation over a 300 day period was 99.26 and one does not reject the hypothesis of a zero intercept and a unit slope on regressing one against the other. This value is denoted $k_{i,t}$ and the simulated cash flow to asset i on the variance swap on path s is then obtained as

$$c_{i,s} = V_{i,s} - k_{i,t}^2.$$

The matrix of cash flows to potential swaps is 51 by 10000 and can be generated on each portfolio design date using this procedure.

5 Optimization Criterion and Portfolio Design

We now consider the formation of portfolios taking positions $a = (a_i, i = 1, \ldots, 51)$ in the variance swap on asset i. The magnitude a_i is a dollar notional amount that may be positive or negative depending on whether the swap is purchased or sold, with the portfolio cash flow $c_s(a)$ on path s being

$$c_s(a) = \sum_{i=1}^{51} a_i c_{is}.$$

The portfolio design strategy is to choose a with a view to optimizing the random variable being accessed. Traditional design criteria include Sharpe ratios, expected utility, *RAROC*, the gain loss ratio among other possibilities. We note that all these performance measurement criteria are just functions of the probability law of the random variable being accessed. Recently, [2] reviewed these criteria for measuring performance and axiomatized and defined indices of acceptability for random cash flows that range from zero for a positive expectation to infinity for an arbitrage. The indices, like other performance measures are scale invariant. The indices that depend solely on the probability law are easily constructed as the highest level of stress such that a stressed expectation is positive at this level. Stressed expectations are expectations under a concave distortion of the distribution function that effectively exaggerates losses and simultaneously discounts gains, much like marginal utility, though unlike a specific utility the reweighting is further exaggerated as one raises the stress level. The criterion delivers a convex set of acceptable risks at each level of acceptability and tractable examples of such indices include *MINVAR* where the stress amounts to computing the expectation of the minimum of a number of independent draws. It is noted that large losses are not heavily discounted in *MINVAR* and the proposed criterion is *MINMAXVAR* that uses the distortion Ψ^γ at level γ of

$$\Psi^\gamma(u) = 1 - (1 - u^{\frac{1}{1+\gamma}})^{1+\gamma}, \quad \cdot \, 0 \le u \le 1.$$

We choose our portfolio of variance swaps to maximize the level of acceptability of the cash flow being accessed. This criterion has been employed in portfolio allocation across stocks by [3]. The computation of the distorted expectation amounts to evaluating

$$\int_{-\infty}^{\infty} c\, d\Psi^{\gamma}(F(c))$$

for a cash flow with distribution function $F(c)$. When a cash flow is sampled by simulation one estimates the distorted expectation by sorting the cash flows in increasing order $c_1 < c_2 < \cdots < c_N$ and evaluating the sum

$$\sum_{n=1}^{N} c_n \left(\Psi^{\gamma}\left(\frac{n}{N} \right) - \Psi^{\gamma}\left(\frac{n-1}{N} \right) \right).$$

The index is given by the largest level of γ for which the distorted expectation is positive and it is clear that this is infinity if there are no negative outcomes and one has a sample space arbitrage or a real arbitrage.

Given that the objective is scale invariant one may restrict the portfolio design search to the unit sphere by requiring that

$$\sum_{i=1}^{51} a_i^2 = 1. \tag{2}$$

We are interested in designing a relative variance trade that goes buys the more volatile and sells the less volatile and so we take an aggregate portfolio that is zero dollar. This constrains the portfolio to satisfy

$$\sum_{i=1}^{51} k_{i,t}^2 a_i = 0. \tag{3}$$

Additionally we avoid an aggregate volatility risk with a zero vega constraint by requiring

$$\sum_{i=1}^{51} k_{i,t} a_i = 0. \tag{4}$$

We therefore design the portfolio to maximize the index of acceptability subject to the constraints (2), (3), and (4).

The optimization may be accomplished on a fixed sample space, or one may regenerate the sample space and treat the problem as one with a stochastic objective. The regeneration is quite expensive and we maximized the criterion on a fixed sample space. A full nonlinear optimization may however unduly leverage the sample space and one may consider instead a stochastic approximation algorithm. Given the dimension of 51 the gradient computation in such an algorithm is expensive and so we coupled stochastic approximation with simultaneous perturbation to get a gradient estimate in just two function evaluations (see [6]). We present results for both algorithms and report on both the in sample and out of sample performance.

Table 1 Robust regression results

Ticker	Constant	Lag 1	Lag 2	Lag 3	Lag 4	Lag 5
xom	−2.7045	−0.0122	0.0449	0.0647	0.0928	0.1192
ge	−1.7717	0.0771	0.0110	0.0551	0.1496	0.2501
msft	−2.6504	0.2050	−0.0059	0.0307	0.0587	0.0308
jpm	−2.5522	0.1024	0.0489	0.0200	0.1089	0.0038
bac	−1.6697	0.1135	0.1250	0.1089	0.1140	0.0474
aapl	−2.3872	−0.0151	0.1460	0.1363	0.1135	−0.0191
goog	−2.3593	0.0975	0.1255	−0.0556	0.1487	0.0697
c	−2.1977	0.1395	0.1332	−0.0750	0.1280	0.0463
mmm	−2.5363	0.0553	−0.0053	0.0219	0.1715	0.1293
spx	−1.9689	0.0314	0.0890	0.0818	0.1448	0.1687

We find that the full nonlinear optimization of the index of acceptability does find sample space arbitrages with the zero dollar and zero vega constraints imposed. In this case we also maximize the minimum cash flow on the sample space.

6 Results on Portfolio Design

Portfolios were formulated once a month in 2008 for 9 months ranging from *February* through *October*. The exact design dates were *February* 27, *March* 27, *April* 25, *May* 26, *June* 24, *July* 23, *August* 21, *September* 19, and *October* 20. We shall summarize the back test results of all nine positionings but describe in detail the components for just the last portfolio for October 20.

We present results for a subset of the 50 companies and the *SPX*. The first step was the robust regression performed on the inverse Hardy Littlewood Gauss transform of daily squared log returns with five lags and a constant run on six months of prior data. The coefficient matrix for the subset is presented in Table 1.

We observe that the constant term for the *SPX* is lower indicative of a lower overall volatility. The lagged coefficients are mainly positive with a substantial relative size at lag five for four of the ten names listed.

The second step is the estimation of marginal laws for the residuals of the regression. This was done on standardized data for the residuals with generated samples being destandardized using the sample means and standard deviations. The probability model used was the variance gamma. We report the parameter estimates for the selected 10 names in Table 2.

We observe significant levels of excess kurtosis and negative skewness indicating that residual densities are skewed down relative to the regression forecast. This is particularly the case for the SPX.

We next use the marginal distributions to transform the residual data to a standard Gaussian density that we then correlate. The eigenvalues of the correlation matrix are reported in Table 3.

Table 2 VG estimates for standardized residuals

Ticker	σ	ν	θ
xom	0.6491	1.2013	−0.2003
ge	0.5139	0.4502	−0.7925
msft	0.4593	0.7754	−0.5303
jpm	0.6269	0.7561	−0.3358
bac	0.7387	1.0487	−0.2548
aapl	0.6173	1.6938	−0.1939
goog	0.5239	0.6704	−0.3807
c	0.6148	0.8940	−0.5860
mmm	0.5278	0.8309	−0.3996
spx	0.0794	0.3201	−1.2305

Table 3 Top 25 eigenvalues of correlation matrix in decreasing order

22.2209	2.9558	2.0834	1.5779	1.3561
1.1752	1.1204	1.0572	1.0271	0.9435
0.9092	0.8961	0.7928	0.7532	0.7460
0.7048	0.6626	0.6148	0.5842	0.5702
0.5178	0.5051	0.4914	0.4654	0.4371

Table 4 Selected residual correlations

	xom	ge	msft	jpm	bac	aapl	goog	c	mmm	spx
xom	1	0.4734	0.4361	0.3369	0.3106	0.4033	0.3278	0.2686	0.4563	0.5165
ge	0.4734	1	0.3821	0.5097	0.4662	0.5495	0.4011	0.5454	0.6269	0.6990
msft	0.4361	0.3821	1	0.3910	0.3252	0.5290	0.6485	0.3809	0.4588	0.5729
jpm	0.3369	0.5097	0.3910	1	0.7459	0.4795	0.4303	0.7082	0.4901	0.5695
bac	0.3106	0.4662	0.3252	0.7459	1	0.4537	0.3859	0.7124	0.5328	0.5680
aapl	0.4033	0.5495	0.5290	0.4795	0.4537	1	0.5408	0.5324	0.4862	0.5748
goog	0.3278	0.4011	0.6485	0.4303	0.3859	0.5408	1	0.4379	0.4869	0.6172
c	0.2686	0.5454	0.3809	0.7082	0.7124	0.5324	0.4379	1	0.4876	0.5694
mmm	0.4563	0.6269	0.4588	0.4901	0.5328	0.4862	0.4869	0.4876	1	0.7070
spx	0.5165	0.6990	0.5729	0.5695	0.5680	0.5748	0.6172	0.5694	0.7070	1

The smallest eigenvalue was 0.0643. The correlations between the ten selected names are provided in Table 4.

We observe that the financial stocks are more correlated with each other that with other assets as are the technology stocks.

We may now simulate correlated Gaussians from the estimated correlation matrix, then transform them to dependent marginals that are fed into updating the regression equations whose output is transformed by the Hardy Littlewood Gauss

Table 5 VGSSD parameters on 20081020

Ticker	σ	ν	θ	γ	RMSE	APE	MOP
xom	0.3879	1.2669	−0.3690	0.4099	0.1959	0.0376	52
ge	0.5399	0.9931	−0.4632	0.5422	0.0961	0.0460	28
msft	0.4566	1.6828	−0.3560	0.4038	0.0505	0.0214	35
jpm	0.5517	0.9707	−0.6450	0.5258	0.1196	0.0288	64
bac	0.5991	0.8029	−0.8299	0.5103	0.0605	0.0187	43
aapl	0.5733	0.3933	−0.7815	0.4249	0.2355	0.0265	105
goog	0.4057	0.3503	−0.6342	0.4422	1.0779	0.0423	260
c	0.7001	0.9206	−0.8064	0.4643	0.0683	0.0272	26
mmm	0.3481	0.5172	−0.3573	0.3505	0.1052	0.0243	26
spx	0.2736	0.8210	−0.3529	0.4141	1.1356	0.0211	194

Table 6 One month variance swaps on 20081020

xom	ge	msft	jpm	bac	aapl	goog	c	mmm	spx
0.6675	0.6101	0.7611	0.7387	0.8696	0.8710	0.6192	1.0479	0.6075	0.5087

transform to a reading on daily squared returns for each of the 51 underliers. To get to the cash flow we need to get the price of the individual variance swap contracts and this is done by calibrating the *VGSSD* Sato process to the option surface of each underlier and we report the results of this calibration for our selected underliers in Table 5.

We used our method of pricing the log contract analytically to derive the prices of the one month variance swap contract and for the selected underliers on 20081020 these are provided in Table 6.

We may now construct simulated cash flows to one month variance swaps for all 51 underliers to generate a matrix of size 51 by 10000 that gives the cash flow to each underlier for each of 10000 simulated scenarios.

This matrix may be used to construct portfolio cash flows given by a simulated vector of 10000 readings for each portfolio. An optimizer is then employed to construct an optimal portfolio that maximizes the level of acceptability (*AI*) of the cash flow using the stress function *MINMAXVAR*. We also maximized the minimum outcome (*MN*) and found that this was positive yielding a sample space arbitrage even after imposing a zero dollar zero vega constraint. In addition we employed *SPSA* to generate optimal portfolios. We present the positions for the selected assets using all three methods in Table 7.

We now report on the results of holding the optimal portfolio as constructed by *AI*, *MN*, or *SPSA* for the following month and recording the actual resulting cash flow. These one month forward realized results are presented in Table 8.

We see that the *June* positions lost money as did *March*. The *August* and *September* trades were very profitable. It appears that *SPSA* performed the best though this

Table 7 Sample positions for the three strategies

Ticker	AI	MN	SPSA
xom	0.2890	0.0867	−0.0198
ge	0.0609	0.0989	0.0114
msft	−0.2665	−0.1816	−0.2117
jpm	0.2493	0.3573	−0.0247
bac	0.3010	0.0687	0.1278
aapl	−0.0268	−0.0200	−0.0754
goog	0.0935	0.1932	0.1015
c	0.0604	0.1446	0.0317
mmm	−0.1653	−0.2522	0.0977
spx	−0.0845	−0.1556	0.2104

Table 8 Realized results one month forward

Design Date	AI	MN	SPSA
20080227	0.1383	0.1383	0.1990
20080327	−0.0893	−0.0716	−0.0842
20080425	−0.0051	−0.0032	0.0167
20080526	0.0701	0.0552	0.0746
20080624	−0.2342	−0.1122	−0.2850
20080723	0.0888	0.0527	0.1120
20080821	0.2108	0.1771	0.3917
20080919	0.4327	0.4180	0.5294
Total	0.6122	0.6544	0.9542

is not a full optimization and therefore it does not excessively leverage the particular path space. From an out of sample perspective this may be a strong point of this procedure, noting that performs better than the full optimization alternatives in six of the eight months. A more extensive study of this issue is called for.

7 Conclusion

We develop procedures for constructing optimal portfolios of variance swaps. This requires a specification of the joint law of squared returns accounting for both autocorrelation and cross asset dependencies. We account for autocorrelation using a linear model on transformed variates focused on maintaining the positivity of squared returns in a subsequent simulation. Cross asset dependence is modeled by a copula approach, specifically by applying a multivariate normal model on marginal transformed to Gaussian variates. Market prices of variance swaps are extracted from option surface calibrations. The methods allow for a simulation of cash flows to portfolios of variance swaps. The portfolio design maximizes the arbitrage consistent recently introduced (see [2]) index of acceptability of random cash flows. Full nonlinear optimizers are contrasted with *SPSA* (Simultaneous Perturbation Stochastic Approximation) algorithms. Preliminary evidence suggests the possible potential superiority of *SPSA*, but further investigations are called for.

The portfolio design program is illustrated on the top 50 stocks of the *S&P*500 index as at November 11 2008. Positions were developed for each month of 2008 from February through October. The positions were generally profitable with losses occurring in *June* and *March*.

References

1. Carr, P., Geman, H., Madan, D., Yor, M.: Self decomposability and option pricing. Math. Finance **17**, 31–57 (2007)
2. Cherny, A., Madan, D.B.: New measures of performance evaluation. Rev. Financ. Stud. Forthcoming, published on line (2009)
3. Eberlein, E., Madan, D.B.: Maximally acceptable portfolios. Working paper, University of Maryland (2009)
4. Malvergne, Y., Sornette, D.: Testing the Gaussian copula hypothesis for financial asset dependencies. Quant. Finance **3**, 231–250 (2003)
5. Markowitz, H.: Portfolio selection. J. Finance **7**, 77–91 (1952)
6. Spall, J.: Multivariate stochastic approximation using a simultaneous perturbation gradient approximation. IEEE Trans. Autom. Control **37**, 332–341 (1992)

Stochastic Partial Differential Equations and Portfolio Choice

Marek Musiela and Thaleia Zariphopoulou

Abstract We introduce a stochastic partial differential equation which describes the evolution of the investment performance process in portfolio choice models. The equation is derived for two formulations of the investment problem, namely, the traditional one (based on maximal expected utility of terminal wealth) and the recently developed forward formulation. The novel element in the forward case is the volatility process which is up to the investor to choose. We provide various examples for both cases and discuss the differences and similarities between the different forms of the equation as well as the associated solutions and optimal processes.

1 Introduction

We introduce a stochastic partial differential equation (SPDE) arising in optimal portfolio selection problems which describes the evolution of the value function process. The SPDE is expected to hold under mild conditions on the asset price dynamics and in general incomplete markets.

The aim herein is not to study questions on the existence, uniqueness and regularity of the solution of the investment performance SPDE. These questions are very challenging due to the possible degeneracy and full nonlinearity of the equation as

M. Musiela
BNP Paribas, 10 Harewood Avenue, London NW1 6AA, UK
e-mail: marek.musiela@bnpparibas.com

T. Zariphopoulou
Oxford-Man Institute and Mathematical Institute, University of Oxford, 2 Walton Well Road, Oxford OX2 6ED, UK

T. Zariphopoulou (✉)
Departments of Mathematics and IROM, McCombs School of Business, The University of Texas at Austin, 1 University Station, C1200, Austin, TX 78712, USA
e-mail: thaleia.zariphopoulou@oxford-man.ox.ac.uk

C. Chiarella, A. Novikov (eds.), *Contemporary Quantitative Finance*,
DOI 10.1007/978-3-642-03479-4_11, © Springer-Verlag Berlin Heidelberg 2010

well as other difficulties stemming from the market incompleteness. They require extensive study and effort and are being currently investigated by the authors and others. We stress that similar questions have not yet been established even for the simplest possible extension of the classical Merton model in incomplete markets, namely, a single factor model with Markovian dynamics (see [44]).

Abstracting from technical considerations, we provide various representative examples with explicit solutions to the SPDE for two different formulations of the optimal investment problem. The first problem is the classical one in which one maximizes the expected utility of terminal wealth. This problem has been extensively analyzed either in its primal formulation via the associated Hamilton-Jacobi-Bellman (HJB) equation in Markovian models or in its dual formulation. However, once one departs from the complete market setting very little, if anything, can be said about the properties of the maximal expected utility, especially with regards to its regularity, the optimal policies and related verification results.

The investment performance SPDE offers an alternative way to examine the evolution of this process beyond the class of Markovian models. One might think of it as the non-Markovian analogue of the HJB equation. Besides providing information for the maximal expected investment performance, it also provides the optimal investment strategy in a generalized stochastic feedback form. Analyzing the SPDE could perhaps lead to a better understanding of the nature and properties of the value function as well as the optimal wealth and optimal investment processes.

The second problem for which we provide the associated SPDE arises in an alternative approach for portfolio choice that is based on the so-called forward investment performance criterion. In this approach, developed by the authors during the last years, the investor does not choose her risk preferences at a single point in time but has the flexibility to revise them dynamically. Recall that in the classical problem once the trading horizon is chosen the investor not only can he not revise his preferences but he cannot extend his utility beyond the initially chosen horizon either. For the new problem, the SPDE plays a very important role, for it exposes in a very transparent way how this flexibility is being modeled. Indeed, this is done via the volatility component of the forward performance process. This input is up to the investor to choose. It represents his uncertainty about the upcoming changes – from one trading period to the next – of the shape of his current risk preferences.

As expected, the SPDEs in the traditional and the forward formulations have similar structure (see (12) and (28)). However, there are fundamental differences. In the first case, a terminal condition is imposed (see (13)) which is in most cases deterministic. In other words, the solution is progressively measurable with regards to the market filtration in $[0, T)$ but degenerates to a deterministic function at the end of the horizon. In contrast, in the second formulation an initial condition is imposed and the solution does not degenerate at any future time. Moreover, as we will discuss later on, in the classical utility problem the investor's volatility is uniquely determined while in the forward case it is not.

A common characteristic of the traditional and forward SPDEs is the form of the drift. Their drifts are uniquely determined once the volatility and the market inputs are specified. One could say that there is similarity between the investment performance SPDEs and the ones appearing in term structure models.

The paper is organized as follows. In Sect. 2 we describe the investment model. In Sect. 3 we recall the classical maximal expected utility problem and derive the associated SPDE. In Sect. 4, we provide examples from models with Markovian dynamics driven by stochastic factors. We also examine the power and logarithmic cases. In Sect. 5, we recall the forward portfolio choice problem and, in analogy to the classical case, we derive the associated SPDE. We finish the section by discussing the connection between the forward investment problem and the traditional expected utility maximization one. In Sect. 6 we provide several examples. We start with the zero volatility case. We then examine two families of non-zero volatility models. The first family incorporates non-zero performance volatilities that model different market views and investment in terms of a benchmark (or different numeraire choice) while the second family corresponds to the forward analogue of the stochastic factor Markovian model.

2 The Investment Model

The market environment consists of one riskless and k risky securities. The risky securities are stocks and their prices are modeled as Itô processes. Namely, for $i = 1, \ldots, k$, the price S_t^i, $t \geq 0$, of the ith risky asset satisfies

$$dS_t^i = S_t^i \left(\mu_t^i dt + \sum_{j=1}^d \sigma_t^{ji} dW_t^j \right), \tag{1}$$

with $S_0^i > 0$, for $i = 1, \ldots, k$. The process $W_t = (W_t^1, \ldots, W_t^d)$, $t \geq 0$, is a standard d-dimensional Brownian motion, defined on a filtered probability space $(\Omega, \mathscr{F}, \mathbb{P})$. For simplicity, it is assumed that the underlying filtration, $\mathscr{F}_t$, coincides with the one generated by the Brownian motion, that is $\mathscr{F}_t = \sigma(W_s : 0 \leq s \leq t)$.

The coefficients μ_t^i and $\sigma_t^i = (\sigma_t^{1i}, \ldots, \sigma_t^{di})$, $i = 1, \ldots, k$, $t \geq 0$, are $\mathscr{F}_t$-progressively measurable processes with values in $\mathbb{R}$ and $\mathbb{R}^d$, respectively. For brevity, we use σ_t to denote the volatility matrix, i.e. the $d \times k$ random matrix (σ_t^{ji}), whose ith column represents the volatility σ_t^i of the ith risky asset. Alternatively, we write (1) as

$$dS_t^i = S_t^i (\mu_t^i dt + \sigma_t^i \cdot dW_t).$$

The riskless asset, the savings account, has the price process B_t, $t \geq 0$, satisfying

$$dB_t = r_t B_t dt \tag{2}$$

with $B_0 = 1$, and for a nonnegative, $\mathscr{F}_t$-progressively measurable interest rate process r_t. Also, we denote by μ_t the $k \times 1$ vector with the coordinates μ_t^i and by $\mathbf{1}$ the k-dimensional vector with every component equal to one. The market coefficients, μ_t, σ_t and r_t, are taken to be bounded (by a deterministic constant).

We assume that the volatility vectors are such that

$$\mu_t - r_t \mathbf{1} \in Lin(\sigma_t^T),$$

where $Lin(\sigma_t^T)$ denotes the linear space generated by the columns of σ_t^T. This implies that $\sigma_t^T(\sigma_t^T)^+(\mu_t - r_t\mathbf{1}) = \mu_t - r_t\mathbf{1}$ and, therefore, the vector

$$\lambda_t = (\sigma_t^T)^+(\mu_t - r_t\mathbf{1}) \tag{3}$$

is a solution to the equation $\sigma_t^T x = \mu_t - r_t\mathbf{1}$. The matrix $(\sigma_t^T)^+$ is the Moore-Penrose pseudo-inverse of the matrix σ_t^T. It easily follows that

$$\sigma_t \sigma_t^+ \lambda_t = \lambda_t \tag{4}$$

and, hence, $\lambda_t \in Lin(\sigma_t)$. We assume throughout that the process λ_t is bounded by a deterministic constant $c > 0$, i.e., for all $t \geq 0$, $|\lambda_t| \leq c$.

Starting at $t = 0$ with an initial endowment $x \in \mathbb{R}^+$, the investor invests at any time $t > 0$ in the risky and riskless assets. The present value of the amounts invested are denoted, respectively, by π_t^0 and π_t^i, $i = 1, \ldots, k$.

The present value of her aggregate investment is, then, given by $X_t^\pi = \sum_{i=0}^k \pi_t^i$, $t > 0$. We will refer to X^π as the discounted wealth generated by the (discounted) strategy $(\pi_t^0, \pi_t^1, \ldots, \pi_t^k)$. The investment strategies will play the role of control processes. Their admissibility set is defined as

$$\mathscr{A} = \left\{ \pi : \pi_t \text{ is self-financing and } \mathscr{F}_t\text{-progressively measurable} \right.$$

$$\left. \text{with } \mathbb{E}\left(\int_0^t |\sigma_s \pi_s|^2 ds < \infty \right) \text{ and } X_t^\pi \geq 0, t \geq 0 \right\}. \tag{5}$$

Using (1) and (2) we deduce that the discounted wealth satisfies, for $t > 0$,

$$X_t^\pi = x + \sum_{i=1}^k \int_0^t \pi_s^i(\mu_s^i - r_s)ds + \sum_{i=1}^k \int_0^t \pi_s^i \sigma_s^i \cdot dW_s.$$

Writing the above in vector notation and using (3) and (4) yields

$$dX_t^\pi = \pi_t \cdot (\mu_t - r_t\mathbf{1})dt + \sigma_t\pi_t \cdot dW_t = \sigma_t\pi_t \cdot (\lambda_t dt + dW_t), \tag{6}$$

where the (column) vector, $\pi_t = (\pi_t^i; i = 1, \ldots, k)$.

3 The Backward Formulation of the Portfolio Choice Problem and the Associated SPDE

The traditional criterion for optimal portfolio choice has been based on maximal expected utility (see the seminal paper [31]). The key ingredients are the choices of the trading horizon $[0, T]$ and the investor's utility, $u_T : \mathbb{R}^+ \to \mathbb{R}$ at terminal time T. The utility function reflects the risk attitude of the investor at time T and is an increasing and concave function of his wealth.

The objective is to maximize the expected utility of terminal wealth over admissible strategies. We will denote the set of such strategies by $\mathscr{A}_T$, a straightforward restriction of $\mathscr{A}$ on $[0, T]$. The maximal expected utility is defined as

$$V(x, t; T) = \sup_{\mathscr{A}_T} E_\mathbb{P}(u_T(X_T^\pi)|\mathscr{F}_t, X_t = x), \tag{7}$$

for $(x, t) \in \mathbb{R}^+ \times [0, T]$. The function u_T satisfies the standard Inada condition (see, for example, [26] and [27]). We introduce the T-notation throughout this section to highlight the dependence of all quantities on the investment horizon at which the investor's risk preferences are chosen.

As solution of a stochastic optimization problem, $V(x, t; T)$ is expected to satisfy the Dynamic Programming Principle (DPP), namely,

$$V(x, t; T) = \sup_{\mathscr{A}_T} E_{\mathbb{P}}(V(X_s^\pi, s; T) | \mathscr{F}_t, X_t = x), \qquad (8)$$

for $t \leq s \leq T$. This is a fundamental result in optimal control and has been proved for a wide class of optimization problems. For a detailed discussion on the validity (and strongest forms) of the DPP in problems with controlled diffusions, we refer the reader to [14] and [42] (see, also, [5, 10, 29]). Key issues are the measurability and continuity of the value function process as well as the compactness of the set of admissible controls. It is worth mentioning that a proof specific to the problem at hand has not been produced to date.[1]

Besides its technical challenges, the DPP exhibits two important properties of the solution. Specifically, $V(X_t^\pi, t; T)$, is a supermartingale for an arbitrary investment strategy and becomes a martingale at an optimum (provided certain integrability conditions hold). Observe also that the DPP yields a backward in time algorithm for the computation of the maximal utility, starting at expiration with u_T and using the martingality property to compute the solution at earlier times. For this, we refer to this formulation of the optimal portfolio choice problem as *backward*.

Regularity results for the process $V(x, t; T)$ have not been produced to date except for special cases. To the best of our knowledge, the most general result for arbitrary utilities can be found in [28].

We continue with the derivation of the SPDE for the value function process. For the moment, the discussion is informal, for general regularity results are lacking. To this end, let us assume that $V(x, t; T)$ admits the Itô decomposition

$$dV(x, t; T) = b(x, t; T)dt + a(x, t; T) \cdot dW_t \qquad (9)$$

for some coefficients $b(x, t; T)$ and $a(x, t; T)$ which are $\mathscr{F}_t$-progressively measurable processes.

Let us also assume that the mapping $x \to V(x, t; T)$ is strictly concave and increasing and that $V(x, t; T)$ is smooth enough so that the Itô-Ventzell formula can be applied to $V(X_t^\pi, t; T)$ for any strategy $\pi \in \mathscr{A}_T$. We then obtain

$$
\begin{aligned}
dV&(X_t^\pi, t; T) \\
&= b(X_t^\pi, t; T)dt + a(X_t^\pi, t; T) \cdot dW_t + V_x(X_t^\pi, t; T)dX_t^\pi \\
&\quad + \frac{1}{2}V_{xx}(X_t^\pi, t; T)d\langle X^\pi \rangle_t + a_x(X_t^\pi, t; T) \cdot d\langle W, X^\pi \rangle_t \\
&= (b(X_t^\pi, t; T) + \sigma_t \pi_t \cdot (V_x(X_t^\pi, t; T)\lambda_t + a_x(X_t^\pi, t; T)))dt
\end{aligned}
$$

---

[1]Recently, a weak version of the DPP was proposed in [6] where conditions related to measurable selection and boundness of controls are relaxed.

$$+ \frac{1}{2} V_{xx}(X_t^\pi, t; T) |\sigma_t \pi_t|^2 dt + (a(X_t^\pi, t; T) + V_x(X_t^\pi, t; T) \sigma_t \pi_t) \cdot dW_t$$

$$= \left(b(X_t^\pi, t; T) + \sigma_t \pi_t \cdot \sigma_t \sigma_t^+ (V_x(X_t^\pi, t; T)\lambda_t + a_x(X_t^\pi, t; T)) \right.$$

$$\left. + \frac{1}{2} V_{xx}(X_t^\pi, t; T)|\sigma_t \pi_t|^2 \right) dt + (a(X_t^\pi, t) + V_x(X_t^\pi, t; T)\sigma_t \pi_t) \cdot dW_t.$$

From the DPP we know that the process $V(X_t^\pi, t; T)$ is a supermartingale for arbitrary admissible policies and becomes a martingale at an optimum.

Let us now choose as control policy the process

$$\pi_t^* = \pi_t^*(X_t^*, t; T)$$

where the feedback process in the right hand side (denoted by a slight abuse of notation by $\pi_t^*(x, t; T)$) is given by

$$\pi_t^*(x, t; T) = -\sigma_t^+ \frac{V_x(x, t; T)\lambda_t + \sigma_t \sigma_t^+ a_x(x, t; T)}{V_{xx}(x, t; T)}, \tag{10}$$

where X_t^* is the wealth process generated by (6) with π_t^* being used. It is assumed that $\pi_t^* \in \mathscr{A}_T$. It is easy to check that π_t^* is the point at which the quadratic expression appearing in the drift above achieves its maximum and, moreover, that the maximum value at this point is given by

$$\frac{1}{2} V_{xx}(X_t^*, t; T)|\sigma_t \pi_t^*|^2 + \sigma_t \pi_t^* \cdot \sigma_t \sigma_t^+ (V_x(X_t^*, t; T)\lambda_t + a_x(X_t^*, t))$$

$$= -\frac{1}{2} \frac{|V_x(X_t^*, t; T)\lambda_t + \sigma_t \sigma_t^+ a_x(X_t^*, t; T)|^2}{V_{xx}(X_t^*, t; T)}.$$

We, then, deduce that the drift coefficient $b(x, t; T)$ must satisfy

$$b(x, t; T) = \frac{1}{2} \frac{|V_x(x, t; T)\lambda_t + \sigma_t \sigma_t^+ a_x(x, t; T)|^2}{V_{xx}(x, t; T)}. \tag{11}$$

Combining the above leads to the SPDE

$$dV(x, t; T) = \frac{1}{2} \frac{|V_x(x, t; T)\lambda_t + \sigma_t \sigma_t^+ a_x(x, t; T)|^2}{V_{xx}(x, t; T)} dt + a(x, t; T) \cdot dW_t \tag{12}$$

with

$$V(x, T; T) = u_T(x). \tag{13}$$

To the best of our knowledge, the above SPDE has not been derived to date. For deterministic terminal utilities, the volatility $a(x, t; T)$ is present because of the stochasticity of the investment opportunity set, as the examples in the next section show.

The optimal feedback portfolio process π_t^* consists of two terms, namely,

$$\pi_t^{*,m} = -\frac{V_x(x, t; T)}{V_{xx}(x, t; T)} \sigma_t^+ \lambda_t \quad \text{and} \quad \pi_t^{*,h} = -\sigma_t^+ \frac{a_x(x, t; T)}{V_{xx}(x, t; T)}.$$

The first component, $\pi_t^{*,m}$, known as the *myopic* investment strategy, resembles the investment policy followed by an investor in markets in which the investment opportunity set remains constant through time. The second term, $\pi_t^{*,h}$, is called the *excess hedging demand* and represents the additional (positive or negative) investment generated by the volatility $a(x, t; T)$ of the performance process V.

4 Examples: Markovian Stochastic Factor Models

Stochastic factors have been used in portfolio choice to model asset predictability and stochastic volatility. The predictability of stock returns was first discussed in [12, 13, 15] (see also [1, 7, 41] and others). The role of stochastic volatility in investment decisions was first studied in [2, 15, 16, 19] (see also [8, 23, 30] and others). There is a vast literature on such models and we refer the reader to the review paper [44] for detailed bibliography, exposition of existing results and open problems.

4.1 Single Stochastic Factor Models

There is only one risky asset whose price S_t, $t \geq 0$, is modeled as a diffusion process solving

$$dS_t = \mu(Y_t) S_t dt + \sigma(Y_t) S_t dW_t^1, \tag{14}$$

with $S_0 > 0$. The stochastic factor Y_t, $t \geq 0$, satisfies

$$dY_t = b(Y_t)dt + d(Y_t)\left(\rho dW_t^1 + \sqrt{1-\rho^2}dW_t^2\right), \tag{15}$$

with $Y_0 = y$, $y \in \mathbb{R}$. It is assumed that $\rho \in (-1, 1)$.

The market coefficients $f = \mu, \sigma, b$ and d satisfy the standard global Lipschitz and linear growth conditions $|f(y) - f(\bar{y})| \leq K|y - \bar{y}|$ and $f^2(y) \leq K(1+y^2)$, for $y, \bar{y} \in \mathbb{R}$. Moreover, it is assumed that the non-degeneracy condition $\sigma(y) \geq l > 0$, $y \in \mathbb{R}$, holds.

It is also assumed that the riskless asset (cf. (2)) offers constant interest rate $r > 0$. The coefficients appearing in (12) take the form

$$\sigma_t = (\sigma(Y_t), 0)^T, \quad \sigma_t^+ = \left(\frac{1}{\sigma(Y_t)}, 0\right) \quad \text{and} \quad \lambda_t = \left(\frac{\mu(Y_t) - r}{\sigma(Y_t)}, 0\right)^T. \tag{16}$$

We easily see that condition (4) is trivially satisfied.

The value function is defined as

$$v(x, y, t; T) = \sup_{\mathscr{A}_T} E_{\mathbb{P}}(u_T(X_T)|X_t = x, Y_t = y), \tag{17}$$

for $(x, y, t) \in \mathbb{R}^+ \times \mathbb{R} \times [0, T]$ and $\mathscr{A}_T$ being the set of admissible strategies.

Regularity results for the value function (17) for general utility functions have not been obtained to date except for the special cases of homothetic preferences (see, for example, [25, 32, 39, 43]). We, thus, proceed with an informal discussion for the associated HJB equation, the form of the process V and the optimal policies. To this end, the HJB turns out to be

$$v_t + \max_\pi \left(\frac{1}{2}\sigma^2(y)\pi^2 v_{xx} + \pi(\mu(y)v_x + \rho\sigma(y)d(y)v_{xy}) \right)$$

$$+ \frac{1}{2}d^2(y)v_{yy} + b(y)v_y = 0, \tag{18}$$

with $v(x, y, T; T) = u_T(x)$, $(x, y, t) \in \mathbb{R}^+ \times \mathbb{R} \times [0, T]$.

Its solution yields the process V, namely, for $0 \le t \le T$,

$$V(x, t; T) = v(x, Y_t, t; T). \tag{19}$$

We next show that $V(x, t; T)$ above solves the SPDE (12) with volatility vector

$$a(x, t; T) = (a_1(x, t; T), a_2(x, t; T))$$

with

$$a_1(x, t; T) = \rho d(Y_t)v_y \quad \text{and} \quad a_2(x, t; T) = \sqrt{1 - \rho^2}d(Y_t)v_y, \tag{20}$$

where the arguments of v_y have been suppressed for convenience.

Indeed, using (19), (16), (20) and Itô's formula yields

$$dV(x, t; T)$$

$$= \left(v_t(x, Y_t, t; T) + \frac{1}{2}d(Y_t)^2 v_{yy}(x, Y_t, t; T) + b(Y_t)v_y(x, Y_t, t; T) \right)dt$$

$$+ \left(\rho d(Y_t)v_y(x, Y_t, t; T), \sqrt{1 - \rho^2}d(Y_t)v_y(x, Y_t, t; T) \right) \cdot dW_t$$

$$= -\frac{1}{2}\max_\pi \left(\frac{1}{2}\sigma^2(Y_t)\pi^2 v_{xx}(x, Y_t, t; T) \right.$$

$$+ \pi(\mu(Y_t)v_x(x, Y_t, t; T) + \rho\sigma(Y_t)d(Y_t)v_{xy}(x, Y_t, t; T)) \Big)dt$$

$$+ \left(\rho d(Y_t)v_y(x, Y_t, t; T), \sqrt{1 - \rho^2}d(Y_t)v_y(x, Y_t, t; T) \right) \cdot dW_t$$

$$= \frac{1}{2}\frac{(\mu(Y_t)v_x(x, Y_t, t; T) + \rho\sigma(Y_t)d(Y_t)v_{xy}(x, Y_t, t; T))^2}{v_{xx}(x, Y_t, t; T)}dt$$

$$+ \left(\rho d(Y_t)v_y(x, Y_t, t; T), \sqrt{1 - \rho^2}d(Y_t)v_y(x, Y_t, t; T) \right) \cdot dW_t$$

$$= \frac{1}{2}\frac{|V_x(x, t; T)\lambda_t + \sigma_t\sigma_t^+ a_x(x, t)|^2}{V_{xx}(x, t; T)}dt + a(x, t; T) \cdot dW_t.$$

The optimal feedback portfolio process for $t \le s \le T$ is obtained from (10). We easily deduce that it coincides with the feedback policy (evaluated at Y_s) derived from the first order conditions in the HJB equation. Indeed, for $t \le s \le T$,

$$\pi_s^*(x, s; T) = -\sigma_s^+ \frac{V_x(x, s; T)\lambda_s + \sigma_s\sigma_s^+ a_x(x, s; T)}{V_{xx}(x, s; T)}$$

$$= -\frac{\lambda(Y_s)}{\sigma(Y_s)} \frac{v_x(x, Y_s, s; T)}{v_{xx}(x, Y_s, s; T)} - \rho\frac{d(Y_s)}{\sigma(Y_s)} \frac{v_{xy}(x, Y_s, s; T)}{v_{xx}(x, Y_s, s; T)}$$

$$= \pi^*(x, Y_s, s; T) \tag{21}$$

where $\pi^* : \mathbb{R}^+ \times \mathbb{R} \times [0, T]$ is given by

$$\pi^*(x, y, s; T) = \arg\max_\pi \left(\frac{1}{2}\sigma^2(y)\pi^2 v_{xx} + \pi(\mu(y)v_x + \rho\sigma(y)d(y)v_{xy})\right).$$

Next, we present two cases for which the above results are rigorous. Specifically, we provide examples for the most frequently used utilities, the power and the logarithmic ones. These utilities have convenient homogeneity properties which, in combination with the linearity of the wealth dynamics in the control policies, enable us to reduce the HJB equation to a quasilinear one. Under a "distortion" transformation (see, for example, [43]) the latter can be linearized and solutions in closed form can be produced using the Feynman-Kac formula. The smoothness of the value function and, in turn, the verification of the optimal feedback policies follows (see, among others, [23–25, 30, 43]).

4.1.1 The CRRA Case: $u_T(x) = \frac{1}{\gamma}x^\gamma, 0 < \gamma < 1, \gamma \neq 0$

The value function v (cf. (17)) is multiplicatively separable and given, for $(x, y, t) \in \mathbb{R}^+ \times \mathbb{R} \times [0, T]$, by

$$v(x, y, t; T) = \frac{1}{\gamma}x^\gamma f(y, t; T)^\delta, \quad \delta = \frac{1-\gamma}{1-\gamma+\rho^2\gamma}, \tag{22}$$

where $f : \mathbb{R} \times [0, T] \to \mathbb{R}^+$ solves the linear parabolic equation

$$f_t + \frac{1}{2}d^2(y)f_{yy} + \left(b(y) + \rho\frac{\gamma}{1-\gamma}\lambda(y)d(y)\right)f_y$$

$$+ \frac{\gamma}{2(1-\gamma)}\frac{\lambda^2(y)}{\delta}f = 0, \tag{23}$$

with $f(x, y, T; T) = 1$. The value function process $V(x, t; T)$ is given by

$$V(x, t; T) = \frac{1}{\gamma}x^\gamma f(Y_t, t; T)^\delta,$$

with f solving (23) and δ as in (22). Direct calculations show that it satisfies the SPDE (12) with volatility components (cf. (20))

$$\begin{cases} a_1(x, t; T) = \rho\frac{\delta}{\gamma}x^\gamma d(Y_t)f_y(Y_t, t; T)f(Y_t, t; T)^{\delta-1}, \\ a_2(x, t; T) = \sqrt{1-\rho^2}\frac{\delta}{\gamma}x^\gamma d(Y_t)f_y(Y_t, t; T)f(Y_t, t; T)^{\delta-1}. \end{cases} \tag{24}$$

The optimal feedback portfolio process is given for $t \leq s \leq T$ by (10), namely,

$$\pi_s^*(x,s;T) = -\sigma_s^+ \frac{V_x(x,s;T)\lambda_s + \sigma_s \sigma_s^+ a_x(x,s;T)}{V_{xx}(x,s;T)}$$

$$= \left(\frac{\lambda(Y_s)}{\sigma(Y_s)(1-\gamma)} + \rho \frac{d(Y_s)}{\sigma(Y_s)(1-\gamma+\rho^2\gamma)} \frac{f_y(Y_s,s;T)}{f(Y_s,s;T)} \right) x,$$

which is in accordance with (21) and (22).

4.1.2 The Logarithmic Case: $u_T(x) = \ln x, \, x > 0$

The value function is additively separable, namely,

$$v(x,y,t;T) = \ln x + h(y,t;T),$$

with $h : \mathbb{R} \times [0,T] \rightarrow \mathbb{R}^+$ solving

$$h_t + \frac{1}{2}d^2(y)h_{yy} + b(y)h_y + \frac{1}{2}\lambda^2(y)h = 0 \tag{25}$$

and $h(y,T;T) = 1$. The value function process $V(x,t;T)$ is, in turn, given by

$$V(x,t;T) = \ln x + h(Y_t,t;T).$$

We easily deduce that it satisfies the SPDE (12) with volatility vector

$$a_1(x,t;T) = \rho h_y(Y_t,t;T) \quad \text{and} \quad a_2(x,t;T) = \sqrt{1-\rho^2}h_y(Y_t,t;T).$$

Observe that because the volatility process does not depend on wealth, the excess risky demand is zero, as (10) indicates. Indeed, the optimal portfolio process is always myopic, given, for $t \leq s \leq T$, by $\pi_s^* = \frac{\lambda(Y_s)}{\sigma(Y_s)} X_s^*$ with

$$X_s^* = x \exp\left(\int_t^s \frac{1}{2}\lambda^2(Y_u)du + \int_t^s \lambda(Y_u)dW_u^1 \right).$$

The logarithmic utility plays a special role in portfolio choice. The optimal portfolio is known as the "growth optimal portfolio" and has been extensively studied in general market settings (see, for example, [3] and [21]). The associated optimal wealth is the so-called "numeraire portfolio." It has also been extensively studied, for it is the numeraire with regards to which all wealth processes are supermartingales under the historical measure (see, among others, [17] and [18]).

4.2 Multi-stochastic Factor Models

Multi-stochastic factor models for homothetic preferences have been analyzed by various authors. The theory of BSDE has been successfully used to characterize and represent the solutions of the reduced HJB equation (see [11]). The regularity of its solutions has been studied using PDE arguments in [32] and [39], for power and exponential utilities, respectively. Finally, explicit solutions for a three factor model can be found in [30]. Working as in the previous examples one obtains that the value function process satisfies the SPDE (12) and that the optimal policies can be expressed in the stochastic feedback form (10). These calculations are routine but tedious and are omitted for the sake of the presentation.

5 The Forward Formulation of the Portfolio Choice Problem and the Associated SPDE

In the classical expected utility models of terminal wealth, discussed in the previous section, one chooses the investment horizon $[0, T]$ and then assigns the utility function $u_T(x)$ at the end of it (cf. (7)). Once these choices are made, the investor's risk preferences cannot be revised. In addition, no investment decisions can be assessed for trading times beyond T.

Recently, the authors proposed an alternative approach to optimal portfolio choice which is based on the so-called forward investment performance criterion (see, for example, [38]). In this approach the investor does not choose her risk preferences at a single point in time but has the flexibility to revise them dynamically for all trading times. Investment strategies are chosen from the set $\mathscr{A}$ defined in (5). A strategy is deemed optimal if it generates a wealth process whose average performance is maintained over time. In other words, the average performance of this strategy at any future date, conditional on today's information, preserves the performance of this strategy up until today. Any strategy that fails to maintain the average performance over time is, then, sub-optimal.

Next, we recall the definition of the forward investment performance. This criterion was first introduced in [33] (see also [37]) in the context of an incomplete binomial model and subsequently studied in [34, 35, 38]. A rich class of such processes which are monotone in time was recently completely analyzed in [36].

We note that the definition below is slightly different than the original one in that the initial condition is not explicitly included. As the example in Sect. 6.1 shows, not all strictly increasing and concave solutions can serve as initial conditions, even for the special class of time monotone investment performance processes (see (40) in Remark 2). Characterizing the set of appropriate initial conditions is a challenging question and is currently being investigated by the authors. Another difference is that, herein, we only allow for policies that keep the wealth nonnegative. Going beyond such strategies raises very interesting questions on possible arbitrage opportunities which are left for future study.

Definition 1 An $\mathscr{F}_t$-progressively measurable process $U(x, t)$ is a forward performance if for $t \geq 0$ and $x \in \mathbb{R}^+$:

i) the mapping $x \to U(x, t)$ is strictly concave and increasing,
ii) for each $\pi \in \mathscr{A}$, $E(U(X_t^\pi, t))^+ < \infty$, and

$$E(U_s(X_s^\pi)|\mathscr{F}_t) \leq U_t(X_t^\pi), \quad s \geq t, \qquad (26)$$

iii) there exists $\pi^* \in \mathscr{A}$, for which

$$E(U(X_s^{\pi^*}, s)|\mathscr{F}_t) = U_t(X_t^{\pi^*}, t), \quad s \geq t. \tag{27}$$

It might seem that all this definition produces is a criterion that is dynamically consistent across time. Indeed, internal consistency is an ubiquitous requirement and needs to be ensured in any proposed criterion. It is satisfied, for example, by the traditional value function process. However, the new criterion allows for much more flexibility as it is manifested by the volatility process $a(x, t)$ introduced below. The volatility process is the *novel* element in the new approach of optimal portfolio choice.

We continue with the derivation of the SPDE associated with the forward investment performance process. As in Sect. 3 we proceed with informal arguments and present rigorous results in the upcoming examples. To this end, we consider a process, say $U(x, t)$, that is $\mathscr{F}_t$-progressively measurable and satisfies condition (i) of the above definition. We also assume that the mapping $x \to U(x, t)$ is smooth enough so that the Itô-Ventzell formula can be applied to $U(X_t^\pi, t)$, for any strategy $\pi \in \mathscr{A}$ and that $E(U(X_t^\pi, t))^+ < +\infty, t \geq 0$.

Let us now assume that $U(x, t)$ satisfies the SPDE

$$dU(x, t) = \frac{1}{2} \frac{|U_x(x, t)\lambda_t + \sigma_t \sigma_t^+ a_x(x, t)|^2}{U_{xx}(x, t)} dt + a(x, t) \cdot dW_t, \tag{28}$$

where the volatility $a(x, t)$ is an $\mathscr{F}_t$-progressively measurable, d-dimensional and continuously differentiable in the spatial argument process.

We fist show that under appropriate integrability conditions $U(X_t^\pi, t)$ is a supermartingale for every admissible portfolio strategy. Indeed, denote the above drift coefficient by

$$b(x, t) = \frac{1}{2} \frac{|U_x(x, t)\lambda_t + \sigma_t \sigma_t^+ a_x(x, t)|^2}{U_{xx}(x, t)}$$

and rewrite (28) as

$$dU(x, t) = b(x, t)dt + a(x, t) \cdot dW_t.$$

Consider the wealth process X^π (cf. (6)) generated using an admissible strategy π. Applying the Itô-Ventzell formula to $U(X_t^\pi, t)$ yields

$$dU(X_t^\pi, t) = b(X_t^\pi, t)dt + a(X_t^\pi, t) \cdot dW_t$$

$$+ U_x(X_t^\pi, t)dX_t^\pi + \frac{1}{2}U_{xx}(X_t^\pi, t)d\langle X^\pi \rangle_t + a_x(X_t^\pi, t) \cdot d\langle W, X^\pi \rangle_t$$

$$= \left(b(X_t^\pi, t) + \sigma_t \pi_t \cdot (U_x(X_t^\pi, t)\lambda_t + a_x(X_t^\pi, t)) \right.$$

$$\left. + \frac{1}{2}U_{xx}(X_t^\pi, t)|\sigma_t \pi_t|^2 \right) dt + (a(X_t^\pi, t) + U_x(X_t^\pi, t)\sigma_t \pi_t) \cdot dW_t$$

$$= \left(b(X_t^\pi, t) + \sigma_t \pi_t \cdot \sigma_t \sigma_t^+ (U_x(X_t^\pi, t)\lambda_t + a_x(X_t^\pi, t)) \right.$$

$$+ \frac{1}{2}U_{xx}(X_t^\pi, t)|\sigma_t \pi_t|^2\Big)dt + (a(X_t^\pi, t) + U_x(X_t^\pi, t)\sigma_t \pi_t) \cdot dW_t$$

$$= \frac{1}{2}U_{xx}(X_t^\pi, t)\left|\sigma_t \pi_t + \sigma_t \sigma_t^+ \frac{U_x(X_t^\pi, t)\lambda_t + a_x(X_t^\pi, t)}{U_{xx}(X_t^\pi, t)}\right|^2 dt,$$

$$+ (a(X_t^\pi, t) + U_x(X_t^\pi, t)\sigma_t \pi_t) \cdot dW_t \tag{29}$$

and we conclude using the concavity assumption on $U(x, t)$.

We next assume that the stochastic differential equation

$$dX_t^* = -\frac{U_x(X_t^*, t)\lambda_t + \sigma_t \sigma_t^+ a_x(X_t^*, t)}{U_{xx}(X_t^*, t)} \cdot (\lambda_t dt + dW_t) \tag{30}$$

has a nonnegative solution $X_t^*, t \ge 0$, with $X_0 = x, x \in \mathbb{R}^+$ and that the strategy π_t^*, $t \ge 0$, defined by

$$\pi_t^* = -\sigma_t^+ \frac{U_x(X_t^*, t)\lambda_t + a_x(X_t^*, t)}{U_{xx}(X_t^*, t)} \tag{31}$$

is admissible. Notice that X_t^* corresponds to the wealth generated by this investment strategy.

From (29) we then see that $U(X_t^*, t)$ is a martingale (under appropriate integrability conditions).

Using the supermartingality property of $U(X_t^\pi, t)$ and the martingality property of $U(X_t^*, t)$ we easily deduce that if $U(x, t)$ solves (28) then it is a forward investment performance process. Moreover, the processes X_t^* and π_t^*, given in (30) and (31), are optimal.

The analysis of the forward performance SPDE (28) is a formidable task. The reasons are threefold. Firstly, it is degenerate and fully nonlinear. Moreover, it is formulated forward in time, which might lead to "ill-posed" behavior. Secondly, one needs to specify the appropriate class of admissible volatility processes, namely, volatility inputs that yield solutions that satisfy the requirements of Definition 1. This question is challenging both from the modeling as well as the technical points of view. Thirdly, as it was mentioned earlier, one also needs to specify the appropriate class of initial conditions.

Addressing these issues is an ongoing research project of the authors and not the scope of this paper. Herein, we only construct explicit solutions for different choices of the volatility process $a(x, t)$. These choices provide a rich class of forward performance processes which exhibit several interesting modeling features.

The initial condition represents the investor's current performance criterion. The volatility process $a(x, t)$ represents the uncertainty about the future shape of this criterion. From the modeling perspective, one can see an analogy to term structure, where the initial condition is the current forward curve as traded in the market, while the volatility captures the way the curve moves from one day to the next. However the analogy stops here. One has to develop different methods to recover the initial condition and to specify the volatility $a(x, t)$ for the investment problem governed by (28).

We stress the fundamental difference between the volatility processes of the investment performance criteria in the backward and the forward formulations. In the backward case, the volatility is uniquely determined through the backward construction of the maximal expected utility. The investor does not have the flexibility to choose this process (see for instance example 4.1.1 and the volatility components (24)). In contrast, in the forward case, the volatility process is chosen by the investor as the examples in Sect. 6.2 show.

5.1 Stochastic Optimization and Forward Investment Performance Process

The intuition behind Definition 1 comes from the analogous martingale and super-martingale properties that the traditional maximal expected utility has, as seen from (8).

However, there are two important observations to make. Firstly, the analogous equivalence between stochastic optimization and the martingality and supermartingality of the solution in the forward formulation of the problem has not yet been established. Specifically, one could define the forward performance process via the (forward) stochastic optimization problem

$$U(x,t) = \sup_{\mathscr{A}} E(U(X_s^\pi, s)|\mathscr{F}_t, X_t^\pi = x),$$

for all $0 \le t \le s$ and with the appropriate initial condition. Characterizing its solutions poses a number of challenging questions, some of them being currently investigated by the authors.[2] From a different perspective, one could seek an *axiomatic* construction of a forward performance process. Results in this direction, as well as on the dual formulation of the problem, can be found in [45] for the exponential case.

The second observation is on the relation between the forward performance process and the classical maximal expected utility. One would expect that in a finite trading horizon, say $[0, T]$, the following holds. Define as utility at terminal time the random variable $u_T(x) \in \mathscr{F}_T$ given by

$$u_T(x) = U(x, T),$$

with $U(x, T)$ being the value of the forward performance process at T. Solve, for $0 \le t \le T$, the stochastic optimization problem of state-dependent utility

$$V(x, t; T) = \sup_{\mathscr{A}_T} E(u_T(X_T)|\mathscr{F}_t, X_t = x) \tag{32}$$

(see, among others, [9, 20, 22, 40]). Then, for $0 \le t \le T$, the classical value function process and the forward investment performance process coincide,

$$U(x, t) = V(x, t; T).$$

---

[2]While preparing this revised version, the authors came across the revised version of [4] where similar questions are studied.

6 Examples

In this section we provide examples of forward investment performance processes which satisfy the SPDE (28). We first look at the case of zero volatility. We then examine two families with non-zero volatility. The examples in the first family (examples 6.2.1, 6.2.2 and 6.2.3) build on the zero volatility case. The second family yields the forward analogue of the stochastic factor model presented for the backward case in example 4.1.

We note that the fact that a process solves the SPDE (28) does not automatically guarantee that it satisfies Definition 1, for there are additional conditions to be verified. One can show that the solutions presented in examples 6.2.1, 6.2.2 and 6.2.3 indeed satisfy these conditions; we refer the reader to [36] for all technical details.

6.1 The Case of Zero Volatility: $a(x, t) \equiv 0$

When $a(x, t) \equiv 0$ the SPDE (28) reduces to

$$dU(x, t) = \frac{1}{2}|\lambda_t|^2 \frac{U_x(x, t)^2}{U_{xx}(x, t)} dt.$$

In [38] it was shown that its solution is given by the time-monotone process

$$U(x, t) = u(x, A_t), \tag{33}$$

where $u : \mathbb{R}^+ \times [0, +\infty) \to \mathbb{R}$ is increasing and strictly concave in x, and satisfies the fully nonlinear equation

$$u_t = \frac{1}{2} \frac{u_x^2}{u_{xx}}. \tag{34}$$

The process A_t is given in (39) below. In a more recent paper, see [36], it was shown that there is a one-to-one correspondence between increasing and strictly concave solution to (34) and strictly increasing positive solutions $h : \mathbb{R} \times [0, +\infty) \to \mathbb{R}^+$ to the heat equation

$$h_t + \frac{1}{2}h_{xx} = 0. \tag{35}$$

Specifically, we have the representation

$$h(x, t) = \int_{0+}^{+\infty} \frac{e^{yx - \frac{1}{2}y^2 t}}{y} v(dy) \tag{36}$$

of strictly increasing solutions of (35) and

$$u(x, t) = -\frac{1}{2} \int_0^t e^{-h^{(-1)}(x, s) + \frac{s}{2}} h_x(h^{(-1)}(x, s), s) ds + \int_0^x e^{-h^{(-1)}(z, 0)} dz \tag{37}$$

for strictly concave and increasing solutions of (34). The measure v appearing above is a positive Borel measure that satisfies appropriate integrability conditions. We refer the reader to [36] for a detailed study of this measure and the interplay between its support and various properties of the functions h and u.

The optimal wealth and portfolio processes are given in closed form, namely,

$$X_t^* = h(h^{(-1)}(x,0) + A_t + M_t, A_t) \quad \text{and} \quad \pi_t^* = h_x(h^{(-1)}(X_t^*, A_t), A_t)\sigma_t^+ \lambda_t, \quad (38)$$

where the market input processes A_t and M_t, $t \geq 0$, are defined as

$$A_t = \int_0^t |\lambda_s|^2 ds \quad \text{and} \quad M_t = \int_0^t \lambda_s \cdot dW_s. \quad (39)$$

Remark 1 Formulae (37) and (36) indicate that not all concave and increasing functions can serve as initial conditions. Indeed, from (33), (37), (36) and (39), we see that the initial condition $U(x,0)$ must be represented as

$$U(x,0) = \int_0^x e^{-h^{(-1)}(z,0)} dz \quad \text{with} \quad h(x,t) = \int_{0^+}^{+\infty} \frac{e^{yx}}{y} v(dy). \quad (40)$$

Remark 2 One could choose the volatility to be $a(x,t) \equiv k_t$, with the process k_t being $\mathscr{F}_t$-progressively measurable and independent of x, U and its derivatives. This case essentially reduces to the one with zero volatility. Indeed, we easily conclude that the process

$$U(x,t) = u(x, A_t) + \int_0^t k_s \cdot dW_s, \quad (41)$$

with u and A_t, $t \geq 0$, as in (34) and (39) solves (28). The optimal investment and wealth processes remain the *same* as in (38) above.

6.1.1 Single Stochastic Factor Models

This is the same model as in Sect. 4.1. Using (33) we see that the forward performance process is given by

$$U(x,t) = u\left(x, \int_0^t \lambda^2(Y_s) ds\right) \quad (42)$$

with u solving (34). Using (38) and (37) one deduces that the optimal portfolio process is given by

$$\pi_t^* = -\frac{u_x(X_t^*, A_t)}{u_{xx}(X_t^*, A_t)} \sigma_t^+ \lambda_t. \quad (43)$$

The calculations are tedious and can be found in Sect. 4 of [36].

Notice that the optimal portfolio (43) is purely myopic even though the investment opportunity set is stochastic. This is because the volatility of the investor's performance process was chosen to be zero. This is in contrast with the one in (21). We also observe that the investment performance process U in (42) is of bounded variation while the one in (19) is not.

6.2 Cases of Non-zero Volatility

We start with three cases of non-zero volatility. Two auxiliary process are involved, φ_t and δ_t, $t \geq 0$. They are both independent of wealth and are taken to be $\mathscr{F}_t$-progressively measurable and bounded by a deterministic constant. In addition, it is assumed that δ_t satisfies, similarly to the process λ_t (cf. (3)), the condition $\sigma_t \sigma_t^+ \delta_t = \delta_t$ (cf. (4)) and, thus, $\delta_t \in Lin(\sigma_t)$. Because the third case is the combination of the first two, we provide the complete calculations therein.

We conclude with a fourth case which is the forward analogue of example 4.1.

6.2.1 The Market View Case: $a(x, t) = U(x, t)\varphi_t$

The forward performance SPDE (28) simplifies to

$$dU(x,t) = \frac{1}{2} \frac{|U_x(x,t)(\lambda_t + \sigma_t\sigma_t^+\varphi_t)|^2}{U_{xx}(x,t)} dt + U(x,t)\varphi_t \cdot dW_t.$$

It turns out (see example 6.2.3 for $\delta_t \equiv 0$, $t \geq 0$) that the process

$$U(x,t) = u(x, A_t)Z_t,$$

with u satisfying (34), Z_t, $t \geq 0$, solving

$$dZ_t = Z_t\varphi_t \cdot dW_t \quad \text{with } Z_0 = 1 \tag{44}$$

and $A_t = \int_0^t |\lambda_s + \sigma_s\sigma_s^+\varphi_s|^2 ds$ satisfies (28).

One may interpret Z_t as a device that offers the flexibility to modify our views on asset returns. For this reason, we call this case the "market-view" case.

Using (31) we obtain that the optimal allocation vector, π_t^*, $t \geq 0$, has the same functional form as (43) but for a different time-rescaling process, namely,

$$\pi_t^* = -\frac{u_x(X_t^*, A_t)}{u_{xx}(X_t^*, A_t)} \sigma_t^+(\lambda_t + \varphi_t),$$

with A_t, $t \geq 0$, as above. It is worth noticing that if the process φ_t is chosen to satisfy $\varphi_t = -\lambda_t$, solutions become static. Specifically, the time-rescaling process vanishes, $A_t = 0$, and, in turn, the forward performance process becomes constant across times. The optimal investment and wealth processes degenerate $\pi_t^* = 0$ and $X_t^{\pi^*} = x$, $t \geq 0$. In other words, even for non-zero λ_t, the optimal policy is to allocate *zero* wealth in every risky asset and at all times.

6.2.2 The Benchmark Case: $a(x, t) = xU_x(x, t)\delta_t$

The forward performance SPDE (28) simplifies to

$$dU(x,t) = \frac{1}{2} \frac{|U_x(x,t)(\lambda_t - \delta_t) - xU_{xx}(x,t)\delta_t|^2}{U_{xx}} dt - xU_x(x,t)\delta_t \cdot dW_t.$$

One can verify (see example 6.2.3 for $\varphi_t \equiv 0$, $t \geq 0$) that the process

$$U(x, t) = u\left(\frac{x}{Y_t}, A_t\right),$$

with Y_t, $t \geq 0$, solving

$$dY_t = Y_t \delta_t \cdot (\lambda_t dt + dW_t) \text{ with } Y_0 = 1 \tag{45}$$

and $A_t = \int_0^t |\lambda_s - \delta_s|^2 ds$ satisfies (28).

One may interpret the auxiliary process Y_t as a benchmark (or numeraire) with respect to which the performance of investment policies is measured.

Next, we define the benchmarked optimal portfolio and wealth processes,

$$\tilde{\pi}_t^* = \frac{\pi_t^*}{Y_t} \quad \text{and} \quad \tilde{X}_t^* = \frac{X_t^*}{Y_t}. \tag{46}$$

Then, $\tilde{\pi}_t^*$ is given by (cf. (49) for $\varphi_t \equiv 0$),

$$\tilde{\pi}_t^* = \tilde{X}_t^* \sigma_t^+ \delta_t - \frac{u_x(\tilde{X}_t^*, A_t)}{u_{xx}(\tilde{X}_t^*, A_t)} \sigma_t^+ (\lambda_t - \delta_t),$$

with A_t as above with $\tilde{X}_t^*$ solving

$$d\tilde{X}_t^* = -\frac{u_x(\tilde{X}_t^*, A_t)}{u_{xx}(\tilde{X}_t^*, A_t)} (\lambda_t - \delta_t) \cdot ((\lambda_t - \delta_t)dt + dW_t).$$

6.2.3 The Combined Market View/Benchmark Case:
$a(x, t) = -xU_x(x, t)\delta_t + U(x, t)\varphi_t$

The forward performance SPDE (28) becomes

$$dU(x, t) = \frac{1}{2} \frac{|U_x(x, t)(\lambda_t + \sigma_t \sigma_t^+ \varphi_t - \delta_t) - xU_{xx}(x, t)\delta_t|^2}{U_{xx}} dt$$
$$+ (-xU_x(x, t)\delta_t + U(x, t)\varphi_t) \cdot dW_t. \tag{47}$$

We introduce the time-rescaling process $A_t = \int_0^t |\lambda_s + \sigma_s \sigma_s^+ \varphi_s - \delta_s|^2 ds$.

We define the process

$$U(x, t) = u\left(\frac{x}{Y_t}, A_t\right) Z_t \tag{48}$$

with u solving (34) and Y_t and Z_t as in (45) and (44). We claim that it solves (47). Indeed, expanding yields

$$dU(x, t) = \left(du\left(\frac{x}{Y_t}, A_t\right)\right) Z_t + u\left(\frac{x}{Y_t}, A_t\right) dZ_t + d\left\langle u\left(\frac{x}{Y}, A\right), Z\right\rangle_t.$$

Moreover,

$$du\left(\frac{x}{Y_t}, A_t\right) = u_x\left(\frac{x}{Y_t}, A_t\right) d\left(\frac{x}{Y_t}\right) + u_t\left(\frac{x}{Y_t}, A_t\right) dA_t + \frac{1}{2} u_{xx}\left(\frac{x}{Y_t}, A_t\right) d\left\langle \frac{x}{Y}\right\rangle_t,$$

and

$$d\left(\frac{x}{Y_t}\right) = -\frac{x}{Y_t}\delta_t \cdot ((\lambda_t - \delta_t)dt + dW_t).$$

Consequently,

$$d\left\langle u\left(\frac{x}{Y}, A\right), Z\right\rangle_t = -\frac{x}{Y_t}u_x\left(\frac{x}{Y_t}, A_t\right)Z_t\delta_t \cdot \varphi_t dt,$$

$$u\left(\frac{x}{Y_t}, A_t\right)dZ_t = u\left(\frac{x}{Y_t}, A_t\right)Z_t\varphi_t \cdot dW_t = U(x,t)\varphi_t \cdot dW_t$$

and

$$\left(du\left(\frac{x}{Y_t}, A_t\right)\right)Z_t$$

$$= -\frac{x}{Y_t}u_x\left(\frac{x}{Y_t}, A_t\right)Z_t\delta_t \cdot ((\lambda_t - \delta_t)dt + dW_t)$$

$$+ u_t\left(\frac{x}{Y_t}, A_t\right)Z_t|\lambda_t + \sigma_t\sigma_t^+\varphi_t - \delta_t|^2 dt + \frac{1}{2}u_{xx}\left(\frac{x}{Y_t}, A_t\right)Z_t\left|-\frac{x}{Y_t}\delta_t\right|^2 dt.$$

We then deduce that

$$dU(x,t) = -\frac{x}{Y_t}u_x\left(\frac{x}{Y_t}, A_t\right)Z_t\delta_t \cdot (\lambda_t + \varphi_t - \delta_t)dt + u_t\left(\frac{x}{Y_t}, A_t\right)Z_t|\lambda_t$$

$$+ \sigma_t\sigma_t^+\varphi_t - \delta_t|^2 dt + \frac{1}{2}u_{xx}\left(\frac{x}{Y_t}, A_t\right)Z_t\left|-\frac{x}{Y_t}\delta_t\right|^2 dt$$

$$+ \left(-\frac{x}{Y_t}u_x\left(\frac{x}{Y_t}, A_t\right)Z_t\delta_t + U(x,t)\varphi_t\right) \cdot dW_t$$

$$= -xU_x(x,t)\delta_t \cdot (\lambda_t + \sigma_t\sigma_t^+\varphi_t - \delta_t)dt$$

$$+ \frac{1}{2}\frac{U_x^2(x,t)}{U_{xx}(x,t)}|\lambda_t + \sigma_t\sigma_t^+\varphi_t - \delta_t|^2 dt + \frac{1}{2}x^2U_{xx}(x,t)|\delta_t|^2 dt$$

$$+ (-xU_x(x,t)\delta_t + U(x,t)\varphi_t) \cdot dW_t$$

$$= \frac{1}{2}\frac{|U_x(x,t)(\lambda_t + \sigma_t\sigma_t^+\varphi_t - \delta_t) - xU_{xx}(x,t)\delta_t|^2}{U_{xx}(x,t)}dt$$

$$+ (-xU_x(x,t)\delta_t + U(x,t)\varphi_t) \cdot dW_t,$$

where we used (34). The optimal benchmarked policies $\tilde{\pi}_t^*$ and $\tilde{X}_t^*$ are defined as in (46). Using (21) and (48) we deduce, after some routine but tedious calculations, that

$$\tilde{\pi}_t^* = \tilde{X}_t^*\sigma_t^+\delta_t - \frac{u_x(\tilde{X}_t^*, A_t)}{u_{xx}(\tilde{X}_t^*, A_t)}\sigma_t^+(\lambda_t + \varphi_t - \delta_t) \tag{49}$$

where $\tilde{X}_t^*$ solves

$$d\tilde{X}_t^* = -\frac{u_x(\tilde{X}_t^*, A_t)}{u_{xx}(\tilde{X}_t^*, A_t)}(\lambda_t + \sigma_t\sigma_t^+\varphi_t - \delta_t) \cdot ((\lambda_t - \delta_t)dt + dW_t). \tag{50}$$

6.2.4 Single Stochastic Factor Models

This is the forward analogue of example 4.1. Consider a single stock and a stochastic factor satisfying (14) and (15). Let $w : \mathbb{R} \times [0, +\infty) \to \mathbb{R}$ satisfying the (forward) HJB equation

$$w_t + \max_\pi \left(\frac{1}{2} \sigma^2(y) \pi^2 w_{xx} + \pi(\mu(y) w_x + \rho \sigma(y) d(y) w_{xy}) \right)$$
$$+ \frac{1}{2} d^2(y) w_{yy} + b(y) w_y = 0, \tag{51}$$

for an appropriate initial condition $w(y, 0)$. Then, for $t \geq 0$, the process

$$U(x, t) = w(x, Y_t, t) \tag{52}$$

satisfies the SPDE (28) with volatility vector $a(x, t) = (a_1(x, t), a_2(x, t))$ with

$$a_1(x, t) = \rho d(Y_t) w_y(y, t) \quad \text{and} \quad a_2(x, t) = \sqrt{1 - \rho^2} d(Y_t) w_y(y, t).$$

Notice the main differences between the forward investment performance processes (52) and (48). Firstly, they are constructed by deterministic functions, w and u, that solve different pdes (see (51) and (34), respectively). Secondly, the investment performance process in (52) does not involve time-rescaling while the one in (48) does.

Acknowledgements   This work was presented, among others, at the AMaMeF Meeting, Vienna (2007), QMF (2007), the 5th World Congress of the Bachelier Finance Society, London (2008) and at conferences and workshops in Oberwolfach (2007 and 2008), Konstanz (2008), Princeton (2008) and UCSB (2009). The authors would like to thank the participants for fruitful comments. They also like to thank G. Zitkovic for his comments as well as an anonymous referee whose suggestions were very valuable in improving the original version of this manuscript. An earlier version of this paper was first posted in September 2007.

References

1. Ait-Sahalia, Y., Brandt, M.: Variable selection for portfolio choice. J. Finance **56**, 1297–1351 (2001)
2. Bates, D.S.: Post-87 crash fears and S&P futures options. J. Econom. **94**, 181–238 (2000)
3. Becherer, D.: The numeraire portfolio for unbounded semimartingales. Finance Stoch. **5**, 327–344 (2001)
4. Berrier, F., Rogers, L.C., Tehranchi, M.: A characterization of forward utility functions. Preprint (2009)
5. Borkar, V.S.: Optimal control of diffusion processes. Pitman Res. Notes Math. Ser., **203** (1983)
6. Bouchard, B., Touzi, N.: Weak dynamic programming principle for viscosity solutions. Preprint (2009)
7. Brandt, M.: Estimating portfolio and consumption choice: A conditional Euler equation approach. J. Finance **54**, 1609–1645 (1999)
8. Chacko, G., Viceira, L.M.: Dynamic consumption and portfolio choice with stochastic volatility in incomplete markets. Rev. Financ. Stud. **18**, 1369–1402 (2005)

9. Cvitanić, J., Schachermayer, W., Wang, H.: Utility maximization in incomplete markets with random endowment. Finance Stoch. **5**, 259–272 (2001)
10. El Karoui, N., Nguyen, D.H., Jeanblanc, M.: Compactification methods in the control of degenerate diffusions: Existence of an optimal control. Stochastics **20**, 169–220 (1987)
11. El Karoui, N., Peng, S., Quenez, M.C.: Backward stochastic differential equations in finance. Math. Finance **7**(1), 1–71 (1997)
12. Fama, W.E., Schwert, G.W.: Asset returns and inflation. J. Financ. Econ. **5**, 115–146 (1977)
13. Ferson, W.E., Harvey, C.R.: The risk and predictability of international equity returns. Rev. Financ. Stud. **6**, 527–566 (1993)
14. Fleming, W.H., Soner, M.H.: Controlled Markov Processes and Viscosity Solutions, 2nd edn. Springer, New York (2005)
15. French, K.R., Schwert, G.W., Stambaugh, R.F.: Expected stock returns and volatility. J. Financ. Econ. **19**, 3–29 (1987)
16. Glosten, L.R., Jagannathan, R., Runkle, D.E.: On the relation between the expected value and the volatility of the nominal excess return of stocks. J. Finance **48**, 1779–1801 (1993)
17. Goll, T., Kallsen, J.: Optimal portfolios for logarithmic utility. Stoch. Process. Appl. **89**, 31–48 (2000)
18. Goll, T., Kallsen, J.: A complete explicit solution to the log-optimal portfolio problem. Ann. Appl. Probab. **12**(2), 774–799 (2003)
19. Heaton, J., Lucas, D.: Market frictions, savings behavior and portfolio choice. Macroecon. Dyn. **1**, 76–101 (1997)
20. Hugonnier, J., Kramkov, D.: Optimal investment with random endowments in incomplete markets. Ann. Appl. Probab. **14**, 845–864 (2004)
21. Karatzas, I., Kardaras, C.: The numeraire portfolio in semimartingale financial models. Finance Stoch. **11**, 447–493 (2007)
22. Karatzas, I., Zitković, G.: Optimal consumption from investment and random endowment in incomplete semimartingale markets. Ann. Appl. Probab. **31**(4), 1821–1858 (2003)
23. Kim, T.S., Omberg, E.: Dynamic nonmyopic portfolio behavior. Rev. Financ. Stud. **9**, 141–161 (1996)
24. Korn, R., Korn, E.: Option Pricing and Portfolio Optimization – Modern Methods of Financial Mathematics. Am. Math. Soc., Providence (2001)
25. Kraft, H.: Optimal portfolios and Heston's stochastic volatility model. Quant. Finance **5**, 303–313 (2005)
26. Kramkov, D., Schachermayer, W.: The asymptotic elasticity of utility functions and optimal investment in incomplete markets. Ann. Appl. Probab. **9**(3), 904–950 (1999)
27. Kramkov, D., Schachermayer, W.: Necessary and sufficient conditions in the problem of optimal investment in incomplete markets. Ann. Appl. Probab. **13**(4), 1504–1516 (2003)
28. Kramkov, D., Sirbu, M.: On the two times differentiability of the value function in the problem of optimal investment in incomplete market. Ann. Appl. Probab. **16**(3), 1352–1384 (2006)
29. Krylov, N.: Controlled Diffusion Processes. Springer, New York (1980)
30. Liu, J.: Portfolio selection in stochastic environments. Rev. Financ. Stud. **20**(1), 1–39 (2007)
31. Merton, R.: Lifetime portfolio selection under uncertainty: The continuous-time case. Rev. Econ. Stat. **51**, 247–257 (1969)
32. Mnif, M.: Portfolio optimization with stochastic volatilities and constraints: An application in high dimension. Appl. Math. Optim. **56**, 243–264 (2007)
33. Musiela, M., Zariphopoulou, T.: The backward and forward dynamic utilities and the associated pricing systems: The case study of the binomial model. Preprint (2003)
34. Musiela, M., Zariphopoulou, T.: Investment and valuation under backward and forward dynamic exponential utilities in a stochastic factor model. Advances in Mathematical Finance. Applied and Numerical Harmonic Analysis Series, 303–334 (2007)
35. Musiela, M., Zariphopoulou, T.: Optimal asset allocation under forward exponential performance criteria (2006). In: Ethier, S., Feng, J., Stockbridge, R. (eds.) Markov Processes and related topics. A Festschrift for T.G. Kurtz. *Lecture Notes-Monograph Series*, 285–300. Institute of Mathematical Statistics, Beachwood (2008)

36. Musiela, M., Zariphopoulou, T.: Portfolio choice under space-time monotone performance criteria. SIAM Journal on Financial Mathematics 1, 326–365 (2010)
37. Musiela, M., Zariphopoulou, T.: The single period binomial model. In: Carmona R. (ed.) Indifference Pricing. Princeton University Press, Princeton, 3–41 (2009)
38. Musiela, M., Zariphopoulou, T.: Portfolio choice under dynamic investment performance criteria. Quant. Finance 9(2), 161–170 (2009)
39. Pham, H.: Smooth solutions to optimal investment models with stochastic volatilities and portfolio constraints. Appl. Math. Optim. 46, 1–55 (2002)
40. Schachermayer, W.: A super-martingale property of the optimal portfolio process. Finance Stoch. 7(4), 433–456 (2003)
41. Wachter, J.: Portfolio and consumption decisions under mean-reverting returns: An exact solution for complete markets. J. Financ. Quant. Anal. 37, 63–91 (2002)
42. Yong, J., Zhou, X.Y.: Stochastic Controls: Hamiltonian Systems and HJB Equations. Springer, New York (1999)
43. Zariphopoulou, T.: A solution approach to valuation with unhedgeable risks. Finance Stoch. 5, 61–82 (2001)
44. Zariphopoulou, T.: Optimal asset allocation in a stochastic factor model – An overview and open problems. In: Hansjorg, A., Runggaldier, W., Schachermayer, W. (eds.) Advanced Financial Modeling. RADON Series on Computational and Applied Mathematics, 8, 427–453 (2009)
45. Žitković, G.: A dual characterization of self-generation and exponential forward performances. Ann. Appl. Probab. 19(6), 2176–2210 (2009)

Issuers' Commitments Would Add More Value than Any Rating Scheme Could Ever Do

Carlos Veiga and Uwe Wystup

Abstract This paper analyzes the evolution of the structured products market focusing on the tools available for private investors, on which they rely for the selection process. The selection process is extremely difficult because there is a myriad of products, because of the dynamic nature of the market and market participants' actions, and because of the complexity of many of the products. We consider the existing types of tools, in particular the rating schemes that have been proposed by industry participants to provide guidance to the investor. We propose a set of properties that a rating scheme should show and check whether the existing schemes carry these properties. Our findings suggest that the existing rating schemes do not carry the desired properties. Furthermore, for the purpose of solving a highly indefinite selection process, an effective rating scheme may not exist. In light of this, we propose the introduction of a new quantity, the *floor*, that has a legal and financial meaning, on which issuers can also compete in addition to price and spread. Its acceptance and use would also yield standardization towards investors' interests by excluding some pricing practices and severely limiting others. Even though very little research has been produced in this area, we believe this to be a topic of high importance in establishing guidelines for healthy industry development and regulation that upholds investors' interests.

1 Introduction

Around the world there is a growing number of securities and contracts issued and written by financial institutions. Their purpose is to offer a customized risk/return

C. Veiga · U. Wystup (✉)
Frankfurt School of Finance & Management, Sonnemannstraße 9-11, 60314 Frankfurt am Main, Germany
e-mail: uwe.wystup@mathfinance.com

C. Veiga
e-mail: veiga.carlos@gmail.com

C. Chiarella, A. Novikov (eds.), *Contemporary Quantitative Finance*,
DOI 10.1007/978-3-642-03479-4_12, © Springer-Verlag Berlin Heidelberg 2010

profile that suits investors' preferences. These so called structured products[1] are linked to diverse underlying assets and are used by private and institutional investors alike. They cover short, medium and long term products from low risk to high risk and leverage.

In Germany and elsewhere this market shows significant activity with the number of issuers surpassing ten in the most liquid underlyings. For example, in June 2008, the most active German exchange for structured products, Börse Stuttgart's Euwax, reported 33 active issuers and more than 300,000 structured products listed. Other countries, specially in Europe and Australia, have also developed structured product markets with several issuers, thousands of products, and whose liquidity is close to 5% of the country's stock market. In the year 2008 the structured products' exchange traded volume, on the European exchanges members of FESE[2], amounted to €213 billion while equities volume amounted to €3,885 billion. In addition to the exchange traded volume, one should also consider the over the counter transactions of listed and unlisted structured products. These are surely a significant percentage of the total structured products trading, but for which, unfortunately, there are no aggregated statistics.

The key difference between structured products and the standardized derivative contracts, i.e. exchange traded options and futures, is the fact that they are issued as securities. This means that a structured product issue has a definite number of "shares" and is bound to the dynamics of securities trading. These dynamics differ strongly from those of standardized derivative contracts specially when selling is concerned. Simply stated, a security can only be sold if it is held (either by previous purchase or borrow), while taking up a selling position in standardized derivative contracts is not hindered by that constraint.

The importance of this difference is clear in light of arbitrage theory. It states that for a claim's price (security or contract) to be coherent with the price(s) of its underlying asset(s) (that again may be securities or contracts), it is necessary that an agent be able to sell the claim short, if it is overpriced with respect to its underlying, and to buy it, if it is underpriced. However, in the case of structured products, borrowing is impossible[3] and, consequently, so is short selling. Thus, there is no market force driving the price of an overpriced security towards its arbitrage theory fair price. That is, the price it would have if short selling were possible.

Hence, the consequence of the impossibility of short selling is that the claim may be overpriced but may never be underpriced. However, arbitrage theory only states the overpricing *can* occur, not that it *does* occur. Though, it should come as no surprise that banks require a reward for going through the costs of issuing and

---

[1] We shall use the terms *products* and *structured products* interchangeably.

[2] Federation of European Securities Exchanges

[3] Borrowing is impossible for several reasons, the most important of which are the unwillingness of the issuers to lend the securities, the dispersion of holders of such products, and the nonexistence of a securities lending market for these securities. Exchange traded funds (ETFs), though being securities, are different from structured products for they have built in, in the fund's by-laws, the borrowing possibility for their market makers.

maintaining these products and that profit is their true *raison d'être*. There is some research corroborating this fact by Stoimenov and Wilkens [4] and [6], and Wilkens et al. [5] that detail the dynamics of the overpricing over the life cycle of a product. These dynamics exhibit overpricing at issuance, overpricing decaying over the life of the product, significant overpricing drops after issuance, and order flow driven price behavior.

One may rightfully ask also why does an investor even consider buying securities that are possibly overpriced. There are certainly several reasons for doing so but here we just state one: many investors do not have the size or will to invest in non-biased securities in a way that would replicate the structured product's payoff. Thus, the trade-offs are ones of price versus size or price versus convenience, which are also present in any other market, financial or not. What is not similar to other markets is the inability of an investor, due to lack of information, to choose the best trade-off available. This is the core subject of this paper.

So far, the efforts to produce the lacking information have been devoted to the development of rating schemes that classify and order products according to a scale. We devote Sect. 2 to assess whether such ratings do produce relevant information, to enable the choice of the best trade-off, and conclude that they do not. In Sect. 3 we develop a formal analysis of the lack of information problem, propose a solution, and show that it produces relevant information. We conclude, in Sect. 4, giving our view of the development of the structured products market in connection with the lack of information problem.

2 Rating Schemes

Before analyzing existing rating schemes, we shall first state what we believe are the properties a rating should have in order to be effective in providing relevant information for the selection problem. Hence a rating should be:

- focused – the rating should measure only one well defined target feature;
- easily perceived – should allow for immediate perception of level and order between products;
- informative – produce additional information to set already available;
- impartial – consider only attributes specific to the product itself;
- current – the rating should be updated to reflect changes of the input data;
- robust – the rating should not be hindered by unusual or complex payoff profiles. It should be applicable to whatever product or contract.

These principles are probably easier to agree upon than to fulfill. Even the well-known and established ratings that classify the credit worthiness of issuers like the Moody's, Standard & Poor's or Fitch's credit ratings do not fulfill all the principles above. Common critics are that (i) ratings react slowly to changes in the environment, (ii) rating agencies choose the timing of rating reviews to be cautious about the political impact on the subject country or company, (iii) the fact that rated subjects pay for the rating service and that sole fact may bias the judgment (similar to an auditor's problem), (iv) rating agencies make significant subjective evaluations, and (v) that rating procedures are not robust enough to be standard across all industries and are not easily applicable to complex structures. This last issue has even been severely highlighted in the course of the current financial crisis. These critics put into question all principles above except the first three; credit ratings are focused solely on measuring the ability to meet future payments, are easily perceived, and do allow for decisions. Nevertheless, they are regarded by industry participants and regulators extremely useful classifications.

Given the success of credit rating schemes, several institutions started to apply the same concept to distill the large quantity of information present in the structured products market. Examples of these schemes are the ones from Institut für ZertifikateAnalyse[4] (IZA), Scope Group[5] and European Derivatives Group[6] (EDG), as are issuers' classification schemes. We shall analyze these rating schemes in general as our analysis is focused on the foundations and concepts that underlie these schemes. By keeping the analysis general, we believe that it remains valid not only for existing schemes but also for future ones that address the same problem. For illustration purposes we do take the mentioned schemes as examples to highlight the problems and implications that arise in connection with principles above.

We shall proceed by taking each principle above individually and examine what sort of procedures it rules out.

Focused excludes:

• a target feature created and defined within the rating process itself.

Examples of such target features are the *quality* of a product or its *appropriateness* to a given investor profile. These concepts are defined within the rating process, they do not mean anything outside of it. Examples of proper target features are the ability to meet future payments or the overall cost of a structured product. These concepts exist *a priori*. When costs, credit rating, investor risk preferences, etc. are aggregated or composed into a single measure, the result is an arbitrary and meaningless concept that cannot be attached to anything outside the scope of the rating process. Furthermore, if such a concept were to be taken as reference, it would, at best, reflect the preferences of a theoretical investor that, for being so individually specific, no other investor could relate to. Any investor, other than the theoretical, with different preferences with respect to any of the attributes, would not rate the products in the same order or scale as the rating would.

Easily Perceived excludes:

• multidimensional rating assessments;
• use of the same symbols to order distinct groups of products.

---

[4]http://www.iza.de

[5]http://www.scope.de

[6]http://www.derivatives-group.com

A multidimensional rating assessment fails to fulfill its very purpose since it does not map the set of products to an ordered scale. For example, a two dimensional rating, e.g. a measure of cost and another of expected return, can be sorted in an infinite number of ways by linearly combining the two measures. Thus, the ordering is left unresolved and hence the investor still lacks a clear basis for a decision. The ease of perceptiveness also excludes the use of the same symbols on several subsets of products that are not comparable with each other. Although the definition of subsets would simplify the rating process, the reuse of the same symbols would yield and implicit comparison that is not intended by the rating itself.

Informative excludes:

• redundant measurements of target features.

A rating that orders by issue date or maturity date also does not add any information to the existing set. A uniform classification of all products also would not carry any information, as it would not order the products.

Impartial excludes:

• the inclusion of investor preferences;
• the use of valuation models;
• estimated parameters;
• arbitrary or subjective assessments.

The key to understand the impartiality concept can be found in measurement theory, see Luce et al. [3] for an in-depth reference and Ahlert et al. [1] for an application. The problem is that the inclusion of measurements of attributes that are not specific to the rated products will change the ordering and evaluation of the products. This inclusion shall never be consensual as it biases the rating towards some of the products. On the other hand, the inclusion of attributes that are specific to the product cannot be argued against for it is the product itself that is being rated. Investor preferences are evidently product non-specific. A valuation model implicitly biases the evaluations towards some products, just consider a barrier and a vanilla option in light of a model that assumes the existence of jumps and one that does not. One can very easily construct an example with two products where the two models yield different orders. The same is true for the inclusion of estimates. Estimates are sample and estimation method dependent and, furthermore, for the calibration of models to market prices, there may be several parameter sets that would calibrate the model. The same is true for arbitrary and subjective assessments that, if changed, would also change the ordering of the products. These assessments include the choice inherent to any aggregation or composition of measurements of different target features.

Current excludes:

• rating revision not linked to input data variability.

An immediate and dramatic consequence of this principle is that the rating should be reassessed every time the input data is refreshed. Thus, if the rating depends on live information, e.g. product price, stock prices or option prices, the rating must also be updated live. Given the nature of the structured products market, where prices are typically overpriced, the price of the structured product is a necessary input to assess the costs embedded in it. The other necessary inputs are the prices of the underlying asset and of related derivatives needed to calculate the theoretical price. Thus, including costs in the rating assessment implies that the rating should be updated as frequently as the product price updates and as often the underlying asset prices updates. For most exchange traded structured products this makes it infeasible to include the cost estimates as an input for the rating process. The same may be said with respect to estimated data, i.e., the rating should be updated as often as the sample that underlies the estimation develops.

Robust excludes:

- rating processes valid only for a specific product type or class;
- any specific model.

The robustness requires that the rating process is a general approach valid for any product that exists or may exist. Different rating processes for different product types raise the problem of comparison across types. Furthermore, the inclusion of future products in the requirement comes from the fact that if they are not included by construction, new products may be created specifically targeted to take advantage of the limitations of the rating process. By the same token, no model may be able to properly evaluate and describe the risk of all types of products. It is, in fact, quite well known that typical models of a given asset class do not preform well when applied to other asset classes.

Still on the robustness principle, one may argue that it is too demanding and should not be considered. Even the well accepted and established credit ratings do not fulfill this principle, so why should the structured products ratings do. We believe it should be upheld for the sole reason that the structured products market has seen a remarkable dynamic in its short history in terms of creation of new types of products. There is also no evidence that this trend is abating.

To complete our analysis of rating schemes for structured products, we check what is left after the exclusions implied by the principles. Although no existing rating scheme belongs to the class of ratings schemes satisfying the principles above, that class may be non-empty. We are unable to produce a formal proof of the existence or the non-existence of rating schemes that fulfill the principles. We can report though that we have not been able to find one. Given the market characteristics, we believe that a rating scheme should consider the overall costs of the products, but that implies the existence of a model to calculate the theoretical price, and that, in turn, is not allowed by the principles. It would also use live data that would imply a live rating. We also fail to see how two products with the same overall costs, but with different payoff profiles, can be ordered in a non-subjective way.

With respect to the rating schemes mentioned above, no single one satisfies the principles above. All three, IZA, Scope and EDG produce scores which do not measure any objective feature of the product. They all rely extensively on aggregation of measurements of both objective and subjective features, e.g., cost, risk, concept of the product, and information produced to describe the product. They consider cost and risk, which in turn require the choice of a model with parameters that need to be estimated and calibrated, and whose assessments are highly ephemeral and not consensual among market participants. Risk is measured by the value at risk only, even though risk may be assessed in multiple ways, yielding each of these its own order. Moreover, computing the value at risk requires a model assumption and possibly parameter assumptions on the distribution of the underlying, which is highly subjective. Though they all exhibit rating reassessement periods that are longer than two weeks. EDG and IZA even consider investor preferences as part of the rating process, as if an investor would know how to describe his or her risk profile in these terms, or check if it would match any of the predefined ones.

Another perspective of the problem is to ask what is the harm in choosing a rating scheme that does not fulfill the principles. Such a choice would foster standardization and all products would still be rated on an equal basis. Even though all principles stand relevant in such a case, the impartiality principle assumes increased importance. If a rating that does not fulfill it is taken as a standard, or even enforced by regulation, that would yield only a standardization towards the (subjective and arbitrary) rating definition and not towards investors' interests. Furthermore, even though investors still need to solve the selection problem on their own, as existing ratings are not effective in ordering products in a meaningful way, they bear their costs. Either payed directly to an agency or embedded in the price of the product (in which the issuer reflects all its costs including the rating related ones), investors end up paying for rating schemes.

Therefore, we believe that a rating scheme is not the answer to bring standardization and informed investor choice to the structured products market. We believe instead that it can be achieved by introducing more tangible information, of the sort of bid-ask spreads and prices.

3 Floor

The proposal we describe in this section builds on the work of Stoimenov and Wilkens [4] and [6], and Wilkens et al. [5] that describe the dynamics of the price of a structured product over its life cycle. This dynamics exhibit overpricing at issuance, overpricing decaying over the life of the product, significant overpricing drops after issuance, and order flow driven price behavior. The authors rely on the concept of theoretical value and super-hedging boundaries to establish a price reference. This price is then compared with market prices to determine the overpricing and its dynamics.

To formalize these observations, without loss of generality, we assume that the issuer determines its bid and ask prices according to the functions

$$Ask(t) = f^A(t) + Markup^A(t), \tag{1}$$

$$Bid(t) = f^B(t) + Markup^B(t), \tag{2}$$

where $f^{A,B}(t)$ is the issuer's estimate of the product's theoretical values, using the relevant spread sides for each variable, and $Markup^{A,B}(t)$ are arbitrary functions.

The markup functions may depend on any factor, including the total quantity sold of the product up to time t.

The price policies described above generate profit for the issuers that can be decomposed in two parcels: interest and capital gains, denoted by P_i and P_{cg} respectively.

The interest is earned on the sale price markup $Markup^A(t_0)$ only, for we assume $f^A(t_0)$ was spent to purchase the issuer's hedge. If we assume a bank account yielding an overnight rate $r(i)$, the profit accumulated up to time t is just

$$P_i(t) = \left(\prod_{i=[t_0]}^{[t]-1} (1 + r(i)) - 1 \right) \times Markup^A(t_0), \qquad (3)$$

where t_0 is the trade time and i running from the day of t_0, $[t_0]$ to the day before t, $[t] - 1$.

It is important to note that this parcel of the profit cannot be controlled by the issuer after the initial transaction. On the investor's perspective, the loss, corresponding to the issuer's profit P_i, is included in his or her overall carry cost of holding the product. That cost is, to a large extent, predictable and/or bounded.

For the capital gains we need to write first the capital gains or losses on the whole structured product transaction, that is

$$Ask(t_0) - Bid(t) = f^A(t_0) - f^B(t) + Markup^A(t_0) - Markup^B(t). \qquad (4)$$

We now assume that $f^A(t_0) - f^B(t)$ is covered by the issuer's hedge. Therefore, the issuer's capital gain attributable to the pricing policy is just

$$P_{cg}(t) = Markup^A(t_0) - Markup^B(t). \qquad (5)$$

Unlike P_i, P_{cg} does depend on the issuer's pricing policy. The issuer is free to change $Markup^B(t)$ at any point; even set it at negative values[7]. On the investor's perspective, an decrease of $Markup^B(t)$ constitutes a loss. Such a loss is unpredictable in its size and moment.

We now claim that investors are better off if the $Markup^B(t)$ is known in advance, that is, before the investor purchases the product. Better off for the sole reason that investors would have enough information to weigh the total costs of the product against the benefits it brings them. Without the knowledge of $Markup^B(t)$ there is a loose end in the costs side until the product's maturity is reached, time when, by definition, $Markup^B(t)$ is zero.

Accordingly, we proceed with our analysis assuming, from this point on, that the issuer has committed to use the function $Markup^B(t)$ and that it stated on the product's term sheet.

However, there is still one open problem. This analysis has assumed that the issuer's estimates of the product's theoretical value, $f^{A,B}(t)$, are not subject to arbitrary revisions. If they are, the commitment is hollow because $f^{A,B}(t)$ may include not only the issuer's estimate of theoretical value but also hide part of the

---

[7]This is equivalent to setting a bid price at a discount to the reference price. Stoimenov and Wilkens [4] provide evidence of this.

$Markup^{A,B}(t)$. If that is allowed to happen, we are back to the initial situation, where there is not enough information to determine in advance the issuer's total profit. However, it is not reasonable to ask the issuer to disclose $f^{A,B}(t)$ for it may include trade secrets, be extremely complex and unusable by other parties.

Therefore, we need to replace $f^{A,B}(t)$, chosen by the issuer, by new function $h^{A,B}(t)$, independent of the issuer's views, such that expression (5) remains valid. In turn, this means $h^A(t_0) - h^B(t)$ is covered by the issuer's hedge.

This is easily accomplished if there is a static hedge for the structured product. Then $h^{A,B}(t)$ are just the prices of that static hedge portfolio, and $h^A(t_0) - h^B(t)$ is just the result from setting up and unwinding the hedge portfolio.

If there is a static super hedge, and $h^{A,B}(t)$ is the price of the super hedge portfolio, expression (5) is still valid for all t before maturity. At maturity time T, $P_{cg}(T) \geq Markup^A(t_0) - Markup^B(T)$ because the payoff of the super hedge portfolio may be greater than the payoff of the product. However, if the investor sells the structured product before maturity, the additional loss is not incurred.

Thus, if there is a static hedge or a super hedge for the structured product, there are functions $h^{A,B}(t)$ to replace $f^{A,B}(t)$ that are independent from the issuer's assessments. Functions $h^{A,B}(t)$ may even track a dynamic hedge (or super hedge) self-financing portfolio that the issuer is able to trade.

For products that cannot be statically super-hedged there may or may not be functions $h^{A,B}(t)$ to replace $f^{A,B}(t)$. However, if a product that can be decomposed as a portfolio, with its elements only taking positive values, the bid price may track only those elements that can be statically hedged (or super hedged). In such a case, the bid and ask functions (1) and (2) would be revised as

$$Ask(t) = h^A(t) + g^A(t) + Markup^A(t), \tag{6}$$

$$Bid(t) = h^B(t) + Markup^B(t), \tag{7}$$

with $h^{A,B}(t)$ the prices of the static hedge portfolio and $g^A(t) \geq 0$ the issuer's estimate of the price of the elements that are not statically hedgeable.

Hence, be $h^B(t)$ an hedge, super-hedge or sub-hedge, its determination is independent from the issuer's will. Furthermore, as $Markup^B(t)$ is defined before issuance, Bid(t) does not depend on the issuer's will at any point in time during the life of the product.

For example, consider a capital guaranteed product composed by a zero coupon bond and an exotic option. Furthermore, assume the issuer considers the Reuters' or Bloomberg's zero coupon bond price estimate as a reference price for $h^{A,B}(t)$. Thus, the product would trade at least at the zero coupon bond price, which is still better than no lower boundary at all. We say at least at zero coupon bond price because, in some situations, the bid will significantly underprice the structured product. The issuer will then, most likely, bid the structured product above the bid commitment to prevent the bid-ask spread from getting too large and to show a more competitive price.

This example shows that issuers may have reasons to bid their structured products above their commitments. It is even likely that issuers do this on a consistent basis on all products, at least by a small amount. The reason being to avoid unintended breaches of the bid price commitments and diminish potential conflicts. This observation is what motivated us to name the issuer's commitment as *floor* and not bid price commitment. From now on we will refer to it only as floor.

The cases we considered so far are cases where it is simple to find a floor and where the floor does not charge the issuer with extra risks. However, the issuer is free to choose the floor, even floors that carry extra risks with them. For the cases we considered above, the term sheet of a structured product should include at least these additional clauses:

- Floor in the Secondary Market: applicable.
- Floor Guarantor: legal name of entity.
- Floor Type: sub-hedge, exact, super-hedge.
- Floor Reference Price: instrument identification and price location.
- Initial Floor Markup: X currency units.
- Floor Markup Daily Decay: Y currency units.

We remark that these rules, on the one hand, exclude some pricing policies reported in Stoimenov and Wilkens [4] and, on the other hand, make some others predictable. Arbitrary pricing policies are excluded as they cannot be described by any function. This is a major difference as the issuer is no longer free to charge investors that hold their structured products in an non-disclosed-in-advance way. Pricing policies that depend on transaction volume or total outstanding quantity would have to be described in advance in a function. Furthermore, its relevant quantities would have to be made public and refreshed at a rate set by the markup definition. This is probably enough to deter issuers from including such rules in the markup definition. Markup functions may still have a non-linear decay, as the reported large decays after issuance. However, as this information is known in advance, investors may postpone the purchase of the product until that period has passed. The floor still leaves room for regular and predictable pricing strategies that are essential for the issuer to be able to profit from its products.

We also make note that, the floor is a new value that should be disseminated through the information network. Just like it is done with the usual set of prices that include the bid and ask prices, the last traded price, the daily maximum and minimum, and the previous sessions' close price.

To conclude this section we cover the most common types of structured products and provide examples of static or dynamic (super) hedges.

We start with a simple example of a very common structured product. The product is called index-tracker and pays off the value of an equity price index on the maturity date. The choice of the index itself as the floor reference is problematic because an equity price index is not a valid static portfolio, for it suffers from cash withdrawals by the amount of the dividends its shares pay. Therefore, if the issuer would choose this index as the Floor Reference Price, the following three clauses should be reviewed to

- Floor Type: super-hedge.
- Initial Floor Markup: implicit in Floor Reference Price.

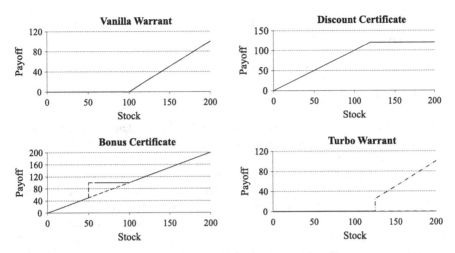

Fig. 1 Payoff profiles of common structured products

- Floor Markup Daily Decay: implicit in Floor Reference Price.

Figure 1 shows payoffs for vanilla warrants, discount certificates, bonus certificates and turbo warrants. The dashed lines represent the several possible values the product may pay off, depending on a barrier monitoring.

The proposals of hedge portfolios that follow, assume the existence of exchange traded options and futures on the same underlying asset and with the same maturity as the product. They also assume the availability of risk-free cash deposits for those maturities. A vanilla warrant has as static hedge an exchange traded option on the same underlying with the same strike and maturity. For call warrants, an exchange traded option with a lower strike constitutes a super hedge. The static hedge of a discount certificate is composed by: a short position on an exchange traded call, a long position on a future and a deposit of the total unused cash. A super hedge is obtained with a higher strike call. The price of bonus a certificate is always higher than the value of a portfolio with a future plus a deposit that pays out the contracted future price. A turbo warrant is a barrier option with the barrier located on the in-the-money side of the strike. There is no static hedge for it using the instruments we assumed. This is a typical case where an issuer may choose to assume a floor that introduces additional risks. Consider a turbo warrant call on a stock that does not pay dividends or on a total return index. Assume also a zero interest rate. A possible floor would be the intrinsic value of the turbo warrant, that is, just the difference of underlying price and strike. It is as if the turbo warrant were of American style, exercisable at any moment. To hedge this new liability, the issuer has to buy one unit of the underlying. If its price never touches the barrier the hedge works. If the price does touch the barrier, the issuer needs to sell the unit of underlying at the barrier level to maximize his result. At least, the issuer needs to sell the hedge above the strike to prevent a loss. However, this may not be possible because stock prices and indexes sometimes evolve in a discontinuous fashion. This is thus the extra risk this floor involves: the risk of not being able to unwind the hedge above the strike price.

This is an example where stating a floor would generate a more valuable product and also justify charging a higher price for it.

4 Conclusion

The goal of this paper is to enable investors to be able to identify the best trade-off available in the structured products market. The trade-off is one of price versus benefit brought to the investor.

To do so we analyze existing tools that claim to contribute to this identification. In particular, we survey the effectiveness of rating schemes to this purpose. We conclude ratings are not effective for they are in essence arbitrary in their definition and, therefore, are only able to produce arbitrary orderings. In order to fairly analyze the schemes, we first present a list of principles we believe every effective rating scheme should have. The rating schemes are then analyzed in light of these principles. We conclude that none of the considered schemes satisfy them. We also report to have failed to develop a rating scheme that would fulfill those principles.

We proceed with a simplification of the problem by removing the benefit to the investor from the analysis. We do so because the investor's benefit is an individual assessment that does not lend itself to modeling. We are then left with the price and realize that, even in price, there is currently no way to have a precise assessment.

Fortunately price is a more tangible concept that allows for modeling and formal description. We offer a framework to study the concept and then develop a proposal that enables a clear assessment of the price side of the trade-off. The proposal is a floor guarantee until the product's maturity. The floor is a quantity that can be freely defined by the issuer but is legally binding. It is the lowest price the issuer can bid at each moment for the structured product. In a way, it constitutes the time continuous counterpart of the discrete time commitments stated in the term sheet. The floor excludes some of the pricing policies referenced in the literature and seriously limits others. However, it still leaves room for regular and predictable pricing strategies that are essential for the issuer to be able to profit from its products. We complete the proposal with examples of application to the most common types of structured products.

Having a guaranteed floor, or the lack of it, quoted by an issuer for a complex product is probably the best rating the investor can get, as the inability or unwillingness to define a floor is itself an indication of the complex balance of risks and rewards underlying the product.

As to the question of whether the floor would make products more expensive, the answer is: not necessarily. On the one hand, any floor is always better than none at all (even if it is theoretically redundant) and thus should imply an extra cost. On the other hand, as explained above, it is impossible to know how expensive existing structured products really are due to their lack of floors. Therefore, the floor introduces a qualitative change that makes the products that bear it incomparable to the ones that do not. Only with a floor is a proper assessment of the cost of a product possible.

Actual rating schemes, as we showed, are not primarily scientific constructions but instead, the result of accumulation of credibility over many years, even centuries, the most elegant scientific formulation would ever replace.

As to the acceptance of these proposals, we do not expect market leader issuers to take them up promptly, as standardization may harm their margins and market share. As in any market, it is more likely that smaller players, that want to grow their market share, use these proposals as a mean to develop products that are objectively superior to the products of their competitors.

Acknowledgements   The author wishes to thank Millennium bcp investimento, S.A. for the financial support being provided during the course of his Ph.D. studies.

References

1. Ahlert, M., Gubernatis, G., Kliemt, H.: Kidney allocation in Eurotransplant. Anal. Krit. **23**, 156–172 (2001)
2. Cabral, L.: Introduction to Industrial Organization. MIT Press, Cambridge (2000)
3. Luce, R., Krantz, D., Suppes, P., Tversky, A.: Foundations of Measurement, vol. III. Academic Press, San Diego (1990)
4. Stoimenov, P., Wilkens, S.: Are structured products 'fairly' priced? An analysis of the German market for equity-linked instruments. J. Bank. Finance **29**, 2971–2993 (2005)
5. Wilkens, S., Erner, C., Röder, K.: The pricing of structured products in Germany. J. Deriv., Fall, 55–69 (2003)
6. Wilkens, S., Stoimenov, P.: The pricing of leverage products: An empirical investigation of the German market for 'long' and 'short' stock index certificates. J. Bank. Finance **31**, 735–750 (2007)

Pricing and Hedging of CDOs: A Top Down Approach

Damir Filipović and Thorsten Schmidt

Abstract This paper considers the pricing and hedging of collateralized debt obligations (CDOs). CDOs are complex derivatives on a pool of credits which we choose to analyse in the top down model proposed in Filipović et al. (Math. Finance, forthcoming, 2009). We reflect on the implied forward rates and bring them in connection with the top-down framework in Lipton and Shelton (Working paper, 2009) and Schönbucher (Working paper, ETH Zurich, 2005). Moreover, we derive variance-minimizing hedging strategies for hedging single tranches with the full index. The hedging strategies are given for the general case. We compute them also explicitly for a parsimonious one-factor affine model.

1 Introduction

In this paper we gradually develop a general formula for the variance-minimizing hedging strategy for a single tranche CDO within the top-down model framework recently developed in Filipović et al. [3].

In the last decade the markets of collateralized debt obligations (CDOs) have witnessed a tremendous activity and many models have been developed for pricing and some for hedging. The current market turmoil however illustrates that mostly used approaches – typically static models, such as the Gaussian copula model – are not able to capture the dynamic nature of the model. This is important for consistent pricing and even more important for hedging. In this paper we concentrate on

D. Filipović (✉)
Swiss Finance Institute, Ecole Polytechnique Fédérale de Lausanne (EPFL) and Swiss Finance Institute, Quartier UNIL-Dorigny, Extranef 218, CH-1015 Lausanne, Switzerland
e-mail: damir.filipovic@epfl.ch

T. Schmidt
Department of Mathematics, Chemnitz University of Technology, D-09126 Chemnitz, Germany
e-mail: thorsten.schmidt@mathematik.tu-chemnitz.de

C. Chiarella, A. Novikov (eds.), *Contemporary Quantitative Finance*,
DOI 10.1007/978-3-642-03479-4_13, © Springer-Verlag Berlin Heidelberg 2010

the dynamic top down model proposed in Filipović et al. [3] and derive variance-minimizing hedging strategies. For an overview of credit risk modelling we refer to the respective chapters in [9].

The most liquidly traded CDOs are those based on so-called indices. In 2004 the CDX in North America and the iTraxx in Europe have been created. For example, the iTraxx Europe consists of the 125 most liquid investment grade corporate credit default swaps. Besides the index *single tranche CDOs* (STCDOs) are liquidly traded. The STCDOs allow to invest in parts of the CDO, so-called tranches, see Sect. 4 below for details. In this paper we analyse the hedging of a STCDO with the index and derive the variance-minimizing hedging strategy. Besides the hedging strategy for the general form, we compute the hedging strategy explicitly in a simple one-factor affine model. This simple model is dynamic and allows to fit any given initial term structure of CDOs perfectly. For related articles on the dynamic hedging of credit portfolio products and CDOs we refer the reader to [1].

We assume a stochastic basis $(\Omega, \mathscr{F}, (\mathscr{F}_t), \mathbb{Q})$ satisfying the usual conditions, and where $\mathbb{Q}$ denotes a risk-neutral pricing measure. We consider a portfolio of credits with an overall outstanding notional normalised to 1. We follow a top-down approach and assume that the loss process

$$L_t = \sum_{s \leq t} \Delta L_s = \int_0^t \int_{(0,1]} x \, \mu(ds, dx)$$

be an $\mathscr{I} := [0, 1]$-valued non-decreasing marked point process with absolutely continuous compensator $v(t, dx)dt$. We let $\mu(dt, dx)$ be the integer-valued random measure associated with the jumps of L. The case of finitely many loss fractions $\mathscr{I} = \{\frac{i}{n} : i = 0, \dots, n\}$ is a special case of the setup.

We then define the (T, x)-bond which pays $1_{\{L_T \leq x\}}$ at maturity T, for $x \in [0, 1]$. In other words, this is a zero-recovery defaultable bond. Its arbitrage-free price at time $t \leq T$ is denoted by $P(t, T, x)$. By construction, $P(t, T) = P(t, T, 1)$ is the risk-free zero coupon bond. As a consequence, $P(t, T, \cdot)/P(t, T) = \mathbb{E}_{\mathbb{Q}^T}[1_{\{L_T \leq x\}} \mid \mathscr{F}_t]$ is just the $\mathscr{F}_t$-conditional cumulative distribution function of L_T under the T-forward measure $\mathbb{Q}^T$. Moreover, any European type T-claim on the loss process with absolutely continuous payoff function $H(L_T)$ can be decomposed into an infinite linear combination of (T, x)-bond payoffs:

$$H(L_T) = H(1) - \int_{(0,1]} H'(x) 1_{\{L_T \leq x\}} \, dx.$$

Here with H' we denote the Dini derivative of H, which is known to be locally integrable in x (see e.g. [12, Theorem 7.29]). As a simple consequence the price π_t of the claim at any time $t \leq T$ is given by

$$\pi_t = H(1)P(t, T) - \int_{(0,1]} H'(x) P(t, T, x) \, dx. \tag{1}$$

This paper is structured as follows. In Sect. 2 we derive the arbitrage free (T, x)-bond dynamics, and obtain a no-arbitrage criterion as proposed in Filipović et al. [3].

In Sect. 3 we reflect on the implied forward rates and bring them in connection with the top-down framework in Lipton and Shelton [8] and Schönbucher [11]. In Sect. 4 we calculate the gains processes from a single tranche CDO with arbitrary detachment points. This includes the entire index in particular. It is understood that the index can be replicated by holding the respective positions in the constituent CDS. In Sect. 5 we then derive the variance-minimizing hedging strategy for any STCDO in terms of the index in general. This is then explicitly computed for a simple one-factor affine model in Sect. 6.

2 (T, x)-Bond Dynamics

In this section we recap the framework for arbitrage-free term structure movements as laid out in [3]. The (T, x)-bond price is decomposed into default event and market risk:

$$P(t, T, x) = 1_{\{L_t \leq x\}} e^{-\int_t^T f(t, u, x) du}.$$

We assume that, for all (T, x), the (T, x)-forward rate process $f(t, T, x)$, $t \leq T$, follows a semimartingale of the form

$$f(t, T, x) = f(0, T, x) + \int_0^t a(s, T, x) ds + \int_0^t b(s, T, x)^\top dW_s$$

$$+ \int_0^t \int_{(0,1]} c(s, T, x; y) \mu(ds, dy) \tag{2}$$

where W is some d-dimensional Brownian motion. To justify the subsequent stochastic analysis we make the following technical assumptions, where $\mathscr{O}$ and $\mathscr{P}$ denote the optional and predictable σ-algebra on $\Omega \times \mathbb{R}_+$, respectively:

(A1) The initial forward curve $f(0, T, x)$ is $\mathscr{B}(\mathbb{R}_+) \otimes \mathscr{B}(\mathscr{I})$-measurable, and locally integrable:

$$\int_0^T |f(0, u, x)| du < \infty \quad \text{for all } (T, x).$$

The drift parameter $a(t, T, x)$ is $\mathbb{R}$-valued $\mathscr{O} \otimes \mathscr{B}(\mathbb{R}_+) \otimes \mathscr{B}(\mathscr{I})$-measurable, and locally integrable:

$$\int_0^T \int_0^T |a(t, u, x)| dt\, du < \infty \quad \text{for all } (T, x).$$

The volatility parameter $b(t, T, x)$ is $\mathbb{R}^d$-valued $\mathscr{O} \otimes \mathscr{B}(\mathbb{R}_+) \otimes \mathscr{B}(\mathscr{I})$-measurable, and locally bounded in the sense that

$$\sup_{0 \leq t \leq u \leq T} \|b(t, u, x)\| < \infty \quad \text{for all } (T, x).$$

The contagion parameter $c(t, T, x; y)$ is $\mathbb{R}$-valued $\mathscr{P} \otimes \mathscr{B}(\mathbb{R}_+) \otimes \mathscr{B}(\mathscr{I}) \otimes \mathscr{B}((0, 1])$-measurable, and locally bounded in the sense that

$$\sup_{0 \leq t \leq u \leq T, y \in (0,1]} |c(t, u, x; y)| < \infty \quad \text{for all } (T, x).$$

Under these conditions, the integrals in (2) are well defined in particular. Moreover, they imply that the risk-free short rate $r_t = f(t, t, 1)$ has a progressive version and satisfies $\int_0^T |r_t| \, dt < \infty$ for all T, see e.g. [4, Corollary 6.3]. Hence the risk-free numeraire $e^{\int_0^t r_s \, ds}$ is well defined. However, these conditions do not imply that spreads $f(t, T, x_1) - f(t, T, x_2)$, for $x_1 < x_2$ are nonnegative in general. Nonnegativity of spreads has to be established from case to case, such as in the affine model in Sect. 6 below.

Denote the discounted (T, x)-bond price by

$$Z(t, T, x) = e^{-\int_0^t r_s \, ds} P(t, T, x).$$

Lemma 1 *Under* (A1) *the implied dynamics of the discounted (T, x)-bond price process is given by*

$$\frac{dZ(t, T, x)}{Z(t-, T, x)} = \alpha(t, T, x) \, dt + \beta(t, T, x) \, dW_t$$

$$+ \int_{(0,1]} \gamma(t, T, x, \xi)(\mu(dt, d\xi) - \nu(t, d\xi)dt)$$

where

$$\alpha(t, T, x) = -r_t - \lambda(t, x) + f(t, t, x) - \int_t^T a(t, u, x) \, du$$

$$+ \frac{1}{2} \left\| \int_t^T b(t, u, x) \, du \right\|^2$$

$$+ \int_{(0,1]} \left(e^{-\int_t^T c(t,u,x;y) \, du} - 1 \right) 1_{\{L_t + y \leq x\}} \nu(t, dy) \quad (3)$$

$$\beta(t, T, x) = -\int_t^T b(t, u, x)^\top du \quad (4)$$

$$\gamma(t, T, x, \xi) = e^{-\int_t^T c(t,u,x;\xi) \, du} 1_{\{L_{t-} + \xi \leq x\}} - 1, \quad (5)$$

and we define

$$\lambda(t, x) = \int_{(0,1]} 1_{\{L_t + y > x\}} \nu(t, dy). \quad (6)$$

The corresponding stochastic exponential representation of $Z(t, T, x)$ reads

$$Z(t, T, x) = Z(0, T, x) \exp \left(\int_0^t \alpha(s, T, x) \, ds \right)$$

$$\times \exp \left(\int_0^t \beta(s, T, x) \, dW_s - \frac{1}{2} \int_0^t \| \beta(s, T, x) \|^2 \, ds \right)$$

$$\times \exp \left(- \int_0^t \gamma(s, T, x, \xi) \, v(s, d\xi) \, ds \right)$$

$$\times \prod_{s \le t} \left(1 + \gamma(s, T, x, \Delta L_s) 1_{\{\Delta L_s > 0\}} \right). \tag{7}$$

In particular, we have $\Delta Z(t, T, x)/Z(t-, T, x) = \gamma(t, T, x, \Delta L_t) 1_{\{\Delta L_t > 0\}}$, which equals -1 if the loss process crosses level x at t, that is, $L_{t-} \le x < L_t$. This is consistent with the fact that $Z(t, T, x) = 1_{\{L_t \le x\}} Z(t, T, x)$ for all t.

Remark 1 Note that $\lambda(t, x)$ in (6) is nothing but the intensity of the x-crossing time $\tau_x = \inf\{t \mid L_t > x\}$ of L. Indeed, this becomes obvious since we can write

$$1_{\{\tau_x \le t\}} = 1_{\{L_t > x\}} = \int_0^t \int_{(0,1]} 1_{\{L_{s-} + y > x\}} 1_{\{L_{s-} \le x\}} \mu(ds, dy). \tag{8}$$

Moreover, conversely to (6), $\lambda(t, x)$ uniquely determines $v(t, dx)$ via

$$v(t, (0, x]) = \lambda(t, L_t) - \lambda(t, L_t + x), \quad x \in (0, 1], \tag{9}$$

where we define $\lambda(t, x) = 0$ for $x \ge 1$. Furthermore, $\lambda(t, x)$ is non-increasing and càdlàg in x for any t by (9).

As a corollary of Lemma 1, we obtain the no-arbitrage drift condition from [3, Theorem 3.2]:

Corollary 1 *No-arbitrage, that is, $Z(t, T, x)$ is a local martingale for all (T, x), holds if and only if*

$$\int_t^T a(t, u, x) \, du = \frac{1}{2} \left\| \int_t^T b(t, u, x) \, du \right\|^2 \tag{10}$$

$$+ \int_{(0,1]} \left(e^{-\int_t^T c(t, u, x; y) \, du} - 1 \right) 1_{\{L_t + y \le x\}} \, v(t, dy),$$

$$r_t + \lambda(t, x) = f(t, t, x) \tag{11}$$

on $\{L_t \le x\}$, $dt \otimes d\mathbb{Q}$-a.s. for all (T, x).

Proof of Lemma 1 As in the proof of [3, Theorem 3.2] we decompose $p(t, T, x) = e^{-\int_t^T f(t,u,x)\,du}$ as

$$\frac{dp(t, T, x)}{p(t-, T, x)} = \left\{ f(t, t, x) - \int_t^T a(t, u, x)\,du + \frac{1}{2} \left\| \int_t^T b(t, u, x)\,du \right\|^2 \right.$$

$$+ \int_{(0,1]} \left(e^{-\int_t^T c(t,u,x;y)\,du} - 1 \right) v(t, dy) \bigg\} dt$$

$$- \int_t^T b(t, u, x)^\top du \cdot dW_t$$

$$+ \int_{(0,1]} \left(e^{-\int_t^T c(t,u,x;y)\,du} - 1 \right) (\mu(dt, dy) - v(t, dy)\,dt).$$

Note that we can write, as in (8)

$$1_{\{L_t \le x\}} = 1 + \int_0^t \int_{(0,1]} (-1_{\{L_{s-}+y>x\}} 1_{\{L_{s-} \le x\}}) \mu(ds, dy).$$

Integration by parts thus gives

$$d(1_{\{L_t \le x\}} p(t, T, x))$$

$$= 1_{\{L_{t-} \le x\}} dp(t, T, x) + p(t-, T, x)\,d1_{\{L_t \le x\}} + d[1_{\{L_t \le x\}}, p(t, T, x)]$$

$$= 1_{\{L_{t-} \le x\}} p(t-, T, x) \frac{dp(t, T, x)}{p(t-, T, x)}$$

$$+ p(t-, T, x) \int_{(0,1]} (-1_{\{L_{t-}+y>x\}} 1_{\{L_{t-} \le x\}}) \mu(dt, dy)$$

$$+ p(t-, T, x) \int_{(0,1]} \left(e^{-\int_t^T c(t,u,x;y)\,du} - 1 \right) (-1_{\{L_{t-}+y>x\}} 1_{\{L_{t-} \le x\}}) \mu(dt, dy)$$

$$= 1_{\{L_{t-} \le x\}} p(t-, T, x) \left(\frac{dp(t, T, x)}{p(t-, T, x)} \right.$$

$$\left. - \int_{(0,1]} e^{-\int_t^T c(t,u,x;y)\,du} 1_{\{L_{t-}+y>x\}} \mu(dt, dy) \right)$$

$$= 1_{\{L_{t-} \le x\}} p(t-, T, x)(D(t)\,dt + dN(t))$$

where the local martingale part is given by

$$dN(t) = - \int_t^T b(t, u, x)^\top du \cdot dW_t$$

$$+ \int_{(0,1]} \left(e^{-\int_t^T c(t,u,x;y)\,du} 1_{\{L_{t-}+y \le x\}} - 1 \right) (\mu(dt, dy) - v(t, dy)\,dt),$$

and the drift part is

$$
D(t) = f(t, t, x) - \int_t^T \dot{a}(t, u, x)\, du + \frac{1}{2}\left\|\int_t^T b(t, u, x)\, du\right\|^2
$$

$$
+ \int_{(0,1]} \left(e^{-\int_t^T c(t,u,x;y)\, du} - 1\right) v(t, dy)
$$

$$
- \int_{(0,1]} e^{-\int_t^T c(t,u,x;y)\, du} 1_{\{L_t+y>x\}}\, v(t, dy)
$$

$$
= f(t, t, x) - \int_t^T a(t, u, x)\, du + \frac{1}{2}\left\|\int_t^T b(t, u, x)\, du\right\|^2
$$

$$
+ \int_{(0,1]} \left(e^{-\int_t^T c(t,u,x;y)\, du} - 1\right) 1_{\{L_t+y\leq x\}}\, v(t, dy) - \int_{(0,1]} 1_{\{L_t+y>x\}}\, v(t, dy).
$$

Note that the last summand equals $\lambda(t, x)$. Discounting by $e^{\int_0^t r_s\, ds}$ yields (3)–(5). The stochastic exponential representation (7) is standard, see e.g. [7, Sect. I.4f]. $\square$

3 (T, x)-Forward Rates

In this intermediary section, we briefly reflect on the corresponding forward rates and discuss their relation to some other top-down approaches. Note that $f(t, T) = f(t, T, 1)$ is the risk-free forward rate.

Equation (11) states that the no-arbitrage property of $P(t, T, x)$ implies that the short rates equal the sum of risk-free short rate plus τ_x-intensity. In a heuristic manner, we can carry this property over to the forward rates:

$$
f(t, T, x) - f(t, T) = \lim_{\Delta T \to 0} \frac{1}{\Delta T}\mathbb{Q}^T[L_{T+\Delta T} > x \mid L_T \leq x, \mathscr{F}_t] \qquad (12)
$$

where $\mathbb{Q}^T$ denotes the T-forward measure. Whence $f(t, T, x) - f(t, T)$ is the $\mathbb{Q}^T$-forward transition rate prevailing at date t for L to jump at T from below or equal to above level x.

Indeed, for the sake of simplicity, let us for the rest of this section assume zero risk-free rates $f(t, T) = r_t = 0$. Then, assuming a continuous term structure $T \mapsto f(t, T, x)$, we obtain

$$
f(t, T, x) = \lim_{\Delta T \to 0} \frac{1}{\Delta T} \frac{P(t, T, x) - P(t, T + \Delta T, x)}{P(t, T, x)}
$$

$$
= \lim_{\Delta T \to 0} \frac{1}{\Delta T} \frac{\mathbb{E}[1_{\{L_T\leq x\}} - 1_{\{L_{T+\Delta T}\leq x\}} \mid \mathscr{F}_t]}{\mathbb{E}[1_{\{L_T\leq x\}} \mid \mathscr{F}_t]}
$$

$$= \lim_{\Delta T \to 0} \frac{1}{\Delta T} \frac{\mathbb{E}[1_{\{L_T \leq x\}} 1_{\{L_{T+\Delta T} > x\}} \mid \mathscr{F}_t]}{\mathbb{E}[1_{\{L_T \leq x\}} \mid \mathscr{F}_t]}$$

$$= \lim_{\Delta T \to 0} \frac{1}{\Delta T} \mathbb{Q}[L_{T+\Delta T} > x \mid L_T \leq x, \mathscr{F}_t],$$

which is (12).

3.1 Relation to the Top-Down Model in Lipton–Shelton [8]

Our approach can be brought in connection with the top-down model in Lipton and Shelton [8, Sect. 6.2], which is based on Schönbucher [11], as follows.

Suppose, as in [8], that the risk-free rates $f(t, T) = r_t = 0$, and that the credit portfolio consists of N_c credits with nominal 1 each, and the loss process can only assume fractions i/N_c, $i = 0, \ldots, N_c$. Then $f(t, T, x) = f(t, T, i/N_c)$ for all $i/N_c \leq x < (i+1)/N_c$. If, moreover, there are no simultaneous defaults, with the sole exception of a systemic default; that is, L_T can either jump from i/N_c to $(i+1)/N_c$ or to 1. Then, view of (12), we obtain with the notation of Lipton and Shelton [8, Sect. 6.2]:

$$f(t, T, x) = \frac{\sum_{l \leq [N_c x]} p_{l|m}(t, T) \left(\sum_{n > [N_c x]} a_{ln}(t, T) \right)}{\sum_{l \leq [N_c x]} p_{l|m}(t, T)}$$

$$= 1_{\{[N_c x] < N_c\}} \frac{p_{[N_c x]|m}(t, T) a_{[N_c x]}(t, T) + \sum_{l \leq [N_c x]} p_{l|m}(t, T) b(t, T)}{\sum_{l \leq [N_c x]} p_{l|m}(t, T)}$$

(13)

given that $L_t = m/N_c$, where $[z] = \max\{i \in \mathbb{Z} \mid i \leq z\}$ denotes the largest integer smaller or equal to z. Here $a_{ln}(t, T)$, $a_l(t, T) = a_{l,l+1}(t, T)$ and $b(t, T)$ denote the forward transition rates prevailing at date t for L to jump at T from l/N_c to n/N_c, $(l+1)/N_c$ and 1 (systemic default), respectively. Moreover, $p_{l|m}(t, T) = \mathbb{Q}[L_T = l/N_c \mid L_t = m/N_c, \mathscr{F}_t]$.

For the short rates, we thus obtain

$$f(t, t, x) = 1_{\{x < 1\}} \left(a_{[N_c x]}(t, t) 1_{\{L_t = [N_c x]/N_c\}} + b(t, t) \right)$$

(14)

on $\{L_t \leq x\}$. Combining this with (9) and (11), we obtain the following obvious relation:

$$v(t, (0, x]) = f(t, t, L_t) - f(t, t, L_t + x)$$

$$= 1_{\{L_t < 1\}} \left(a_{[N_c L_t]}(t, t) 1_{\{x \geq 1/N_c\}} + b(t, t) 1_{\{L_t + x \geq 1\}} \right).$$

4 Single Tranche CDOs (STCDOs)

STCDO are the most liquidly traded CDO derivatives. For the iTraxx Europe, the tranches 0–3%, 3–6%, 6–9%, 9–12%, and 12–22% are traded. Also special subindices exist, e.g. the iTraXX Europe HiVol which contains the 30 highest spread entities from the iTraxx Europe. We first describe the payment structure of STCDOs and then compute the dynamics of the related gains processes. A STCDO issued at $t = 0$ is specified by

- a number of coupon payment dates $0 < T_1 < \cdots < T_n$ (T_n is the maturity of the STCDO),
- a *tranche* with lower and upper detachment points $x_1 < x_2$ in $\mathscr{I}$,
- a fixed swap rate s_0.

We write

$$H(x) := (x_2 - x)^+ - (x_1 - x)^+ = \int_{(x_1, x_2]} 1_{\{x \leq y\}} dy.$$

An investor in this STCDO

- receives $s_0 H(L_{T_i})$ at T_i, $i = 1, \ldots, n$ (payment leg),
- pays $-dH(L_t) = H(L_{t-}) - H(L_t)$ at any time $t \leq T_n$ where $\Delta L_t \neq 0$ (default leg).

The accumulated discounted cash flow by time t thus equals

$$A_t = s_0 \sum_{T_i \leq t} e^{-\int_0^{T_i} r_s \, ds} H(L_{T_i}) + \int_0^t e^{-\int_0^u r_s \, ds} dH(L_u)$$

$$= s_0 \sum_{T_i \leq t} e^{-\int_0^{T_i} r_s \, ds} H(L_{T_i})$$

$$+ e^{-\int_0^t r_s \, ds} H(L_t) - H(L_0) + \int_0^t r_u e^{-\int_0^u r_s \, ds} H(L_u) \, du$$

$$= \int_{(x_1, x_2]} \left\{ s_0 \sum_{T_i \leq t} e^{-\int_0^{T_i} r_s \, ds} 1_{\{L_{T_i} \leq x\}} \right.$$

$$\left. + e^{-\int_0^t r_s \, ds} 1_{\{L_t \leq x\}} - 1_{\{L_0 \leq x\}} + \int_0^t r_u e^{-\int_0^u r_s \, ds} 1_{\{L_u \leq x\}} \, du \right\} dx \qquad (15)$$

where we have integrated by parts the default leg cash flow.

Throughout we shall assume that the accumulated discounted cash flow is square integrable:

(A2) $A_t \in L^2$ for all $t \leq T_n$.

The discounted time t spot value of the STCDO is given by the expectation of future discounted cash-flows and computes to

$$\Gamma_t = \mathbb{E}\left[A_{T_n} - A_t \mid \mathscr{F}_t \right]$$

$$= \int_{(x_1, x_2]} \left\{ s_0 \sum_{t < T_i} Z(t, T_i, x) - e^{-\int_0^t r_s \, ds} 1_{\{L_t \leq x\}} + Z(t, T_n, x) + \delta(t, x) \right\} dx$$

where

$$\delta(t,x) = \int_t^{T_n} \mathbb{E}\left[r_u e^{-\int_0^u r_s\,ds} 1_{\{L_u \leq x\}} \mid \mathscr{F}_t\right] du.$$

The par swap rate at time t, which is quoted in the market, is defined as the swap rate which gives the STCDO zero value if entered at t, that is, $\Gamma_t = 0$. It computes to

$$s_t = \frac{\int_{(x_1,x_2]} \left\{ e^{-\int_0^t r_s\,ds} 1_{\{L_t \leq x\}} - Z(t,T_n,x) - \delta(t,x) \right\} dx}{\int_{(x_1,x_2]} \sum_{t < T_i} Z(t,T_i,x)\,dx}.$$

The discounted t spot value Γ_t can thus be expressed as

$$\Gamma_t = (s_0 - s_t) \int_{(x_1,x_2]} \sum_{t<T_i} Z(t,T_i,x)\,dx.$$

The gains process from holding the STCDO equals

$$G_t = A_t + \Gamma_t = \mathbb{E}\left[A_{T_n} \mid \mathscr{F}_t\right]$$

$$= \int_{(x_1,x_2]} \left\{ s_0 \left(\sum_{T_i \leq t} e^{-\int_0^{T_i} r_s\,ds} 1_{\{L_{T_i} \leq x\}} + \sum_{t<T_i} Z(t,T_i,x) \right) \right.$$

$$\left. - 1_{\{L_0 \leq x\}} + Z(t,T_n,x) + \int_0^t r_u e^{-\int_0^u r_s\,ds} 1_{\{L_u \leq x\}}\,du + \delta(t,x) \right\} dx.$$

In view of (A2), G is a square integrable martingale, and it satisfies

$$dG_t = \int_{(x_1,x_2]} \left\{ s_0 \sum_{t<T_i} dZ(t,T_i,x) \right.$$

$$\left. + dZ(t,T_n,x) + r_t e^{-\int_0^t r_s\,ds} 1_{\{L_t \leq x\}}\,dt + d\delta(t,x) \right\} dx. \qquad (16)$$

The above analysis, including assumption (A2), prevails for any tranche $(x_1,x_2]$. We shall write accordingly

$$s_t = s_t^{(x_1,x_2]}, \quad A_t = A_t^{(x_1,x_2]}, \quad \Gamma_t = \Gamma_t^{(x_1,x_2]}, \quad G_t = G_t^{(x_1,x_2]}$$

and formulate the modified version of (A2).

(A2′) Assume that for each $x_1 < x_2$ with $x_1, x_2 \in \mathscr{I}$ it holds that $A_t^{(x_1,x_2]} \in L^2$ for all $t \leq T_n$.

Here is a sufficient condition:

Lemma 2 *If* $\sup_{t \leq T_n} e^{-\int_0^t r_s\,ds} \in L^2$ *then (A2′) is satisfied. This holds in particular if r is nonnegative.*

Proof Follows directly from the representation of A_t in Eq. (15).   □

Note that $x_1 = 0$ and $x_2 = 1$ corresponds to the entire index, which is composed by some balanced portfolio of single name CDS. It is understood that the index can be replicated by holding the respective positions in the constituent CDS. In the following section, we provide a risk minimizing hedging strategy for any STCDO in terms of the index.

5 Variance-Minimizing Hedging

As proposed by Cont and Kan [1], we now derive the variance-minimizing hedging strategy of the (x_1, x_2)-tranche STCDO with the index. Recall that, by assumption (A2'), the gains process $G_t^{(x_1,x_2]}$ for any tranche $(x_1, x_2]$ is a square integrable martingale.

For a pair of square integrable martingales M and N we denote by $\langle M, N \rangle$ their predictable quadratic covariation (see [7, Sect. I.4a]). Notice that

$$\mathbb{E}[M_{T_n} N_{T_n}] = \mathbb{E}[\langle M, N \rangle_{T_n}] + \mathbb{E}[M_0 N_0]$$

defines a scalar product on the space of square integrable martingales on $[0, T_n]$. The following can thus be seen as an orthogonal projection statement.

Theorem 1 *Assume that (A2') holds. For any time interval $0 \le t < T \le T_n$, the self-financing strategy*

$$\phi^* = \frac{d \langle G^{(x_1,x_2]}, G^{(0,1]} \rangle}{d \langle G^{(0,1]} \rangle}$$

along with the initial capital $c^* = G_t^{(x_1,x_2]}$ is the unique minimizer of the quadratic hedging error

$$\text{ess inf}_{c,\phi} \, \mathbb{E}\left[\left(c + \int_t^T \phi_s \, dG_s^{(0,1]} - G_T^{(x_1,x_2]}\right)^2 \mid \mathscr{F}_t\right].$$

Here the essential infimum is taken over all $c \in L^2(\mathscr{F}_t)$ and predictable processes ϕ with

$$\mathbb{E}\left[\int_0^{T_n} \phi_s^2 \, d \langle G^{(0,1]} \rangle_s\right] < \infty. \tag{17}$$

This strategy is referred to as the variance-minimizing strategy.

Proof By Assumption (A2') the process given by $G_t^{(x_1,x_2]} = \mathbb{E}(A_{T_n}^{(x_1,x_2]} | \mathscr{F}_t)$ is a square-integrable martingale. Hence, by Proposition 10.4 in [2] it can be decomposed in the so-called Goultchuk-Kunita-Watanabe decomposition:

$$G_t^{(x_1,x_2]} = G_0^{(x_1,x_2]} + \int_0^t \xi_s \, dG_s^{(0,1]} + G_t'. \tag{18}$$

Here G' is a square integrable martingale with mean zero and orthogonal to $G^{(0,1]}$ in the sense that $\langle G', G^{(0,1]} \rangle = 0$. Theorem 2.1 in Møller [10] states that the variance-minimizing strategy is given by the term ξ in Eq. (18). Orthogonality of G' and $G^{(0,1]}$ yields

$$d \langle G^{(x_1,x_2]}, G^{(0,1]} \rangle_t = \xi_t \, d \langle G^{(0,1]} \rangle_t$$

and the representation of $\xi = \phi^*$ follows. The initial cost, given by Eq. (2.3) in Møller [10], equals $A_t^{(x_1,x_2]} + \Gamma_t^{(x_1,x_2]} = G_t^{(x_1,x_2]}$. $\qquad \square$

Remark 2 The variance-minimizing hedging strategy minimizes a quadratic risk directly under the risk-neutral measure Q. This approach was also pursued in a number of different papers [1, 5, 6]. It is particularly useful when the distribution under the objective measure is difficult to obtain. As markets of CDOs are quite young, only few historical data is available which makes statistical estimation difficult.

Note that the variance-minimizing strategy ϕ^* does not depend on the reference time interval $[t, T]$, while it does on the tranche $(x_1, x_2]$ of course. Intuitively speaking, ϕ_t^* minimizes, locally for all t,

$$\mathbb{E}\big[(dG_t^{(x_1,x_2]} - \phi_t \, dG_t^{(0,1]})^2 \mid \mathscr{F}_t \big]$$

among all predictable ϕ_t which satisfy (17).

In the following, we compute ϕ^* for model specifications in various degrees of generality.

5.1 Deterministic Risk Free Rates

In this section we assume deterministic interest rates and derive the respective variance-minimizing hedging strategy in detail. We first compute the necessary terms of Eq. (16). The gains process and the hedging strategy follow.

Lemma 3 *If the risk free interest rates r_t are deterministic, then*

$$d\delta(t, x) = \mathscr{B}(t, x) \, dW_t$$

$$+ \int_{(0,1]} \mathscr{C}(t, x, \xi) \, (\mu(dt, d\xi) - \nu(t, d\xi)dt) - r_t e^{-\int_0^t r_s} 1_{\{L_t \le x\}} \, dt$$

where

$$\mathscr{B}(t,x) = \int_t^{T_n} r_u Z(t,u,x)\beta(t,u,x)\,du$$

$$\mathscr{C}(t,x,\xi) = \int_t^{T_n} r_u Z(t-,u,x)\gamma(t,u,x,\xi)\,du.$$

Proof If the risk free interest rates r_t are deterministic, we obtain

$$\delta(t,x) = \int_t^{T_n} r_u Z(t,u,x)\,du.$$

Using a stochastic Fubini argument as in the proof of [3, Theorem 3.2], we transform in the following

$$\int_t^{T_n}\int_0^t \cdots ds\,du = \int_0^t \int_t^{T_n} \cdots du\,ds = \int_0^t \int_s^{T_n} \cdots du\,ds - \int_0^t \int_0^u \cdots ds\,du,$$

and similarly for $dW_s\,du$ and $(\mu(ds,d\xi) - v(s,d\xi)ds)\,du$. We thus obtain

$$\int_t^{T_n} r_u Z(t,u,x)\,du$$

$$= \int_t^{T_n} r_u \Bigg(Z(0,u,x) + \int_0^t Z(s,u,x)\beta(s,u,x)\,dW_s$$

$$+ \int_0^t Z(s-,u,x)\int_{(0,1]} \gamma(s,u,x,\xi)(\mu(ds,d\xi) - v(s,d\xi)ds)\Bigg)du$$

$$= \int_0^{T_n} r_u Z(0,u,x)\,du$$

$$+ \int_0^t \mathscr{B}(s,x)\,dW_s + \int_0^t \int_{(0,1]} \mathscr{C}(s,x,\xi)(\mu(ds,d\xi) - v(s,d\xi)ds)$$

$$- \int_0^t r_u Z(u,u,x)\,du,$$

which yields the claim. $\qquad\square$

The gains process (16) accordingly simplifies to

$$dG_t^{(x_1,x_2]} = \int_{(x_1,x_2]} \Bigg\{ s_0^{(x_1,x_2]} \sum_{t<T_i} dZ(t,T_i,x) + dZ(t,T_n,x)$$

$$+ \int_t^{T_n} r_u Z(t,u,x)\beta(t,u,x)\,du\,dW_t$$

$$+ \int_{(0,1]} \int_t^{T_n} r_u Z(t-, u, x) \gamma(t, u, x, \xi) \, du \, (\mu(dt, d\xi) - v(t, d\xi)dt) \Big\} dx$$

$$= e^{-\int_0^t r_u du} \left(B_t^{(x_1, x_2]} dW_t + \int_{(0,1]} C_t^{(x_1, x_2]}(\xi) \, (\mu(dt, d\xi) - v(t, d\xi)dt) \right)$$

where

$$B_t^{(x_1, x_2]} = \int_{(x_1, x_2]} \left\{ s_0^{(x_1, x_2]} \sum_{t < T_i} P(t, T_i, x) \beta(t, T_i, x) \right.$$

$$\left. + P(t, T_n, x)\beta(t, T_n, x) + \int_t^{T_n} r_u P(t, u, x)\beta(t, u, x) \, du \right\} dx \quad (19)$$

$$C_t^{(x_1, x_2]}(\xi) = \int_{(x_1, x_2]} \left\{ s_0^{(x_1, x_2]} \sum_{t < T_i} P(t-, T_i, x)\gamma(t, T_i, x, \xi) \right.$$

$$+ P(t-, T_n, x)\gamma(t, T_n, x, \xi)$$

$$\left. + \int_t^{T_n} r_u P(t-, u, x)\gamma(t, u, x, \xi) \, du \right\} dx. \quad (20)$$

The predictable quadratic covariation thus computes to

$$\frac{d\langle G^{(x_1, x_2]}, G^{(0,1]} \rangle}{dt}$$

$$= e^{-2 \int_0^t r_u du} \left(B_t^{(x_1, x_2]} B_t^{(0,1]} - \int_{(0,1]} C_t^{(x_1, x_2]}(\xi) C_t^{(0,1]}(\xi) \, f(t, t, L_t + d\xi) \right)$$

where we have used $v(t, d\xi) = -f(t, t, L_t + d\xi)$, which follows from (11) and (6). Hence the variance-minimizing strategy given by

$$\phi_t^* = \frac{B_t^{(x_1, x_2]} B_t^{(0,1]} - \int_{(0,1]} C_t^{(x_1, x_2]}(\xi) C_t^{(0,1]}(\xi) \, f(t, t, L_t + d\xi)}{(B_t^{(0,1]})^2 - \int_{(0,1]} (C_t^{(0,1]}(\xi))^2 \, f(t, t, L_t + d\xi)} \quad (21)$$

can be computed at any time t by the observables

$$s_0^{(x_1, x_2]}, \quad P(t, u, x), \quad t \le u \le T_n, \quad x \in \mathscr{I},$$

and the model parameters

$$r_u, \quad \beta(t, u, x), \quad \gamma(t, u, x, \cdot), \quad t \le u \le T_n, \quad x \in \mathscr{I}.$$

The model parameters can be calibrated to the prevailing market data which could be either time series or option prices.

6 Affine Term Structure

In this section we consider a one-factor affine model proposed in Sect. 7.1 in [3]. This simple model is able to calibrate perfectly to any given initial term structure in the market and also allows for the explicit computation of the variance-minimizing hedging strategy as we show now.

Assume a constant risk-free short rate r. The factor Y is assumed to be a Feller square root process:

$$dY_t = (\mu_0 + \mu_1 Y_t)dt + \sigma\sqrt{Y_t}dW_t, \quad Y_0 = y \in \mathbb{R}_+ \tag{22}$$

and the forward rate follows an *affine term structure model*

$$f(t, T, x) = A'(t, T, x) + B'(t, T, x)Y_t$$

for some functions $A'(t, T, x)$ and $B'(t, T, x)$ with values in $\mathbb{R}$ and $\mathbb{R}^d$, respectively. We denote

$$A(t, T, x) = \int_t^T A'(t, u, x)du, \quad B(t, T, x) = \int_t^T B'(t, u, x)du.$$

The functions A and B are determined in terms of Riccati equations, which under (22) can be solved explicitly. From Sect. 7.1 in [3] we obtain that

$$\lambda(t, x) = \alpha_0(t, x) - r + \beta_0(x)Y_t, \tag{23}$$

with some $\mathbb{R}_+$-valued bounded measurable functions $\alpha_0(t, x)$ and $\beta_0(x)$ which are non-increasing and càdlàg, and $\alpha_0(t, 1) = r \geq 0$ and $\beta_0(1) = 0$. These functions can be used to calibrate the model to an initial term structure of STCDO prices. The Riccati equations become

$$A(t, T, x) = \int_t^T (\alpha_0(s, x) + \mu_0 B(s, T, x)) \, ds$$

$$-\partial_t B(t, T, x) = \beta_0(x) + \mu_1 B(t, T, x) - \frac{\sigma^2}{2}B(t, T, x)^2, \quad B(T, T, x) = 0. \tag{24}$$

The equation for B has the solution

$$B(t, T, x) \equiv B(T - t, x) = \frac{2\beta_0(x)\left(e^{\rho(x)(T-t)} - 1\right)}{\rho(x)\left(e^{\rho(x)(T-t)} + 1\right) - \mu_1\left(e^{\rho(x)(T-t)} - 1\right)} \tag{25}$$

where $\rho(x) = \sqrt{\mu_1^2 + 2\sigma^2\beta_0(x)}$. Note that

$$A'(t, T, x) = \partial_T A(t, T, x) = \alpha_0(T, x) + \mu_0 B(T - t, x),$$

and therefore the forward rate takes the following form

$$f(t, T, x) = \alpha_0(T, x) + \mu_0 B(T - t, x) + \partial_T B(T - t, x)Y_t. \tag{26}$$

Moreover,

$$P(t, T, x) = 1_{\{L_t \le x\}} e^{-A(t,T,x)-B(T-t,x)Y_t}.$$

In the following we compute the variance-minimizing hedging strategy in this model. We do not assume that the STCDO-prices are observed for any level x, but rather fix the attachment point structure $0 = x_0 < x_1 < \cdots < x_M = 1$. For simplicity we consider only $L_t = 0$.

In the following, we compute the essential terms for the hedging strategy. We assume that α_0 is piecewise linear and β_0 is piecewise constant:

(A3) Assume that $L_t = 0$ and

$$\alpha_0(s, x) = r + \sum_{i=1}^{M} (\alpha_{1,i}(s) + \alpha_{2,i}(s)x)1_{\{x \in [x_{i-1}, x_i)\}},$$

$$\beta_0(x) = \sum_{i=1}^{M} \beta_i 1_{\{x \in [x_{i-1}, x_i)\}}.$$

Equation (23) shows that α_0 is the intensity of the loss process crossing level x when the factor Y equals zero. Under (A3) this is interpolated linearly, e.g. from the observed tranche prices. The factor β_i determines the (linear) influence of Y on the intensity.

Remark 3 The assumption $L_t = 0$ is taken for simplicity of the notation. It is straightforward to extend the following results to the general case $L_t \ge 0$. By (23) and Remark 1, α_0 is non-increasing in x with $\alpha_0(t, 1) = r$ such that $\alpha_{2,i}(s) \le 0$. Moreover, as $Y \ge 0$ also β_0 is non-increasing in x and $\beta_0(1) = 0$. Hence $\beta_1 \ge \cdots \ge \beta_M \ge 0$. Continuity of α_0 in x eases the expressions at some places; continuity of α_0 in x is equivalent to

$$\alpha_{1,l+1}(t) + \alpha_{2,l+1}(t)x_l = \alpha_{1,l}(t) + \alpha_{2,l}(t)x_l,$$

for all $t \ge 0$ and $0 \le l \le M$.

Under (A3), we obtain that A is piecewise linear and B piecewise constant in x:

$$A(t, T, x) = \sum_{i=1}^{n} (A_{1,i}(t, T) + A_{2,i}(t, T)x)1_{\{x \in [x_{i-1}, x_i)\}}$$

$$B(t, T, x) = \sum_{i=1}^{n} B_i(T - t)1_{\{x \in [x_{i-1}, x_i)\}}$$

with

$$A_{1,i}(t, T) = \int_t^T \left(r + \alpha_{1,i}(s) + B_i(T - s) \right) ds$$

$$A_{2,i}(t, T) = \int_t^T \alpha_{2,i}(s)ds$$

$$B_i(T) = \frac{2\beta_i \left(e^{\rho(i)T} - 1\right)}{\rho(i) \left(e^{\rho(i)T} + 1\right) - \mu_1 \left(e^{\rho(i)T} - 1\right)}$$

and $\rho(i) = \sqrt{\mu_1^2 + 2\sigma^2 \beta_i}$.

Lemma 4 *Assume that (A3) holds. Then, in the affine one-factor model we have*

$$B_t^{(x_{i-1}, x_i]} = \sigma \sqrt{Y_t} \int_t^{T_n} \left(P(t, u, x_{i-1}) - P(t, u, x_i-)\right) \frac{B_i(u - t)}{A_{2,i}(t, u)} dw_u^{(x_{i-1}, x_i]}$$

where

$$dw_u^{(x_{i-1}, x_i]} = r \, du + \sum_{j=1}^M \left(s_0^{(x_{i-1}, x_i]} + 1_{\{j=n\}}\right) \delta_{T_j}(du)$$

where δ_T is the Dirac measure at T.

Proof From (26) and (22) we obtain that $b(t, T, x) = B'(T - t, x)\sigma\sqrt{Y_t}$. As $B(0, x) = 0$, inserting this in (4) gives

$$\beta(t, T, x) = -B(T - t, x)\sigma\sqrt{Y_t}.$$

Denote $w(i, j) := s_0^{(x_{i-1}, x_i]} + 1_{\{j=n\}}$. Equation (19) yields together with $L_t = 0$ that

$$B_t^{(x_{i-1}, x_i]} = -\sigma\sqrt{Y_t} \int_{(x_{i-1}, x_i]} \left\{ s_0^{(x_{i-1}, x_i]} \sum_{t < T_j} P(t, T_j, x)B(T_j - t, x) \right.$$

$$+ P(t, T_n, x)B(T_n - t, x) + r \int_0^{T_n} P(t, u, x)B(u - t, x) \, du \Bigg\} dx$$

$$= -\sigma\sqrt{Y_t} \int_{(x_{i-1}, x_i]} \left\{ \sum_{t < T_j} w(i, j)P(t, T_j, x)B(T_j - t, x) \right.$$

$$+ r \int_t^{T_n} P(t, u, x)B(u - t, x) \, du \Bigg\} dx.$$

The affine structure and (A3) allow to compute

$$\int_{(x_{i-1}, x_i]} P(t, T, x)B(T - t, x)dx = \int_{(x_{i-1}, x_i)} e^{-A(t, T, x) - B(T-t, x)Y_t} B(T - t, x)dx$$

$$= B_i(T - t)e^{-A_{1,i}(t, T) - B_i(T-t)Y_t} \int_{(x_{i-1}, x_i)} e^{-A_{2,i}(t, T)x} dx$$

$$= \frac{B_i(T-t)e^{-A_{1,i}(t,T)-B_i(T-t)Y_t}}{A_{2,i}(t,T)} \left(e^{-A_{2,i}(t,T)x_{i-1}} - e^{-A_{2,i}(t,T)x_i} \right)$$

$$= \frac{B_i(T-t)}{A_{2,i}(t,T)} \left(P(t,T,x_{i-1}) - P(t,T,x_i-) \right). \tag{27}$$

Hence

$$B_t^{(x_{i-1},x_i]} = -\sigma\sqrt{Y_t}\left\{ \sum_{t<T_j} w(i,j)\frac{B_i(T_j-t)}{A_{2,i}(t,T_j)} \left(P(t,T_j,x_{i-1}) - P(t,T_j,x_i-) \right) \right.$$

$$\left. + r\int_t^{T_n} \frac{B_i(u-t)}{A_{2,i}(t,u)} \left(P(t,u,x_{i-1}) - P(t,u,x_i-) \right) du \right\},$$

which is exactly the claim. $\qquad\qquad\square$

The following result gives the remaining part of the hedging strategy. Let

$$V(i,l,u) := \frac{x_l - x_{l-1}}{A_{2,i}(u)} - 1_{\{i=l\}}\frac{1}{(A_{2,i}(u))^2} \tag{28}$$

as well as $W(i,l,u) := V(i,l,u)$ for $1 \leq i < l$ and $W(l,l,u) := -\frac{1}{(A_{2,l}(u))^2}$ and set

$$w^k(i,u) := \sum_{l=(k+1)\wedge i}^{M} \alpha_{2,i}(t)W(i,l,u), \qquad v^k(i,u) := \sum_{l=(k+1)\wedge i}^{M} \alpha_{2,i}(t)V(i,l,u). \tag{29}$$

Proposition 1 *Under* (A3) *we have that*

$$\int_{(0,1]} C_t^{(x_{k-1},x_k]}(\xi)C_t^{(0,1]}(\xi)\,f(t,t,L_t+d\xi)$$

$$= -C_t^{(x_{k-1},x_k]}(x_{k-1}) \cdot \left(\sum_{i=1}^{M}\int_t^{T_n} \left(v^k(i,u)P(t,u,x_i) - w^k(i,u)P(t,u,x_i-) \right)dv_u^{(0,1]} \right)$$

$$+ \alpha_{2,k}(t)I_k$$

$$+ \sum_{l=1}^{M} \left(f(t,t,x_l) - f(t,t,x_l-) \right)C_t^{(x_{k-1},x_k]}(x_l)C_t^{(0,1]}(x_l),$$

where I_k *is given in Eq.* (32) *below and*

$$f(t,t,x_l) - f(t,t,x_l-) = \alpha_{1,l+1}(t) - \alpha_{1,l}(t) + x_l(\alpha_{2,l+1}(t) - \alpha_{2,l}(t))$$

$$+ Y_t(\beta_{l+1} - \beta_l).$$

If α_0 *is chosen to be continuous, then*

$$f(t,t,x_l) - f(t,t,x_l-) = Y_t(\beta_{l+1} - \beta_l).$$

Proof of Proposition 1

This section contains the proof of Proposition 1. For the proof we make use of the following result.

Lemma 5 *In the affine one-factor model under* (A3),

$$C_t^{(x_{i-1},x_i]}(\xi) = -1_{\{\xi > x_{i-1}\}} \int_t^{T_n} \frac{P(t, u, x_{i-1}) - P(t, u, (x_i \wedge \xi)-)}{A_{2,i}(t, u)} dv_u^{(x_{i-1},x_i]}$$

and

$$C_t^{(0,1]}(\xi) = -\sum_{i=1}^M 1_{\{\xi > x_{i-1}\}} \int_t^{T_n} \frac{P(t, u, x_{i-1}) - P(t, u, (x_i \wedge \xi)-)}{A_{2,i}(t, u)} dv_u^{(0,1]}$$

with

$$v_u^{(x_1,x_2]} = \left(ru + \sum_{u < T_j} \left(s_0^{(x_1,x_2]} + 1_{\{j=n\}} \right) \right).$$

Proof With $w(i, j) := s_0^{(x_{i-1},x_i]} + 1_{\{j=n\}}$ we obtain from Eq. (20) that

$$C_t^{(x_{i-1},x_i]}(\xi) = \int_{(x_{i-1},x_i]} \left\{ \sum_{t < T_j} w(i, j) P(t-, T_j, x) \gamma(t, T_j, x, \xi) \right.$$

$$\left. + \int_t^{T_n} r P(t-, u, x) \gamma(t, u, x, \xi) du \right\} dx.$$

First, $c = 0$ in (5) yields $\gamma(t, u, x, \xi) = -1_{\{L_{t-} + \xi > x\}} = -1_{\{\xi > x\}}$. Moreover, as in (27)

$$\int_{(x_{i-1},x_i]} P(t-, T, x) 1_{\{\xi > x\}} dx = 1_{\{\xi > x_{i-1}\}} \frac{P(t, T, x_{i-1}) - P(t, T, (x_i \wedge \xi)-)}{A_{2,i}(t, T)} \tag{30}$$

and we obtain

$$C_t^{(x_{i-1},x_i]}(\xi) = -1_{\{\xi > x_{i-1}\}} \left\{ \sum_{t < T_j} w(i, j) \frac{P(t, T_j, x_{i-1}) - P(t, T_j, (x_i \wedge \xi)-)}{A_{2,i}(t, T_j)} \right.$$

$$\left. + r \int_t^{T_n} \frac{P(t, u, x_{i-1}) - P(t, u, (x_i \wedge \xi)-)}{A_{2,i}(t, u)} du \right\}. \tag{31}$$

The expression for $C_t^{(0,1]}(\xi)$ follows in a similar way. $\qquad\square$

Proof of Proposition 1 Under (A3) f is piecewise linear but not necessarily continuous. With $\xi \in (0, 1]$ we obtain

$$f(t, t, d\xi) = \sum_{l=1}^{M} \left(1_{\{\xi \in [x_{l-1}, x_l)\}} f_x(t, t, x_{l-1}) d\xi + (f(t, t, x_l) - f(t, t, x_l-)) \delta_{x_l}(d\xi) \right),$$

where δ_x denotes the Dirac measure at x. We have that

$$f_x(t, t, x_{l-1}) = \alpha_{2,l}(t),$$

$$f(t, t, x_l) - f(t, t, x_l-) = \alpha_{1,l+1}(t) - \alpha_{1,l}(t) + x_l(\alpha_{2,l+1}(t) - \alpha_{2,l}(t))$$
$$+ Y_t(\beta_{l+1} - \beta_l).$$

Next,

$$\int_{(0,1]} C_t^{(x_{k-1}, x_k]}(\xi) C_t^{(0,1]}(\xi) f(t, t, d\xi)$$

$$= \sum_{l=1}^{M} \alpha_{2,l}(t) \underbrace{\int_{[x_{l-1}, x_l)} C_t^{(x_{k-1}, x_k]}(\xi) C_t^{(0,1]}(\xi) d\xi}_{=:I_l}$$

$$+ \sum_{l=1}^{M} \left(f(t, t, x_l) - f(t, t, x_l-) \right) C_t^{(x_{k-1}, x_k]}(x_l) C_t^{(0,1]}(x_l).$$

For $l < k$ the integral vanishes. For $l > k$ we have from (30) that

$$C_t^{(x_{k-1}, x_k]}(\xi) = -\int_t^{T_n} \frac{P(t, u, x_{k-1}) - P(t, u, x_k-)}{A_{2,k}(u)} dv_u^{(x_{k-1}, x_k]} = C_t^{(x_{k-1}, x_k]}(1) \quad (1)$$

and

$$\int_{(x_{l-1}, x_l]} C_t^{(0,1]}(\xi) d\xi$$

$$= -\int_t^{T_n} \sum_{i=1}^{l} \int_{(x_{l-1}, x_l]} \frac{P(t, u, x_{i-1}) - P(t, u, (x_i \wedge \xi)-)}{A_{2,i}(u)} d\xi \, dv_u^{(0,1]}$$

$$= -\sum_{i=1}^{l-1} (x_l - x_{l-1}) \int_t^{T_n} \frac{P(t, u, x_{i-1}) - P(t, u, x_i-)}{A_{2,i}(u)} dv_u^{(0,1]}$$

$$- \int_t^{T_n} \left\{ \frac{P(t, u, x_{l-1})}{A_{2,l}(u)} ((x_l - x_{l-1}) - (A_{2,l}(u))^{-1}) + \frac{P(t, u, x_l-)}{(A_{2,l}(u))^2} \right\} dv_u^{(0,1]}$$

$$= -\sum_{i=1}^{l} \int_t^{T_n} \left(V(i, l, u) P(t, u, x_{i-1}) - W(i, l, u) P(t, u, x_i-) \right) dv_u^{(0,1]}$$

with W, V given in (28). Hence

$$
\sum_{l=k+1}^{M} I_l = \sum_{l=k+1}^{M} \alpha_{2,l}(t) C_t^{(x_{k-1}, x_k]}(1)
$$

$$
\cdot \left(-\sum_{i=1}^{l} \int_t^{T_n} \big(V(i,l,u) P(t,u,x_{i-1}) - W(i,l,u) P(t,u,x_i-) \big) dv_u^{(0,1]} \right)
$$

$$
= -C_t^{(x_{k-1}, x_k]}(1) \cdot \left(\sum_{i=1}^{M} \int_t^{T_n} \big(v^k(i,u) P(t,u,x_i) \right.
$$

$$
\left. - w^k(i,u) P(t,u,x_i-) \big) dv_u^{(0,1]} \right)
$$

where w, v is as in (29). Finally, we consider the case where $l = k$. Then

$$
I_k = \int_{x_{k-1}}^{x_k} \int_t^{T_n} \frac{P(t,u,x_{k-1}) - P(t,u,\xi)}{A_{2,k}(u)} dv_u^{(x_{k-1},x_k]}
$$

$$
\cdot \int_t^{T_n} \sum_{i=1}^{k} \frac{P(t,z,x_{i-1}) - P(t,z,x_i \wedge \xi)}{A_{2,i}(z)} dv_z^{(0,1]} d\xi
$$

$$
= \int_t^{T_n} \frac{P(t,u,x_{k-1})\big((A_{2,k}(u))(x_k - x_{k-1}) - 1 \big) + P(t,u,x_k-)}{(A_{2,k}(u))^2} dv_u^{(x_{k-1},x_k]}
$$

$$
\cdot \left(\sum_{i=1}^{k-1} \int_t^{T_n} \frac{P(t,z,x_{i-1}) - P(t,z,x_i-)}{A_{2,i}(z)} dv_z^{(0,1]} \right)
$$

$$
+ \int_t^{T_n} \int_t^{T_n} \int_{x_{k-1}}^{x_k} \frac{(P(t,u,x_{k-1}) - P(t,u,\xi))(P(t,z,x_{k-1}) - P(t,z,\xi))}{A_{2,k}(u) A_{2,k}(z)} d\xi \, dv_u^{(x_{k-1},x_k]} dv_z^{(0,1]}.
$$

Note that

$$
\sum_{i=1}^{k-1} \int_t^{T_n} \frac{P(t,u,x_{i-1}) - P(t,u,x_i)}{A_{2,i}(u)} dv_u^{(0,1]} = C_t^{(0,1]}(x_{k-1})
$$

and

$$
\int_{(x_{k-1},x_k]} \big((P(t,u,x_{k-1}) - P(t,u,\xi))(P(t,z,x_{k-1}) - P(t,z,x_k)) \big) d\xi
$$

$$
= P(t,u,x_{k-1}) P(t,z,x_{k-1}) \left((x_k - x_{k-1}) - \frac{1}{A_{2,k}(u)} - \frac{1}{A_{2,k}(z)} \right.
$$

$$
\left. + \frac{1}{A_{2,k}(u) + A_{2,k}(z)} \right) + P(t,u,x_{k-1}) P(t,z,x_k-) \frac{1}{A_{2,k}(z)}
$$

$$+ P(t, u, x_k-) P(t, z, x_{k-1}) \frac{1}{A_{2,k}(u)}$$

$$- P(t, u, x_k-) P(t, z, x_k-) \frac{1}{A_{2,k}(u) A_{2,k}(z)}.$$

Summarising,

$$I_k = \int_t^{T_n} \frac{P(t, u, x_{k-1})\big((A_{2,k}(u))(x_k - x_{k-1}) - 1\big) + P(t, u, x_k-)}{(A_{2,k}(u))^2} dv_u^{(x_{k-1}, x_k]}$$

$$\cdot C_t^{(0,1]}(x_{k-1}) + \int_t^{T_n} \int_t^{T_n} \frac{1}{A_{2,k}(u) A_{2,k}(z)} \bigg\{ P(t, u, x_{k-1}) P(t, z, x_k-) \frac{1}{A_{2,k}(z)}$$

$$+ P(t, u, x_{k-1}) P(t, z, x_{k-1}) \bigg((x_k - x_{k-1}) - \frac{1}{A_{2,k}(u)} - \frac{1}{A_{2,k}(z)}$$

$$+ \frac{1}{A_{2,k}(u) + A_{2,k}(z)} \bigg) + P(t, u, x_k-) P(t, z, x_{k-1}) \frac{1}{A_{2,k}(u)}$$

$$- P(t, u, x_k-) P(t, z, x_k-) \frac{1}{A_{2,k}(u) A_{2,k}(z)} \bigg\} dv_u^{(x_{k-1}, x_k]} dv_z^{(0,1]}. \tag{32}$$

$\square$

7 Conclusion and Outlook

This paper derives dynamic hedging strategies for a large class of top-down models for CDO markets. The goal is to hedge single-tranche CDOs with the CDO index. Explicit formulas are provided for a simple one-factor affine model. Further studies shall analyse the empirical performance of the model and the hedging strategies; of particular importance is the comparison to other approaches in the literature.

Acknowledgements Damir Filipović gratefully acknowledges financial support from WWTF (Vienna Science and Technology Fund). We thank an anonymous referee for helpful remarks.

References

1. Cont, R., Kan, Y.-H.: Dynamic hedging of portfolio credit derivatives. Financial Engineering Report No. 2008-08, Columbia University Center for Financial Engineering (2008)
2. Cont, R., Tankov, P.: Financial Modelling with Jump Processes. Chapman & Hall, London (2004)
3. Filipović, D., Overbeck, L., Schmidt, T.: Dynamic modelling of CDO term structures. Math. Finance, forthcoming (2009)
4. Filipović, D.: Term Structure Models. Springer, Berlin (2009)

5. Frey, R., Backhaus, J.: Pricing and hedging of portfolio credit derivatives with interacting default intensities. Int. J. Theor. Appl. Financ. **11**, 611–634 (2008)
6. Föllmer, H., Sondermann, D.: Hedging of non-redundant contingent claims. In: Hildenbrand, W., Mas-Colell, A. (Eds.) Contributions to Mathematical Economics, pp. 147–160. North-Holland, Amsterdam (1986)
7. Jacod, J., Shiryaev, A.N.: Limit Theorems for Stochastic Processes. Springer, New York (1987)
8. Lipton, A., Shelton, D.: Single and multi-name credit derivatives: Theory and practice. Working paper (2009)
9. McNeil, A., Frey, R., Embrechts, P.: Quantitative Risk Management: Concepts, Techniques and Tools. Princeton University Press, Princeton (2005)
10. Møller, T.: Risk-minimizing hedging strategies for insurance payment processes. Finance Stoch. **5**, 419–446 (2001)
11. Schönbucher, P.: Portfolio losses and the term structure of loss transition rates: A new methodology for the pricing of portfolio credit derivatives. Working Paper, ETH Zurich (2005)
12. Wheeden, R.L., Zygmund, A.: Measure and Integral. Dekker, New York (1977)

Constructing Random Times with Given Survival Processes and Applications to Valuation of Credit Derivatives

Pavel V. Gapeev, Monique Jeanblanc, Libo Li, and Marek Rutkowski

Abstract We provide an explicit construction of a random time when the associated Azéma semimartingale (also known as the survival process) is given in advance. Our approach hinges on the use of a variant of Girsanov's theorem combined with a judicious choice of the Radon-Nikodým density process. The proposed solution is also partially motivated by the classic example arising in the filtering theory.

1 Introduction

The goal of this work is to address the following problem:

Problem (P) Let $(\Omega, \mathscr{G}, \mathbb{F}, \mathbb{P})$ be a probability space endowed with the filtration $\mathbb{F} = (\mathscr{F}_t)_{t \in \mathbb{R}_+}$. Assume that we are given a strictly positive, càdlàg, $(\mathbb{P}, \mathbb{F})$-local

This paper is dedicated to Professor Eckhard Platen on the occasion of his 60th birthday. Even though the topic of this work is not related to his exciting benchmark approach, we hope he will find some interest in this research.

P.V. Gapeev
Department of Mathematics, London School of Economics, Houghton Street, London WC2A 2AE, UK
e-mail: p.v.gapeev@lse.ac.uk

M. Jeanblanc (✉)
Département de Mathématiques, Université d'Évry Val d'Essonne, 91025 Évry Cedex, France
e-mail: monique.jeanblanc@univ-evry.fr

L. Li · M. Rutkowski
School of Mathematics and Statistics, University of Sydney, Sydney, NSW 2006, Australia

L. Li
e-mail: l.li@maths.usyd.edu.au

M. Rutkowski
e-mail: m.rutkowski@maths.usyd.edu.au

C. Chiarella, A. Novikov (eds.), *Contemporary Quantitative Finance*,
DOI 10.1007/978-3-642-03479-4_14, © Springer-Verlag Berlin Heidelberg 2010

martingale N with $N_0 = 1$ and an $\mathbb{F}$-adapted, continuous, increasing process Λ, with $\Lambda_0 = 0$ and $\Lambda_\infty = \infty$, and such that $G_t := N_t e^{-\Lambda_t} \leq 1$ for every $t \in \mathbb{R}_+$. The goal is to construct a random time τ on an extended probability space and a probability measure $\mathbb{Q}$ on this space such that:

(i) $\mathbb{Q}$ is equal to $\mathbb{P}$ when restricted to $\mathbb{F}$, that is, $\mathbb{Q}|_{\mathscr{F}_t} = \mathbb{P}|_{\mathscr{F}_t}$ for every $t \in \mathbb{R}_+$,

(ii) the Azéma supermartingale $G_t^{\mathbb{Q}} := \mathbb{Q}(\tau > t \,|\, \mathscr{F}_t)$ of τ under $\mathbb{Q}$ with respect to the filtration $\mathbb{F}$ satisfies

$$G_t^{\mathbb{Q}} = N_t e^{-\Lambda_t}, \quad \forall t \in \mathbb{R}_+. \tag{1}$$

In that case, the pair $(\tau, \mathbb{Q})$ is called a *solution* to Problem (P).

We will sometimes refer to the Azéma supermartingale $G^{\mathbb{Q}}$ as the *survival process* of τ under $\mathbb{Q}$ with respect to $\mathbb{F}$. The solution to this problem is well known if $N_t = 1$ for every $t \in \mathbb{R}_+$ (see Sect. 3) and thus we will focus in what follows on the case where N is not equal identically to 1.

Condition (i) implies that the postulated inequality $G_t := N_t e^{-\Lambda_t} \leq 1$ is necessary for the existence of a solution $(\tau, \mathbb{Q})$. Note also that in view of (i), the joint distribution of (N, Λ) is set to be identical under $\mathbb{P}$ and $\mathbb{Q}$ for any solution $(\tau, \mathbb{Q})$ to Problem (P). In particular, N is not only a $(\mathbb{P}, \mathbb{F})$-local martingale, but also a $(\mathbb{Q}, \mathbb{F})$-local martingale. However, in the construction of a solution to Problem (P) provided in this work, the so-called H-hypothesis is not satisfied under $\mathbb{Q}$ by the filtration $\mathbb{F}$ and the enlarged filtration $\mathbb{G}$ generated by $\mathbb{F}$ and the observations of τ. Hence the process N is not necessarily a $(\mathbb{Q}, \mathbb{G})$-local martingale.

In the approach proposed in this work, in the first step we construct a finite random time τ on an extended probability space using the *canonical construction* in such a way that

$$G_t^{\mathbb{P}} := \mathbb{P}(\tau > t \,|\, \mathscr{F}_t) = e^{-\Lambda_t}, \quad \forall t \in \mathbb{R}_+.$$

To avoid the need for an extension of Ω, it suffices to postulate, without loss of generality, that there exists a random variable Θ defined on Ω such that Θ is exponentially distributed under $\mathbb{P}$ and it is independent of $\mathscr{F}_\infty$. In the second step, we propose a change of the probability measure by making use of a suitable version of Girsanov's theorem. Since we purportedly identify the extended space with Ω, it make sense to compare the probability measures $\mathbb{P}$ and $\mathbb{Q}$. Let us mention in this regard that the probability measures $\mathbb{P}$ and $\mathbb{Q}$ are not necessarily equivalent. However, for any solution $(\tau, \mathbb{Q})$ to Problem (P), the equality $\mathbb{Q}(\tau < \infty) = 1$ is satisfied (see Lemma 3) and thus τ is necessarily a finite random time under $\mathbb{Q}$.

In the existing literature, one can find easily examples where the Doob-Meyer decomposition of the Azéma supermartingale is given, namely, $G_t = M_t - A_t$ (see, e.g., Mansuy and Yor [5]). It is then straightforward to deduce the multiplicative decomposition by setting $N_t = \int_0^t e^{\Lambda_s} \, dM_s$ and $\Lambda_t = \int_0^t \frac{dA_s}{G_{s-}}$. However, to the best of our knowledge, a complete solution to the problem stated above is not yet available, though some partial results were obtained. Nikeghbali and Yor [6] study a similar problem for a particular process Λ, namely, $\Lambda_t = \ln(\sup_{s \leq t} N_s)$ for a local martingale N which converges to 0 as t goes to infinity. It is worth stressing that in [6] the process G can take the value one for some $t > 0$. We will conduct the first part of our study under the standing assumption that $G_t \leq 1$. However, to provide an explicit construction of a probability measure $\mathbb{Q}$, we will work in Sect. 5 under the stronger assumption that the inequality $G_t < 1$ holds for every $t > 0$.

The paper is organized as follows. We start by presenting in Sect. 2 an example of a random time τ, which is not a stopping time with respect to the filtration $\mathbb{F}$, such that the Azéma supermartingale of τ with respect to $\mathbb{F}$ can be computed explicitly. In fact, we revisit here a classic example arising in the non-linear filtering theory. In the present context, it can be seen as a motivation for the problem stated at the beginning. In addition, some typical features of the Azéma supermartingale, which are apparent in the filtering example, are later rediscovered in a more general set-up, which is examined in the subsequent sections.

The goal of Sect. 3 is to furnish some preliminary results on Girsanov's change of a probability measure in the general set-up. In Sect. 4, the original problem is first reformulated and then reduced to a more tractable analytical problem (see Problems (P.1)–(P.3) therein). In Sect. 5, we analyze in some detail the case of a Brownian filtration. Under the assumption that $G_t < 1$ for every $t \in \mathbb{R}_+$, we identify a solution to the original problem in terms of the Radon-Nikodým density process.

Section 6 discusses the relevance of the multiplicative decomposition of survival process of a default time τ for the risk-neutral valuation of credit derivatives. From this perspective, it is important to observe that a random time τ constructed in this work has the same intensity under $\mathbb{P}$ and $\mathbb{Q}$, but it has different conditional probability distributions with respect to $\mathbb{F}$ under $\mathbb{P}$ and $\mathbb{Q}$. This illustrates the important fact that the default intensity does not contain enough information to price credit derivatives (in this regard, we refer to El Karoui et al. [3]). In several papers in the financial literature, the modeling of credit risk is based on the postulate that the process

$$M_t := \mathbb{1}_{\{\tau \leq t\}} - \int_0^{t \wedge \tau} \lambda_u \, du$$

is a martingale with respect to a filtration $\mathbb{G}$ such that τ is a (totally inaccessible) $\mathbb{G}$-stopping time. However, we will argue in Sect. 6 that this information is insufficient for the computation of prices of credit derivatives. Indeed, it appears that, except for the most simple examples of models and credit derivatives, the martingale component in the multiplicative decomposition of the Azéma supermartingale of τ has a non-negligible impact on risk-neutral values of credit derivatives.

2 Filtering Example

The starting point for this research was a well-known problem arising in the filtering theory. The goal of this section is to recall this example and to examine some interesting features of the conditional distributions of a random time, which will be later rediscovered in a different set-up.

2.1 Azéma Supermartingale

Let $W = (W_t, t \in \mathbb{R}_+)$ be a Brownian motion on the probability space $(\Omega, \mathcal{G}, \mathbb{P})$, and τ be a random time, independent of W and such that $\mathbb{P}(\tau > t) = e^{-\lambda t}$ for every $t \in \mathbb{R}_+$ and a constant $\lambda > 0$. We define the process $U = (U_t, t \in \mathbb{R}_+)$ by setting

$$U_t = \exp\left(\left(a + b - \frac{\sigma^2}{2}\right)t - b(t - \tau)^+ + \sigma W_t\right),$$

where a, b and σ are strictly positive constants.

It is not difficult to check that the process U solves the following stochastic differential equation

$$dU_t = U_t\,(a + b\,\mathbb{1}_{\{\tau > t\}})\,dt + U_t\,\sigma\,dW_t. \tag{2}$$

In the filtering problem, the goal is to assess the conditional probability that the moment τ has already occurred by a given date t, using the observations of the process U driven by (2).

Let us take as $\mathbb{F}$ the natural filtration generated by the process U, that is,

$$\mathcal{F}_t = \sigma(U_s \mid 0 \le s \le t)$$

for every $t \in \mathbb{R}_+$. By means of standard arguments (see, for instance, [8, Chapt. IV, Sect. 4] or [4, Chapt. IX, Sect. 4]), it can be shown that the process U admits the following semimartingale decomposition with respect to its own filtration

$$dU_t = U_t(a + bG_t)\,dt + U_t\sigma\,d\overline{W}_t,$$

where $G = (G_t, t \in \mathbb{R}_+)$ is the Azéma supermartingale, given by $G_t = \mathbb{P}(\tau > t \mid \mathcal{F}_t)$, and the innovation process $\overline{W} = (\overline{W}_t, t \in \mathbb{R}_+)$, defined by

$$\overline{W}_t = W_t + \frac{b}{\sigma}\int_0^t (\mathbb{1}_{\{\tau > u\}} - G_u)\,du,$$

is a standard Brownian motion with respect to $\mathbb{F}$. It is easy to show, using the arguments based on the notion of strong solutions of stochastic differential equations (see, for instance, Liptser and Shiryaev [4, Chapt. IV, Sect. 4]), that the natural filtration of $\overline{W}$ coincides with $\mathbb{F}$. It follows from [4, Chapt. IX, Sect. 4] (see also Shiryaev [8, Chapt. IV, Sect. 4]) that G solves the following SDE

$$dG_t = -\lambda G_t\,dt + \frac{b}{\sigma}G_t(1 - G_t)\,d\overline{W}_t. \tag{3}$$

Consequently, the process $N = (N_t, t \in \mathbb{R}_+)$, given by $N_t = e^{\lambda t}G_t$, satisfies

$$dN_t = \frac{b}{\sigma}e^{\lambda t}G_t(1 - G_t)\,d\overline{W}_t. \tag{4}$$

Since $G(1 - G)$ is bounded, it is clear that N is a strictly positive $(\mathbb{P}, \mathbb{F})$-martingale with $N_0 = 1$. We conclude that the Azéma supermartingale G of τ with respect to the filtration $\mathbb{F}$ admits the following representation

$$G_t = N_t e^{-\lambda t}, \quad \forall t \in \mathbb{R}_+, \tag{5}$$

where the $(\mathbb{P}, \mathbb{F})$-martingale N is given by (4).

Let us observe that equality (3) provides the (additive) Doob-Meyer decomposition of the bounded $(\mathbb{P}, \mathbb{F})$-supermartingale G, whereas equality (5) yields its multiplicative decomposition.

2.2 Conditional Distributions

From the definition of the Azéma supermartingale G and the fact that $(G_t e^{\lambda t}, t \in \mathbb{R}_+)$ is a $(\mathbb{P}, \mathbb{F})$-martingale it follows that, for a fixed $u > 0$ and every $t \in [0, u]$,

$$\mathbb{P}(\tau > u \mid \mathscr{F}_t) = \mathbb{E}_\mathbb{P}(\mathbb{P}(\tau > u \mid \mathscr{F}_u) \mid \mathscr{F}_t) = e^{-\lambda u} \mathbb{E}_\mathbb{P}(N_u \mid \mathscr{F}_t) = e^{-\lambda u} N_t. \tag{6}$$

Standard arguments given in Shiryaev [7, Chapt. IV, Sect. 4] (which are also summarized in [8, Chapt. IV, Sect. 4]), based on an application of the Bayes formula, yield the following result, which extends formula (6) to $t \in [u, \infty)$.

Proposition 1 *The conditional survival probability equals, for every $t, u \in \mathbb{R}_+$,*

$$\mathbb{P}(\tau > u \mid \mathscr{F}_t) = 1 - \frac{X_t}{X_{u \wedge t}} + X_t Y_{u \wedge t} e^{-\lambda u}, \tag{7}$$

where the process Y is given by

$$Y_t = \exp\left(\frac{b}{\sigma^2} \left(\ln U_t - \frac{2a + b - \sigma^2}{2} t \right) \right) \tag{8}$$

and the process X satisfies

$$\frac{1}{X_t} = 1 + \int_0^t e^{-\lambda u} \, dY_u. \tag{9}$$

It follows immediately from (7) that $G_t = X_t Y_t e^{-\lambda t}$ so that the equality $N = XY$ is valid, and thus (7) coincides with (6) when $t \in [0, u]$. Moreover, by standard computations, we see that

$$dY_t = \frac{b}{\sigma} Y_t \, dW_t + \frac{b^2}{\sigma^2} G_t Y_t \, dt. \tag{10}$$

Using (4) and (10), we obtain

$$dX_t = d\left(\frac{N_t}{Y_t}\right) = -\frac{b}{\sigma} G_t X_t d\overline{W}_t,$$

and thus X is a strictly positive $(\mathbb{P}, \mathbb{F})$-martingale with $X_0 = 1$. Finally, it is interesting to note that we deal here with the model where

$$\tau = \inf\{t \in \mathbb{R}_+ : \Lambda_t \geq \Theta\}$$

with $\Lambda_t = \lambda t$ (so that $\lambda \tau = \Theta$) and the barrier Θ is an exponentially distributed random variable, which is not independent of the σ-field $\mathscr{F}_\infty$. Indeed, we have that, for every $u > 0$ and $0 \leq t < u/\lambda$,

$$\mathbb{P}(\Theta > u \mid \mathscr{F}_t) = \mathbb{P}(\tau > u/\lambda \mid \mathscr{F}_t) = N_t e^{-u} \neq e^{-u}.$$

3 Preliminary Results

We start by introducing notation. Let τ be a random time, defined the probability space $(\Omega, \mathscr{G}, \mathbb{F}, \mathbb{P})$ endowed with the filtration $\mathbb{F}$, and such that $\mathbb{P}(\tau > 0) = 1$. We denote by $\mathbb{G} = (\mathscr{G}_t)_{t \in \mathbb{R}_+}$ the $\mathbb{P}$-completed and right-continuous version of the progressive enlargement of the filtration $\mathbb{F}$ by the filtration $\mathbb{H} = (\mathscr{H}_t)_{t \in \mathbb{R}_+}$ generated by the process $H_t = \mathbb{1}_{\{\tau \leq t\}}$. It is assumed throughout that the H-*hypothesis* (see, for instance, Elliott et al. [2]) is satisfied under $\mathbb{P}$ by the filtrations $\mathbb{F}$ and $\mathbb{G}$ so that, for every $u \in \mathbb{R}_+$,

$$\mathbb{P}(\tau > u \mid \mathscr{F}_t) = \mathbb{P}(\tau > u \mid \mathscr{F}_u), \quad \forall t \in [u, \infty).$$

The main tool in our construction of a random time with a given Azéma supermartingale is a locally equivalent change of a probability measure. For this reason, we first present some results related to Girsanov's theorem in the present set-up.

3.1 Properties of $(\mathbb{P}, \mathbb{G})$-Martingales

It order to define the Radon-Nikodým density process, we first analyze the properties of $(\mathbb{P}, \mathbb{G})$-martingales. The following auxiliary result is based on El Karoui et al. [3], in the sense that it can be seen as a consequence of Theorem 5.7 therein. For the sake of completeness, we provide a simple proof of Proposition 2.

In what follows, Z stands for a càdlàg, $\mathbb{F}$-adapted, $\mathbb{P}$-integrable process, whereas $Z_t(u)$ denotes an $\mathscr{O}(\mathbb{F}) \otimes \mathscr{B}(\mathbb{R}_+)$-measurable map, where $\mathscr{O}(\mathbb{F})$ stands for the $\mathbb{F}$-optional σ-field in $\Omega \times \mathbb{R}_+$ (for details, see [3]).

Proposition 2 *Assume that the H-hypothesis is satisfied under* $\mathbb{P}$ *by the filtrations* $\mathbb{F}$ *and* $\mathbb{G}$. *Let the* $\mathbb{G}$-*adapted,* $\mathbb{P}$-*integrable process* $Z^{\mathbb{G}}$ *be given by the formula*

$$Z_t^{\mathbb{G}} = Z_t \mathbb{1}_{\{\tau > t\}} + Z_t(\tau)\mathbb{1}_{\{\tau \le t\}}, \quad \forall t \in \mathbb{R}_+, \tag{11}$$

where:

(i) *the projection of* $Z^{\mathbb{G}}$ *onto* $\mathbb{F}$, *which is defined by*

$$Z_t^{\mathbb{F}} := \mathbb{E}_{\mathbb{P}}(Z_t^{\mathbb{G}} \mid \mathscr{F}_t) = Z_t \, \mathbb{P}(\tau > t \mid \mathscr{F}_t) + \mathbb{E}_{\mathbb{P}}(Z_t(\tau)\mathbb{1}_{\{\tau \le t\}} \mid \mathscr{F}_t)$$

is a $(\mathbb{P}, \mathbb{F})$-*martingale*,

(ii) *for any fixed* $u \in \mathbb{R}_+$, *the process* $(Z_t(u), \, t \in [u, \infty))$ *is a* $(\mathbb{P}, \mathbb{F})$-*martingale*.

Then the process $Z^{\mathbb{G}}$ *is a* $(\mathbb{P}, \mathbb{G})$-*martingale*.

Proof Let us take $s < t$. Then

$$\mathbb{E}_{\mathbb{P}}(Z_t^{\mathbb{G}} \mid \mathscr{G}_s) = \mathbb{E}_{\mathbb{P}}(Z_t \mathbb{1}_{\{\tau > t\}} \mid \mathscr{G}_s) + \mathbb{E}_{\mathbb{P}}(Z_t(\tau)\mathbb{1}_{\{s < \tau \le t\}} \mid \mathscr{G}_s)$$
$$+ \mathbb{E}_{\mathbb{P}}(Z_t(\tau)\mathbb{1}_{\{\tau \le s\}} \mid \mathscr{G}_s)$$
$$= I_1 + I_2 + I_3.$$

An application of the standard formula for the conditional expectation yields

$$I_1 + I_2 = \mathbb{1}_{\{\tau > s\}} \frac{1}{G_s^{\mathbb{P}}} \mathbb{E}_{\mathbb{P}}(Z_t G_t^{\mathbb{P}} \mid \mathscr{F}_s) + \mathbb{1}_{\{\tau > s\}} \frac{1}{G_s^{\mathbb{P}}} \mathbb{E}_{\mathbb{P}}(Z_t(\tau)\mathbb{1}_{\{s < \tau \le t\}} \mid \mathscr{F}_s),$$

whereas for I_3, we obtain

$$I_3 = \mathbb{E}_{\mathbb{P}}(Z_t(\tau)\mathbb{1}_{\{\tau \le s\}} \mid \mathscr{G}_s) = \mathbb{1}_{\{\tau \le s\}} \mathbb{E}_{\mathbb{P}}(Z_t(u) \mid \mathscr{F}_s)_{u=\tau}$$
$$= \mathbb{1}_{\{\tau \le s\}} \mathbb{E}_{\mathbb{P}}(Z_s(u) \mid \mathscr{F}_s)_{u=\tau} = \mathbb{1}_{\{\tau \le s\}} Z_s(\tau),$$

where the first equality holds under the H-hypothesis[1] (see Sect. 3.2 in El Karoui et al. [3]) and the second follows from (ii). It thus suffices to show that $I_1 + I_2 = Z_s \mathbb{1}_{\{\tau > s\}}$. Condition (i) yields

$$\mathbb{E}_{\mathbb{P}}(Z_t G_t^{\mathbb{P}} \mid \mathscr{F}_s) + \mathbb{E}_{\mathbb{P}}(Z_t(\tau)\mathbb{1}_{\{\tau \le t\}} \mid \mathscr{F}_s) - \mathbb{E}_{\mathbb{P}}(Z_s(\tau)\mathbb{1}_{\{\tau \le s\}} \mid \mathscr{F}_s) = Z_s G_s^{\mathbb{P}}.$$

Therefore,

$$I_1 + I_2 = \mathbb{1}_{\{\tau > s\}} \frac{1}{G_s^{\mathbb{P}}} \left(Z_s G_s^{\mathbb{P}} + \mathbb{E}_{\mathbb{P}}((Z_s(\tau) - Z_t(\tau))\mathbb{1}_{\{\tau \le s\}} \mid \mathscr{F}_s) \right) = Z_s \mathbb{1}_{\{\tau > s\}},$$

---

[1]Essentially, this equality holds, since under the H-hypothesis the σ-fields $\mathscr{F}_t$ and $\mathscr{G}_s$ are conditionally independent given $\mathscr{F}_s$.

where the last equality holds since

$$\mathbb{E}_{\mathbb{P}}((Z_s(\tau) - Z_t(\tau))\mathbb{1}_{\{\tau \le s\}} \,|\, \mathscr{F}_s) = \mathbb{1}_{\{\tau \le s\}} \mathbb{E}_{\mathbb{P}}((Z_s(u) - Z_t(u)) \,|\, \mathscr{F}_s)_{u=\tau} = 0.$$

For the last equality in the formula above, we have again used condition (ii) in Proposition 2.   □

In order to define a probability measure $\mathbb{Q}$ locally equivalent to $\mathbb{P}$ under which (1) holds, we will search for a process $Z^{\mathbb{G}}$ satisfying the following set of assumptions.

Assumption (A) *The process $Z^{\mathbb{G}}$ is a $\mathbb{G}$-adapted and $\mathbb{P}$-integrable process given by*

$$Z_t^{\mathbb{G}} = Z_t \mathbb{1}_{\{\tau > t\}} + Z_t(\tau) \mathbb{1}_{\{\tau \le t\}}, \quad \forall t \in \mathbb{R}_+, \tag{12}$$

such that the following properties are valid:

(A.1) *the projection of $Z^{\mathbb{G}}$ onto $\mathbb{F}$ is equal to one, that is, $Z_t^{\mathbb{F}} := \mathbb{E}_{\mathbb{P}}(Z_t^{\mathbb{G}} \,|\, \mathscr{F}_t) = 1$ for every $t \in \mathbb{R}_+$,*
(A.2) *the process $Z^{\mathbb{G}}$ is a strictly positive $(\mathbb{P}, \mathbb{G})$-martingale.*

Remark 1 Since $\mathbb{P}(\tau > 0) = 1$ is clear that $Z_0^{\mathbb{G}} = Z_0 = 1$, so that $\mathbb{E}_{\mathbb{P}}(Z_t^{\mathbb{G}}) = 1$ for every $t \in \mathbb{R}_+$. We will later define a probability measure $\mathbb{Q}$ using the process $Z^{\mathbb{G}}$ as the Radon-Nikodým density. Then condition (A.1) will imply that the restriction of $\mathbb{Q}$ to $\mathbb{F}$ equals $\mathbb{P}$ and, together with the equality $Z = N$, will give us control over the Azéma supermartingale of τ under $\mathbb{Q}$. Let us also note that assumption (A.1) implies that condition (i) of Proposition 2 is trivially satisfied.

The following lemma provides a simple condition, which is equivalent to property (A.1).

Lemma 1 *The projection of $Z^{\mathbb{G}}$ on $\mathbb{F}$ equals $Z_t^{\mathbb{F}} := \mathbb{E}_{\mathbb{P}}(Z_t^{\mathbb{G}} \,|\, \mathscr{F}_t) = 1$ if and only if the processes Z and $Z_t(\tau)$ satisfy the following relationship*

$$Z_t = \frac{1 - \mathbb{E}_{\mathbb{P}}(Z_t(\tau)\mathbb{1}_{\{\tau \le t\}} \,|\, \mathscr{F}_t)}{\mathbb{P}(\tau > t \,|\, \mathscr{F}_t)}. \tag{13}$$

Proof Straightforward calculations yield

$$\mathbb{E}_{\mathbb{P}}(Z_t^{\mathbb{G}} \,|\, \mathscr{F}_t) = \mathbb{E}_{\mathbb{P}}(Z_t \mathbb{1}_{\{\tau > t\}} + Z_t(\tau)\mathbb{1}_{\{\tau \le t\}} \,|\, \mathscr{F}_t)$$

$$= Z_t \, \mathbb{P}(\tau > t \,|\, \mathscr{F}_t) + \mathbb{E}_{\mathbb{P}}(Z_t(\tau)\mathbb{1}_{\{\tau \le t\}} \,|\, \mathscr{F}_t) = 1.$$

The last equality is equivalent to formula (13).   □

We find it convenient to work with the following assumption, which is more explicit and slightly stronger than Assumption (A).

Assumption (B) *We postulate that the process Z and the map $Z_t(u)$ are such that:*

(B.1) *equality* (13) *is satisfied,*
(B.2) *for every $u \in \mathbb{R}_+$, the process $(Z_t(u), t \in [u, \infty))$ is a strictly positive $(\mathbb{P}, \mathbb{F})$-martingale.*

Lemma 2 *Assumption* (B) *implies Assumption* (A).

Proof In view of Lemma 1, the conditions (A.1) and (B.1) are equivalent. In view of Proposition 2, conditions (B.1) and (B.2) imply (A.2).   □

Remark 2 It is not true, in general, that Assumption (A) implies Assumption (B), since it is not true that Assumption (A) implies condition (B.2). However, if the intensity $(\lambda_u, u \in \mathbb{R}_+)$ of τ under $\mathbb{P}$ exists then one can show that for any $u \in \mathbb{R}_+$ the process $(Z_t(u)\lambda_u G_u, t \in [u, \infty))$ is a $(\mathbb{P}, \mathbb{F})$-martingale (see El Karoui et al. [3]). This property implies in turn condition (B.2) provided that the intensity process λ does not vanish.

3.2 Girsanov's Theorem

To establish a suitable version of Girsanov's theorem, we need to specify a set-up in which the H-hypothesis is satisfied. Let Λ be an $\mathbb{F}$-adapted, continuous, increasing process with $\Lambda_0 = 0$ and $\Lambda_\infty = \infty$. We define the random time τ using the *canonical construction*, that is, by setting

$$\tau = \inf\{t \in \mathbb{R}_+ : \Lambda_t \geq \Theta\}, \tag{14}$$

where Θ is an exponentially distributed random variable with parameter 1, independent of $\mathscr{F}_\infty$, and defined on a suitable extension of the space Ω.

In fact, we will formally identify the probability space Ω with its extension, so that the probability measures $\mathbb{P}$ and $\mathbb{Q}$ will be defined on the same space.

It is easy to check that in the case of the canonical construction, the H-hypothesis is satisfied under $\mathbb{P}$ by the filtrations $\mathbb{F}$ and $\mathbb{G}$. Moreover, the Azéma supermartingale of τ with respect to $\mathbb{F}$ under $\mathbb{P}$ equals

$$G_t^{\mathbb{P}} := \mathbb{P}(\tau > t \,|\, \mathscr{F}_t) = e^{-\Lambda_t}, \quad \forall t \in \mathbb{R}_+.$$

Under Assumption (B), the strictly positive $(\mathbb{P}, \mathbb{G})$-martingale $Z^{\mathbb{G}}$ given by (cf. (12))

$$Z_t^{\mathbb{G}} = Z_t \mathbb{1}_{\{\tau > t\}} + Z_t(\tau) \mathbb{1}_{\{\tau \leq t\}}, \quad \forall t \in \mathbb{R}_+, \tag{15}$$

can be used to define a probability measure $\mathbb{Q}$ on $(\Omega, \mathscr{G}_\infty)$, locally equivalent to $\mathbb{P}$, such that the Radon-Nikodým density of $\mathbb{Q}$ with respect to $\mathbb{P}$ equals

$$\frac{d\mathbb{Q}}{d\mathbb{P}}\Big|\mathcal{G}_t = Z_t^G \qquad (16)$$

for every $t \in \mathbb{R}_+$.

The next result describes the conditional distributions of τ under a locally equivalent probability measure $\mathbb{Q}$.

Proposition 3 *Under Assumption* (B), *let the probability measure* $\mathbb{Q}$ *locally equivalent to* $\mathbb{P}$ *be given by* (15)–(16). *Then the following properties hold:*

(i) *the restriction of* $\mathbb{Q}$ *to the filtration* $\mathbb{F}$ *is equal to* $\mathbb{P}$,
(ii) *the Azéma supermartingale of* τ *under* $\mathbb{Q}$ *satisfies*

$$G_t^{\mathbb{Q}} := \mathbb{Q}(\tau > t \mid \mathcal{F}_t) = Z_t e^{-\Lambda_t}, \quad \forall t \in \mathbb{R}_+, \qquad (17)$$

(iii) *for every* $u \in \mathbb{R}_+$,

$$\mathbb{Q}(\tau > u \mid \mathcal{F}_t) = \begin{cases} \mathbb{E}_{\mathbb{P}}(Z_u e^{-\Lambda_u} \mid \mathcal{F}_t), & t \le u, \\ Z_t e^{-\Lambda_t} + \mathbb{E}_{\mathbb{P}}(Z_t(\tau)\mathbb{1}_{\{u < \tau \le t\}} \mid \mathcal{F}_t), & t \ge u. \end{cases} \qquad (18)$$

Proof Part (i) follows immediately the fact that condition (B.1) is equivalent to condition (A.1). For part (ii), it suffices to check that the second equality in (17) holds. By the abstract Bayes formula

$$\mathbb{Q}(\tau > t \mid \mathcal{F}_t) = \frac{\mathbb{E}_{\mathbb{P}}(Z_t^G \mathbb{1}_{\{\tau > t\}} \mid \mathcal{F}_t)}{\mathbb{E}_{\mathbb{P}}(Z_t^G \mid \mathcal{F}_t)} = \frac{\mathbb{E}_{\mathbb{P}}(Z_t \mathbb{1}_{\{\tau > t\}} \mid \mathcal{F}_t)}{Z_t^F}$$

$$= Z_t \mathbb{P}(\tau > t \mid \mathcal{F}_t) = Z_t e^{-\Lambda_t},$$

as desired. Using the abstract Bayes formula, we can also find expressions for the conditional probabilities $\mathbb{Q}(\tau > u \mid \mathcal{F}_t)$ for every $u, t \in \mathbb{R}_+$. For a fixed $u \in \mathbb{R}_+$ and every $t \in [0, u]$, we have that

$$\mathbb{Q}(\tau > u \mid \mathcal{F}_t) = \mathbb{Q}(\mathbb{Q}(\tau > u \mid \mathcal{F}_u) \mid \mathcal{F}_t) = \mathbb{E}_{\mathbb{Q}}(Z_u e^{-\Lambda_u} \mid \mathcal{F}_t) = \mathbb{E}_{\mathbb{P}}(Z_u e^{-\Lambda_u} \mid \mathcal{F}_t).$$

For a fixed $u \in \mathbb{R}_+$ and every $t \in [u, \infty)$, we obtain

$$\mathbb{Q}(\tau > u \mid \mathcal{F}_t) = \frac{\mathbb{E}_{\mathbb{P}}(Z_t^G \mathbb{1}_{\{\tau > u\}} \mid \mathcal{F}_t)}{\mathbb{E}_{\mathbb{P}}(Z_t^G \mid \mathcal{F}_t)}$$

$$= \mathbb{E}_{\mathbb{P}}((Z_t \mathbb{1}_{\{\tau > t\}} + Z_t(\tau)\mathbb{1}_{\{\tau \le t\}})\mathbb{1}_{\{\tau > u\}} \mid \mathcal{F}_t)$$

$$= \mathbb{E}_{\mathbb{P}}(Z_t \mathbb{1}_{\{\tau > t\}} + Z_t(\tau)\mathbb{1}_{\{u < \tau \le t\}} \mid \mathcal{F}_t)$$

$$= Z_t \mathbb{P}(\tau > t \mid \mathcal{F}_t) + \mathbb{E}_{\mathbb{P}}(Z_t(\tau)\mathbb{1}_{\{u < \tau \le t\}} \mid \mathcal{F}_t)$$

$$= Z_t G_t^{\mathbb{P}} + \mathbb{E}_{\mathbb{P}}(Z_t(\tau)\mathbb{1}_{\{u < \tau \le t\}} \mid \mathcal{F}_t)$$

$$= Z_t e^{-\Lambda_t} + \mathbb{E}_{\mathbb{P}}(Z_t(\tau)\mathbb{1}_{\{u < \tau \le t\}} \mid \mathcal{F}_t).$$

This completes the proof. □

Remark 3 (i) It is worth stressing that the H-hypothesis does not hold under $\mathbb{Q}$. This property follows immediately from (17), since under the H-hypothesis the Azéma supermartingale is necessarily a decreasing process.

(ii) It is not claimed that τ is finite under $\mathbb{Q}$. Indeed, this property holds if and only if

$$\lim_{t \to \infty} \mathbb{Q}(\tau > t) = \lim_{t \to \infty} \mathbb{E}_{\mathbb{Q}}(Z_t e^{-\Lambda_t}) =: c = 0,$$

otherwise, we have that $\mathbb{Q}(\tau = \infty) = c$. Of course, if $\mathbb{Q}$ is equivalent to $\mathbb{P}$ then necessarily $\mathbb{Q}(\tau < \infty) = 1$ since from (14) we deduce that $\mathbb{P}(\tau < \infty) = 1$ (this is a consequence of the assumption that $\Lambda_\infty = \infty$).

(iii) We observe that the formula

$$G_t^{\mathbb{Q}} := \mathbb{Q}(\tau > t \mid \mathscr{F}_t) = Z_t e^{-\Lambda_t}, \quad \forall t \in \mathbb{R}_+, \tag{19}$$

represents the multiplicative decomposition of the Azéma supermartingale $G^{\mathbb{Q}}$ if and only if the process Z is a $(\mathbb{Q}, \mathbb{F})$-local martingale or, equivalently, if Z is a $(\mathbb{P}, \mathbb{F})$-local martingale (it is worth stressing that Assumption (B) does not imply that Z is a $(\mathbb{P}, \mathbb{F})$-martingale; see Example 1 below). In other words, an equivalent change of a probability measure may result in a change of the decreasing component in the multiplicative decomposition as well. The interested reader is referred to Sect. 6 in El Karoui et al. [3] for a more detailed analysis of the change of a probability measure in the framework of the so-called *density approach* to the modeling of a random time.

Example 1 To illustrate the above remark, let us set $Z_t(u) = 1/2$ for every $u \in \mathbb{R}_+$ and $t \in [0, u]$. Then the process $Z^{\mathbb{G}}$, which is given by the formula

$$Z_t^{\mathbb{G}} = \frac{1 - (1/2)\mathbb{P}(\tau \leq t \mid \mathscr{F}_t)}{\mathbb{P}(\tau > t \mid \mathscr{F}_t)} \mathbb{1}_{\{\tau > t\}} + (1/2)\mathbb{1}_{\{\tau \leq t\}},$$

satisfies Assumption (B). The process Z is not a $(\mathbb{P}, \mathbb{F})$-local martingale, however, since

$$Z_t = \frac{1 + \mathbb{P}(\tau > t \mid \mathscr{F}_t)}{2\mathbb{P}(\tau > t \mid \mathscr{F}_t)} = \frac{1 + e^{-\Lambda_t}}{2e^{-\Lambda_t}}.$$

Moreover, the Azéma supermartingale of τ under $\mathbb{Q}$ equals, for every $t \in \mathbb{R}_+$,

$$\mathbb{Q}(\tau > t \mid \mathscr{F}_t) = Z_t e^{-\Lambda_t} = \frac{1}{2}(1 + e^{-\Lambda_t}) = e^{-\widehat{\Lambda}_t},$$

where $\widehat{\Lambda}$ is an $\mathbb{F}$-adapted, continuous, increasing process different from Λ. Note, however, that $Z^{\mathbb{G}}$ is not a uniformly integrable $(\mathbb{P}, \mathbb{G})$-martingale and $\mathbb{Q}$ is not equivalent to $\mathbb{P}$ on $\mathscr{G}_\infty$ since $\widehat{\Lambda}_\infty = \ln 2 < \infty$, so that $\mathbb{Q}(\tau < \infty) < 1$.

Let us also observe that the martingale $Z^{\mathbb{G}}$ has a jump at time τ and this in fact implies the processes Λ and $\widehat{\Lambda}$ do not coincide. We will see in the sequel (cf. Lemma 4) that, under mild technical assumptions, it is necessary to set $Z_t(t) = Z_t$ when solving Problem (P), so that the density process $Z^{\mathbb{G}}$ is continuous at τ. This is by no means surprising, since this equality was identified in El Karoui et al. [3] within the density approach as the crucial condition for the preservation of the $\mathbb{F}$-intensity of a random time τ under an equivalent change of a probability measure.

4 Construction Through a Change of Measure

We are in the position to address the issue of finding a solution $(\tau, \mathbb{Q})$ to Problem (P). Let N be a strictly positive $(\mathbb{P}, \mathbb{F})$-local martingale with $N_0 = 1$ and let Λ be an $\mathbb{F}$-adapted, continuous, increasing process with $\Lambda_0 = 0$ and $\Lambda_\infty = \infty$. We postulate, in addition, that $G = N e^{-\Lambda} \leq 1$. It is well known that a strictly positive local martingale is a supermartingale; this implies, in particular, that the process N is $\mathbb{P}$-integrable.

Before we proceed to an explicit construction of a random time τ and a probability measure $\mathbb{Q}$, let us show that for any solution $(\tau, \mathbb{Q})$ to Problem (P), we necessarily have that $\mathbb{Q}(\tau < \infty) = 1$.

Lemma 3 *For any solution $(\tau, \mathbb{Q})$ to Problem* (P), *we have that* $\mathbb{Q}(\tau = \infty) = 0$.

Proof Note first that

$$\mathbb{Q}(\tau = \infty) = \lim_{t \to \infty} \mathbb{Q}(\tau > t) = \lim_{t \to \infty} \mathbb{E}_{\mathbb{Q}}(\mathbb{Q}(\tau > t \mid \mathscr{F}_t)) = \lim_{t \to \infty} \mathbb{E}_{\mathbb{P}}(\mathbb{Q}(\tau > t \mid \mathscr{F}_t))$$
$$= \lim_{t \to \infty} \mathbb{E}_{\mathbb{P}}(N_t e^{-\Lambda_t}).$$

Since N is a strictly positive $(\mathbb{P}, \mathbb{F})$-local martingale, and thus a positive supermartingale, we have that $\lim_{t \to \infty} N_t =: N_\infty < \infty$, $\mathbb{P}$-a.s. Recall that, by assumption, we also have that $0 \leq N_t e^{-\Lambda_t} \leq 1$. Hence by applying the dominated convergence theorem, we obtain

$$\mathbb{Q}(\tau = \infty) = \lim_{t \to \infty} \mathbb{E}_{\mathbb{P}}(N_t e^{-\Lambda_t}) = \mathbb{E}_{\mathbb{P}}\left(\lim_{t \to \infty} N_t e^{-\Lambda_t}\right) = \mathbb{E}_{\mathbb{P}}\left(N_\infty \lim_{t \to \infty} e^{-\Lambda_t}\right) = 0,$$

where the last equality follows from the assumption that $\Lambda_\infty = \infty$. $\qquad\square$

In the first step, using the canonical construction, we define a random time τ by formula (14). In the second step, we propose a suitable change of a probability measure.

In order to define a probability measure $\mathbb{Q}$ locally equivalent to $\mathbb{P}$ under which (1) holds, we wish to employ Proposition 3 with $Z = N$. To this end, we postulate that Assumption (B) is satisfied by $Z = N$ and a judiciously selected $\mathscr{O}(\mathbb{F}) \otimes \mathscr{B}(\mathbb{R}_+)$-measurable map $Z_t(u)$. The choice of a map $Z_t(u)$, for a given in advance process N, is studied in what follows.

Let $\mathbb{Q}$ be defined by (15) with Z replaced by N. Then, by Proposition 3, we conclude that the Azéma supermartingale of τ under $\mathbb{Q}$ equals

$$G_t^{\mathbb{Q}} := \mathbb{Q}(\tau > t \mid \mathscr{F}_t) = N_t e^{-\Lambda_t}, \quad \forall t \in \mathbb{R}_+. \tag{20}$$

By the same token, formula (18) remains valid when Z is replaced by N.

Our next goal is to investigate Assumption (B), which was crucial in the proof of Proposition 3 and thus also in obtaining equality (20). For this purpose, let us formulate the following auxiliary problem, which combines Assumption (B) with the assumption that $N = Z$.

Problem (P.1) Let a strictly positive $(\mathbb{P}, \mathbb{F})$-local martingale N with $N_0 = 1$ be given. Find an $\mathcal{O}(\mathbb{F}) \otimes \mathcal{B}(\mathbb{R}_+)$-measurable map $Z_t(u)$ such that the following conditions are satisfied:

(i) for every $t \in \mathbb{R}_+$,

$$1 - N_t \, \mathbb{P}(\tau > t \mid \mathcal{F}_t) = \mathbb{E}_{\mathbb{P}}(Z_t(\tau)\mathbb{1}_{\{\tau \le t\}} \mid \mathcal{F}_t), \tag{21}$$

(ii) for any fixed $u \in \mathbb{R}_+$, the process $(Z_t(u), t \in [u, \infty))$ is a strictly positive $(\mathbb{P}, \mathbb{F})$-martingale.

Of course, equality (21) is obtained by combining (B.1) with the equality $Z = N$. If, for a given process N, we can find a solution $(Z_t(u), t \in [u, \infty))$ to Problem (P.1) then the pair $(Z_t, Z_t(u)) = (N_t, Z_t(u))$ will satisfy Assumption (B).

To examine the existence of a solution to Problem (P.1), we note first that formula (21) can be represented as follows

$$1 - G_t^{\mathbb{P}} N_t = \mathbb{E}_{\mathbb{P}}(Z_t(\tau)\mathbb{1}_{\{\tau \le t\}} \mid \mathcal{F}_t) = \int_0^t Z_t(u) \, d\mathbb{P}(\tau \le u \mid \mathcal{F}_t).$$

Since the H-hypothesis is satisfied under $\mathbb{P}$, the last formula is equivalent to

$$1 - G_t^{\mathbb{P}} N_t = \int_0^t Z_t(u) \, d\mathbb{P}(\tau \le u \mid \mathcal{F}_u) = \int_0^t Z_t(u) \, d(1 - G_u^{\mathbb{P}}) = -\int_0^t Z_t(u) \, dG_u^{\mathbb{P}}.$$

Since we work under the standing assumption that $G_t^{\mathbb{P}} = e^{-\Lambda_t}$, we thus obtain the following equation, which is equivalent to (21)

$$N_t e^{-\Lambda_t} = 1 + \int_0^t Z_t(u) \, de^{-\Lambda_u}. \tag{22}$$

We conclude that, within the present set-up, Problem (P.1) is equivalent to the following one.

Problem (P.2) Let a strictly positive $(\mathbb{P}, \mathbb{F})$-local martingale N with $N_0 = 1$ be given. Find an $\mathcal{O}(\mathbb{F}) \otimes \mathcal{B}(\mathbb{R}_+)$-measurable map $Z_t(u)$ such that the following conditions hold:

(i) for every $t \in \mathbb{R}_+$

$$N_t e^{-\Lambda_t} = 1 + \int_0^t Z_t(u) \, de^{-\Lambda_u}, \tag{23}$$

(ii) for any fixed $u \in \mathbb{R}_+$, the process $(Z_t(u), t \in [u, \infty))$ is a strictly positive $(\mathbb{P}, \mathbb{F})$-martingale.

Assume that $\widehat{Z}_t := Z_t(t)$ is an $\mathbb{F}$-optional process. We will now show that the equality $N = \widehat{Z}$ necessarily holds for any solution to (23), in the sense made precise in Lemma 4. In particular, it follows immediately from this result that the processes N and $\widehat{Z}$ are indistinguishable when $\widehat{Z}$ is an $\mathbb{F}$-adapted, càdlàg process and the process Λ has strictly increasing sample paths (for instance, when $\Lambda_t = \int_0^t \lambda_u \, du$ for some strictly positive *intensity process* λ).

Lemma 4 *Suppose that $Z_t(u)$ solves Problem (P.2) and the process $(\widehat{Z}_t, t \in \mathbb{R}_+)$ given by $\widehat{Z}_t = Z_t(t)$ is $\mathbb{F}$-optional. Then $N = \widehat{Z}$, ν-a.e., where the measure ν on $(\Omega \times \mathbb{R}_+, \mathcal{O}(\mathbb{F}))$ is generated by the increasing process Λ, that is, for every $s < t$ and any bounded, $\mathbb{F}$-optional process V*

$$\nu(\mathbb{1}_{[s,t[} V) = \mathbb{E}_{\mathbb{P}}\left(\int_s^t V_u \, d\Lambda_u\right).$$

Proof The left-hand side in (23) has the following Doob-Meyer decomposition

$$N_t e^{-\Lambda_t} = 1 + \int_0^t e^{-\Lambda_u} \, dN_u + \int_0^t N_u \, de^{-\Lambda_u}, \qquad (24)$$

whereas the right-hand side in (23) can be represented as follows

$$1 + \int_0^t Z_t(u) \, de^{-\Lambda_u} = 1 + \int_0^t (Z_t(u) - \widehat{Z}_u) \, de^{-\Lambda_u} + \int_0^t \widehat{Z}_u \, de^{-\Lambda_u}$$

$$= 1 + I_1(t) + I_2(t), \qquad (25)$$

where I_2 is an $\mathbb{F}$-adapted, continuous process of finite variation. We will show that I_1 is a $(\mathbb{P}, \mathbb{F})$-martingale, so that right-hand side in (25) yields the Doob-Meyer decomposition as well. To this end, we need to show that the equality

$$\mathbb{E}_{\mathbb{P}}(I_1(t) \,|\, \mathscr{F}_s) = I_1(s)$$

holds for every $s < t$ or, equivalently,

$$\mathbb{E}_{\mathbb{P}}\left(\int_0^t Z_t(u) \, de^{-\Lambda_u} - \int_0^s Z_s(u) \, de^{-\Lambda_u} \,\bigg|\, \mathscr{F}_s\right) = \mathbb{E}_{\mathbb{P}}\left(\int_s^t \widehat{Z}_u \, de^{-\Lambda_u} \,\bigg|\, \mathscr{F}_s\right). \qquad (26)$$

We first observe that, for every $s < t$,

$$\mathbb{E}_{\mathbb{P}}\left(\int_0^s Z_s(u) \, de^{-\Lambda_u} \,\bigg|\, \mathscr{F}_s\right) = \mathbb{E}_{\mathbb{P}}\left(\int_0^s \mathbb{E}_{\mathbb{P}}(Z_t(u) \,|\, \mathscr{F}_s) \, de^{-\Lambda_u} \,\bigg|\, \mathscr{F}_s\right)$$

$$= \mathbb{E}_{\mathbb{P}}\left(\int_0^s Z_t(u) \, de^{-\Lambda_u} \,\bigg|\, \mathscr{F}_s\right),$$

and thus the right-hand side in (26) satisfies

$$\mathbb{E}_\mathbb{P}\left(\int_s^t Z_t(u)\,de^{-\Lambda_u}\,\Big|\,\mathscr{F}_s\right) = \mathbb{E}_\mathbb{P}\left(\int_s^t \mathbb{E}_\mathbb{P}(Z_t(u)\,|\,\mathscr{F}_u)\,de^{-\Lambda_u}\,\Big|\,\mathscr{F}_s\right)$$

$$= \mathbb{E}_\mathbb{P}\left(\int_s^t \widehat{Z}_u\,de^{-\Lambda_u}\,\Big|\,\mathscr{F}_s\right),$$

where we have used the following equality, which holds for every $\mathbb{F}$-adapted, continuous process A of finite variation and every càdlàg process V (not necessarily $\mathbb{F}$-adapted)

$$\mathbb{E}_\mathbb{P}\left(\int_s^t V_u\,dA_u\,\Big|\,\mathscr{F}_s\right) = \mathbb{E}_\mathbb{P}\left(\int_s^t \mathbb{E}_\mathbb{P}(V_u\,|\,\mathscr{F}_u)\,dA_u\,\Big|\,\mathscr{F}_s\right).$$

We thus see that (26) holds, so that I_1 is a $(\mathbb{P},\mathbb{F})$-martingale. By comparing the right-hand sides in (24) and (25) and using the uniqueness of the Doob-Meyer decomposition, we conclude that

$$\int_0^t (N_u - \widehat{Z}_u)\,de^{-\Lambda_u} = 0, \quad \forall t \in \mathbb{R}_+.$$

The formula above implies that $N = \widehat{Z}$, $\widetilde{\nu}$-a.e., where the measure $\widetilde{\nu}$ on $(\Omega \times \mathbb{R}_+, \mathcal{O}(\mathbb{F}))$ is generated by the decreasing process $e^{-\Lambda_u}$. It is easily see that the measures $\widetilde{\nu}$ and ν are equivalent and thus $N = \widehat{Z}$, ν-a.e. $\qquad\square$

To address the issue of existence of a solution to Problem (P.2) (note that it is not claimed that a solution $Z_t(u)$ to Problem (P.2) is unique), we start by postulating that, as in the filtering case described in Sect. 2, an $\mathcal{O}(\mathbb{F}) \otimes \mathcal{B}(\mathbb{R}_+)$-measurable map $Z_t(u)$ satisfies: $Z_t(u) = X_t Y_{t \wedge u}$ for some $\mathbb{F}$-adapted, continuous, strictly positive processes X and Y. It is then easy to check that condition (ii) implies that the process X is necessarily a $(\mathbb{P},\mathbb{F})$-martingale. Moreover, (23) becomes

$$N_t e^{-\Lambda_t} = 1 + X_t \int_0^t Y_u\,de^{-\Lambda_u}. \tag{27}$$

Note that at this stage we are searching for a pair (X, Y) of strictly positive, $\mathbb{F}$-adapted processes such that X is a $(\mathbb{P},\mathbb{F})$-martingale and equality (27) holds for every $t \in \mathbb{R}_+$. In view of Lemma 4, it is also natural to postulate that $N = XY$. We will then be able to find a simple relation between processes X and Y (see formula (28) below) and thus to reduce the dimensionality of the problem.

Lemma 5 (i) *Assume that a pair* (X, Y) *of strictly positive processes is such that the process* $Z_t(u) = X_t Y_{t \wedge u}$ *solves Problem* (P.2) *and the equality* $N = XY$ *holds. Then the process* X *is a* $(\mathbb{P},\mathbb{F})$*-martingale and the process* Y *equals*

$$Y_t = Y_0 + \int_0^t e^{\Lambda_u}\,d\left(\frac{1}{X_u}\right). \tag{28}$$

(ii) Conversely, if X is a strictly positive $(\mathbb{P}, \mathbb{F})$*-martingale and the process Y given by* (28) *is strictly positive, then the process* $Z_t(u) = X_t Y_{t \wedge u}$ *solves Problem* (P.2) *for the process* $N = XY$.

Proof For part (i), we observe that under the assumption that $N = XY$, equation (27) reduces to

$$X_t Y_t e^{-\Lambda_t} = 1 + X_t \int_0^t Y_u \, de^{-\Lambda_u}, \tag{29}$$

which in turn is equivalent to

$$Y_t e^{-\Lambda_t} = \frac{1}{X_t} + \int_0^t Y_u \, de^{-\Lambda_u}. \tag{30}$$

The integration by parts formula yields

$$\frac{1}{X_t} = Y_0 + \int_0^t e^{-\Lambda_u} \, dY_u$$

and this in turn is equivalent to (28). To establish part (ii), we first note that (28) implies (29), which means that (23) is satisfied by the processes $N = XY$ and $Z_t(u) = X_t Y_{t \wedge u}$. It is also clear that for any fixed $u \in \mathbb{R}_+$, the process $(Z_t(u), t \in [u, \infty))$ is a strictly positive $(\mathbb{P}, \mathbb{F})$-martingale. Let us also note that, by the Itô formula, the process XY satisfies

$$d(X_t Y_t) = Y_t \, dX_t - e^{\Lambda_t}(1/X_t) \, dX_t, \tag{31}$$

and thus it is a strictly positive $(\mathbb{P}, \mathbb{F})$-local martingale. □

We conclude that in order to find a solution $Z_t(u) = X_t Y_{t \wedge u}$ to Problem (P.2), it suffices to solve the following problem.

Problem (P.3) Assume that we are given a strictly positive $(\mathbb{P}, \mathbb{F})$-local martingale N with $N_0 = 1$ and an $\mathbb{F}$-adapted, continuous, increasing process Λ with $\Lambda_0 = 0$ and $\Lambda_\infty = \infty$. Find a strictly positive $(\mathbb{P}, \mathbb{F})$-martingale X such that for the process Y given by (28) we have that $N = XY$.

The following corollary is an easy consequence of part (ii) in Lemma 5 and thus its proof is omitted.

Corollary 1 *Assume that a process X solves Problem* (P.3) *and let Y be given by* (28). *Then the processes* $Z = N$ *and* $Z_t(u) = X_t Y_{t \wedge u}$ *solve Problem* (P.2) *and thus they satisfy Assumption* (B).

5 Case of the Brownian Filtration

The aim of this section is to examine the existence of a solution to Problem (P.3) under the following standing assumptions:

(i) the filtration $\mathbb{F}$ is generated by a Brownian motion W,
(ii) we are given an $\mathbb{F}$-adapted, continuous, increasing process Λ with $\Lambda_0 = 0$ and $\Lambda_\infty = \infty$ and a strictly positive $(\mathbb{P}, \mathbb{F})$-local martingale N satisfying

$$N_t = 1 + \int_0^t v_u N_u \, dW_u, \quad \forall t \in \mathbb{R}_+, \tag{32}$$

for some $\mathbb{F}$-predictable process v,

(iii) the inequality $G_t := N_t e^{-\Lambda_t} < 1$ holds for every $t > 0$, so that $N_t < e^{\Lambda_t}$ for every $t > 0$.

We start by noting that X is postulated to be a strictly positive $(\mathbb{P}, \mathbb{F})$-martingale and thus it is necessarily given by

$$X_t = \exp\left(\int_0^t x_s \, dW_s - \frac{1}{2} \int_0^t x_s^2 \, ds \right), \quad \forall t \in \mathbb{R}_+, \tag{33}$$

where the process x is yet unknown. The goal is to specify x in terms of N and Λ in such a way that the equality $N = XY$ will hold for Y given by (28).

Lemma 6 *Let X be given by (33) with the process x satisfying*

$$x_t = \frac{v_t N_t}{N_t - e^{\Lambda_t}}, \quad \forall t \in \mathbb{R}_+. \tag{34}$$

Assume that x is a square-integrable process. Then the equality $N = XY$ holds, where the process Y given by (28) with $Y_0 = 1$.

Proof Using (31) and (33), we obtain

$$d(X_t Y_t) = Y_t \, dX_t - e^{\Lambda_t}(1/X_t) \, dX_t = Y_t x_t X_t \, dW_t - e^{\Lambda_t}(1/X_t) \, x_t X_t \, dW_t,$$

and thus

$$d(X_t Y_t) = x_t (X_t Y_t - e^{\Lambda_t}) \, dW_t. \tag{35}$$

Let us denote $V = XY$. Then V satisfies the following SDE

$$dV_t = x_t (V_t - e^{\Lambda_t}) \, dW_t. \tag{36}$$

In view of (34), it is clear that the process N solves this equation as well. Hence to show that the equality $N = XY$ holds, it suffices to show that a solution to the SDE

(36) is unique. We note that we deal here with the integral equation of the form

$$V_t = H_t + \int_0^t x_u V_u \, dW_u. \tag{37}$$

We will show that a solution to (37) is unique. To this end, we argue by contradiction. Suppose that V^i, $i = 1, 2$ are any two solutions to (37). Then the process $U = V^1 - V^2$ satisfies

$$dU_t = x_t U_t \, dW_t, \quad U_0 = 0, \tag{38}$$

which admits the obvious solution $U = 0$. Suppose that $\widehat{U}$ is a non-null solution to (38). Then the Doléans-Dade equation $dX_t = x_t X_t \, dW_t$, $X_0 = 1$ would admit the usual solution X given by (33) and another solution $X + \widehat{U} \neq X$, and this is well known to be false. We conclude that (38) admits a unique solution, and this in turn implies the uniqueness of a solution to (37). This shows that $N = XY$, as was stated.   □

The following result is an immediate consequence of Corollary 1 and Lemma 6.

Corollary 2 *Let the filtration $\mathbb{F}$ be generated by a Brownian motion W given on the probability space $(\Omega, \mathscr{G}, \mathbb{F}, \mathbb{P})$. Assume that we are given an $\mathbb{F}$-adapted, continuous supermartingale G such that $0 < G_t < 1$ for every $t > 0$ and*

$$G_t = N_t e^{-\Lambda_t}, \quad \forall t \in \mathbb{R}_+, \tag{39}$$

where Λ is an $\mathbb{F}$-adapted, continuous, increasing process with $\Lambda_0 = 0$ and $\Lambda_\infty = \infty$, and N is a strictly positive $\mathbb{F}$-local martingale, so that there exists an $\mathbb{F}$-predictable process v such that

$$N_t = 1 + \int_0^t v_u N_u \, dW_u, \quad \forall t \in \mathbb{R}_+. \tag{40}$$

Let X be given by (33) with the process x satisfying

$$x_t = \frac{v_t G_t}{G_t - 1}, \quad \forall t \in \mathbb{R}_+. \tag{41}$$

Then:

(i) *the equality $N = XY$ holds for the process Y given by (28),*
(ii) *the processes $Z = N$ and $Z_t(u) = X_t Y_{t \wedge u}$ satisfy Assumption (B).*

In the next result, we denote by τ the random time defined by the canonical construction on a (possibly extended) probability space $(\Omega, \mathscr{G}, \mathbb{F}, \mathbb{P})$. By construction, the Azéma supermartingale of τ with respect to $\mathbb{F}$ under $\mathbb{P}$ equals

$$G_t^{\mathbb{P}} := \mathbb{P}(\tau > t \mid \mathscr{F}_t) = e^{-\Lambda_t}, \quad \forall t \in \mathbb{R}_+.$$

Let us note that, since the H-hypothesis is satisfied, the Brownian motion W remains a Brownian motion with respect to the enlarged filtration $\mathbb{G}$ under $\mathbb{P}$. It is still a Brownian motion under $\mathbb{Q}$ with respect to the filtration $\mathbb{F}$, since the restriction of $\mathbb{Q}$ to $\mathbb{F}$ is equal to $\mathbb{P}$. However, the process W is not necessarily a Brownian motion under $\mathbb{Q}$ with respect to the enlarged filtration $\mathbb{G}$.

The following result furnishes a solution to Problem (P) within the set-up described at the beginning of this section.

Proposition 4 *Under the assumptions of Corollary 2, we define a probability measure $\mathbb{Q}$ locally equivalent to $\mathbb{P}$ by the Radon-Nikodým density process $Z^{\mathbb{G}}$ given by formula (12) with $Z_t = X_t Y_t = N_t$ and $Z_t(u) = X_t Y_{t \wedge u}$ or, more explicitly,*

$$Z_t^{\mathbb{G}} = N_t \mathbb{1}_{\{\tau > t\}} + X_t Y_{t \wedge \tau} \mathbb{1}_{\{\tau \le t\}}, \quad \forall t \in \mathbb{R}_+. \tag{42}$$

Then the Azéma supermartingale of τ with respect to $\mathbb{F}$ under $\mathbb{Q}$ satisfies

$$\mathbb{Q}(\tau > t \mid \mathscr{F}_t) = X_t Y_t e^{-\Lambda_t} = N_t e^{-\Lambda_t}, \quad \forall t \in \mathbb{R}_+. \tag{43}$$

Moreover, the conditional distribution of τ given $\mathscr{F}_t$ satisfies

$$\mathbb{Q}(\tau > u \mid \mathscr{F}_t) = \begin{cases} \mathbb{E}_{\mathbb{P}}(N_u e^{-\Lambda_u} \mid \mathscr{F}_t), & t < u, \\ N_t e^{-\Lambda_t} + X_t \mathbb{E}_{\mathbb{P}}(Y_\tau \mathbb{1}_{\{u < \tau \le t\}} \mid \mathscr{F}_t), & t \ge u. \end{cases}$$

Proof In view of Corollary 2, Assumption (B) is satisfied and thus the probability measure $\mathbb{Q}$ is well defined by the Radon-Nikodým density process $Z^{\mathbb{G}}$ given by (15), which is now equivalent to (42). Therefore, equality (43) is an immediate consequence of Proposition 3. Using (18), for every $u \in \mathbb{R}_+$, we obtain, for every $t \in [0, u]$

$$\mathbb{Q}(\tau > u \mid \mathscr{F}_t) = \mathbb{E}_{\mathbb{P}}(Z_u e^{-\Lambda_u} \mid \mathscr{F}_t) = \mathbb{E}_{\mathbb{P}}(N_u e^{-\Lambda_u} \mid \mathscr{F}_t),$$

whereas for every $t \in [u, \infty)$, we get

$$\begin{aligned} \mathbb{Q}(\tau > u \mid \mathscr{F}_t) &= Z_t e^{-\Lambda_t} + \mathbb{E}_{\mathbb{P}}(Z_t(\tau) \mathbb{1}_{\{u < \tau \le t\}} \mid \mathscr{F}_t) \\ &= X_t Y_t e^{-\Lambda_t} + \mathbb{E}_{\mathbb{P}}(X_t Y_{t \wedge \tau} \mathbb{1}_{\{u < \tau \le t\}} \mid \mathscr{F}_t) \\ &= N_t e^{-\Lambda_t} + X_t \mathbb{E}_{\mathbb{P}}(Y_\tau \mathbb{1}_{\{u < \tau \le t\}} \mid \mathscr{F}_t), \end{aligned}$$

as required. $\qquad \square$

Example 2 The set-up considered in this example is related to the filtering problem examined in Sect. 2. Let $W = (W_t, t \in \mathbb{R}_+)$ be a Brownian motion defined on the probability space $(\Omega, \mathscr{G}, \mathbb{P})$ and let $\mathbb{F}$ be its natural filtration. We wish to model a random time with the Azéma semimartingale with respect to the filtration $\mathbb{F}$ given by the solution to the following SDE (cf. (3))

$$dG_t = -\lambda G_t \, dt + \frac{b}{\sigma} G_t (1 - G_t) \, dW_t, \quad G_0 = 1. \tag{44}$$

A comparison theorem for SDEs implies that $0 < G_t < 1$ for every $t > 0$. Moreover, by an application of the Itô formula, we obtain

$$G_t = N_t e^{-\lambda t}, \quad \forall t \in \mathbb{R}_+,$$

where the martingale N satisfies

$$dN_t = \frac{b}{\sigma}(1 - G_t)N_t \, dW_t. \tag{45}$$

As in Sect. 3, we start the construction of τ by first defining a random variable Θ, with exponential distribution with parameter 1 and independent of $\mathscr{F}_\infty$ under $\mathbb{P}$, by setting

$$\tau = \inf\{t \in \mathbb{R}_+ : \lambda t \geq \Theta\}.$$

In the second step, we propose an equivalent change of a probability measure. For this purpose, we note that the process x is here given by (cf. (34))

$$x_t = \frac{v_t N_t}{N_t - e^{\lambda t}} = -\frac{b}{\sigma}G_t,$$

and thus x is a bounded process. Next, in view of (33) and (45), the process X solves the SDE

$$dX_t = -\frac{b}{\sigma}G_t X_t \, dW_t,$$

and the process Y satisfies (cf. (28))

$$dY_t = e^{\lambda t} \, d\left(\frac{1}{X_t}\right) = \frac{N_t}{X_t}\left(\frac{b}{\sigma} \, dW_t + \frac{b^2}{\sigma^2} G_t \, dt\right).$$

The integration by parts formula yields

$$\begin{aligned}
d\left(\frac{N_t}{X_t}\right) &= \frac{1}{X_t} \, dN_t + N_t \, d\left(\frac{1}{X_t}\right) + d\left[\frac{1}{X}, N\right]_t \\
&= \frac{N_t}{X_t}\frac{b}{\sigma}(1 - G_t) \, dW_t + \frac{N_t}{X_t}\left(\frac{b}{\sigma} G_t \, dW_t + \frac{b^2}{\sigma^2} G_t^2 \, dt\right) \\
&\quad + \frac{N_t}{X_t}\frac{b^2}{\sigma^2} G_t(1 - G_t) \, dt \\
&= \frac{N_t}{X_t}\left(\frac{b}{\sigma} \, dW_t + \frac{b^2}{\sigma^2} G_t \, dt\right).
\end{aligned}$$

It is now easy to conclude that $N = XY$, as was expected. Under the probability measure $\mathbb{Q}$ introduced in Proposition 4, we have that

$$G_t^{\mathbb{Q}} := \mathbb{Q}(\tau > t \mid \mathscr{F}_t) = G_t = N_t e^{-\lambda t}, \quad \forall t \in \mathbb{R}_+.$$

Moreover, the conditional distribution of τ given $\mathscr{F}_t$ satisfies

$$\mathbb{Q}(\tau > u \mid \mathscr{F}_t) = \begin{cases} \mathbb{E}_\mathbb{P}(N_u e^{-\lambda u} \mid \mathscr{F}_t) = N_t e^{-\lambda u}, & t < u, \\ N_t e^{-\lambda t} + X_t \, \mathbb{E}_\mathbb{P}(Y_\tau \mathbb{1}_{\{u < \tau \leq t\}} \mid \mathscr{F}_t), & t \geq u. \end{cases}$$

Since τ is here independent of $\mathscr{F}_\infty$ under $\mathbb{P}$, we obtain, for $t \geq u$,

$$\mathbb{E}_\mathbb{P}(Y_\tau \mathbb{1}_{\{u < \tau \leq t\}} \mid \mathscr{F}_t) = \int_u^t Y_v \lambda e^{-\lambda v} \, dv = - \int_u^t Y_v \, de^{-\lambda v}$$

$$= Y_u e^{-\lambda u} - Y_t e^{-\lambda t} + \left(\frac{1}{X_t} - \frac{1}{X_u} \right),$$

where the last equality can be deduced, for instance, from (30). Therefore, for every $t \geq u$

$$\mathbb{Q}(\tau > u \mid \mathscr{F}_t) = 1 - \frac{X_t}{X_u} + X_t Y_u e^{-\lambda u}.$$

We conclude that, for every $t, u \in \mathbb{R}_+$,

$$\mathbb{Q}(\tau > u \mid \mathscr{F}_t) = 1 - \frac{X_t}{X_{u \wedge t}} + X_t Y_{u \wedge t} e^{-\lambda u}.$$

It is interesting to note that this equality agrees with the formula (7), which was established in Proposition 1 in the context of filtering problem using a different technique.

6 Applications to Valuation of Credit Derivatives

We will examine very succinctly the importance of the multiplicative decomposition of the Azéma supermartingale (i.e., the survival process) of a random time for the risk-neutral valuation of credit derivatives. Unless explicitly stated otherwise, we assume that the interest rate is null. This assumption is made for simplicity of presentation and, obviously, it can be easily relaxed.

As a risk-neutral probability, we will select either the probability measure $\mathbb{P}$ or the equivalent probability measure $\mathbb{Q}$ defined here on $(\Omega, \mathscr{G}_T)$, where T stands for the maturity date of a credit derivative. Recall that within the framework considered in this paper the survival process of a random time τ is given under $\mathbb{P}$ and $\mathbb{Q}$ by the following formulae

$$G_t^\mathbb{P} := \mathbb{P}(\tau > t \mid \mathscr{F}_t) = e^{-\Lambda_t},$$

and

$$G_t^\mathbb{Q} := \mathbb{Q}(\tau > t \mid \mathscr{F}_t) = N_t e^{-\Lambda_t},$$

respectively. The random time τ is here interpreted as the *default time* of a reference entity of a credit derivative. We assume from now on that the increasing process Λ

satisfies $\Lambda_t = \int_0^t \lambda_u \, du$ for some non-negative, $\mathbb{F}$-progressively measurable process λ. Then we have the following well known result.

Lemma 7 *The process M, given by the formula*

$$M_t = \mathbb{1}_{\{\tau \le t\}} - \int_0^{t \wedge \tau} \lambda_u \, du,$$

is a $\mathbb{G}$-martingale under the probability measures $\mathbb{P}$ and $\mathbb{Q}$.

The property established in Lemma 7 is frequently adopted in the financial literature as the definition of the *default intensity* λ. In the present set-up, Lemma 7 implies that the default intensity is the same under the equivalent probability measures $\mathbb{P}$ and $\mathbb{Q}$, despite the fact that the corresponding survival processes $G^{\mathbb{P}}$ and $G^{\mathbb{Q}}$ are different (recall that we postulate that N is a non-trivial local martingale). Hence the following question arises: is the specification of the default intensity λ sufficient for the risk-neutral valuation of credit derivatives related to a reference entity? Similarly as in El Karoui et al. [3], we will argue that the answer to this question is negative. To support this claim, we will show that the risk-neutral valuation of credit derivatives requires the full knowledge of the survival process, and thus the knowledge of the decreasing component Λ of the survival process in not sufficient for this purpose, in general.

To illustrate the importance of martingale component N for the valuation of credit derivatives, we first suppose $\Lambda = (\Lambda_t, \, t \in \mathbb{R}_+)$ is deterministic. We will argue that in that setting the process N has no influence on the prices of some simple credit derivatives, such as: a defaultable zero-coupon bond with zero recovery or a stylized credit default swap (CDS) with a deterministic protection payment. This means that the model calibration based on these assets will only allow us to recover the function Λ, but will provide no information regarding the local martingale component N of the survival process $G^{\mathbb{Q}}$. However, if the assumptions of the deterministic character of Λ and/or protection payment of a CDS are relaxed, then the corresponding prices will depend on the choice of N as well, and thus the explicit knowledge of N becomes important (for an example, see Corollary 3). As expected, this feature becomes even more important when we deal with a credit risk model in which the default intensity λ is stochastic, as is typically assumed in the financial literature.

6.1 Defaultable Zero-Coupon Bonds

By definition, the risk-neutral price under $\mathbb{Q}$ of the T-maturity defaultable zero-coupon bond with zero recovery equals, for every $t \in [0, T]$,

$$D^{\mathbb{Q}}(t, T) := \mathbb{Q}(\tau > T \mid \mathscr{G}_t) = \mathbb{1}_{\{\tau > t\}} \frac{\mathbb{Q}(\tau > T \mid \mathscr{F}_t)}{G_t^{\mathbb{Q}}} = \mathbb{1}_{\{\tau > t\}} \frac{\mathbb{E}_{\mathbb{Q}}(N_T e^{-\Lambda_T} \mid \mathscr{F}_t)}{G_t^{\mathbb{Q}}}.$$

Assuming that Λ is deterministic, we obtain the pricing formulae independent of N. Indeed, the risk-neutral price of the bond under $\mathbb{P}$ equals, for every $t \in [0, T]$,

$$D^{\mathbb{P}}(t, T) := \mathbb{P}(\tau > T \mid \mathscr{G}_t) = \mathbb{1}_{\{\tau > t\}} e^{-(\Lambda_T - \Lambda_t)}.$$

Using the fact that the restriction of $\mathbb{Q}$ to $\mathbb{F}$ is equal to $\mathbb{P}$, we obtain

$$D^{\mathbb{Q}}(t, T) := \mathbb{Q}(\tau > T \mid \mathscr{G}_t) = \mathbb{1}_{\{\tau > t\}} \frac{1}{N_t e^{-\Lambda_t}} \, \mathbb{E}_{\mathbb{Q}}(N_T e^{-\Lambda_T} \mid \mathscr{F}_t)$$

$$= \mathbb{1}_{\{\tau > t\}} \frac{1}{N_t e^{-\Lambda_t}} \, e^{-\Lambda_T} \, \mathbb{E}_{\mathbb{P}}(N_T \mid \mathscr{F}_t) = \mathbb{1}_{\{\tau > t\}} e^{-(\Lambda_T - \Lambda_t)},$$

where we have assumed that N is a (true) $(\mathbb{P}, \mathbb{F})$-martingale. If N is a strict local martingale then it is a strict supermartingale and thus $D^{\mathbb{Q}}(t, T) \leq D^{\mathbb{P}}(t, T)$.

If we allow for a stochastic process Λ, then the role of N in the valuation of defaultable zero-coupon bonds becomes important, as can be seen from the following expressions

$$D^{\mathbb{P}}(t, T) = \mathbb{1}_{\{\tau > t\}} \frac{1}{e^{-\Lambda_t}} \, \mathbb{E}_{\mathbb{P}}(e^{-\Lambda_T} \mid \mathscr{F}_t)$$

and

$$D^{\mathbb{Q}}(t, T) = \mathbb{1}_{\{\tau > t\}} \frac{1}{N_t e^{-\Lambda_t}} \, \mathbb{E}_{\mathbb{Q}}(N_T e^{-\Lambda_T} \mid \mathscr{F}_t) = \mathbb{1}_{\{\tau > t\}} \frac{1}{N_t e^{-\Lambda_t}} \, \mathbb{E}_{\mathbb{P}}(N_T e^{-\Lambda_T} \mid \mathscr{F}_t),$$

where the last equality follows from the standing assumption that the restriction of $\mathbb{Q}$ to $\mathbb{F}$ is equal to $\mathbb{P}$. We thus see that the inequality $D^{\mathbb{P}}(t, T) \neq D^{\mathbb{Q}}(t, T)$ is likely to hold when Λ is stochastic.

In the case of a stochastic interest rate, the bond valuation problem is more difficult. Let the discount factor $\beta = (\beta_t, \, t \in \mathbb{R}_+)$ be defined by

$$\beta_t = \exp\left(-\int_0^t r_s \, ds\right),$$

where the short-term interest rate process $r = (r_t, \, t \in \mathbb{R}_+)$ is assumed to be $\mathbb{F}$-adapted. Then the risk-neutral prices under the probability measures $\mathbb{P}$ and $\mathbb{Q}$ of the T-maturity defaultable zero-coupon bond with zero recovery are given by the following expressions

$$D^{\mathbb{P}}(t, T) = \mathbb{1}_{\{\tau > t\}} \frac{1}{e^{-\Lambda_t} \beta_t} \, \mathbb{E}_{\mathbb{P}}(e^{-\Lambda_T} \beta_T \mid \mathscr{F}_t)$$

and

$$D^{\mathbb{Q}}(t, T) = \mathbb{1}_{\{\tau > t\}} \frac{1}{N_t e^{-\Lambda_t} \beta_t} \, \mathbb{E}_{\mathbb{P}}(N_T e^{-\Lambda_T} \beta_T \mid \mathscr{F}_t).$$

Once again, it is clear that the role of the martingale component of the survival process is non-trivial, even in the case when the default intensity is assumed to be deterministic.

6.2 Credit Default Swaps

Let us now consider a stylized CDS with the protection payment process R and fixed spread κ, which gives protection over the period $[0, T]$. It is known that the risk-neutral price under $\mathbb{P}$ of this contract is given by the formula, for every $t \in [0, T]$,

$$S_t^{\mathbb{P}}(\kappa) := \mathbb{E}_{\mathbb{P}}\big(\mathbb{1}_{\{t < \tau \leq T\}} R_\tau - \kappa((T \wedge \tau) - (t \vee \tau)) \,|\, \mathscr{G}_t\big). \tag{46}$$

Let us make an additional assumption that the process Λ satisfies $\Lambda_t = \int_0^t \lambda_u \, du$ for some intensity process λ.

Then one can show that (see, for instance, Bielecki et al. [1])

$$S_t^{\mathbb{P}}(\kappa) = \mathbb{1}_{\{\tau > t\}} \frac{1}{G_t^{\mathbb{P}}} \mathbb{E}_{\mathbb{P}}\left(\int_t^T G_u^{\mathbb{P}}(R_u \lambda_u - \kappa) \, du \,\bigg|\, \mathscr{F}_t\right). \tag{47}$$

Analogous formulae are valid under $\mathbb{Q}$, if we decide to choose $\mathbb{Q}$ as a risk-neutral probability measure. To analyze the impact of N on the value of the CDS, let us first consider the special case when the default intensity λ and the protection payment R are assumed to be deterministic. In that case, the risk-neutral price of the CDS under $\mathbb{P}$ can be represented as follows

$$S_t^{\mathbb{P}}(\kappa) = \mathbb{1}_{\{\tau > t\}} e^{\Lambda_t} \int_t^T e^{-\Lambda_u}(R_u \lambda_u - \kappa) \, du.$$

For the risk-neutral price under $\mathbb{Q}$, we obtain

$$\begin{aligned}
S_t^{\mathbb{Q}}(\kappa) &= \mathbb{1}_{\{\tau > t\}} \frac{1}{G_t^{\mathbb{Q}}} \mathbb{E}_{\mathbb{Q}}\left(\int_t^T G_u^{\mathbb{Q}}(R_u \lambda_u - \kappa) \, du \,\bigg|\, \mathscr{F}_t\right) \\
&= \mathbb{1}_{\{\tau > t\}} \frac{1}{G_t^{\mathbb{Q}}} \int_t^T \mathbb{E}_{\mathbb{P}}(G_u^{\mathbb{Q}} \,|\, \mathscr{F}_t)(R_u \lambda_u - \kappa) \, du \\
&= \mathbb{1}_{\{\tau > t\}} \frac{1}{N_t e^{-\Lambda_t}} \int_t^T \mathbb{E}_{\mathbb{P}}(N_u \,|\, \mathscr{F}_t) e^{-\Lambda_u}(R_u \lambda_u - \kappa) \, du \\
&= \mathbb{1}_{\{\tau > t\}} e^{\Lambda_t} \int_t^T e^{-\Lambda_u}(R_u \lambda_u - \kappa) \, du,
\end{aligned}$$

where in the second equality we have used once again the standing assumption that the restriction of $\mathbb{Q}$ to $\mathbb{F}$ is equal to $\mathbb{P}$ and the last one holds provided that N is a $(\mathbb{P}, \mathbb{F})$-martingale. It is thus clear that the equality $S_t^{\mathbb{P}}(\kappa) = S_t^{\mathbb{Q}}(\kappa)$ holds for every $t \in [0, T]$, so that the price of the CDS does not depend on the particular choice of N when λ and R are deterministic.

Let us now consider the case when the intensity λ is assumed to be deterministic, but the protection payment R is allowed to be stochastic. Since our goal is to provide an explicit example, we postulate that $R = 1/N$, though we do not pretend that this is a natural choice of the protection payment process R.

Corollary 3 *Let us set* $R = 1/N$ *and let us assume that the default intensity* λ *is deterministic. If the process N is a* $(\mathbb{P}, \mathbb{F})$-*martingale then the fair spread* $\kappa_0^{\mathbb{P}}$ *of the CDS under* $\mathbb{P}$ *equals*

$$\kappa_0^{\mathbb{P}} = \frac{\int_0^T \mathbb{E}_{\mathbb{P}}((N_u)^{-1})\lambda_u e^{-\Lambda_u}\, du}{\int_0^T e^{-\Lambda_u}\, du} \tag{48}$$

and the fair spread $\kappa_0^{\mathbb{Q}}$ *under* $\mathbb{Q}$ *satisfies*

$$\kappa_0^{\mathbb{Q}} = \frac{1 - e^{-\Lambda_T}}{\int_0^T e^{-\Lambda_u}\, du} < \kappa_0^{\mathbb{P}}. \tag{49}$$

Proof Recall that the fair spread at time 0 is defined as the level of κ for which the value of the CDS at time 0 equals zero, that is, $S_0^{\mathbb{P}}(\kappa) = 0$ or $S_0^{\mathbb{Q}}(\kappa) = 0$, depending on the choice of a risk-neutral probability measure. By applying formula (47) with $t = 0$ to risk-neutral probability measures $\mathbb{P}$ and $\mathbb{Q}$, we thus obtain

$$\kappa_0^{\mathbb{P}} = \frac{\mathbb{E}_{\mathbb{P}}\left(\int_0^T G_u^{\mathbb{P}} R_u \lambda_u\, du\right)}{\mathbb{E}_{\mathbb{P}}\left(\int_0^T G_u^{\mathbb{P}}\, du\right)}$$

and

$$\kappa_0^{\mathbb{Q}} = \frac{\mathbb{E}_{\mathbb{Q}}\left(\int_0^T G_u^{\mathbb{Q}} R_u \lambda_u\, du\right)}{\mathbb{E}_{\mathbb{Q}}\left(\int_0^T G_u^{\mathbb{Q}}\, du\right)}.$$

Let us first observe that the denominators in the two formulae above are in fact equal since

$$\mathbb{E}_{\mathbb{P}}\left(\int_0^T G_u^{\mathbb{P}}\, du\right) = \int_0^T e^{-\Lambda_u}\, du$$

and

$$\mathbb{E}_{\mathbb{Q}}\left(\int_0^T G_u^{\mathbb{Q}}\, du\right) = \mathbb{E}_{\mathbb{Q}}\left(\int_0^T N_t e^{-\Lambda_u}\, du\right) = \int_0^T \mathbb{E}_{\mathbb{Q}}(N_t) e^{-\Lambda_u}\, du$$

$$= \int_0^T \mathbb{E}_{\mathbb{P}}(N_t) e^{-\Lambda_u}\, du = \int_0^T e^{-\Lambda_u}\, du$$

since the restriction of $\mathbb{Q}$ to $\mathbb{F}$ equals $\mathbb{P}$ and N is a $(\mathbb{P}, \mathbb{F})$-martingale, and thus we have that $\mathbb{E}_{\mathbb{P}}(N_t) = 1$ for every $t \in [0, T]$. For the numerators, using the postulated equality $R = 1/N$, we obtain

$$\mathbb{E}_{\mathbb{P}}\left(\int_0^T G_u^{\mathbb{P}} R_u \lambda_u\, du\right) = \mathbb{E}_{\mathbb{P}}\left(\int_0^T (N_u)^{-1} \lambda_u e^{-\Lambda_u}\, du\right)$$

$$= \int_0^T \mathbb{E}_{\mathbb{P}}((N_u)^{-1}) \lambda_u e^{-\Lambda_u}\, du$$

and

$$\mathbb{E}_{\mathbb{Q}}\left(\int_0^T G_u^{\mathbb{Q}} R_u \lambda_u \, du\right) = \int_0^T \lambda_u e^{-\Lambda_u} \, du = 1 - e^{-\Lambda_T}.$$

This proves equalities in (48) and (49). The inequality in formula (49) follows from the fact that $1/N$ is a strict submartingale and thus $\mathbb{E}_{\mathbb{P}}((N_t)^{-1}) > 1$ for every t. □

Formulae (48) and (49) make it clear that the fair spreads $\kappa_0^{\mathbb{P}}$ and $\kappa_0^{\mathbb{Q}}$ are not equal when the $(\mathbb{P}, \mathbb{F})$-martingale N is not constant. This example supports our claim that the knowledge of the default intensity λ, even in the case when the default intensity λ is deterministic, is not sufficient for the determination of the risk-neutral price of a credit derivative, in general. To conclude, a specific way in which the default time of the underlying credit entity is modeled should always be scrutinized in detail and, whenever possible, the multiplicative decomposition of the associated survival process should be computed explicitly.

Acknowledgements This research was initiated when the first author was visiting Université d'Evry Val d'Essonne in November 2008. The warm hospitality from the Département de Mathématiques and financial support from the Europlace Institute of Finance and the European Science Foundation (ESF) through the grant number 2500 of the programme Advanced Mathematical Methods for Finance (AMaMeF) are gratefully acknowledged. It was continued during the stay of the second author at CMM, Universidad de Chile, Santiago de Chile. The warm hospitality of Jaime San Martin and Soledad Torres is gratefully acknowledged.

The research of P.V. Gapeev was supported by ESF AMaMeF Short Visit Grant 2500. The research of Monique Jeanblanc benefited from the support of the "Chaire Risque de Crédit", Fédération Bancaire Française and Institut Europlace de Finance. The research of M. Rutkowski was supported under Australian Research Council's Discovery Projects funding scheme (project number DP0881460).

References

1. Bielecki, T.R., Jeanblanc, M., Rutkowski, M.: Pricing and trading credit default swaps. Ann. Appl. Probab. **18**, 2495–2529 (2008)
2. Elliott, R.J., Jeanblanc, M., Yor, M.: On models of default risk. Math. Financ. **10**, 179–196 (2000)
3. El Karoui, N., Jeanblanc, M., Jiao, Y.: Stoch. Process. Appl. **120**, 1011–1032 (2010)
4. Liptser, R.S., Shiryaev, A.N.: Statistics of Random Processes I. Springer, Berlin (1977)
5. Mansuy, R., Yor, M.: Random Times and Enlargements of Filtrations in a Brownian Setting. *Lecture Notes in Mathematics*, vol. 1873, Springer, Berlin (2004)
6. Nikeghbali, A., Yor, M.: Doob's maximal identity, multiplicative decompositions and enlargements of filtrations. Ill. J. Math. **50**, 791–814 (2006)
7. Shiryaev, A.N.: Statistical Sequential Analysis. Am. Math. Soc., Providence (1973)
8. Shiryaev, A.N.: Optimal Stopping Rules. Springer, New York (1978)

Representation of American Option Prices Under Heston Stochastic Volatility Dynamics Using Integral Transforms

Carl Chiarella, Andrew Ziogas, and Jonathan Ziveyi

Abstract We consider the evaluation of American options on dividend paying stocks in the case where the underlying asset price evolves according to Heston's stochastic volatility model in (Heston, Rev. Financ. Stud. 6:327–343, 1993). We solve the Kolmogorov partial differential equation associated with the driving stochastic processes using a combination of Fourier and Laplace transforms and so obtain the joint transition probability density function for the underlying processes. We then use this expression in applying Duhamel's principle to obtain the expression for an American call option price, which depends upon an unknown early exercise surface. By evaluating the pricing equation along the free surface boundary, we obtain the corresponding integral equation for the early exercise surface.

1 Introduction

In this paper[1] we seek to generalise the constant volatility analysis of American option pricing to the stochastic volatility case, focusing in particular on Heston's square root model in [13]. We derive the solution by first posing the problem in a form proposed by Jamshidian in [15] for American option pricing. This involves converting the American option pricing problem from one involving the solution of a homogeneous partial differential equation (PDE) on a finite domain (determined

C. Chiarella (✉) · J. Ziveyi
School of Finance and Economics, University of Technology, Sydney, Australia
e-mail: carl.chiarella@uts.edu.au

J. Ziveyi
e-mail: jonathan.ziveyi@student.uts.edu.au

A. Ziogas
Bank of Scotland Treasury, Sydney, Australia
e-mail: andrew.ziogas@bostreasury.com.au

[1] This is a shortened version of the working paper [9].

C. Chiarella, A. Novikov (eds.), *Contemporary Quantitative Finance*,
DOI 10.1007/978-3-642-03479-4_15, © Springer-Verlag Berlin Heidelberg 2010

by the early exercise boundary) to one involving the solution to an inhomogeneous partial differential equation in an infinite domain. We represent the solution of the inhomogeneous PDE as the sum of the solution to the homogeneous part and the corresponding inhomogeneous term by taking advantage of Duhamel's principle. We then solve the PDE associated with the density function using Fourier and Laplace transforms. In this way we incidentally demonstrate how Heston's results for the European call, derived by him using characteristic functions, may also be found by means of Fourier and Laplace transforms, without the need to assume the form of the solution.

Stochastic volatility models for pricing derivative securities have been developed as an extension to the original, constant volatility model of Black and Scholes in [4]. Many studies of market prices for option contracts have consistently found that implied volatilities vary with respect to both maturity and the moneyness of the option. Directly modelling the volatility of the underlying asset using an additional stochastic process provides one means by which this empirical feature can be incorporated into the Black-Scholes pricing framework.

There exist a number of methods that one can use to model volatility stochastically. In this paper we focus on Heston's square root model, which seems to have become the dominant stochastic volatility model in the literature, one reason perhaps being that Heston in [13] was able to provide a convenient analytic expression for European option prices.

Pricing American options under stochastic volatility has much in common with the standard geometric Brownian motion case. In the constant volatility case, it is well known that the price of an American call option can be decomposed into the sum of a corresponding European call and an early exercise premium term. Kim in [17], Jacka in [14] and Carr, Jarrow and Myneni in [5] demonstrate this result using a variety of approaches. The American call price takes the form of an integral equation involving the unknown early exercise boundary. By evaluating this equation at the free boundary, a corresponding integral equation for the early exercise condition is produced. It turns out that this characterisation also holds when the volatility is allowed to evolve randomly.

Since the cash flows arising from early exercise are not explicit functions of the volatility, the main complexity arising in generalising American option pricing theory from the constant volatility case to that of stochastic volatility is related to the early exercise boundary. Lewis in [18] indicates that the free boundary becomes a two-dimensional free surface, in which the early exercise value of the underlying is a function of time to maturity and the volatility level. This result is reiterated in a more general context by Detemple and Tian in [10]. Further analysis is provided by Touzi in [22], who proves a number of fundamental properties for the free boundary and option price under stochastic volatility for the American put, focusing on how the surface changes with respect to the volatility.

The task of solving the two-dimensional PDE for the early exercise premium is obviously more involved than in the one-dimensional case. For European calls, Heston in [13] uses characteristic functions to derive the pricing equation under stochastic volatility. This is closely related to the Fourier transform approach, which was applied to American option pricing, originally by McKean in [20], and used by Chiarella and Ziogas in [8] to consider American strangles. While the Fourier transform, taken with respect to the log of the underlying asset price, is sufficient to solve the problem under constant volatility, it turns out that when the volatility follows a square-root process in addition one needs to take the Laplace transform with respect to the volatility variable. Feller in [12] used the Laplace transform to solve the Fokker-Planck equation for the transition density of the square-root process. Here we use the Laplace transform in conjunction with the Fourier transform to derive the integral equation for the price of an American call under Heston's stochastic volatility model. Subsequently, a corresponding integral equation for the early exercise surface is readily determined.

Since the expression for the early exercise premium for the American call involves a three-dimensional integral, it is generally considered too cumbersome for numerical solution methods. Here we shall simply indicate some possible approaches to solving directly the integral equation, which are developed fully in a companion paper by Adolfsson et al. in [2].

The remainder of this paper is structured as follows. Section 2 outlines the pricing problem for an American call option under Heston's square root process. We also outline the Jamshidian formulation for the American call, and use Duhamel's principle to determine the form of the solution as the sum of the corresponding European price and the early exercise premium. These expressions are integrals involving the transition density function for the joint stock price and variance processes, so Sect. 3 uses integral transform methods to provide a general solution to the PDE for this quantity. We then derive an analytic integral equation for the option price and early exercise boundary in Sect. 4. In Sect. 5 we briefly discuss some possible numerical procedures that may be developed from the structure proposed in this paper and make some concluding remarks. Most of the lengthy mathematical derivations are given in appendices.

2 Problem Statement – the Heston Model

Let $C_E(S, v, \tau)$ and $C_A(S, v, \tau)$ be respectively the price of a European and an American call option written on a dividend paying underlying asset of price S with time-to-maturity τ[2] and strike price K. The underlying asset is assumed to pay a continuously compounded dividend yield at a rate q. The payoff at maturity of the contract is denoted as $(S - K)^+$. The dynamics of S under the real world measure, $\mathbb{P}$, are driven by the stochastic differential equation,

$$dS = \mu S dt + \sqrt{v} S dZ_1, \tag{1}$$

where μ is the instantaneous return per unit time, v is the instantaneous variance (squared volatility) per unit time of the return on the underlying asset and Z_1 is a

---

[2]Here, $\tau = T - t$, where T is the maturity time of the option contract and t is current time.

standard Wiener process. We also allow v to evolve stochastically using the square root process proposed by Heston in [13], so that under the real world measure, the dynamics of v are governed by

$$dv = \kappa(\theta - v)dt + \sigma\sqrt{v}dZ_2, \qquad (2)$$

where θ is the long-run mean of v, κ is the speed of mean-reversion, σ is the instantaneous volatility of v per unit time, and Z_2 is a standard Wiener process correlated with Z_1 such that $\mathbb{E}(dZ_1 dZ_2) = \rho dt$.

It is more convenient to convert the correlated Wiener processes to independent Wiener processes. This is accomplished by applying the Cholesky decomposition to the Wiener processes, Z_1 and Z_2 to obtain independent Wiener processes W_1 and W_2 such that,

$$dZ_1 = dW_1,$$
$$dZ_2 = \rho dW_1 + \sqrt{1 - \rho^2}dW_2. \qquad (3)$$

Applying this transformation to (1) and (2) we obtain,

$$dS = \mu Sdt + \sqrt{v}SdW_1, \qquad (4)$$
$$dv = \kappa(\theta - v)dt + \sigma\rho\sqrt{v}dW_1 + \sigma\sqrt{1 - \rho^2}\sqrt{v}dW_2. \qquad (5)$$

Equations (4) and (5) are expressed under the real world probability measure, $\mathbb{P}$. However, for fair valuation of derivative securities, we need to work under the risk-neutral measure that we denote as $\mathbb{Q}$. The transformation from $\mathbb{P}$ to $\mathbb{Q}$ is accomplished by applying Girsanov's Theorem for Wiener processes, the application of which to (4) and (5) gives,

$$dS = (r - q)Sdt + \sqrt{v}Sd\tilde{W}_1, \qquad (6)$$
$$dv = \left[\kappa(\theta - v) - \lambda(S, v, t)\sigma\sqrt{1 - \rho^2}\sqrt{v}\right]dt + \rho\sigma\sqrt{v}d\tilde{W}_1$$
$$\quad + \sigma\sqrt{1 - \rho^2}\sqrt{v}d\tilde{W}_2, \qquad (7)$$

where r is the risk-free rate of interest, $\lambda(t, S, v)$ is the market price of volatility risk, and $\tilde{W}_1$ and $\tilde{W}_2$ are Wiener processes under the measure $\mathbb{Q}$. The market price of volatility risk must be strictly positive as investors expect a positive risk premium in order for them to hold a risky security. To ensure positivity of v the constant parameters must be chosen such that,

$$2\kappa\theta \geq \sigma^2, \quad \text{and} \quad -1 < \rho < \min\left(\frac{\kappa}{\sigma}, 1\right). \qquad (8)$$

These conditions also ensure that the stochastic variance process, v neither explodes nor attains zero and hence there will always exist a positive risk premium associated with the non-traded factor.[3] In determining the market prices of

---

[3] For a proof of these assertions we refer the reader to Appendix 1 in [7].

the volatility risk, we follow the arguments of Heston in [13] and assume that, $\lambda(S, v, t) = \lambda\sqrt{v}/(\sigma\sqrt{1-\rho^2})$, where λ is a constant. Substituting this into (7) we obtain,

$$dS = (r - q)S dt + \sqrt{v} S d\tilde{W}_1, \tag{9}$$

$$dv = [\kappa\theta - (\kappa + \lambda)v] dt + \rho\sigma\sqrt{v} d\tilde{W}_1 + \sigma\sqrt{1 - \rho^2}\sqrt{v} d\tilde{W}_2. \tag{10}$$

Given (4) and (5) we can use the well established hedging techniques and Ito's lemma to obtain the pricing PDE for the value of an option $C(S, v, \tau)$ written on the stock as,

$$\frac{\partial C}{\partial \tau} = (r - q)S\frac{\partial C}{\partial S} + \frac{1}{2}vS^2\frac{\partial^2 C}{\partial S^2} + [\kappa\theta - (\kappa + \lambda)v]\frac{\partial C}{\partial v}$$
$$+ \frac{1}{2}\sigma^2 v\frac{\partial^2 C}{\partial v^2} + \rho\sigma vS\frac{\partial^2 C}{\partial S\partial v} - rC. \tag{11}$$

The value of the European call option $C_E(S, v, \tau)$ is determined by solving (11) on $0 \le S < \infty$ and $0 < v < \infty$ subject to,

$$C_E(S, v, 0) = (S - K)^+. \tag{12}$$

The value of the American call option $C_A(S, v, \tau)$ is obtained by solving (11) on the restricted interval $0 \le S \le b(v, \tau)$ and $0 < v < \infty$, where $b(v, \tau)$ is the early exercise surface at variance, v and time-to-maturity, τ, subject to the initial and boundary conditions,

$$C_A(S, v, 0) = (S - K)^+, \quad 0 \le S < \infty, \tag{13}$$

$$C_A(b(v, \tau), v, \tau) = b(v, \tau) - K, \tag{14}$$

$$\lim_{S \to b(v,\tau)} \frac{\partial C_A}{\partial S}(S, v, \tau) = 1. \tag{15}$$

The conditions (12) and (13) give the payoff at maturity of the option contract. Equation (14) is the value matching condition which ensures continuity of the American option value function at the early exercise surface, $b(v, \tau)$. Equation (15) is called the smooth pasting condition which together with the value matching condition are imposed on the American option to avoid arbitrage opportunities.[4]

As noted above the PDE (11) for the American call option is to be solved in a region where the underlying asset is bounded above by the early exercise boundary, $b(v, \tau)$. Jamshidian in [15] shows how the homogeneous PDE (11) for the American option can be transformed so that S can have an unrestricted domain with $0 \le S < \infty$ by introducing an indicator function and transforming it to (see Appendix 1),

---

[4]In fact, (15) is not used in deriving the expressions for the price and free boundary below. However it is required when solving the option pricing partial differential equation numerically, for instance using the method of lines.

$$\frac{\partial C_A}{\partial \tau} = (r-q)S\frac{\partial C_A}{\partial S} + \frac{1}{2}vS^2\frac{\partial^2 C_A}{\partial S^2} + [\kappa\theta - (\kappa+\lambda)v]\frac{\partial C_A}{\partial v}$$

$$+ \frac{1}{2}\sigma^2 v\frac{\partial^2 C_A}{\partial v^2} + \rho\sigma v S\frac{\partial^2 C_A}{\partial S\partial v} - rC_A + \mathbb{1}_{S\geq b(v,\tau)}(qS - rK), \quad (16)$$

where $0 \leq S < \infty$, $0 < v < \infty$, and $\mathbb{1}_{S\geq b(S,v,\tau)}$ is an indicator function which is equal to one if $S \geq b(v,\tau)$ or zero otherwise.

Also associated with the two stochastic differential equations (9) and (10) is a bivariate transition density function $G(S, v, \tau; S_0, v_0)$ which denotes the transition from S, v at time-to-maturity τ to S_0, v_0 at maturity. This function is also known as the Green's function and will be used to represent the solution of the PDE (11). The transition density function satisfies the backward Kolmogorov PDE associated with S and v which is,

$$\frac{\partial G}{\partial \tau} = (r-q)S\frac{\partial G}{\partial S} + \frac{1}{2}vS^2\frac{\partial^2 G}{\partial S^2} + [\kappa\theta - (\kappa+\lambda)v]\frac{\partial G}{\partial v} + \frac{1}{2}\sigma^2 v\frac{\partial^2 G}{\partial v^2}$$

$$+ \rho\sigma v S\frac{\partial^2 G}{\partial S\partial v}. \quad (17)$$

Equation (17) is to be solved subject to the initial condition,

$$G(S, v, 0; S_0, v_0) = \delta(S - S_0)\delta(v - v_0),$$

where $\delta(\cdot)$ is the Dirac delta function.

Equation (16) is an inhomogeneous PDE the solution of which can be represented by use of Duhamel's principle (see [19]). For convenience, we let $S = e^x$, $V_E(x, v, \tau) \equiv C_E(e^x, v, \tau)$, $V_A(x, v, \tau) \equiv C_A(e^x, v, \tau)$ and $H(x, v, \tau; x_0, v_0) \equiv G(e^x, v, \tau; e^{x_0}, v_0)$ so that the PDE (16) becomes,

$$\frac{\partial V_A}{\partial \tau} = \left(r - q - \frac{1}{2}v\right)\frac{\partial V_A}{\partial x} + \frac{1}{2}v\frac{\partial^2 V_A}{\partial x^2} + [\kappa\theta - (\kappa+\lambda)v]\frac{\partial V_A}{\partial v}$$

$$+ \frac{1}{2}\sigma^2 v\frac{\partial^2 V_A}{\partial v^2} + \rho\sigma v\frac{\partial^2 V_A}{\partial x\partial v} - rV_A + \mathbb{1}_{x\geq\ln b(v,\tau)}(qe^x - rK). \quad (18)$$

Proposition 1 *Let $V_E(x, v, \tau)$ denote the solution to the homogeneous part of (18), which is the European option value. It can then be shown that,*

$$V_E(x, v, \tau) = e^{-r\tau}\int_0^\infty\int_{-\infty}^\infty (e^u - K)^+ H(x, v, \tau; u, w)du dw. \quad (19)$$

Application of Duhamel's principle shows that if the solution to the homogeneous part of the PDE (18) is given by (19), then the American call option value can be represented as,

$$V_A(x, v, \tau) = V_E(x, v, \tau) + V_P(x, v, \tau), \quad (20)$$

where,

$$V_P(x, v, \tau)$$

$$= \int_0^\tau e^{-r(\tau-\xi)}\int_0^\infty\int_{\ln b(w,\xi)}^\infty (qe^u - rK)H(x, v, \tau - \xi; u, w)du dw d\xi, \quad (21)$$

is the early exercise premium.

Proof Refer to Appendix 2. ☐

By effecting the transformations indicated just before Proposition 1, the Kolmogorov PDE becomes,

$$\frac{\partial H}{\partial \tau} = \frac{v}{2}\frac{\partial^2 H}{\partial x^2} + \rho\sigma v\frac{\partial^2 H}{\partial x \partial v} + \frac{\sigma^2 v}{2}\frac{\partial^2 H}{\partial v^2} + \left(r - q - \frac{v}{2}\right)\frac{\partial H}{\partial x}$$

$$+ (\alpha - \beta v)\frac{\partial H}{\partial v}, \tag{22}$$

where $-\infty < x < \infty$, $0 < v < \infty$, $0 \le \tau \le T$, with $\alpha = \kappa\theta$ and $\beta = \kappa + \lambda$. The PDE (22) is to be solved subject to the initial condition,

$$H(x, v, 0; x_0, v_0) = \delta(x - x_0)\delta(v - v_0). \tag{23}$$

For convenience and when there is no risk of confusion, we will just write $H(x, v, 0)$ for $H(x, v, 0; x_0, v_0)$. Since x does not appear in the coefficients for any terms in (22), we can apply the Fourier transform to the x-variable and the result is presented in Proposition 2 below. It should be noted that integral transforms require knowledge of the behaviour of the functions being transformed at the extremities of the domain. In applying the Fourier transform, we require that $H(x, v, \tau)$ and $\partial H/\partial x$ tend to zero as $x \to \pm\infty$. By their nature, density functions tend to zero as their state variables tend to infinity thus it seems reasonable to impose these conditions on the density function. It is of course also possible to apply the integral transform approach directly to the option pricing equation, but then in order to handle the extremities of the domain we would have to consider the damped option price approach proposed by Carr and Madan in [6].

3 Finding the Density Function Using Integral Transforms

In this section we derive the explicit form of the density function by solving the PDE, (22). As indicated, this will be accomplished by applying the Fourier transform to the logarithm of the underlying asset variable and the Laplace transform to the stochastic variance variable. Applying these transform techniques transforms the PDE to a corresponding system of ordinary differential equations, known as the characteristic equations, which are then solved by the method of characteristics. These steps are given by the propositions below.

Proposition 2 *Let $\mathcal{F}_x\{H(x, v, \tau)\}$ be the Fourier transform of $H(x, v, \tau)$ taken with respect to x, defined as*

$$\mathcal{F}_x\{H(x, v, \tau)\} = \int_{-\infty}^{\infty} e^{i\phi x} H(x, v, \tau)dx \equiv \hat{H}(\phi, v, \tau). \tag{24}$$

Applying the Fourier transform to (22) we find that $\hat{H}$ satisfies the PDE

$$\frac{\partial \hat{H}}{\partial \tau} = \frac{\sigma^2 v}{2} \frac{\partial^2 \hat{H}}{\partial v^2} + (\alpha - \Theta v)\frac{\partial \hat{H}}{\partial v} + \left(\frac{\Lambda}{2}v - i\phi(r-q)\right)\hat{H}, \qquad (25)$$

where

$$\Theta = \Theta(\phi) \equiv \beta + \rho \sigma i\phi \quad and \quad \Lambda = \Lambda(\phi) \equiv i\phi - \phi^2. \qquad (26)$$

The PDE (25) is to be solved subject to the initial condition,

$$\hat{H}(\phi, v, 0) = e^{i\phi x_0}\delta(v - v_0) \equiv \hat{u}(\phi, v). \qquad (27)$$

Proof By use of (24), and the assumptions that

$$\lim_{x \to \pm\infty} H(x, v, \tau) = \lim_{x \to \pm\infty} \frac{\partial H}{\partial x}(x, v, \tau) = 0$$

we have

$$\mathscr{F}_x\left\{\frac{\partial H}{\partial x}\right\} = -i\phi\hat{H}, \qquad \mathscr{F}_x\left\{\frac{\partial^2 H}{\partial x^2}\right\} = -\phi^2\hat{H},$$

$$\mathscr{F}_x\left\{\frac{\partial^2 H}{\partial x \partial v}\right\} = -i\phi\frac{\partial \hat{H}}{\partial v}, \qquad \mathscr{F}_x\left\{\frac{\partial H}{\partial \tau}\right\} = \frac{\partial \hat{H}}{\partial \tau}.$$

Thus a direct application of the transform (24) to the PDE (22) for H yields the PDE (25) for $\hat{H}$. □

To reduce (25) to a first order PDE, we require the use of one more integral transform in the v dimension, namely the Laplace transform. The Laplace transform has been used in [12] to solve the Fokker-Planck equation associated with the diffusion process (5) for v. Here we use the same approach, the only difference being that we are now considering the joint transition density of the underlying asset driven by the mean-reverting process, v. By the nature of transition density functions, we make certain assumptions about the behaviour of $\hat{H}(\phi, v, \tau)$. Specifically, we assume that $e^{-sv}\hat{H}(\phi, v, \tau)$ and $e^{-sv}\partial\hat{H}(\phi, v, \tau)/\partial v$ tend to zero as $v \to \infty$, where s is the Laplace transform variable.[5] Proposition 3 provides the Laplace transform of (25) with respect to v.

Proposition 3 *Let $\mathscr{L}_v\{\hat{H}(\phi, v, \tau)\}$ be the Laplace transform of $\hat{H}(\phi, v, \tau)$ taken with respect to v, defined as*

$$\mathscr{L}_v\{\hat{H}(\phi, v, \tau)\} = \int_0^\infty e^{-sv}\hat{H}(\phi, v, \tau)dv \equiv \tilde{H}(\phi, s, \tau), \qquad (28)$$

where the Laplace transform variable s could be complex whenever an improper integral exists. Applying the Laplace transform to (25) we find that $\tilde{H}$ satisfies the PDE,

$$\frac{\partial \tilde{H}}{\partial \tau} + \left(\frac{\sigma^2}{2}s^2 - \Theta s + \frac{\Lambda}{2}\right)\frac{\partial \tilde{H}}{\partial s} = [(\alpha - \sigma^2)s + \Theta - i\phi(r-q)]\tilde{H} + f_L(\tau), \qquad (29)$$

---

[5]These conditions impose growth constraints on the H function in the v-direction.

with initial condition $\tilde{H}(\phi, s, 0) = e^{i\phi x_0 - s v_0}$, *and where* $f_L(\tau) \equiv (\sigma^2/2 - \alpha) \times$
$\hat{H}(\phi, 0, \tau)$ *is determined by the condition that*

$$\lim_{s \to \infty} \tilde{H}(\phi, s, \tau) = 0. \tag{30}$$

Proof Refer to Appendix 3.   □

Equation (29) is now a first order PDE which can be solved using the method of characteristics. The unknown function $f_L(\tau)$ on the right hand side of (29) is then found by applying the condition (30) and the exact form will be determined in the proof of Proposition 4 in Appendix 4. In this way we are able to solve (29) for $\tilde{H}(\phi, s, \tau)$, and the result is given in Proposition 4.

Proposition 4 *Using the method of characteristics, and subsequently applying condition* (30) *to determine* $f_L(\tau)$, *the solution to the first order PDE* (29) *is*

$$\tilde{H}(\phi, s, \tau)$$
$$= \exp\left\{ \left[\frac{(\alpha - \sigma^2)(\Theta - \Omega)}{\sigma^2} + \Theta - i\phi(r - q) \right] \tau \right\}$$
$$\times \left(\frac{2\Omega}{(\sigma^2 s - \Theta + \Omega)(e^{\Omega\tau} - 1) + 2\Omega} \right)^{2 - \frac{2\alpha}{\sigma^2}}$$
$$\times e^{i\phi x_0} e^{-\left(\frac{\Theta - \Omega}{\sigma^2} \right) v_0} \exp\left\{ \frac{-2\Omega v_0 (\sigma^2 s - \Theta + \Omega) e^{\Omega\tau}}{\sigma^2 [(\sigma^2 s - \Theta + \Omega)(e^{\Omega\tau} - 1) + 2\Omega]} \right\}$$
$$\times \Gamma\left(\frac{2\alpha}{\sigma^2} - 1; \frac{2\Omega v_0 e^{\Omega\tau}}{\sigma^2 (e^{\Omega\tau} - 1)} \times \frac{2\Omega}{(\sigma^2 s - \Theta + \Omega)(e^{\Omega\tau} - 1) + 2\Omega} \right), \tag{31}$$

where

$$\Omega = \Omega(\phi) \equiv \sqrt{\Theta^2(\phi) - \sigma^2 \Lambda(\phi)}, \tag{32}$$

$\Gamma(n; z)$ *is an incomplete (lower) gamma function, defined as*

$$\Gamma(n; z) = \frac{1}{\Gamma(n)} \int_0^z e^{-\xi} \xi^{n-1} d\xi, \tag{33}$$

and $\Gamma(n)$ *is the (complete) gamma function given by*

$$\Gamma(n) = \int_0^\infty e^{-\xi} \xi^{n-1} d\xi. \tag{34}$$

Proof Refer to Appendix 4.   □

Having determined $\tilde{H}(\phi, s, \tau)$, we now seek to revert back to the original variables S and v, and thus obtain the density function $G(S, v, \tau)$. We begin this process by inverting the Laplace transform[6] in Proposition 5, again using the techniques of [12].

---

[6]The inverse Laplace transform is formally defined as

Proposition 5 *The inverse Laplace transform of $\tilde{H}(\phi, s, \tau)$ in (31) is*

$$\hat{H}(\phi, v, \tau) = \exp\left\{\frac{(\Theta - \Omega)}{\sigma^2}(v - v_0 + \alpha\tau)\right\} e^{i\phi x_0 - i\phi(r-q)\tau}$$

$$\times \frac{2\Omega e^{\Omega\tau}}{\sigma^2(e^{\Omega\tau} - 1)}\left(\frac{v_0 e^{\Omega\tau}}{v}\right)^{\frac{\alpha}{\sigma^2} - \frac{1}{2}} \exp\left\{\frac{-2\Omega}{\sigma^2(e^{\Omega\tau} - 1)}(v_0 e^{\Omega\tau} + v)\right\}$$

$$\times I_{\frac{2\alpha}{\sigma^2} - 1}\left(\frac{4\Omega}{\sigma^2(e^{\Omega\tau} - 1)}(v_0 v e^{\Omega\tau})^{\frac{1}{2}}\right),\tag{35}$$

where $I_k(z)$ is the modified Bessel function of the first kind, defined as

$$I_k(z) = \sum_{n=0}^{\infty} \frac{\left(\frac{z}{2}\right)^{2n+k}}{\Gamma(n+k+1)n!}.\tag{36}$$

Proof Refer to Appendix 5. □

All that remains now is to find the inverse Fourier transform of (35), which we give in the next proposition.[7]

Proposition 6 *Given the definition of the Fourier transform $\mathcal{F}_x$ in (24), the inverse Fourier transform of $\hat{H}(\phi, v, \tau)$ is*

$$\mathcal{F}_x^{-1}\{\hat{H}(\phi, v, \tau)\} = \frac{1}{2\pi}\int_{-\infty}^{\infty} e^{-i\phi x}\hat{H}(\phi, v, \tau)d\phi = H(x, v, \tau).\tag{37}$$

Using (37), the inverse Fourier transform of (35) evaluates to

$$H(x, v, \tau; x_0, v_0) = \frac{1}{2\pi}\int_{-\infty}^{\infty} e^{i\phi x_0}U(x, v, \tau; -\phi, v_0)d\phi,\tag{38}$$

where

$$U(x, v, \tau; \psi, v_0)$$
$$= e^{\frac{(\Theta - \Omega)}{\sigma^2}(v - v_0 + \alpha\tau)}e^{i\psi(r-q)\tau}e^{i\psi x}$$
$$\times \frac{2\Omega e^{\Omega\tau}}{\sigma^2(e^{\Omega\tau} - 1)}\left(\frac{v_0 e^{\Omega\tau}}{v}\right)^{\frac{\alpha}{\sigma^2} - \frac{1}{2}} \exp\left\{\frac{-2\Omega}{\sigma^2(e^{\Omega\tau} - 1)}(v_0 e^{\Omega\tau} + v)\right\}$$
$$\times I_{\frac{2\alpha}{\sigma^2} - 1}\left(\frac{4\Omega}{\sigma^2(e^{\Omega\tau} - 1)}(v_0 v e^{\Omega\tau})^{\frac{1}{2}}\right).\tag{39}$$

$$\mathcal{L}_v^{-1}\{\tilde{H}(\phi, s, \tau)\} = \hat{H}(\phi, v, \tau) = \frac{1}{2\pi i}\int_{c-i\infty}^{c+i\infty} e^{sv}\tilde{H}(\phi, s, \tau)ds,$$

where $c > 0$. For our purpose we do not need to evaluate the contour integral as the inverse Laplace transforms that we shall require are known and tabulated.

[7] We recall that the quantities Θ and Ω that appear below were defined by (26) and (32) respectively.

Proof The result follows by simply substituting (35) into (37) and rearranging. □

Now that we have managed to solve for $H(x, v, \tau)$, we revert back to the original underlying variable S and the original density function $G(S, v, \tau; S_0, v_0)$.

Proposition 7 *The density function in terms of the original state variables can be represented as,*

$$G(S, v, \tau; S_0, v_0) = \frac{1}{2\pi} \int_{-\infty}^{\infty} e^{i\phi \ln S_0} U(\ln S, v, \tau; -\phi, v_0) d\phi. \tag{40}$$

Proof Recall that $S = e^x$ and $H(x, v, \tau; S_0, v_0) = G(e^x, v, \tau; e^{x_0}, v_0)$. Substituting these into (38) we obtain the result in the above proposition. □

With the knowledge of the density function, we can apply Proposition 1 to determine the European option component and the early exercise premium.

4 Solution for the American Call Option

Having established the density function we can now carry out the calculations in the full representation of the American call option value given by (20). The solution involves two linear components: the price of the corresponding European call option on the underlying, plus the early exercise premium. We begin by finding the price of the European call option component.

The payoff for the European call is independent of v, and this allows us to simplify (35) before taking the inverse Fourier transform to find the European call price, the result of which we present in Proposition 8.

Proposition 8 *Evaluating (19) by applying the inverse Fourier transform (37) and transforming to the stock price variable, the price of a European call option, $C_E(S, v, \tau)$, is given by*

$$C_E(S, v, \tau) = Se^{-q\tau} P_1^H(S, v, \tau; K) - Ke^{-r\tau} P_2^H(S, v, \tau; K), \tag{41}$$

where

$$P_j^H(S, v, \tau; K) = \frac{1}{2} + \frac{1}{\pi} \int_0^{\infty} Re\left(\frac{f_j(S, v, \tau; \phi)e^{-i\phi \ln K}}{i\phi}\right) d\phi, \tag{42}$$

for $j = 1, 2$, with

$$f_j(S, v, \tau; \phi) = \exp\{B_j(\phi, \tau) + D_j(\phi, \tau)v + i\phi \ln S\}, \tag{43}$$

$$B_j(\phi, \tau) = i\phi(r - q)\tau + \frac{\alpha}{\sigma^2}\left\{(\Theta_j + \Omega_j)\tau - 2\ln\left(\frac{1 - Q_j e^{\Omega_j \tau}}{1 - Q_j}\right)\right\},$$

$$D_j(\phi, \tau) = \frac{(\Theta_j + \Omega_j)}{\sigma^2}\left(\frac{1 - e^{\Omega_j \tau}}{1 - Q_j e^{\Omega_j \tau}}\right),$$

and $Q_j = (\Theta_j + \Omega_j)/(\Theta_j - \Omega_j)$, where we define $\Theta_1 = \Theta(i - \phi)$, $\Omega_1 = \Omega(i - \phi)$, and $\Theta_2 = \Theta(-\phi)$, $\Omega_2 = \Omega(-\phi)$.

Proof Refer to Appendix 6. □

We note that (41) is the solution derived by Heston in [13] using characteristic functions. Here we arrive at Heston's result by using the Fourier transform approach. In addition, we have extended the technique used by Feller in [12] (namely the Laplace transform to find the transition probability density function for the square root process) to find the transition probability density function for the joint processes S and v. In particular, it is clear that one does not need to assume that the solution is of the form given by Heston in [13], but rather this can be derived directly using the Fourier and Laplace transform technique.

Proposition 9 *The early exercise premium for the American call, $C_P(S, v, \tau)$, is given by*

$$C_P(S, v, \tau) = \int_0^\tau \int_0^\infty \left[q S e^{-q(\tau - \xi)} P_1^A(S, v, \tau - \xi; w, b(w, \xi)) \right.$$
$$\left. - rK e^{-r(\tau - \xi)} P_2^A(S, v, \tau - \xi; w, b(w, \xi)) \right] dw d\xi, \qquad (44)$$

where

$$P_j^A(S, v, \tau - \xi; w, b(w, \xi))$$
$$= \frac{1}{2} + \frac{1}{\pi} \int_0^\infty \mathrm{Re} \left(\frac{g_j(S, v, \tau - \xi; \phi, w) e^{-i\phi \ln b(w, \xi)}}{i\phi} \right) d\phi, \qquad (45)$$

for $j = 1, 2$, with

$$g_j(S, v, \tau - \xi; \phi, w) = e^{\frac{(\Theta_j - \Omega_j)}{\sigma^2}(v - w + \alpha(\tau - \xi))} e^{i\phi(r - q)(\tau - \xi)} e^{i\phi \ln S}$$
$$\times \frac{2\Omega_j e^{\Omega_j(\tau - \xi)}}{\sigma^2 (e^{\Omega_j(\tau - \xi)} - 1)} \left(\frac{w e^{\Omega_j(\tau - \xi)}}{v} \right)^{\frac{\alpha}{\sigma^2} - \frac{1}{2}}$$
$$\times \exp \left\{ \frac{-2\Omega_j}{\sigma^2 (e^{\Omega_j(\tau - \xi)} - 1)} (w e^{\Omega_j(\tau - \xi)} + v) \right\}$$
$$\times I_{\frac{2\alpha}{\sigma^2} - 1} \left(\frac{4\Omega_j}{\sigma^2 (e^{\Omega_j(\tau - \xi)} - 1)} (wv e^{\Omega_j(\tau - \xi)})^{\frac{1}{2}} \right), \qquad (46)$$

with $I_k(z)$ given by (36), and Θ_j and Ω_j are given in Proposition 8.

Proof Refer to Appendix 7. □

Having established the European call price and early exercise premium in Propositions 8 and 9, we can now obtain the integral expression for the American call price, along with the integral equation for the early exercise surface $b(v, \tau)$, which is required to evaluate the integrals in (44) and (45).

Proposition 10 *Substituting* (41) *and* (44) *into* (20) *and using the fact that* $x = \ln S$ *the price of the American call option,* $C_A(S, v, \tau)$, *is given by the integral expression*

$$C_A(S, v, \tau) = Se^{-q\tau} P_1^H(S, v, \tau; K) - Ke^{-r\tau} P_2^H(S, v, \tau; K)$$
$$+ \int_0^\tau \int_0^\infty \left[qSe^{-q(\tau-\xi)} P_1^A(S, v, \tau - \xi; w, b(w, \xi)) \right.$$
$$\left. - rKe^{-r(\tau-\xi)} P_2^A(S, v, \tau - \xi; w, b(w, \xi)) \right] dw d\xi, \qquad (47)$$

where P_j^H *and* P_j^A *are given by* (42) *and* (45) *respectively. Furthermore, evaluating* (47) *at* $S = b(v, \tau)$ *and using the boundary condition* (14), *we find that* $b(v, \tau)$ *satisfies the integral equation,*

$$b(v, \tau) - K = b(v, \tau)e^{-q\tau} P_1^H(b(v, \tau), v, \tau; K) - Ke^{-r\tau} P_2^H(b(v, \tau), v, \tau; K)$$
$$+ \int_0^\tau \int_0^\infty \left[qb(v, \tau)e^{-q(\tau-\xi)} P_1^A(b(v, \tau), v, \tau - \xi; w, b(w, \xi)) \right.$$
$$\left. - rKe^{-r(\tau-\xi)} P_2^A(b(v, \tau), v, \tau - \xi; w, b(w, \xi)) \right] dw d\xi. \qquad (48)$$

Proof Direct substitution of (41) and (44) into (20) (recalling that $x = \ln S$ and $V_A(\ln S, v, \tau) = C_A(S, v, \tau)$) yields (47), and using (14) with direct substitution of $S = b(v, \tau)$ into (47) produces (48). $\qquad \square$

In deriving (47) we have provided a generalisation of Heston's derivation for the European call price under the dynamics (9) and (10). Our approach focuses on application of the transforms directly to the PDE, and use of Duhamel's principle to obtain the early exercise premium. This generalises the solution of Heston to the American case, although without direct use of the characteristic function for the joint and marginal densities of S and v.

The solution (47)–(48) has also been derived by Tzavalis and Wang in [23] using the Snell envelope approach of Karatzas in [16]. In their solution, Tzavalis and Wang use expectation operators to represent the early exercise premium, whereas here we provide explicit expressions for these expectations over the joint density of S and v. We also derive the joint density and rewrite these expectations in a form that is analogous to the Heston solution for the European call.

Equations (47) and (48) are presented in a form that matches the American call price presented by Kim in [17] and Jamshidian in [15] in the constant volatility case. Thus (47) is a generalisation of the early exercise premium decomposition of an American call option to incorporate Heston's stochastic volatility dynamics. Furthermore, we see that (47) is an integral expression for the American call price that depends upon the unknown early exercise surface $b(v, \tau)$. As in the constant volatility case, we can readily price the American call using (47) with numerical integration once $b(v, \tau)$ has been determined.

The free surface satisfies the two-dimensional Volterra integral equation (48). Since $0 < v < \infty$, one must find a way to efficiently estimate the boundary for all values of v. Once this has been established, standard techniques can be used to solve

(48) numerically. Hence the two-pass process observed in the constant volatility case, where one first finds $b(v, \tau)$ and then $C_A(S, v, \tau)$, is still valid for the stochastic volatility case. The representation (47) and (48) for the American call value under stochastic volatility may seem difficult to implement as it involves triple integration. It turns out that it is possible to develop a quite efficient numerical scheme by approximating the free surface by a form that is log-linear in the stochastic variance as suggested by Tzavalis and Wang in [23]. This leads to a system of two integral equations for the free surface, which involves a two-dimensional integral. It is then possible to develop a numerical algorithm for solving this integral equation system which employs standard ideas for the solution of Volterra integral equations. Full details of this numerical implementation may be found in the companion paper by Adolfsson et al. in [2].

5 Conclusion

This paper presents an analysis of the American call option pricing problem under stochastic volatility. Taking Heston's square root model proposed in [13] for the random evolution of the asset volatility, Lewis in [18] and Detemple and Tian in [10] show that the option price is the solution to a two-dimensional free boundary value problem. Following the techniques of Jamshidian in [15], we show that the homogeneous partial differential equation (PDE) for the option price that needs to be solved in a restricted domain is replaced by an equivalent inhomogeneous PDE which has to be solved in an unrestricted domain. We then present the general solution of the resulting PDE by using Duhamel's principle. The general solution involves finding the transition density function, which satisfies the Kolmogorov PDE associated with the driving stochastic processes. To our knowledge an explicit expression for the joint density of the stock price and variance has not been given in the literature. We then present a natural, systematic approach to solving this PDE. Once solved, the full integral representation of the American option price is presented. Here we have focused on the American call option but clearly the technique could be applied to American puts, American strangles and other similar standard payoffs.

The general approach adopted here of transforming the American option pricing homogeneous free boundary value problem to an inhomogeneous boundary problem on an unrestricted region and then solving this by use of Duhamel's principle and transform methods can be extended to a number of interesting American option pricing problems. First, the dynamics of the underlying asset could be extended to also allow for jump-diffusions, for example the Bates model in [3]. Some preliminary results in this direction have been obtained by Cheang et al. in [7]. Second, the technique could be extended to certain types of American exotic options under stochastic volatility such as basket options and compound options. Work in this area is ongoing.

Appendix 1: Deriving the Jamshidian Formulation Under Stochastic Volatility

In order to derive the inhomogeneous PDE in the region $0 \leq S < \infty$ that corresponds to the free boundary value problem (11)–(15), we must find the inhomogeneous term such that the PDE (11) is true for all values of S. Noting that $C_A(S, v, \tau) = S - K$ when $S \geq b(v, \tau)$ we find by direct substitution that the PDE in the stopping region evaluates to

$$
-\frac{\partial C_A}{\partial \tau} + \frac{v S^2}{2} \frac{\partial^2 C_A}{\partial S^2} + \rho \sigma v S \frac{\partial^2 C_A}{\partial S \partial v} + \frac{\sigma^2 v}{2} \frac{\partial^2 C_A}{\partial v^2}
$$
$$
+ (r - q) S \frac{\partial C_A}{\partial S} + (\kappa \theta - (\kappa + \lambda) v) \frac{\partial C_A}{\partial v} - r C_A = (r K - q S).
$$

Thus the inhomogeneous PDE that holds for $0 \leq S < \infty$ is

$$
\frac{\partial C_A}{\partial \tau} = \frac{v S^2}{2} \frac{\partial^2 C_A}{\partial S^2} + \rho \sigma v S \frac{\partial^2 C_A}{\partial S \partial v} + \frac{\sigma^2 v}{2} \frac{\partial^2 C_A}{\partial v^2} + (r - q) S \frac{\partial C_A}{\partial S}
$$
$$
+ (\kappa \theta - (\kappa + \lambda) v) \frac{\partial C_A}{\partial v} - r C_A + \mathbb{1}_{\{S - b(v,\tau) > 0\}} (q S - r K).
$$

Appendix 2: Proof of Proposition 1

The inhomogeneous PDE (18), which may be written as

$$
\frac{\partial V}{\partial \tau} = \mathscr{D} V - r V + I(x, v, \tau) \tag{49}
$$

with initial condition $V(x, v, 0) = (e^x - K)$, is to be solved in the region $\tau \geq 0$, $-\infty < x < \infty$ and $0 < v < \infty$, where we define the operator $\mathscr{D}$ as

$$
\mathscr{D} \equiv \frac{v}{2} \frac{\partial^2}{\partial x^2} + \rho \sigma v \frac{\partial^2}{\partial x \partial v} + \frac{\sigma^2 v}{2} \frac{\partial^2}{\partial v^2} + \left(r - q - \frac{1}{2} v \right) \frac{\partial}{\partial x} + (\kappa \theta - (\kappa + \lambda) v) \frac{\partial}{\partial v},
$$

and we set $I(x, v, \tau) = \mathbb{1}_{x \geq \ln b(v, \tau)} (q e^x - r K)$.

When $I(x, v, \tau) = 0$ the solution of the homogeneous PDE may be written as

$$
V_E(x, v, \tau) = e^{-r \tau} \int_0^\infty \int_{-\infty}^\infty (e^u - K)^+ H(x, v, \tau; u, w) du \, dw,
$$

where the Green's function[8] $H(x, v, \tau; u, w)$ satisfies the PDE,

$$
\frac{\partial H}{\partial \tau} = \mathscr{D} H.
$$

The last equation is to be solved subject to the initial condition

---

[8] For more on the theory of Green's functions, we refer the reader to [11].

$$H(x, v, 0; u, w) = \delta(v - w)\delta(x - u),$$

where $\delta(x)$ is the Dirac delta function.

According to Duhamel's principle (see [19]), the solution for $I(x, v, \tau) \neq 0$ is given by,

$$V(x, v, \tau) = e^{-r\tau} \int_0^\infty \int_{-\infty}^\infty (e^u - K)^+ H(x, v, \tau; u, w) du dw$$

$$+ \int_0^\tau e^{-r(\tau-\xi)} \int_0^\infty \int_{-\infty}^\infty I(u, w, \xi) H(x, v, \tau - \xi; u, w) du dw d\xi$$

$$\equiv V_E(x, v, \tau) + V_P(x, v, \tau). \tag{50}$$

We verify that this is the correct solution by showing that (50) satisfies the PDE (49). It is straightforward to demonstrate this by direct substitution of $V(x, v, \tau) = V_E(x, v, \tau) + V_P(s, v, \tau)$ in to (49).[9]

Appendix 3: Proof of Proposition 3

We apply the Laplace transform (28) to the PDE (25). First we note that

$$\mathscr{L}_v\left\{\frac{\Lambda v}{2}\hat{H} - i\phi(r - q)\hat{H}\right\} = \frac{\Lambda}{2}\int_0^\infty v e^{-sv}\hat{H}dv - i\phi(r - q)\tilde{H}$$

$$= -\frac{\Lambda}{2}\frac{\partial\tilde{H}}{\partial s} - i\phi(r - q)\tilde{H}.$$

For the term involving the first order derivative with respect to v we have

$$\mathscr{L}_v\left\{(\alpha - \Theta v)\frac{\partial\hat{H}}{\partial v}\right\} = \int_0^\infty (\alpha - \Theta v)e^{-sv}\frac{\partial\hat{H}}{\partial v}dv$$

$$= \alpha\left\{-\hat{H}(\phi, 0, \tau) + s\int_0^\infty e^{-sv}\hat{H}dv\right\}$$

$$+ \Theta\frac{\partial}{\partial s}\int_0^\infty e^{-sv}\frac{\partial\hat{H}}{\partial v}dv$$

$$= \alpha(-\hat{H}(\phi, 0, \tau) + s\tilde{H}) + \Theta\frac{\partial}{\partial s}\{-\hat{H}(\phi, 0, \tau) + s\tilde{H}\}$$

$$= \Theta s\frac{\partial\tilde{H}}{\partial s} + (\alpha s + \Theta)\tilde{H} - \alpha\hat{H}(\phi, 0, \tau).$$

Finally, for the term involving the second order derivative we calculate

---

[9]For details refer to Appendix 2 of [9].

$$\mathcal{L}_v\left\{\frac{\sigma^2 v}{2}\frac{\partial^2 \hat{H}}{\partial v^2}\right\} = -\frac{\sigma^2}{2}\frac{\partial}{\partial s}\int_0^\infty e^{-sv}\frac{\partial^2 \hat{H}}{\partial v^2}dv = -\frac{\sigma^2}{2}\frac{\partial}{\partial s}\left\{s\int_0^\infty e^{-sv}\frac{\partial \hat{H}}{\partial v}dv\right\}$$

$$= -\frac{\sigma^2}{2}\frac{\partial}{\partial s}\{-s\tilde{H}(\phi,0,\tau) + s^2\tilde{H}\}$$

$$= \frac{-\sigma^2 s^2}{2}\frac{\partial \tilde{H}}{\partial s} - \sigma^2 s\tilde{H} + \frac{\sigma^2}{2}\hat{H}(\phi,0,\tau).$$

Thus the Laplace transform of (25) is the solution of

$$\frac{\partial \tilde{H}}{\partial \tau} + \left(\frac{\sigma^2}{2}s^2 - \Theta s + \frac{\Lambda}{2}\right)\frac{\partial \tilde{H}}{\partial s}$$

$$= [(\alpha - \sigma^2)s + \Theta - i\phi(r-q)]\tilde{H} + \left(\frac{\sigma^2}{2} - \alpha\right)\hat{H}(\phi,0,\tau).$$

Finally we set $f_L(\tau) \equiv (\sigma^2/2 - \alpha)\hat{H}(\phi,0,\tau)$, and note that since $\tilde{H}(\phi,s,\tau)$ must be finite for all $s > 0$, $f_L(\tau)$ must be determined such that $\tilde{H}(\phi,s,\tau) \to 0$ as $s \to \infty$.[10]

Appendix 4: Proof of Proposition 4

Since (29) is a first order PDE, it may be solved using the method of characteristics. We do this in three steps.

(i) *First we express the solution in terms of the so far unknown function $f_L(\tau)$.*
The characteristic equation for (29) is

$$d\tau = \frac{ds}{\left(\frac{\sigma^2}{2}s^2 - \Theta s + \frac{\Lambda}{2}\right)} = \frac{d\tilde{H}}{[(\alpha - \sigma^2)s + \Theta - i\phi(r-q)]\tilde{H} + f_L(\tau)}. \quad (51)$$

Integrating the first pair in (51), we obtain[11]

$$\tau + c_1 = \frac{1}{\Omega}\int\left(\frac{1}{s - (\frac{\Theta+\Omega}{\sigma^2})} - \frac{1}{s - (\frac{\Theta-\Omega}{\sigma^2})}\right)ds,$$

where we use the notation c_j to denote an undetermined constant term, and set

$$\Omega = \Omega(\phi) \equiv \sqrt{\Theta^2 - \Lambda\sigma^2}. \quad (52)$$

Integrating the RHS gives a relation between the transform variable s and time-to-maturity τ, namely

$$\Omega\tau + c_2 = \ln\left(\frac{\sigma^2 s - \Theta - \Omega}{\sigma^2 s - \Theta + \Omega}\right),$$

---

[10]This growth condition on $\hat{H}$ ensures that the Laplace transform is well defined on $0 \le s < \infty$.
[11]Note that $x^2 - \frac{2\Theta}{\sigma^2}x + \frac{\Lambda}{\sigma^2} = 0$ has solution $x = (\Theta \pm \Omega)/\sigma^2$, with Ω defined in (52).

which may be rewritten as

$$c_3 = \frac{(\sigma^2 s - \Theta - \Omega)e^{-\Omega\tau}}{\sigma^2 s - \Theta + \Omega}. \tag{53}$$

We also note that (53) may be re-expressed as

$$s = \frac{(\Theta - \Omega)}{\sigma^2} - \frac{2\Omega e^{-\Omega\tau}}{\sigma^2(c_3 - e^{-\Omega\tau})}. \tag{54}$$

We next consider the last pair in (51), which can be rearranged to give the first order ordinary differential equation (ODE)

$$\frac{d\tilde{H}}{d\tau} + [(\sigma^2 - \alpha)s - \Theta + i\phi(r - q)]\tilde{H} = f_L(\tau). \tag{55}$$

The integrating factor, $R(\tau)$, for this ODE is the solution to

$$\frac{dR}{d\tau} = [(\sigma^2 - \alpha)s - \Theta + i\phi(r - q)]R.$$

Using the expression (54) for s and integrating with respect to τ gives

$$\ln R = \left[\frac{(\sigma^2 - \alpha)(\Theta - \Omega)}{\sigma^2} - \Theta + i\phi(r - q)\right]\tau$$
$$- (\sigma^2 - \alpha)\int \frac{2\Omega e^{-\Omega\xi}}{\sigma^2(c_3 - e^{-\Omega\xi})}d\xi.$$

Setting $u = c_3 - e^{-\Omega\xi}$, the last integral becomes

$$\int \frac{e^{-\Omega\xi}}{c_3 - e^{-\Omega\xi}}d\xi = \frac{1}{\Omega}\ln|u|,$$

and hence

$$R(\tau) = \exp\left\{\left[\frac{(\sigma^2 - \alpha)(\Theta - \Omega)}{\sigma^2} - \Theta \right.\right.$$
$$\left.\left. + i\phi(r - q)\right]\tau\right\}\left|\frac{1}{c_3 - e^{-\Omega\tau}}\right|^{\frac{2}{\sigma^2}(\sigma^2 - \alpha)}. \tag{56}$$

Applying the method of variation of parameters to the ODE (55) we find that

$$R(\tau)\tilde{H}(\phi, s, \tau) = \int_0^\tau R(\xi)f_L(\xi)d\xi + c_4,$$

which on use of the expression (56) for $R(\tau)$ becomes

$$\tilde{H}(\phi, s, \tau)$$
$$= \exp\left\{\left[\frac{(\alpha - \sigma^2)(\Theta - \Omega)}{\sigma^2} + \Theta - i\phi(r - q)\right]\tau\right\}|c_3 - e^{-\Omega\tau}|^{\frac{2}{\sigma^2}(\sigma^2 - \alpha)}$$
$$\times \left\{c_4 + \int_0^\tau f_L(\xi)\exp\left\{\left[\frac{(\sigma^2 - \alpha)(\Theta - \Omega)}{\sigma^2} - \Theta + i\phi(r - q)\right]\xi\right\}\right.$$

$$\times \left| \frac{1}{c_3 - e^{-\Omega \xi}} \right|^{\frac{2}{\sigma^2}(\sigma^2 - \alpha)} d\xi \bigg\}. \tag{57}$$

Next we determine the constant c_4 that appears in (57). We anticipate that we can find a function A such that $c_4 = A(c_3)$, where c_3 is given by (53). When $\tau = 0$, we have from (53) and (57) that

$$\tilde{H}(\phi, s, 0) = \left| \frac{\sigma^2 s - \Theta - \Omega}{\sigma^2 s - \Theta + \Omega} - 1 \right|^{\frac{2}{\sigma^2}(\sigma^2 - \alpha)} A(c_3). \tag{58}$$

Note from (54) that at $\tau = 0$ we have

$$s = \frac{\Theta - \Omega}{\sigma^2} - \frac{2\Omega}{(c_3 - 1)\sigma^2}. \tag{59}$$

Thus from (58) we find that $A(c_3)$ is given by

$$A(c_3) = |c_3 - 1|^{-\frac{2}{\sigma^2}(\sigma^2 - \alpha)} \tilde{H}\left(\phi, \frac{\Theta - \Omega}{\sigma^2} - \frac{2\Omega}{(c_3 - 1)\sigma^2}, 0\right).$$

In (57) consider the term

$$|c_3 - e^{-\Omega \tau}|^{\frac{2}{\sigma^2}(\sigma^2 - \alpha)} A(c_3)$$

$$= \left| \frac{2\Omega e^{-\Omega \tau}}{(\sigma^2 s - \Theta + \Omega)(1 - e^{-\Omega \tau}) + 2\Omega e^{-\Omega \tau}} \right|^{\frac{2}{\sigma^2}(\sigma^2 - \alpha)}$$

$$\times \tilde{H}\left(\phi, \frac{\Theta - \Omega}{\sigma^2} - \frac{2\Omega}{(c_3 - 1)\sigma^2}, 0\right),$$

and note from (53) that

$$\frac{2\Omega}{(c_3 - 1)\sigma^2} = \frac{(\sigma^2 s - \Theta + \Omega)2\Omega}{\sigma^2[(e^{-\Omega \tau} - 1)(\sigma^2 s - \Theta - \Omega) - 2\Omega e^{-\Omega \tau}]}.$$

Also, we note that for $0 \leq \xi \leq \tau$ we have

$$\left| \frac{c_3 - e^{-\Omega \tau}}{c_3 - e^{-\Omega t}} \right|^{\frac{2}{\sigma^2}(\sigma^2 - \alpha)}$$

$$= \left| \frac{(\sigma^2 s - \Theta - \Omega)e^{-\Omega \tau} - (\sigma^2 s - \Theta + \Omega)e^{-\Omega \tau}}{(\sigma^2 s - \Theta - \Omega)e^{-\Omega \tau} - (\sigma^2 s - \Theta + \Omega)e^{-\Omega t}} \right|^{\frac{2}{\sigma^2}(\sigma^2 - \alpha)}$$

$$= \left| \frac{2\Omega e^{-\Omega \tau}}{(\sigma^2 s - \Theta + \Omega)(e^{-\Omega t} - e^{-\Omega \tau}) + 2\Omega e^{-\Omega \tau}} \right|^{\frac{2}{\sigma^2}(\sigma^2 - \alpha)},$$

and make the observation that all real arguments in $|\cdot|$ are positive. Thus by substituting the last equation into (57) we obtain

$$\tilde{H}(\phi, s, \tau)$$

$$= \exp\left\{ \left[\frac{(\alpha - \sigma^2)(\Theta - \Omega)}{\sigma^2} + \Theta - i\phi(r - q) \right] \tau \right\}$$

$$\times \tilde{H}\left(\phi, \frac{\Theta - \Omega}{\sigma^2} - \frac{(\sigma^2 s - \Theta + \Omega)2\Omega}{\sigma^2[(e^{-\Omega\tau} - 1)(\sigma^2 s - \Theta - \Omega) - 2\Omega e^{-\Omega\tau}]}, 0\right)$$

$$\times \left(\frac{2\Omega e^{-\Omega\tau}}{(\sigma^2 s - \Theta + \Omega)(1 - e^{-\Omega\tau}) + 2\Omega e^{-\Omega\tau}}\right)^{\frac{2}{\sigma^2}(\sigma^2 - \alpha)}$$

$$+ \int_0^\tau f_L(\xi) \exp\left\{\left[\frac{(\alpha - \sigma^2)(\Theta - \Omega)}{\sigma^2} + \Theta - i\phi(r - q)\right](\tau - \xi)\right\}$$

$$\times \left(\frac{2\Omega e^{-\Omega\tau}}{(\sigma^2 s - \Theta + \Omega)(e^{-\Omega\xi} - e^{-\Omega\tau}) + 2\Omega e^{-\Omega\tau}}\right)^{\frac{2}{\sigma^2}(\sigma^2 - \alpha)} d\xi. \quad (60)$$

(ii) *Now we determine the function $f_L(\xi)$:* We achieve this by applying the condition (30) to (60). Taking the limit of (60) as $s \to \infty$ by use of l'Hôpital's rule, we require that

$$\int_0^\tau f_L(\xi) \exp\left\{-\left[\frac{(\alpha - \sigma^2)(\Theta - \Omega)}{\sigma^2} + \Theta - i\phi(r - q)\right]\xi\right\}$$

$$\times \left(\frac{1 - e^{-\Omega\tau}}{e^{-\Omega\xi} - e^{-\Omega\tau}}\right)^{\frac{2}{\sigma^2}(\sigma^2 - \alpha)} d\xi$$

$$= -\tilde{H}\left(\phi, \frac{\Theta - \Omega}{\sigma^2} - \frac{2\Omega}{\sigma^2(e^{-\Omega\tau} - 1)}, 0\right). \quad (61)$$

In (61) make the changes of variable

$$\zeta^{-1} = 1 - e^{-\Omega\xi}, \qquad z^{-1} = 1 - e^{-\Omega\tau}, \quad (62)$$

so that

$$\int_z^\infty g(\zeta)(\zeta - z)^{\frac{2}{\sigma^2}(\alpha - \sigma^2)} d\zeta = -\Omega\tilde{H}\left(\phi, \frac{\Theta - \Omega}{\sigma^2} + \frac{2\Omega z}{\sigma^2}, 0\right), \quad (63)$$

where we set

$$g(\zeta) = f_L(\xi) \exp\left\{-\left[\frac{(\alpha - \sigma^2)(\Theta - \Omega)}{\sigma^2} + \Theta - i\phi(r - q)\right]\xi\right\} \frac{\zeta^{\frac{2}{\sigma^2}(\sigma^2 - \alpha)}}{\zeta(\zeta - 1)}. \quad (64)$$

Thus our task is to solve (63) for $g(\zeta)$, and hence we will obtain the function $f_L(\xi)$.

Firstly, by definition (28) for the Laplace transform,

$$\tilde{H}\left(\phi, \frac{\Theta - \Omega}{\sigma^2} + \frac{2\Omega z}{\sigma^2}, 0\right)$$

$$= \int_0^\infty \hat{H}(\phi, w, 0) \exp\left\{-\left(\frac{\Theta - \Omega}{\sigma^2} + \frac{2\Omega z}{\sigma^2}\right)w\right\} dw.$$

Introducing a gamma function, as defined by (34), we have[12]

---

[12]The choice of the term $\Gamma(\frac{2\alpha}{\sigma^2} - 1)$ below may seem arbitrary and it seems we could have chosen $\Gamma(\beta)$ for β arbitrary. However it turns out that to make (65) below match with (63) we need to take $\beta = \frac{2\alpha}{\sigma^2} - 1$.

$$\tilde{H}\left(\phi, \frac{\Theta - \Omega}{\sigma^2} + \frac{2\Omega z}{\sigma^2}, 0\right)$$

$$= \frac{\Gamma\left(\frac{2\alpha}{\sigma^2} - 1\right)}{\Gamma\left(\frac{2\alpha}{\sigma^2} - 1\right)} \int_0^\infty \hat{H}(\phi, w, 0) \exp\left\{-\left(\frac{\Theta - \Omega}{\sigma^2} + \frac{2\Omega z}{\sigma^2}\right) w\right\} dw$$

$$= \frac{1}{\Gamma\left(\frac{2\alpha}{\sigma^2} - 1\right)}$$

$$\times \int_0^\infty \int_0^\infty e^{-a} a^{\frac{2\alpha}{\sigma^2} - 2} \hat{H}(\phi, w, 0) \exp\left\{-\left(\frac{\Theta - \Omega}{\sigma^2} + \frac{2\Omega z}{\sigma^2}\right) w\right\} da\, dw.$$

The change of integration variable $a = (2\Omega w/\sigma^2) y$ gives

$$\tilde{H}\left(\phi, \frac{\Theta - \Omega}{\sigma^2} + \frac{2\Omega z}{\sigma^2}, 0\right)$$

$$= \frac{1}{\Gamma\left(\frac{2\alpha}{\sigma^2} - 1\right)} \int_0^\infty \hat{H}(\phi, w, 0) \exp\left\{-\left(\frac{\Theta - \Omega}{\sigma^2}\right) w\right\}$$

$$\times \left[\int_0^\infty e^{-\frac{2\Omega w}{\sigma^2} y} \left(\frac{2\Omega w}{\sigma^2} y\right)^{\frac{2\alpha}{\sigma^2} - 2} e^{-\frac{2\Omega z}{\sigma^2} w} dy\right] \left(\frac{2\Omega w}{\sigma^2}\right) dw$$

$$= \frac{1}{\Gamma\left(\frac{2\alpha}{\sigma^2} - 1\right)} \int_0^\infty \hat{H}(\phi, w, 0) \exp\left\{-\left(\frac{\Theta - \Omega}{\sigma^2}\right) w\right\} \left(\frac{2\Omega w}{\sigma^2}\right)^{\frac{2\alpha}{\sigma^2} - 1}$$

$$\times \left(\int_0^\infty e^{-\frac{2\Omega w}{\sigma^2} (y + z)} y^{\frac{2\alpha}{\sigma^2} - 2} dy\right) dw.$$

Making one further change of variable, namely $\zeta = z + y$, we obtain

$$\tilde{H}\left(\phi, \frac{\Theta - \Omega}{\sigma^2} + \frac{2\Omega z}{\sigma^2}, 0\right)$$

$$= \frac{1}{\Gamma\left(\frac{2\alpha}{\sigma^2} - 1\right)} \int_0^\infty \hat{H}(\phi, w, 0) \exp\left\{-\left(\frac{\Theta - \Omega}{\sigma^2}\right) w\right\} \left(\frac{2\Omega w}{\sigma^2}\right)^{\frac{2\alpha}{\sigma^2} - 1}$$

$$\times \int_z^\infty e^{-\frac{2\Omega w}{\sigma^2} \zeta} (\zeta - z)^{\frac{2\alpha}{\sigma^2} - 2} d\zeta\, dw$$

$$= \int_z^\infty (\zeta - z)^{\frac{2}{\sigma^2}(\alpha - \sigma^2)} \left[\int_0^\infty \frac{\hat{H}(\phi, w, 0)}{\Gamma\left(\frac{2\alpha}{\sigma^2} - 1\right)} \left(\frac{2\Omega w}{\sigma^2}\right)^{\frac{2\alpha}{\sigma^2} - 1}\right.$$

$$\left. \times \exp\left\{-\left(\frac{\Theta - \Omega}{\sigma^2} + \frac{2\Omega \zeta}{\sigma^2}\right) w\right\} dw\right] d\zeta. \tag{65}$$

Comparing (65) with (63), we can conclude that

$$g(z) = \frac{-\Omega}{\Gamma\left(\frac{2\alpha}{\sigma^2} - 1\right)} \int_0^\infty \hat{H}(\phi, w, 0) \left(\frac{2\Omega w}{\sigma^2}\right)^{\frac{2\alpha}{\sigma^2} - 1}$$

$$\times \exp\left\{-\left(\frac{\Theta - \Omega}{\sigma^2} + \frac{2\Omega z}{\sigma^2}\right) w\right\} dw, \tag{66}$$

and hence $f_L(\xi)$ can be readily found by expressing $f_L(\xi)$ as a function of $g(\zeta)$ using (64). Using the initial condition

$$\hat{H}(\phi, w, 0) = e^{i\phi x_0} \delta(v - w), \tag{67}$$

equation (66) then becomes

$$g(z) = \frac{-\Omega}{\Gamma\left(\frac{2\alpha}{\sigma^2} - 1\right)} \int_0^\infty e^{i\phi x_0} \delta(v - w) \left(\frac{2\Omega w}{\sigma^2}\right)^{\frac{2\alpha}{\sigma^2} - 1}$$

$$\times \exp\left\{-\left(\frac{\Theta - \Omega}{\sigma^2} + \frac{2\Omega z}{\sigma^2}\right) w\right\} dw, \tag{68}$$

which simplifies to,

$$g(z) = \frac{-\Omega}{\Gamma\left(\frac{2\alpha}{\sigma^2} - 1\right)} e^{i\phi x_0} \left(\frac{2\Omega v_0}{\sigma^2}\right)^{\frac{2\alpha}{\sigma^2} - 1} \exp\left\{-\left(\frac{\Theta - \Omega}{\sigma^2} + \frac{2\Omega z}{\sigma^2}\right) v_0\right\}. \tag{69}$$

Given the explicit representation of $g(z)$, we note that,

$$f_L(\xi) = \frac{-\Omega}{\Gamma\left(\frac{2\alpha}{\sigma^2} - 1\right)} e^{i\phi x_0} \left(\frac{2\Omega v_0}{\sigma^2}\right)^{\frac{2\alpha}{\sigma^2} - 1} \exp\left\{-\left(\frac{\Theta - \Omega}{\sigma^2} + \frac{2\Omega z}{\sigma^2}\right) v_0\right\}$$

$$\times \exp\left\{\left[\frac{(\alpha - \sigma^2)(\Theta - \Omega)}{\sigma^2} + \Theta - i\phi(r - q)\right]\xi\right\} \frac{\zeta(\zeta - 1)}{\zeta^{\frac{2}{\sigma^2}(\sigma^2 - \alpha)}}. \tag{70}$$

(iii) *Having found $f_L(\xi)$, all that remains is to substitute it into (60) to obtain $\tilde{H}(\phi, s, \tau)$; this requires us to consider the following expressions. Firstly we have*

$$J_1 = \exp\left\{\left[\frac{(\alpha - \sigma^2)(\Theta - \Omega)}{\sigma^2} + \Theta - i\phi(r - q)\right]\tau\right\}$$

$$\times \tilde{H}\left(\phi, \frac{\Theta - \Omega}{\sigma^2} + \frac{(\sigma^2 s - \Theta + \Omega)2\Omega}{\sigma^2[(1 - e^{-\Omega\tau})(\sigma^2 s - \Theta + \Omega) - 2\Omega e^{-\Omega\tau}]}, 0\right)$$

$$\times \left(\frac{2\Omega e^{-\Omega\tau}}{(\sigma^2 s - \Theta + \Omega)(1 - e^{-\Omega\tau}) + 2\Omega e^{-\Omega\tau}}\right)^{\frac{2}{\sigma^2}(\sigma^2 - \alpha)}$$

$$= \exp\left\{\left[\frac{(\alpha - \sigma^2)(\Theta - \Omega)}{\sigma^2} + \Theta - i\phi(r - q)\right]\tau\right\}$$

$$\times \tilde{H}\left(\phi, \frac{\Theta - \Omega}{\sigma^2} + \frac{2\Omega e^{-\Omega\tau}(\sigma^2 s - \Theta + \Omega)z}{\sigma^2[(\sigma^2 s - \Theta + \Omega) + 2\Omega(z - 1)]}, 0\right)$$

$$\times \left(\frac{2\Omega(z-1)}{(\sigma^2 s - \Theta + \Omega) + 2\Omega(z-1)} \right)^{2 - \frac{2\alpha}{\sigma^2}}.$$

Next we consider

$$J_2 = \int_z^\infty f_L(\xi) \exp\left\{ \left[\frac{(\alpha - \sigma^2)(\Theta - \Omega)}{\sigma^2} + \Theta - i\phi(r-q) \right](\tau - \xi) \right\}$$

$$\times \frac{1}{\Omega \zeta^2 e^{-\Omega \xi}} \left(\frac{2\Omega(z-1)\zeta}{(\sigma^2 s - \Theta + \Omega)(\zeta - 2) + 2\Omega(z-1)\zeta} \right)^{2 - \frac{2\alpha}{\sigma^2}} d\zeta,$$

which by use of (64) becomes

$$J_2 = \frac{1}{\Omega} \exp\left\{ \left[\frac{(\alpha - \sigma^2)(\Theta - \Omega)}{\sigma^2} + \Theta - i(r-q)\phi \right] \tau \right\}$$

$$\times \int_z^\infty g(\zeta) \left(\frac{2\Omega(z-1)}{(\sigma^2 s - \Theta + \Omega)(\zeta - z) + 2\Omega(z-1)\zeta} \right)^{2 - \frac{2\alpha}{\sigma^2}} d\zeta,$$

and substituting for $g(\zeta)$ from (69) we have

$$J_2 = \frac{1}{\Omega} \exp\left\{ \left[\frac{(\alpha - \sigma^2)(\Theta - \Omega)}{\sigma^2} + \Theta - i\phi(r-q) \right] \tau \right\}$$

$$\times \int_z^\infty \frac{-\Omega}{\Gamma\left(\frac{2\alpha}{\sigma^2} - 1\right)} e^{i\phi x_0} \left(\frac{2\Omega v_0}{\sigma^2} \right)^{\frac{2\alpha}{\sigma^2} - 1} \exp\left\{ -\left(\frac{\Theta - \Omega}{\sigma^2} + \frac{2\Omega \zeta}{\sigma^2} \right) v_0 \right\}$$

$$\times \left(\frac{2\Omega(z-1)}{(\sigma^2 s - \Theta + \Omega)(\zeta - z) + 2\Omega(z-1)\zeta} \right)^{2 - \frac{2\alpha}{\sigma^2}} d\zeta$$

$$= \frac{-[2\Omega(z-1)]^{2 - \frac{2\alpha}{\sigma^2}}}{\Gamma\left(\frac{2\alpha}{\sigma^2} - 1\right)} \exp\left\{ \left[\frac{(\alpha - \sigma^2)(\Theta - \Omega)}{\sigma^2} + \Theta - i\phi(r-q) \right] \tau \right\}$$

$$\times e^{i\phi x_0} \left(\frac{2\Omega v_0}{\sigma^2} \right)^{\frac{2\alpha}{\sigma^2} - 1} \exp\left\{ -\left(\frac{\Theta - \Omega}{\sigma^2} \right) v_0 \right\} J_3(v_0), \tag{71}$$

where for convenience we set

$$J_3(v_0) = \int_z^\infty e^{-\frac{2\Omega v_0}{\sigma^2}\zeta} [(\sigma^2 s - \Theta + \Omega)(\zeta - z) + 2\Omega(z-1)\zeta]^{\frac{2\alpha}{\sigma^2} - 2} d\zeta.$$

Before proceeding further, we perform some extensive manipulations on $J_3(v_0)$. Firstly, make the change of integration variable $y = (\sigma^2 s - \Theta + \Omega)(\zeta - z) + 2\Omega(z - 1)\zeta$ to obtain

$$J_3(v_0) = \int_{2\Omega(z-1)z}^\infty \exp\left\{ \frac{-2\Omega v_0}{\sigma^2} \left(\frac{y + (\sigma^2 s - \Theta + \Omega)z}{(\sigma^2 s - \Theta + \Omega) + 2\Omega(z-1)} \right) \right\} y^{\frac{2\alpha}{\sigma^2} - 2}$$

$$\times \frac{dy}{(\sigma^2 s - \Theta + \Omega) + 2\Omega(z-1)}$$

$$= \frac{1}{(\sigma^2 s - \Theta + \Omega) + 2\Omega(z-1)} \exp\left\{ \frac{-2\Omega v_0(\sigma^2 s - \Theta + \Omega)z}{\sigma^2[(\sigma^2 s - \Theta + \Omega) + 2\Omega(z-1)]} \right\}$$

$$\times \int_{2\Omega(z-1)z}^{\infty} \exp\left\{\frac{-2\Omega v_0 y}{\sigma^2[(\sigma^2 s - \Theta + \Omega) + 2\Omega(z-1)]}\right\} y^{\frac{2\alpha}{\sigma^2}-2} dy.$$

By making a further change of integration variable, namely

$$\xi = \frac{2\Omega v_0 y}{\sigma^2[(\sigma^2 s - \Gamma + \Omega) + 2\Omega(z-1)]},$$

we have

$$J_3(v_0) = \frac{\sigma^2}{2\Omega v_0}\left(\frac{\sigma^2[(\sigma^2 s - \Theta + \Omega) + 2\Omega(z-1)]}{2\Omega v_0}\right)^{\frac{2\alpha}{\sigma^2}-2}$$

$$\times \exp\left\{\frac{-2\Omega v_0(\sigma^2 s - \Theta + \Omega)z}{\sigma^2[(\sigma^2 s - \Theta + \Omega) + 2\Omega(z-1)]}\right\}$$

$$\times \int_{\frac{4\Omega^2(z-1)zv_0}{\sigma^2[(\sigma^2 s - \Theta + \Omega) + 2\Omega(z-1)]}}^{\infty} e^{-\xi} \xi^{(\frac{2\alpha}{\sigma^2}-1)-1} d\xi,$$

which in terms of the gamma functions (33) and (34) may be written

$$J_3(v_0) = [(\sigma^2 s - \Theta + \Omega) + 2\Omega(z-1)]^{\frac{2\alpha}{\sigma^2}-2}$$

$$\times \exp\left\{\frac{-2\Omega v_0(\sigma^2 s - \Theta + \Omega)z}{\sigma^2[(\sigma^2 s - \Theta + \Omega) + 2\Omega(z-1)]}\right\}\left(\frac{\sigma^2}{2\Omega v_0}\right)^{\frac{2\alpha}{\sigma^2}-1}$$

$$\times \left[\Gamma\left(\frac{2\alpha}{\sigma^2} - 1\right) - \int_0^{\frac{4\Omega^2(z-1)zv_0}{\sigma^2[(\sigma^2 s - \Theta + \Omega) + 2\Omega(z-1)]}} e^{-\xi} \xi^{(\frac{2\alpha}{\sigma^2}-1)-1} d\xi\right]. \quad (72)$$

Substituting (72) into (71) we find that

$$J_2 = \frac{-1}{\Gamma\left(\frac{2\alpha}{\sigma^2} - 1\right)} \exp\left\{\left[\frac{(\alpha - \sigma^2)(\Theta - \Omega)}{\sigma^2} + \Theta - i\phi(r-q)\right]\tau\right\}$$

$$\times \left(\frac{2\Omega(z-1)}{(\sigma^2 s - \Theta + \Omega) + 2\Omega(z-1)}\right)^{2-\frac{2\alpha}{\sigma^2}}$$

$$\times e^{i\phi x_0} e^{-\left(\frac{\Theta-\Omega}{\sigma^2}\right)v_0} \exp\left\{\frac{-2\Omega v_0(\sigma^2 s - \Theta + \Omega)z}{\sigma^2[(\sigma^2 s - \Theta + \Omega) + 2\Omega(z-1)]}\right\}$$

$$\times \Gamma\left(\frac{2\alpha}{\sigma^2} - 1\right)\left[1 - \Gamma\left(\frac{2\alpha}{\sigma^2} - 1; \frac{4\Omega^2(z-1)zv_0}{\sigma^2[(\sigma^2 s - \Theta + \Omega) + 2\Omega(z-1)]}\right)\right],$$

and since $\tilde{H}(\phi, s, \tau) = J_1 + J_2$, we have

$$\tilde{H}(\phi, s, \tau) = \exp\left\{\left[\frac{(\alpha - \sigma^2)(\Theta - \Omega)}{\sigma^2} + \Theta - i\phi(r-q)\right]\tau\right\}$$

$$\times \left(\frac{2\Omega(z-1)}{(\sigma^2 s - \Theta + \Omega) + 2\Omega(z-1)}\right)^{2-\frac{2\alpha}{\sigma^2}}$$

$$\times e^{i\phi x_0} e^{-\left(\frac{\Theta-\Omega}{\sigma^2}\right)v_0} \exp\left\{\frac{-2\Omega v_0(\sigma^2 s - \Theta + \Omega)z}{\sigma^2[(\sigma^2 s - \Theta + \Omega) + 2\Omega(z-1)]}\right\}$$

$$\times \Gamma\left(\frac{2\alpha}{\sigma^2} - 1; \frac{4\Omega^2(z-1)zv_0}{\sigma^2[(\sigma^2 s - \Theta + \Omega) + 2\Omega(z-1)]}\right),$$

which, after substituting for z from (62) becomes

$$\tilde{H}(\phi, s, \tau) = \exp\left\{\left[\frac{(\alpha - \sigma^2)(\Theta - \Omega)}{\sigma^2} + \Theta - i\phi(r - q)\right]\tau\right\}$$

$$\times \left(\frac{2\Omega}{(\sigma^2 s - \Theta + \Omega)(e^{\Omega\tau} - 1) + 2\Omega}\right)^{2 - \frac{2\alpha}{\sigma^2}}$$

$$\times e^{i\phi x_0} e^{-\left(\frac{\Theta - \Omega}{\sigma^2}\right)v_0} \exp\left\{\frac{-2\Omega v_0(\sigma^2 s - \Theta + \Omega)e^{\Omega\tau}}{\sigma^2[(\sigma^2 s - \Theta + \Omega)(e^{\Omega\tau} - 1) + 2\Omega]}\right\}$$

$$\times \Gamma\left(\frac{2\alpha}{\sigma^2} - 1; \frac{2\Omega v_0 e^{\Omega\tau}}{\sigma^2(e^{\Omega\tau} - 1)} \times \frac{2\Omega}{(\sigma^2 s - \Theta + \Omega)(e^{\Omega\tau} - 1) + 2\Omega}\right),$$

which is the result in Proposition 4.

Appendix 5: Proof of Proposition 5 – The Inverse Laplace Transform

The inverse Laplace transform of (31) is most easily found by using the new variables

$$A = \frac{2\Omega v_0}{\sigma^2(1 - e^{-\Omega\tau})}, \qquad z = \frac{1}{2\Omega}\{(\sigma^2 s - \Theta + \Omega)(e^{\Omega\tau} - 1) + 2\Omega\}. \quad (73)$$

If we set

$$h(\phi, v_0, \tau) = \exp\left\{\left[\frac{(\alpha - \sigma^2)(\Theta - \Omega)}{\sigma^2} + \Theta - i\phi(r - q)\right]\tau\right\}$$

$$\times \exp\left\{-\left(\frac{\Theta - \Omega}{\sigma^2}\right)v_0 + i\phi x_0\right\}, \quad (74)$$

then under the change of variables (73) and making use of (33), (31) becomes

$$\tilde{H}(\phi, s(z), \tau) = h(\phi, v_0, \tau)\exp\left\{-\frac{A}{z}(z-1)\right\}z^{\frac{2\alpha}{\sigma^2} - 2}\Gamma\left(\frac{2\alpha}{\sigma^2} - 1; \frac{A}{z}\right)$$

$$= h(\phi, v_0, \tau)\exp\left\{-\frac{A}{z}(z-1)\right\}\frac{z^{\frac{2\alpha}{\sigma^2} - 2}}{\Gamma\left(\frac{2\alpha}{\sigma^2} - 1\right)}\int_0^{\frac{A}{z}} e^{-\beta}\beta^{\frac{2\alpha}{\sigma^2} - 1}d\beta. \quad (75)$$

Changing the integration variable according to $\xi = 1 - \frac{z}{A}\beta$, (75) becomes

$$\tilde{H}(\phi, s(z), \tau) = h(\phi, v_0, \tau)e^{-A}\frac{A^{\frac{2\alpha}{\sigma^2} - 1}}{\Gamma\left(\frac{2\alpha}{\sigma^2} - 1\right)}\int_0^1 (1 - \xi)^{\frac{2\alpha}{\sigma^2} - 2}z^{-1}e^{\frac{A\xi}{z}}d\xi. \quad (76)$$

In (28), the Laplace transform is defined with respect to the parameter s. In order to invert (76), we must first establish the relationship between the Laplace transform with respect to parameter s, and the inverse Laplace transform with respect to the parameter z which is a function of s as defined in the second part of (73).

From (73) we see that

$$s = \frac{2\Omega(z-1)}{\sigma^2(e^{\Omega\tau}-1)} + \frac{\Theta-\Omega}{\sigma^2}.$$

Substituting this into (28) gives

$$\mathscr{L}_v\{\hat{H}(\phi, v, \tau)\} = \int_0^\infty \exp\left\{-\left[\frac{2\Omega(z-1)}{\sigma^2(e^{\Omega\tau}-1)} + \frac{\Theta-\Omega}{\sigma^2}\right]v\right\}\hat{H}(\phi, v, \tau)dv.$$

By letting

$$y = \frac{2\Omega v}{\sigma^2(e^{\Omega\tau}-1)}, \tag{77}$$

we have

$$\mathscr{L}_v\{\hat{H}(\phi, v(y), \tau)\} = \frac{\sigma^2(e^{\Omega\tau}-1)}{2\Omega}\int_0^\infty e^{-zy}\exp\left\{-\left(\frac{(\Theta-\Omega)(e^{\Omega\tau}-1)}{2\Omega}-1\right)y\right\}$$
$$\times \hat{H}(\phi, v(y), \tau)dy,$$

which, by use of the definition (28) can be written as

$$\mathscr{L}_v\{\hat{H}(\phi, v(y), \tau)\} = \frac{\sigma^2(e^{\Omega\tau}-1)}{2\Omega}$$
$$\times \mathscr{L}_y\left\{\exp\left\{-\left(\frac{(\Theta-\Omega)(e^{\Omega\tau}-1)}{2\Omega}-1\right)y\right\}\hat{H}(\phi, v(y), \tau)\right\}$$

or alternatively (again using (28))

$$\tilde{H}(\phi, s(z), \tau) = \frac{\sigma^2(e^{\Omega\tau}-1)}{2\Omega}$$
$$\times \mathscr{L}_y\left\{\exp\left\{-\left(\frac{(\Theta-\Omega)(e^{\Omega\tau}-1)}{2\Omega}-1\right)y\right\}\hat{H}(\phi, v(y), \tau)\right\},$$

where we set

$$\mathscr{L}_y\{f(y)\} = \int_0^\infty e^{-zy}f(y)dy, \tag{78}$$

and recall that y is given by (77), and z is defined in (73). It then follows that

$$\mathscr{L}_y^{-1}\left\{\frac{2\Omega}{\sigma^2(e^{\Omega\tau}-1)}\tilde{H}(\phi, s(z), \tau)\right\}$$
$$= \exp\left\{-\left(\frac{(\Theta-\Omega)(e^{\Omega\tau}-1)}{2\Omega}-1\right)y\right\}\hat{H}(\phi, v(y), \tau).$$

Then by noting that

$$\hat{H}(\phi, v(y), \tau) = \mathscr{L}_v^{-1}\{\tilde{H}(\phi, s(z), \tau)\}$$

we find that

$$\mathcal{L}_v^{-1}\{\tilde{H}(\phi, s(z), \tau)\} = \frac{2\Omega}{\sigma^2(e^{\Omega\tau} - 1)} \exp\left\{\left[\frac{(\Theta - \Omega)(e^{\Omega\tau} - 1)}{2\Omega} - 1\right]y\right\}$$
$$\times \mathcal{L}_y^{-1}\{\tilde{H}(\phi, s(z), \tau)\}. \tag{79}$$

Applying the inverse transform (79) to (76), we have

$$\hat{H}(\phi, v(y), \tau) = h(\phi, v_0, \tau)\frac{\sigma^2(1 - e^{-\Omega\tau})}{2\Omega}e^{-A}\frac{A^{\frac{2\alpha}{\sigma^2} - 1}}{\Gamma(\frac{2\alpha}{\sigma^2} - 1)}\frac{2\Omega}{\sigma^2(e^{\Omega\tau} - 1)}$$
$$\times \exp\left\{\left[\frac{(\Theta - \Omega)(e^{\Omega\tau} - 1)}{2\Omega} - 1\right]y\right\}$$
$$\times \int_0^1 (1 - \xi)^{\frac{2\alpha}{\sigma^2} - 2}\mathcal{L}_y^{-1}\{z^{-1}e^{\frac{A\xi}{z}}\}d\xi. \tag{80}$$

Referring to [1] we find that

$$\mathcal{L}_y\{I_0(2\sqrt{A\xi y})\} = \frac{1}{z}e^{\frac{A\xi}{z}},$$

where $I_k(x)$ is the modified Bessel function defined by (36). Thus the inverse Laplace transform of $\tilde{H}(\phi, s, \tau)$ becomes

$$\hat{H}(\phi, v(y), \tau) = h(\phi, v_0, \tau)\frac{\sigma^2(1 - e^{-\Omega\tau})}{2\Omega}$$
$$\times e^{-A}\frac{A^{\frac{2\alpha}{\sigma^2} - 1}}{\Gamma(\frac{2\alpha}{\sigma^2} - 1)}\frac{2\Omega}{\sigma^2(e^{\Omega\tau} - 1)}\exp\left\{\left[\frac{(\Theta - \Omega)(e^{\Omega\tau} - 1)}{2\Omega} - 1\right]y\right\}$$
$$\times \int_0^1 (1 - \xi)^{\frac{2\alpha}{\sigma^2} - 2}I_0(2\sqrt{A\xi y})d\xi.$$

We can further simplify this result by noting that[13]

$$\int_0^1 (1 - \xi)^{\frac{2\alpha}{\sigma^2} - 2}I_0(2\sqrt{A\xi y})d\xi = \Gamma\left(\frac{2\alpha}{\sigma^2} - 1\right)(Ay)^{\frac{1}{2} - \frac{\alpha}{\sigma^2}}I_{\frac{2\alpha}{\sigma^2} - 1}(2\sqrt{Ay}),$$

and therefore

$$\hat{H}(\phi, v, \tau) = h(\phi, v_0, \tau)\frac{\sigma^2(1 - e^{-\Omega\tau})}{2\Omega}\frac{2\Omega}{\sigma^2(e^{\Omega\tau} - 1)}$$
$$\times e^{-A-y}\left(\frac{A}{y}\right)^{\frac{\alpha}{\sigma^2} - \frac{1}{2}}\exp\left\{\frac{(\Theta - \Omega)(e^{\Omega\tau} - 1)}{2\Omega}y\right\}I_{\frac{2\alpha}{\sigma^2} - 1}(2\sqrt{Ay}).$$

Recalling the definitions for A and y, from (73) and (77) respectively, we conclude that

---

[13] One way to obtain this result is to simply expand both terms under the integral in power series.

$$\hat{H}(\phi, v, \tau) = h(\phi, v_0, \tau) \frac{2\Omega}{\sigma^2(e^{\Omega\tau} - 1)} \exp\left\{-\frac{2\Omega}{\sigma^2(e^{\Omega\tau} - 1)}(v_0 e^{\Omega\tau} + v)\right\}$$

$$\times \left(\frac{v_0 e^{\Omega\tau}}{v}\right)^{\frac{\alpha}{\sigma^2} - \frac{1}{2}} \exp\left\{\frac{(\Theta - \Omega)}{\sigma^2}v\right\} I_{\frac{2\alpha}{\sigma^2} - 1}\left(\frac{4\Omega}{\sigma^2(e^{\Omega\tau} - 1)}(v_0 v e^{\Omega\tau})^{\frac{1}{2}}\right).$$

Finally, substituting for $h(\phi, v_0, \tau)$ from (74) we obtain the result stated in Proposition 5.

Appendix 6: Proof of Proposition 8 – The European Option Price

We note that the payoff of the European option, $(e^x - K)^+$, is independent of v. From (19) we have that

$$V_E(x, v, \tau) = e^{-r\tau} \int_0^\infty \int_{\ln K}^\infty (e^y - K) H(x, v, \tau; y, w) dy dw.$$

which upon use of (38) may be written

$$V_E(x, v, \tau) = \frac{e^{-r\tau}}{2\pi} \int_0^\infty \int_{\ln K}^\infty \int_{-\infty}^\infty (e^y - K) e^{i\phi y} U(x, v, \tau; -\phi, w) dy dw d\phi. \quad (81)$$

Switching back to the original variable $S(\equiv e^x)$ in (81) and re-arranging the order of integration we have

$$C_E(S, v, \tau) = \frac{e^{-r\tau}}{2\pi} \int_{-\infty}^\infty \int_{\ln K}^\infty (e^y - K) e^{i\phi y} \int_0^\infty U(\ln S, v, \tau; -\phi, w) dw dy d\phi. \quad (82)$$

We denote the integral with respect to w as a separate function by setting

$$f_2(S, v, t, \psi) = \int_0^\infty U(\ln S, v, \tau; \psi, w) dw.$$

Applying the transformations given by (73) and (77) we can write

$$f_2(S, v, \tau, \psi) = e^{i\psi \ln S + i\psi(r-q)\tau} \exp\left\{\left[\frac{\alpha(\Theta - \Omega)}{\sigma^2}\right]\tau\right\}$$

$$\times \int_0^\infty e^{-A-y} \exp\left\{-\left(\frac{\Theta - \Omega}{\sigma^2}\right) \frac{\sigma^2(1 - e^{-\Omega\tau})}{2\Omega} A\right\}$$

$$\times \left(\frac{A}{y}\right)^{\frac{\alpha}{\sigma^2} - \frac{1}{2}} \exp\left\{\frac{(\Theta - \Omega)(e^{\Omega\tau} - 1)}{2\Omega} y\right\} I_{\frac{2\alpha}{\sigma^2} - 1}(2\sqrt{Ay}) dA. \quad (83)$$

To undertake the integration with respect to A, we write (83) as,

$$f_2(S, v, \tau, \psi) = e^{i\psi \ln S + i\psi(r-q)\tau} \exp\left\{\left[\frac{(\alpha - \sigma^2)(\Theta - \Omega)}{\sigma^2} + \Theta - \Omega\right]\tau\right\}$$

$$\times e^{-y} \exp\left\{\frac{(\Theta - \Omega)(e^{\Omega\tau} - 1)y}{2\Omega}\right\} J(y),$$

where, by use of (36),

$$J(y) = \sum_{n=0}^{\infty} y^n \int_0^{\infty} \exp\left\{ -\left[\left(\frac{\Theta - \Omega}{2\Omega} \right)(1 - e^{-\Omega\tau}) + 1 \right] A \right\} \frac{A^{n + \frac{2\alpha}{\sigma^2} - 1}}{n! \Gamma\left(\frac{2\alpha}{\sigma^2} + n \right)} dA.$$

Making the change of variable $\xi = [(\Theta - \Omega)(1 - e^{-\Omega\tau})/2\Omega + 1]A$, and using the definition of the gamma function from (34), we have

$$J(y) = \sum_{n=0}^{\infty} \frac{y^n}{n! \Gamma\left(\frac{2\alpha}{\sigma^2} + n \right)} \frac{2\Omega}{(\Theta - \Omega)(1 - e^{-\Omega\tau}) + 2\Omega}$$

$$\times \int_0^{\infty} e^{-\xi} \left(\frac{2\Omega}{(\Theta - \Omega)(1 - e^{-\Omega\tau}) + 2\Omega} \right)^{n + \frac{2\alpha}{\sigma^2} - 1} \xi^{n + \frac{2\alpha}{\sigma^2} - 1} d\xi$$

$$= \sum_{n=0}^{\infty} \frac{y^n}{n!} \left(\frac{2\Omega}{(\Theta - \Omega)(1 - e^{-\Omega\tau}) + 2\Omega} \right)^{n + \frac{2\alpha}{\sigma^2}} \frac{\Gamma\left(\frac{2\alpha}{\sigma^2} + n \right)}{\Gamma\left(\frac{2\alpha}{\sigma^2} + n \right)}.$$

Recalling the Taylor expansion for e^x, we can express $J(y)$ as

$$J(y) = \left(\frac{2\Omega}{(\Theta - \Omega)(1 - e^{-\Omega\tau}) + 2\Omega} \right)^{\frac{2\alpha}{\sigma^2}} \exp\left\{ \frac{2\Omega y}{(\Theta - \Omega)(1 - e^{-\Omega\tau}) + 2\Omega} \right\},$$

and thus $f_2(S, v, \tau, \psi)$ becomes

$$f_2(S, v, \tau, \psi) = e^{i\psi \ln S + i\psi(r-q)\tau} e^{\frac{(\Theta - \Omega)}{\sigma^2}(\alpha\tau + v)} \left(\frac{2\Omega}{(\Theta - \Omega)(1 - e^{-\Omega\tau}) + 2\Omega} \right)^{\frac{2\alpha}{\sigma^2}}$$

$$\times \exp\left\{ \frac{-2\Omega v}{\sigma^2 (e^{\Omega\tau} - 1)} \right\}$$

$$\times \exp\left\{ \frac{2\Omega}{(\Theta - \Omega)(1 - e^{-\Omega\tau}) + 2\Omega} \times \frac{2\Omega v}{\sigma^2 (e^{\Omega\tau} - 1)} \right\}. \tag{84}$$

We now take the inverse Fourier transform of (84). Before we apply the inverse transform, we will rewrite (84) in the form presented by Heston in [13] for ease of comparison.

We begin by seeking to write $f_2(S, v, \tau, \psi)$ in the form

$$f_2(S, v, \tau; \psi) = \exp\{i\psi \ln S + B_2(\psi, \tau) + D_2(\psi, \tau)v\}, \tag{85}$$

where $B(\psi, \tau)$ and $D(\psi, \tau)$ are determined by equating (84) with (85). Thus for $B_2(\psi, \tau)$ we have

$$B_2(\psi, \tau)$$
$$= i\psi(r - q)\tau + \frac{\alpha}{\sigma^2} \left\{ (\Theta_2 - \Omega_2)\tau - 2\ln\left(\frac{(\Theta_2 - \Omega_2)(1 - e^{-\Omega_2\tau}) + 2\Omega_2}{2\Omega_2} \right) \right\},$$

where $\Theta_2 = \Theta_2(\psi) \equiv \Theta(-\psi)$ and $\Omega_2 = \Omega_2(\psi) \equiv \Omega(-\psi)$. Defining $Q_2 = Q_2(\psi) \equiv (\Theta_2 + \Omega_2)/(\Theta_2 - \Omega_2)$, the expression for $B_2(\psi, \tau)$ becomes

$$B_2(\psi, \tau) = i\psi(r - q)\tau + \frac{\alpha}{\sigma^2} \left\{ (\Theta_2 + \Omega_2)\tau - 2\ln\left(\frac{1 - Q_2 e^{\Omega_2\tau}}{1 - Q_2} \right) \right\}. \tag{86}$$

For $D_2(\psi, \tau)$ we find that

$$D_2(\psi, \tau) = \frac{(\Theta_2 - \Omega_2)}{\sigma^2} - \frac{2\Omega_2}{\sigma^2(e^{\Omega_2\tau} - 1)}$$
$$+ \frac{2\Omega_2}{(\Theta_2 - \Omega_2)(1 - e^{-\Omega_2\tau}) + 2\Omega_2} \times \frac{2\Omega_2}{\sigma^2(e^{\Omega_2\tau} - 1)},$$

which simplifies to

$$D_2(\psi, \tau) = \frac{(\Theta_2 + \Omega_2)}{\sigma^2}\left[\frac{1 - e^{\Omega_2\tau}}{1 - Q_2 e^{\Omega_2\tau}}\right]. \tag{87}$$

Taking the inverse Fourier transform of (85), we find that

$$C_E(S, v, \tau) = \frac{e^{-r\tau}}{2\pi}\int_{-\infty}^{\infty}\left(\int_{\ln K}^{\infty}(e^y - K)e^{i\phi y}dy\right)$$
$$\times \exp\{B_2(-\phi, \tau) + D_2(-\phi, \tau)v\}e^{-i\phi \ln S}d\phi$$
$$= \frac{e^{-r\tau}}{2\pi}\left[\int_{-\infty}^{\infty}f_2(S, v, \tau; -\phi)\int_{\ln K}^{\infty}e^y e^{i\phi y}dyd\phi\right.$$
$$\left. - K\int_{-\infty}^{\infty}f_2(S, v, \tau; -\phi)\int_{\ln K}^{\infty}e^{i\phi y}dyd\phi\right]. \tag{88}$$

We can evaluate the integrals in (88) using (95) and (96) from Appendix 8, provided that $f_2(S, v, \tau; -\phi)$ satisfies the appropriate assumptions. The first assumption we must verify is that $f_2(S, v, \tau; \phi - i)$ can be expressed as a function of ϕ. This assumption is satisfied, since

$$f_2(S, v, \tau; \phi - i) = Se^{(r-q)\tau}f_1(S, v, \tau; \phi),$$

where we define

$$f_1(S, v, \tau; \phi) \equiv \exp\{B_1(\phi, \tau) + D_1(\phi, \tau) + i\phi \ln S\},$$

with

$$B_1(\phi, \tau) = i\phi(r - q)\tau + \frac{\alpha}{\sigma^2}\left\{(\Theta_1 + \Omega_1)\tau - 2\ln\left(\frac{1 - Q_1 e^{\Omega_2\tau}}{1 - Q_1}\right)\right\}, \tag{89}$$

$$D_1(\phi, \tau) = \frac{(\Theta_1 + \Omega_1)}{\sigma^2}\left[\frac{1 - e^{\Omega_1\tau}}{1 - Q_1 e^{\Omega_1\tau}}\right], \tag{90}$$

and where $Q_1 = Q_1(\phi) \equiv (\Theta_1 + \Omega_1)/(\Theta_1 - \Omega_1)$, $\Theta_1 = \Theta_1(\phi) \equiv \Theta(i - \phi)$, and $\Omega_1 = \Omega_1(\phi) \equiv \Omega(i - \phi)$.

Furthermore, we can readily show that $f_j(S, v, \tau; -\phi) = \overline{f_j(S, v, \tau; \phi)}$ (here $\bar{f}_j$ denotes the complex conjugate of f_j) for $j = 1, 2$, and therefore all the assumptions required to carry out the calculations (95) and (96) of Appendix 8 are satisfied. Thus (88) becomes

$$C_E(S, v, \tau) = Se^{-q\tau} \left[\frac{1}{2} + \frac{1}{\pi} \int_0^\infty \mathrm{Re}\left(\frac{f_1(S, v, \tau; \phi)e^{-i\phi \ln K}}{i\phi} \right) d\phi \right]$$
$$- Ke^{-r\tau} \left[\frac{1}{2} + \frac{1}{\pi} \int_0^\infty \mathrm{Re}\left(\frac{f_2(S, v, \tau; \phi)e^{-i\phi \ln K}}{i\phi} \right) d\phi \right],$$

which is the result stated in Proposition 8.

Appendix 7: Proof of Proposition 9

Substituting (38) into (21), changing variables to $S = \ln x$ and noting that $V_P(\ln S, v, \tau) = C_P(S, v, \tau)$ we find that

$$C_P(S, v, \tau) = \int_0^\tau e^{-r(\tau-\xi)} \int_0^\infty \int_{\ln b(w,\xi)}^\infty [qe^y - rK]$$
$$\times \left[\frac{1}{2\pi} \int_{-\infty}^\infty e^{i\phi y} g_2(S, v, \tau - \xi; -\phi, w) d\phi \right] dy\,dw\,d\xi,$$

where we define

$$g_2(S, v, \tau; \phi, w) \equiv e^{\frac{(\Theta_2 - \Omega_2)}{\sigma^2}(v - w + \alpha\tau)} e^{i\phi(r-q)\tau} e^{i\phi \ln S}$$
$$\times \frac{2\Omega_2 e^{\Omega_2 \tau}}{\sigma^2(e^{\Omega_2 \tau} - 1)} \left(\frac{we^{\Omega_2 \tau}}{v} \right)^{\frac{\alpha}{\sigma^2} - \frac{1}{2}}$$
$$\times \exp\left\{ \frac{-2\Omega_2}{\sigma^2(e^{\Omega_2 \tau} - 1)}(we^{\Omega_2 \tau} + v) \right\}$$
$$\times I_{\frac{2\alpha}{\sigma^2} - 1}\left(\frac{4\Omega_2}{\sigma^2(e^{\Omega_2 \tau} - 1)}(wve^{\Omega_2 \tau})^{\frac{1}{2}} \right).$$

Thus we have

$$C_P(S, v, \tau)$$
$$= \int_0^\tau e^{-r(\tau-\xi)} \int_0^\infty \frac{1}{2\pi} \left[q \int_{-\infty}^\infty g_2(S, v, \tau - \xi; -\phi, w) \int_{\ln b(w,\xi)}^\infty e^y e^{i\phi y} dy\,d\phi \right.$$
$$\left. - rK \int_{-\infty}^\infty g_2(S, v, \tau - \xi; -\phi, w) \int_{\ln b(w,\xi)}^\infty e^{i\phi y} dy\,d\phi \right] dw\,d\xi, \qquad (91)$$

which can be evaluated using (95) and (96) from Appendix 8. It turns out to be useful to define a function g_1 according to[14]

$$g_2(S, v, \tau - \xi; \phi - i, w) = Se^{(r-q)(\tau-\xi)} g_1(S, v, \tau - \xi; \phi, w),$$

where we define

---

[14] This transformation is analogous to the one between f_1 and f_2 in Appendix 6.

$$g_1(S, v, \tau - \xi; \phi, w) \equiv e^{\frac{(\Theta_1 - \Omega_1)}{\sigma^2}(v - w + \alpha(\tau - \xi))} e^{i\phi(r-q)(\tau-\xi)} e^{i\phi \ln S}$$

$$\times \frac{2\Omega_1 e^{\Omega_1(\tau-\xi)}}{\sigma^2(e^{\Omega_1(\tau-\xi)} - 1)} \left(\frac{w e^{\Omega_1(\tau-\xi)}}{v}\right)^{\frac{\alpha}{\sigma^2} - \frac{1}{2}}$$

$$\times \exp\left\{\frac{-2\Omega_1}{\sigma^2(e^{\Omega_1(\tau-\xi)} - 1)}(w e^{\Omega_1(\tau-\xi)} + v)\right\}$$

$$\times I_{\frac{2\alpha}{\sigma^2} - 1}\left(\frac{4\Omega_1}{\sigma^2(e^{\Omega_1(\tau-\xi)} - 1)}(w v e^{\Omega_1(\tau-\xi)})^{\frac{1}{2}}\right).$$

It is also simple to demonstrate that $g_j(S, v, \tau - \xi; -\phi, w) = \overline{g_j(S, v, \tau - \xi; \phi, w)}$ for $j = 1, 2$. Hence by use of (95) and (96) of Appendix 8, (91) becomes

$$C_P(S, v, \tau) = \int_0^\tau e^{-r(\tau-\xi)} \int_0^\infty [q S e^{(r-q)(\tau-\xi)} P_1^A(S, v, \tau - \xi; w, b(w, \xi))$$
$$- r K P_2^A(S, v, \tau - \xi; w, b(w, \xi))] dw d\xi,$$

where we set

$$P_j^A(S, v, \tau - \xi; w, b(w, \xi))$$
$$= \frac{1}{2} + \frac{1}{\pi} \int_0^\infty \text{Re}\left(\frac{g_j(S, v, \tau - \xi; \phi, w) e^{-i\phi \ln b(w, \xi)}}{i\phi}\right) d\phi,$$

for $j = 1, 2$, which readily simplifies to (44) in Proposition 9.

Appendix 8: The Evaluation of Some Complex Integral Terms Occurring in Appendices 6 and 7

In this appendix we evaluate some integrals involving a complex function g that arise in the calculations in Appendices 6 and 7. First we set up some notation that will be required in the calculations below. For the complex function g define the function h such that,

$$h(\phi) \equiv g(\phi - i).$$

It is not difficult to show that for the given form of the function h there holds $h(-\phi) = \overline{h(\phi)}$. The integral terms that we need to consider are of the general form (see (88))

$$J_1^H = \frac{1}{2\pi} \int_{-\infty}^\infty g(-\phi) \int_a^\infty e^y e^{i\phi y} dy d\phi, \tag{92}$$

and

$$J_2^H = \frac{1}{2\pi} \int_{-\infty}^\infty g(-\phi) \int_a^\infty e^{i\phi y} dy d\phi. \tag{93}$$

Beginning with J_1^H we have

$$J_1^H = \frac{1}{2\pi} \int_{-\infty}^{\infty} g(-\phi) \int_a^{\infty} e^{i(\phi-i)y} dy d\phi$$

$$= \frac{1}{2\pi} \int_{-\infty}^{\infty} g(-\psi - i) \int_a^{\infty} e^{i\psi y} dy d\psi$$

$$= \frac{1}{2\pi} \int_{-\infty}^{\infty} g(\phi - i) \int_a^{\infty} e^{-i\phi y} dy d\phi$$

$$= \frac{1}{2\pi} \int_{-\infty}^{\infty} g(\phi - i) \left[\lim_{b \to \infty} \frac{e^{-i\phi a} - e^{-i\phi b}}{i\phi} \right] d\phi.$$

Next we rewrite J_1^H as

$$J_1^H = \frac{1}{2\pi} \lim_{b \to \infty} \left[\int_0^{\infty} g(\phi - i) \left(\frac{e^{-i\phi a} - e^{-i\phi b}}{i\phi} \right) d\phi \right.$$

$$\left. + \int_0^{\infty} g(-\phi - i) \left(\frac{e^{i\phi a} - e^{i\phi b}}{-i\phi} \right) d\phi \right]$$

$$= \frac{1}{2\pi} \int_0^{\infty} \frac{g(\phi - i)e^{-i\phi a} - g(-\phi - i)e^{i\phi a}}{i\phi} d\phi$$

$$- \frac{1}{2\pi} \lim_{b \to \infty} \left[\int_0^{\infty} \frac{g(\phi - i)e^{-i\phi b} - g(-\phi - i)e^{i\phi b}}{i\phi} d\phi \right]. \qquad (94)$$

Referring to [21] we can evaluate the integral terms in (94) by way of the relation[15]

$$F(x) = \frac{1}{2} - \frac{1}{2\pi} \int_0^{\infty} \frac{f(\phi)e^{-i\phi x} - f(-\phi)e^{i\phi x}}{i\phi} d\phi,$$

where $F(x)$ denotes a cumulative density function. Defining $h(\phi) \equiv g(\phi - i)$, we can show that

$$\lim_{b \to \infty} \frac{1}{2\pi} \int_0^{\infty} \frac{g(\phi - i)e^{-i\phi b} - g(-\phi - i)e^{i\phi b}}{i\phi} d\phi$$

$$= \lim_{b \to \infty} \frac{1}{2\pi} \int_0^{\infty} \frac{h(\phi)e^{-i\phi b} - h(-\phi)e^{i\phi b}}{i\phi} d\phi = \lim_{b \to \infty} \left(\frac{1}{2} - F(b) \right) = -\frac{1}{2}.$$

Hence (94) becomes

$$J_1^H = \frac{1}{2} + \frac{1}{2\pi} \int_0^{\infty} \frac{h(\phi)e^{-i\phi a} - h(-\phi)e^{i\phi a}}{i\phi} d\phi.$$

Finally, given that $h(\phi)$ and $h(-\phi)$ are complex conjugates, so that $h(-\phi) = \overline{h(\phi)}$, it follows that,

$$\frac{h(-\phi)e^{i\phi a}}{-i\phi} = \overline{\left(\frac{h(\phi)e^{-i\phi a}}{i\phi} \right)},$$

---

[15] The function F is defined by $F(x) = \int_{-\infty}^{x} f(u) du$

and thus we find that

$$J_1^H = \frac{1}{2} + \frac{1}{\pi} \int_0^\infty \mathrm{Re}\left(\frac{h(\phi)e^{-i\phi a}}{i\phi}\right) d\phi. \tag{95}$$

For J_2^H, we make the change of variable let $\psi = -\phi$ to produce

$$J_2^H = \frac{1}{2\pi} \int_{-\infty}^\infty g(\psi) \int_a^\infty e^{-i\psi y} dy d\psi,$$

and similarly we can show that

$$J_2^H = \frac{1}{2} + \frac{1}{\pi} \int_0^\infty \mathrm{Re}\left(\frac{g(\phi)e^{-i\phi a}}{i\phi}\right) d\phi, \tag{96}$$

where we assume that $g(-\phi) = \overline{g(\phi)}$.

References

1. Abramowitz, M., Stegun, I.A.: Handbook of Mathematical Functions. Dover, New York (1970)
2. Adolfsson, T., Chiarella, C., Ziogas, A., Ziveyi, J.: The numerical approximation of American option prices under Heston stochastic volatility dynamics. QFRC Working Paper Series. The University of Tehcnology, Sydney (2010)
3. Bates, D.S.: Jumps and stochastic volatility: Exchange rate processes implicit in deutsche mark options. Rev. Financ. Stud. **9**, 69–107 (1996)
4. Black, F., Scholes, M.: The pricing of corporate liabilities. J. Polit. Econ. **81**, 637–659 (1973)
5. Carr, P., Jarrow, R., Myneni, R.: Alternative characterizations of American put options. Math. Finance **2**, 87–106 (1992)
6. Carr, P., Madan, D.: Option pricing and the fast Fourier transform. J. Comput. Finance **2**(4), 61–73 (1999)
7. Cheang, G., Chiarella, C., Ziogas, A.: The representation of American options prices under stochastic volatility and jump-diffusion dynamics. QFRC Working Paper Series, No. 256. The University of Technology, Sydney (2009)
8. Chiarella, C., Ziogas, A.: Evaluation of American strangles. J. Econ. Dyn. Control **29**, 31–62 (2005)
9. Chiarella, C., Ziogas, A., Ziveyi, J.: American option prices under stochastic volatility dynamics; A representation. QFRC Working Paper Series. The University of Tehcnology, Sydney (2010)
10. Detemple, J., Tian, W.: The valuation of American options for a class of diffusion processes. Manag. Sci. **48**(7), 917–937 (2002)
11. Duffy, D.G.: Green's Functions with Applications. CRC Press, Boca Raton (2001)
12. Feller, W.: Two singular diffusion problems. Ann. Math. **54**, 173–182 (1951)
13. Heston, S.: A closed-form solution for options with stochastic volatility with applications to bond and currency options. Rev. Financ. Stud. **6**, 327–343 (1993)
14. Jacka, S.D.: Optimal stopping and the American put. Math. Finance **1**, 1–14 (1991)
15. Jamshidian, F.: An analysis of American options. Rev. Futures Mark. **11**, 72–80 (1992)
16. Karatzas, I.: On the pricing of American options. Appl. Math. Optim. **17**, 37–60 (1988)
17. Kim, I.J.: The analytic valuation of American options. Rev. Financ. Stud. **3**, 547–572 (1990)
18. Lewis, A.L.: Option Valuation Under Stochastic Volatility. Finance Press, California (2000)
19. Logan, D.: Applied Partial Differential Equations. Springer, New York (2004)

20. McKean, H.P.: Appendix: A free boundary value problem for the heat equation arising from a problem in mathematical economics. Ind. Manag. Rev. **6**, 32–39 (1965)
21. Shephard, N.G.: From characteristic function to distribution function: A simple framework for the theory. Econom. Theory **7**, 519–529 (1991)
22. Touzi, N.: American options exercise boundary when the volatility changes randomly. Appl. Math. Optim. **39**, 411–422 (1999)
23. Tzavalis, E., Wang, S.: Pricing American options under stochastic volatility: A new method using Chebyshev polynomials to approximate the early exercise boundary. Working Paper, No. 488. Department of Economics, Queen Mary, University of London (2003)

Buy Low and Sell High

Min Dai, Hanqing Jin, Yifei Zhong,
and Xun Yu Zhou

Abstract In trading stocks investors naturally aspire to "buy low and sell high (BLSH)". This paper formalizes the notion of BLSH by formulating stock buying/selling in terms of four optimal stopping problems involving the global maximum and minimum of the stock prices over a given investment horizon. Assuming that the stock price process follows a geometric Brownian motion, all the four problems are solved and buying/selling strategies completely characterized via a free-boundary PDE approach.

Min Dai is an affiliated member of Risk Management Institute of NUS.

M. Dai
Department of Mathematics, National University of Singapore (NUS), Singapore, Singapore
e-mail: matdm@nus.edu.sg

H. Jin · Y. Zhong
Mathematical Institute and Nomura Centre for Mathematical Finance, and Oxford-Man Institute
of Quantitative Finance, The University of Oxford, 24–29 St Giles, Oxford OX1 3LB, UK

H. Jin
e-mail: jinh@maths.ox.ac.uk

Y. Zhong
e-mail: yifei.zhong@maths.ox.ac.uk

X.Y. Zhou (✉)
Mathematical Institute and Nomura Centre for Mathematical Finance, and Oxford-Man Institute
of Quantitative Finance, The University of Oxford, 24–29 St Giles, Oxford OX1 3LB, UK
e-mail: zhouxy@maths.ox.ac.uk

X.Y. Zhou
Department of Systems Engineering and Engineering Management, The Chinese University of
Hong Kong, Shatin, Hong Kong

C. Chiarella, A. Novikov (eds.), *Contemporary Quantitative Finance*,
DOI 10.1007/978-3-642-03479-4_16, © Springer-Verlag Berlin Heidelberg 2010

1 Introduction

Assume that a discounted stock price, S_t, evolves according to

$$dS_t = \mu S_t dt + \sigma S_t dB_t,$$

where constants $\mu \in (-\infty, +\infty)$ and $\sigma > 0$ are the excess rate of return and the volatility rate, respectively, and $\{B_t; t > 0\}$ is a standard 1-dimension Brownian motion on a filtered probability space $(\mathbb{S}, \mathscr{F}, \{\mathscr{F}_t\}_{t \geq 0}, \mathbb{P})$ with $B_0 = 0$ almost surely.

We are interested in the following optimal decisions to buy or sell the stock over a given investment horizon $[0, T]$:

$$\text{Buying:} \quad \min_{\tau \in \mathscr{T}} \mathbb{E}\left(\frac{S_\tau}{M_T}\right), \tag{1}$$

$$\min_{\tau \in \mathscr{T}} \mathbb{E}\left(\frac{S_\tau}{m_T}\right); \tag{2}$$

$$\text{Selling:} \quad \max_{\tau \in \mathscr{T}} \mathbb{E}\left(\frac{S_\tau}{M_T}\right), \tag{3}$$

$$\max_{\tau \in \mathscr{T}} \mathbb{E}\left(\frac{S_\tau}{m_T}\right); \tag{4}$$

where $\mathbb{E}$ stands for the expectation, $\mathscr{T}$ is the set of all $\mathscr{F}_t$-stopping time $\tau \in [0, T]$, and M_T and m_T are respectively the global maximum and minimum of the stock price on $[0, T]$, namely,

$$\begin{cases} M_T = \max_{0 \leq v \leq T} S_v, \\ m_T = \min_{0 \leq v \leq T} S_v. \end{cases} \tag{5}$$

Some discussions on the motivations of the above problems are in order. When an investor trades a stock she naturally hopes to "buy low and sell high (BLSH)". Since one could never be able to "buy at the lowest and sell at the highest", there could be different interpretations on the maxim BLSH depending on the meanings of the "low" and the "high". Problems (1)–(4) make precise these in terms of optimal stopping (timing). Specifically, Problem (1) is equivalent to

$$\max_{\tau \in \mathscr{T}} \mathbb{E}\left(\frac{M_T - S_\tau}{M_T}\right),$$

i.e., the investor attempts to *maximize* the expected *relative* error between the buying price and the highest possible stock price by choosing a proper time to buy. This is motivated by a typical investment sentiment that an investor wants to stay away, as far as possible, from the highest price when she is buying. Similarly, Problem (2) is to *minimize* the expected relative error between the buying price and the lowest possible stock price. In the same spirit, Problem (3) (resp. (4)) is to minimize (resp. maximize) the expected relative error between the selling price and the highest (resp. lowest) possible stock price when selling. Therefore, Problems (1)–(4) capture BLSH from various perspectives.

Optimal stopping problems involving the global maximum or minimum of the stock prices were formulated for the first time by Shiryaev in [6]. Optimal stock trading involving relative error was first formulated and solved in [8], where Problem (3) was investigated using a purely probabilistic approach. There is, however, a case that remains unsolved in [8].[1] In this paper, we will take a PDE approach, which enables us to solve not only all the cases associated with Problem (3), but also all the other problems (1), (2) and (4) simultaneously. Moreover, we will derive completely the buying/selling regions for all the problems, leading to optimal *feedback* trading strategies, ones that would respond to all the scenarios in time and in (certain) well-defined states (rather than to the ones at $t = 0$ only).

The results derived from our models are quite intuitive and consistent with the common investment practice. For the two buying problems (1) and (2), our results dictate that one ought to buy immediately or never buy depending on whether the underlying stock is "good" or "bad" (which will be defined precisely). If, on the other hand, the stock is intermediate between good and bad, then one should buy if and only if either the current stock price is sufficiently cheap compared with the historical high (for Problem (1)) or it has sufficiently risen from the historical low (for Problem (2)). For the selling problems (3) and (4), one should never sell (i.e. hold until the final date) or immediately sell depending, again, on whether the stock is "good" or "bad" (which will however be defined differently from the buying problems). If the stock is in between good and bad, then one should sell if and only if either the current stock price is sufficiently close to the historical high (for Problem (3)) or it is sufficiently close to the historical low (for Problem (4)). In particular, at time $t = 0$ the stock price is trivially both historical high and low; hence the selling strategy would be to sell at $t = 0$, suggesting that this intermediate case is "bad" after all.

The remainder of the paper is organized as follows. In Sect. 2 we give mathematical preliminaries needed in solving the four problems, and in Sect. 3 we present the main results. Some concluding remarks are given in Sect. 4, while the proofs are relegated to an appendix.

2 Preliminaries

As they stand Problems (1)–(4) are not standard optimal stopping time problems since they all involve the global maximum and minimum of a stochastic process which are not adapted. In this section we first turn these problems into standard ones, and then present the corresponding free-boundary PDEs for solving them. We do these in two sub-sections for buying and selling respectively.

---

[1]It is argued in [7] and [8] that this missing case is economically insignificant. The case is subsequently covered by du Toit and Peskir in [4] employing the same probabilistic approach.

2.1 Buying Problems

We start with the buying problem (1). Denote by M_t the running maximum stock price over $[0, t]$, i.e., $M_t = \max_{0 \le v \le t} S_v$. Then, we have

$$
\mathbb{E}\left(\frac{S_\tau}{M_T}\right) = \mathbb{E}\left(\frac{S_\tau}{\max\{M_\tau, \max_{\tau \le s \le T} S_s\}}\right) = \mathbb{E}\left(\max\left\{\frac{M_\tau}{S_\tau}, \max_{\tau \le s \le T} \frac{S_s}{S_\tau}\right\}\right)^{-1}
$$

$$
= \mathbb{E}\left[\left[\mathbb{E}\left(\max\left\{\frac{M_\tau}{S_\tau}, \max_{\tau \le s \le T} \frac{S_s}{S_\tau}\right\}\right)^{-1}\Big| \mathscr{F}_\tau\right]\right]
$$

$$
= \mathbb{E}\left[\mathbb{E}\left[\min\left\{e^{-x}, e^{-\max_{t \le s \le T}\{(\mu - \frac{\sigma^2}{2})(s-t) + \sigma B_{(s-t)}\}}\right\}\right]\Big|_{x = \log \frac{M_\tau}{S_\tau}, \, t = \tau}\right]
$$

$$
= \mathbb{E}\left[\Psi\left(\log \frac{M_\tau}{S_\tau}, \tau\right)\right] \tag{6}
$$

where

$$
\Psi(x, t) = \mathbb{E}\left[\min\left\{e^{-x}, e^{-\max_{t \le s \le T}\{(\mu - \frac{\sigma^2}{2})(s-t) + \sigma B_{(s-t)}\}}\right\}\right], \quad \forall (x, t) \in \tilde{\Omega}^+,
$$

and $\Omega^+ = (0, +\infty) \times [0, T)$, $\tilde{\Omega}^+ = \Omega^+ \cup \{x = 0\}$.

The expression of $\Psi(x, t)$ is as follows [cf. Shiryaev, Xu and Zhou in [8]]:

$$
\Psi(x, t) = \begin{cases}
\dfrac{3\sigma^2 - 2\mu}{2(\sigma^2 - \mu)} e^{(\sigma^2 - \mu)(T-t)} \Phi(d_1) + e^{-x}\Phi(d_2) & \\
\quad + \dfrac{\sigma^2}{2(\mu - \sigma^2)} e^{\frac{2(\mu - \sigma^2)x}{\sigma^2}} \Phi(d_3) & \text{if } \mu \ne \sigma^2, \\
e^{-x}\Phi(d_2) + (1 + x + \dfrac{\sigma^2(T-t)}{2})\Phi(d_3) & \\
\quad - \sigma\sqrt{\dfrac{T-t}{2\pi}} e^{-\frac{d_3^2}{2}} & \text{if } \mu = \sigma^2,
\end{cases} \tag{7}
$$

where $d_1 = \dfrac{-x + (\mu - \frac{3}{2}\sigma^2)(T-t)}{\sigma\sqrt{T-t}}$, $d_2 = \dfrac{x - (\mu - \frac{1}{2}\sigma^2)(T-t)}{\sigma\sqrt{T-t}}$, $d_3 = \dfrac{-x - (\mu - \frac{1}{2}\sigma^2)(T-t)}{\sigma\sqrt{T-t}}$, $\Phi(\cdot) = \int_{-\infty}^{\cdot} \frac{1}{\sqrt{2\pi}} e^{-\frac{s^2}{2}} ds$.

Equation (6) implies that (1) is equivalent to a standard optimal stopping problem with a terminal payoff Ψ and an underlying adapted state process

$$
X_t = \log \frac{M_t}{S_t}, \quad X_0 = 0.
$$

In view of the dynamic programming approach, we need to consider the following problem

$$
V(x, t) \doteq \min_{0 \le \tau \le T - t} \mathbb{E}_{t,x}\left(\Psi(X_{\tau+t}, \tau + t)\right), \tag{8}
$$

where $X_t = x$ under $\mathbb{P}_{t,x}$ with $(x, t) \in \Omega^+$ given and fixed. Obviously, the original problem is $V(0, 0) = \min_{0 \le \tau \le T} \mathbb{E}(\frac{S_\tau}{M_T})$.

It is a standard exercise to show that $V(\cdot, \cdot)$, the value function, satisfies the following free-boundary PDE (also known as the variational inequalities)

$$\begin{cases} \max\{\mathscr{L}V, V - \Psi\} = 0, & \text{in } \Omega^+, \\ V_x(0, t) = 0, & V(x, T) = \Psi(x, T), \end{cases} \tag{9}$$

where the operator $\mathscr{L}$ is defined by

$$\mathscr{L} = -\partial_t - \frac{\sigma^2}{2}\partial_{xx} - \left(\frac{1}{2}\sigma^2 - \mu\right)\partial_x. \tag{10}$$

Therefore, the buying region for Model (1) is

$$BR = \{(x, t) \in \widetilde{\Omega^+} : V(x, t) = \Psi(x, t)\}. \tag{11}$$

Let us now turn to the alternative buying model (2). Denote by m_t the running minimum stock price over $[0, t]$, i.e., $m_t = \min_{0 \leq v \leq t} S_v$. Then, we have

$$\mathbb{E}\left(\frac{S_\tau}{m_T}\right) = \mathbb{E}\left(\frac{S_\tau}{\min\{m_\tau, \min_{\tau \leq s \leq T} S_s\}}\right) = \mathbb{E}\left(\min\left\{\frac{m_\tau}{S_\tau}, \min_{\tau \leq s \leq T}\frac{S_s}{S_\tau}\right\}\right)^{-1}$$

$$= \mathbb{E}\left[\left[\mathbb{E}\left(\min\left\{\frac{m_\tau}{S_\tau}, \min_{\tau \leq s \leq T}\frac{S_s}{S_\tau}\right\}\right)^{-1}\bigg|\mathscr{F}_\tau\right]\right]$$

$$= \mathbb{E}\left[\mathbb{E}\left[\max\left\{e^{-x}, e^{-\min_{\tau \leq s \leq T}\{(\mu - \frac{\sigma^2}{2})(s-t) + \sigma B_{(s-t)}\}}\right\}\right]\bigg|_{x = \log\frac{m_\tau}{S_\tau}, t = \tau}\right]$$

$$= \mathbb{E}\left[\psi\left(\log\frac{m_\tau}{S_\tau}, \tau\right)\right]$$

where

$$\psi(x, t) = \mathbb{E}\left[\max\left\{e^{-x}, e^{-\min_{t \leq s \leq T}\{(\mu - \frac{\sigma^2}{2})(s-t) + \sigma B_{(s-t)}\}}\right\}\right], \quad \forall(x, t) \in \widetilde{\Omega}^-,$$

and $\Omega^- = (-\infty, 0) \times [0, T)$, $\widetilde{\Omega}^- = \Omega^- \cup \{x = 0\}$.

Similar to (7), we can find the expression of $\psi(x, t)$, $\forall(x, t) \in \Omega^-$, as follows

$$\psi(x, t) = \begin{cases} \frac{3\sigma^2 - 2\mu}{2(\sigma^2 - \mu)}e^{(\sigma^2 - \mu)(T-t)}\Phi(d_1') + e^{-x}\Phi(d_2') \\ \quad + \frac{\sigma^2}{2(\mu - \sigma^2)}e^{\frac{2(\mu - \sigma^2)x}{\sigma^2}}\Phi(d_3') & \text{if } \mu \neq \sigma^2, \\ e^{-x}\Phi(d_2') + (1 + x + \frac{\sigma^2(T-t)}{2})\Phi(d_3') \\ \quad - \sigma\sqrt{\frac{T-t}{2\pi}}e^{-\frac{(d_3')^2}{2}} & \text{if } \mu = \sigma^2, \end{cases} \tag{12}$$

where $d_1' = \frac{x - (\mu - \frac{3}{2}\sigma^2)(T-t)}{\sigma\sqrt{T-t}}$, $d_2' = \frac{-x + (\mu - \frac{1}{2}\sigma^2)(T-t)}{\sigma\sqrt{T-t}}$, $d_3' = \frac{x + (\mu - \frac{1}{2}\sigma^2)(T-t)}{\sigma\sqrt{T-t}}$.

Thus, we define the associated value function as

$$v(x, t) = \min_{0 \leq \tau \leq T - t} \mathbb{E}_{t,x}\left(\psi(X_{\tau+t}, \tau + t)\right),$$

where $X_t = \log \frac{m_t}{S_t} = x$ under $\mathbb{P}_{t,x}$ with $(x,t) \in \Omega^-$ given and fixed. The variational inequalities that $v(x,t)$ satisfies are given as follows

$$\begin{cases} \max\{\mathscr{L}v, \ v - \psi\} = 0, & \text{in } \Omega^-, \\ v_x(0,t) = 0, & v(x,T) = \psi(x,T), \end{cases} \tag{13}$$

where $\mathscr{L}$ is defined by (10).

The buying region for Model (2) is, therefore

$$BR = \{(x,t) \in \widetilde{\Omega^-} : v(x,t) = \psi(x,t)\}, \tag{14}$$

where $\widetilde{\Omega^-} = \Omega^- \cup \{x = 0\}$.

2.2 Selling Problems

We now consider the selling problems. In a similar manner, we introduce the value function associated with problem (3) as follows:

$$U(x,t) \doteq \max_{0 \le \tau \le T-t} \mathbb{E}(\Psi(X_{\tau+t}, \tau+t)), \quad \forall (x,t) \in \Omega^+.$$

It is also easy to see that U satisfies

$$\begin{cases} \min\{\mathscr{L}U, U - \Psi\} = 0, & \text{in } \Omega^+, \\ U_x(0,t) = 0, & U(x,T) = \Psi(x,T), \end{cases} \tag{15}$$

where $\mathscr{L}$ and Ψ are as given in (10) and (7) respectively. The corresponding selling region is

$$SR = \{(x,t) \in \widetilde{\Omega^+} : U(x,t) = \Psi(x,t)\}. \tag{16}$$

For Problem (4), we introduce the value function

$$u(x,t) \doteq \max_{0 \le \tau \le T-t} \mathbb{E}(\psi(X_{\tau+t}, \tau+t)), \quad \forall (x,t) \in \Omega^-.$$

It is easy to see that u satisfies

$$\begin{cases} \min\{\mathscr{L}u, u - \psi\} = 0, & \text{in } \Omega^-, \\ u_x(0,t) = 0, & u(x,T) = \psi(x,T), \end{cases} \tag{17}$$

where $\mathscr{L}$ and ψ are as given in (10) and (12) respectively. The corresponding selling region is as follows:

$$SR = \{(x,t) \in \widetilde{\Omega^-} : u(x,t) = \psi(x,t)\}. \tag{18}$$

3 Optimal Buying and Selling Strategies

In this section we present the main results of our paper. Again, we divide the section into two sub-sections dealing with the buying and selling decisions respectively.

The following "goodness index" of the stock is crucial in defining whether the stock is "good", "bad" or "intermediate" for the four problems under consideration:

$$\alpha = \frac{\mu}{\sigma^2}.$$

3.1 Buying Strategies

The following result characterizes the optimal buying strategy for Problem (1).

Theorem 1 (Optimal Buying Strategies against the Highest Price) *Let BR be the buying region as defined in* (11).

i) *If* $\alpha \leq 0$, *then* $BR = \emptyset$;

ii) *If* $0 < \alpha < 1$, *then there is a monotonically decreasing boundary* $x_b^*(t)$: $[0, T) \to (0, +\infty)$ *such that*

$$BR = \{(x, t) \in \Omega^+ : x \geq x_b^*(t), 0 \leq t < T\}. \tag{19}$$

Moreover, $\lim_{t \to T^-} x_b^*(t) = 0$, *and*

$$\lim_{T-t \to \infty} x_b^*(t) = \begin{cases} \frac{8(1-\alpha)}{(3-2\alpha)(2\alpha-1)} & \text{if } \frac{1}{2} < \alpha < 1, \\ +\infty & \text{if } 0 < \alpha \leq \frac{1}{2}; \end{cases} \tag{20}$$

iii) *If* $\alpha \geq 1$, *then* $BR = \widetilde{\Omega^+}$.

We place the proof in Appendix.

So, if the buying criterion is to stay away as much as possible from the highest price, then one should never buy if the stock is "bad" ($\alpha \leq 0$), and immediately buy if the stock is "good" ($\alpha \geq 1$). If the stock is somewhat between good and bad ($0 < \alpha < 1$), then one should buy as soon as the ratio between the historical high and the current stock price, M_t/S_t, exceeds certain time-dependent level (or equivalently the current stock price is sufficiently – depending on when the time is – away in proportion from the historical high). Moreover, in this case when the terminal date is sufficiently near one should always buy.

The other buying problem (2) is solved in the following theorem.

Theorem 2 (Optimal Buying Strategies against the Lowest Price) *Let BR be the buying region as defined in* (14).

i) *If* $\alpha \leq 0$, *then* $BR = \emptyset$;

ii) *If* $0 < \alpha < 1$, *then there is a monotonically increasing boundary* $x_b^*(t) : [0, T) \to (-\infty, 0)$ *such that*

$$BR = \{(x, t) \in \Omega^- : x \leq x_b^*(t), 0 \leq t < T\}. \tag{21}$$

Moreover, $\lim_{t \to T^-} x_b^*(t) = 0$, *and*

$$\lim_{T-t \to \infty} x_b^*(t) = -\infty.$$

iii) *If $\alpha \geq 1$, then $BR = \widetilde{\Omega^-}$.*

We place the proof in Appendix.

The above results suggest that, if the buying criterion is to buy at a price as close to the lowest price as possible, then one should never buy if the stock is "bad" ($\alpha \leq 0$), and immediately buy if the stock is "good" ($\alpha \geq 1$). If the stock is intermediate between good and bad ($0 < \alpha < 1$), then one should buy as soon as the ratio between the current price and the historical low, S_t/m_t, exceeds certain time-dependent level (or the current stock price is sufficiently away in proportion from the historical low). Moreover, in this case when the terminal date is sufficiently near one should always buy. On the other hand, one tends not to buy if the duration of the investment horizon is exceedingly long.

Comparing Theorems 1 and 2, we find that the two buying models (1) and (2), albeit different in formulation, produce quite similar trading behaviors. The only difference lies in the (endogenous) criteria (those in terms of M_t/S_t and S_t/m_t) to be used to trigger buying for the intermediate case $0 < \alpha < 1$.

3.2 Selling Problems

The first selling problem (3) is solved in the following theorem.

Theorem 3 (Optimal Selling Strategies against the Highest Price) *Let SR be the optimal selling region as defined in* (16).

i) *If $\alpha > \frac{1}{2}$, then $SR = \emptyset$.*

ii) *If $\alpha = \frac{1}{2}$, then $SR = \{x = 0\}$.*

iii) *If $0 < \alpha < \frac{1}{2}$, then $\{x = 0\} \subset SR$. Moreover, there exists a boundary $x_s^*(t)$:*
$[0, T) \to [0, +\infty)$ *such that*

$$SR = \left\{ (x, t) \in \widetilde{\Omega^+} : x \leq x_s^*(t) \right\}, \tag{22}$$

and $x_s^*(t) \leq x_b^*(t) < \infty$, where $x_b^*(t)$ is defined in (19).

iv) *If $\alpha \leq 0$, then $SR = \widetilde{\Omega^+}$.*

The proof is placed in Appendix.

The above results indicate that, apart from the definitely holding case $\alpha > \frac{1}{2}$ and the immediately selling case $\alpha \leq 0$, there is an intermediate case $0 < \alpha \leq \frac{1}{2}$ where one should sell only when M_t/S_t is sufficiently small. However, at $t = 0$ this quantity is automatically the smallest; hence one should also sell at $t = 0$ if $0 < \alpha \leq \frac{1}{2}$. This therefore fills the gap case $0 < \alpha \leq \frac{1}{2}$ missing in [8].

On the other hand, extensive numerical results have shown consistently that $x_s^*(t)$ is monotonically decreasing and $x_s^*(t) > 0$ when $0 < \alpha < \frac{1}{2}$, although we are yet to be able to establish these analytically. Moreover, $\lim_{T-t \to \infty} x_s^*(t) = \frac{1}{2\alpha-1} \log(1 - (2\alpha - 1)^2)$, which can be shown as (20), is confirmed by our numerical results.

The following is on Problem (4).

Theorem 4 (Optimal Selling Strategies against the Lowest Price) *Let SR be the optimal selling region as defined in* (18).

i) *If* $\alpha > \frac{1}{2}$, *then* $SR = \emptyset$.

ii) *If* $\alpha = \frac{1}{2}$, *then* $SR = \{x = 0\}$.

iii) *If* $0 < \alpha < \frac{1}{2}$, *then* $\{x = 0\} \subset SR$. *Moreover, there exists a boundary* $x_s^*(t)$: $[0, T) \to (-\infty, 0]$ *such that*

$$SR = \left\{ (x, t) \in \widetilde{\Omega^-} : x \geq x_s^*(t) \right\}, \tag{23}$$

and $x_s^*(t) \geq x_b^*(t) > -\infty$, *where* $x_b^*(t)$ *is defined in* (21).

iv) *If* $\alpha \leq 0$, *then* $SR = \widetilde{\Omega^-}$.

The proof is the same as that for Theorem 3. We omit it here.

The trading behavior derived from this model is virtually the same as the other selling model at time $t = 0$.

Incidentally, numerical results show that $x_s^*(t)$ is monotonically increasing. However, it is an open problem to establish $\lim_{T-t \to \infty} x_s^*(t)$ since the solution to the stationary problem is not unique.

4 Concluding Remarks

In this paper four stock buying/selling problems are formulated as optimal stopping problems so as to capture the investment motto "buy low and sell high". The free boundary PDE approach, as opposed to the probabilistic approach taken by Shiryaev, Xu and Zhou in [8], is employed to solve all the problems thoroughly. The optimal trading strategies derived are simple and consistent with the normal investment behaviors. For buying problems, apart from the straightforward extreme cases (depending on the quality of the underlying stock) where one should always or never buy, one ought to buy so long as the stock price has declined sufficiently from the historical high or risen sufficiently from the historical low. For selling problems, optimal strategies exhibit similar (or indeed opposite) patterns.

The paper certainly (or at least we hope to) suggest more problems than solutions. An immediate question would be to extend the geometric Brownian stock price to more complex and realistic processes.

Acknowledgements Dai is partially supported by Singapore MOE AcRF grant (No. R-146-000-096-112) and NUS RMI grant (No. R-146-000-124-720/646). Zhou acknowledges financial support from the Nomura Centre for Mathematical Finance and a start-up fund of the University of Oxford, and Jin, Zhong and Zhou acknowledge research grants from the Oxford–Man Institute of Quantitative Finance.

Appendix: Proofs

Some Transformations

Before proving the results in the paper, let us present some transformations, first introduced by Dai and Zhong in [3], which play a critical role in our analysis:

$$F(x,t) = \log(\Psi(x,t)), \qquad \overline{V}(x,t) = \log\left(\frac{V(x,t)}{\Psi(x,t)}\right), \qquad \text{and}$$

$$\overline{U}(x,t) = \log\left(\frac{U(x,t)}{\Psi(x,t)}\right), \tag{24}$$

and introduce two lemmas which are useful for both buying and selling cases. Without loss of generality, we assume that $\sigma = 1$, as one could make a change of time $\bar{t} = \sigma^2 t$ if otherwise. Thus, we use α instead of μ in the following.

A direct calculation [cf. Shiryaev, Xu and Zhou in [8]] shows that $\Psi(x,t)$ satisfies

$$\begin{cases} \mathcal{L}\Psi = \Psi_x + (1-\alpha)\Psi, & \text{in } \Omega^+, \\ \Psi_x(0,t) = 0, & \Psi(x,T) = e^{-x}. \end{cases} \tag{25}$$

Then, $F(x,t)$ satisfies

$$\begin{cases} -F_t - \frac{1}{2}(F_{xx} + F_x^2) + (\alpha - \frac{3}{2})F_x + (\alpha - 1) = 0, & \text{in } \Omega^+, \\ F_x(0,t) = 0, & F(x,T) = -x. \end{cases} \tag{26}$$

Accordingly, (9) and (15) reduce to

$$\begin{cases} \max\{\mathcal{L}_0\overline{V} + F_x - (\alpha - 1), \overline{V}\} = 0, & \text{in } \Omega^+, \\ \overline{V}_x(0,t) = 0, & \overline{V}(x,T) = 0, \end{cases} \tag{27}$$

and

$$\begin{cases} \min\{\mathcal{L}_0\overline{U} + F_x - (\alpha - 1), \overline{U}\} = 0, & \text{in } \Omega^+, \\ \overline{U}_x(0,t) = 0, & \overline{U}(x,T) = 0, \end{cases} \tag{28}$$

respectively, where $\mathcal{L}_0 = -\partial_t - \frac{1}{2}[\partial_{xx} + (\partial_x)^2 + 2F_x\partial_x] + (\alpha - \frac{1}{2})\partial_x$.

As a result, BR (11) and SR (16) can be rewritten as

$$BR = \{(x,t) \in \widetilde{\Omega^+} : \overline{V} = 0\} \quad \text{and} \quad SR = \{(x,t) \in \widetilde{\Omega^+} : \overline{U} = 0\}.$$

Lemma 1 *Let $\overline{V}(x,t)$ and $\overline{U}(x,t)$ be the solutions to Problems (27) and (28), respectively. Then,*

$$\overline{V}(x,t) < \overline{U}(x,t) \text{ in } \Omega^+.$$

Proof It is easy to see that $\mathcal{L}_0\overline{V} \leq \mathcal{L}_0\overline{U}$ in Ω^+. Applying the strong maximum principle gives the result. □

Lemma 2 *Suppose* $F(x,t)$ *is the solution to* (26). *Then* $F(x,t)$ *has the following properties*:

i) $-1 < F_x(x,t) < 0, \forall (x,t) \in \Omega^+$;
ii) $F_{xx}(x,t) \leq 0, \forall (x,t) \in \Omega^+$;
iii) $F_{xt}(x,t) \leq 0, \forall (x,t) \in \Omega^+$;
iv) $F_x(x,t;\alpha+\delta) \leq F_x(x,t;\alpha)+\delta, \forall \delta > 0, (x,t) \in \Omega^+$, *where* $F(x,t;\beta)$ *denotes* $F(x,t)$ *corresponding to the goodness index* β;
v) $\lim_{x\to\infty} F_x(x,t) = -1, \forall t \in [0,T)$.

Proof Denote $\widetilde{F}(x,t) \doteq F_x(x,t)$, and $F^{xx}(x,t) \doteq F_{xx}(x,t)$. It is easy to verify that $\widetilde{F}$ and F^{xx} satisfy

$$\begin{cases} -\widetilde{F}_t - \frac{1}{2}(\widetilde{F}_{xx} + 2\widetilde{F}\widetilde{F}_x) + (\alpha - \frac{3}{2})\widetilde{F}_x = 0, & \text{in } \Omega^+, \\ \widetilde{F}(0,t) = 0, \ \widetilde{F}(x,T) = -1, \end{cases}$$

and

$$\begin{cases} -F_t^{xx} - \frac{1}{2}(F_{xx}^{xx} + 2F_x F_x^{xx} + 2(F^{xx})^2) + (\alpha - \frac{3}{2})F_x^{xx} = 0, & \text{in } \Omega^+, \\ F^{xx}(0,t) \leq 0, \ F^{xx}(x,T) = 0, \end{cases}$$

respectively. By virtue of the (strong) maximum principle, we obtain parts i) and ii).

To show part iii), let us define $F_x^{-\delta}(x,t) \doteq F_x(x,t-\delta)$. It suffices to show $G(x,t) \doteq F_x^{-\delta}(x,t) - F_x(x,t) \geq 0, \forall \delta \geq 0$. It is easy to verify that $G(x,t)$ satisfies

$$\begin{cases} -G_t - \frac{1}{2}\left(G_{xx} + 2F_x^{-\delta}G_x + 2F_{xx}G\right) + (\alpha - \frac{3}{2})G_x = 0, & \text{in } (-\infty, 0) \times [\delta, T), \\ G(0,t) = 0, \ G(x,T) = F_x(x,T-\delta) + 1 \geq 0. \end{cases}$$

Thanks to the minimum principle, we see $G(x,t) \geq 0$, in $(-\infty, 0] \times [\delta, T)$, which results in $F_{xt}(x,t) \leq 0$.

Next we prove part iv). Denote $\widetilde{F}^\delta(x,t) \doteq F_x(x,t;\alpha+\delta)$ and $\widehat{F}(x,t) \doteq F_x(x,t;\alpha) + \delta$. Let $P = \widetilde{F}^\delta - \widehat{F}$. Then, it suffices to show $P < 0$ in Ω^+. It is easy to check that $\widetilde{F}^\delta(x,t)$ and $\widehat{F}(x,t)$ satisfy

$$\begin{cases} -\widetilde{F}_t^\delta - \frac{1}{2}(\widetilde{F}_{xx}^\delta + 2\widetilde{F}^\delta\widetilde{F}_x^\delta) + (\alpha + \delta - \frac{3}{2})\widetilde{F}_x^\delta = 0, & \text{in } \Omega^+, \\ \widetilde{F}^\delta(0,t) = 0, \ \widetilde{F}^\delta(x,T) = -1 \end{cases} \tag{29}$$

and

$$\begin{cases} -\widehat{F}_t - \frac{1}{2}(\widehat{F}_{xx} + 2\widehat{F}\widehat{F}_x) + (\alpha + \delta - \frac{3}{2})\widehat{F}_x = 0, & \text{in } \Omega^+, \\ \widehat{F}(0,t) = \delta, \ \widehat{F}(x,T) = -1 + \delta, \end{cases} \tag{30}$$

respectively. Subtracting (30) from (29), we obtain

$$\begin{cases} -P_t - \frac{1}{2}(P_{xx} + \widehat{F}P_x + \widetilde{F}_x^\delta P) + (\alpha + \delta - \frac{3}{2})P_x = 0, & \text{in } \Omega^+, \\ P(0,t) = -\delta, \ P(x,T) = -\delta. \end{cases}$$

Applying the maximum principle gives the desired result.

To show part v), note that

$$\Phi\left(-\frac{x}{a}\right) \sim O\left(\frac{1}{x}e^{-\frac{x^2}{2a^2}}\right), \quad \text{as } x \to +\infty, \forall a > 0.$$

It follows

$$\lim_{x \to +\infty} F_x(x,t) = \lim_{x \to +\infty} \frac{\Psi_x(x,t)}{\Psi(x,t)} = \lim_{x \to +\infty} \frac{-e^{-x}\Phi(d_2) + e^{2(\alpha-1)x}\Phi(d_3)}{\Psi(x,t)} = -1,$$

which completes the proof.   □

Due to Lemma 2 part ii) and iii), we can have the following proposition.

Proposition 1 *The variational inequality problem* (27) *has a unique solution* $\overline{V}(x,t) \in W_p^{2,1}(\Omega_N^+)$, $1 < p < +\infty$, *where* Ω_N^+ *is any bounded set in* Ω^+. *Moreover,* $\forall(x,t) \in \Omega^+$, *we have*

i) $0 \le \overline{V}_x \le 1$;
ii) $\overline{V}_t \ge 0$;
iii) $\overline{V}(x,t;\alpha) \le \overline{V}(x,t;\alpha+\delta)$ *for* $\delta > 0$, *where* $\overline{V}(x,t;\beta)$ *denotes* $\overline{V}(x,t)$ *corresponding to the goodness index* β.

Proof Using the penalized approach [cf. Friedman in [5]], it is not hard to show that (27) has a unique solution $\overline{V}(x,t) \in W_p^{2,1}(\Omega_N^+)$, $1 < p < +\infty$, where Ω_N^+ is any bounded set in Ω^+. To prove part i), we only need to confine to the noncoincidence set $\Lambda = \{(x,t) \in \Omega^+ : \overline{V} < 0\}$. Denote $w = \overline{V}_x$ and $\omega = \overline{V}_x - 1$, then w and ω satisfy

$$\begin{cases} -w_t - \frac{1}{2}(w_{xx} + 2ww_x + 2F_{xx}w + 2F_xw_x) + (\alpha - \frac{1}{2})w_x = -F_{xx}, & \text{in } \Lambda, \\ w|_{\partial\Lambda} = 0, \end{cases}$$

and

$$\begin{cases} -\omega_t - \frac{1}{2}(\omega_{xx} + 2\omega\omega_x + 2F_{xx}\omega + 2F_x\omega_x) + (\alpha - \frac{3}{2})\omega_x = 0, & \text{in } \Lambda, \\ \omega|_{\partial\Lambda} = -1, \end{cases}$$

respectively. Since $-F_{xx} \ge 0$, one can deduce $w \ge 0$ and $\omega \le 0$ in Λ by the maximum principle, which is desired.

To show part ii), we denote $\widetilde{V}(x,t) = \overline{V}(x,t-\delta)$. It suffices to show $Q(x,t) \doteq \widetilde{V}(x,t) - \overline{V}(x,t) \le 0$ in Ω^+, $\forall \delta \ge 0$. If it is false, then

$$\Delta = \{(x,\tau) \in \Omega : Q(x,t) > 0\} \ne \emptyset.$$

It is easy to verify that $Q(x,t)$ satisfies

$$\begin{cases} -Q_t - \frac{1}{2}(Q_{xx} + (\widetilde{V}_x + \overline{V}_x)Q_x + 2F_xQ_x) + (\alpha - \frac{1}{2})Q_x \\ \quad \le -\delta F_{xt}(\cdot,\cdot)(\widetilde{V}_x - 1) \le 0, & \text{in } \Delta, \\ Q|_{\partial\Delta} = 0, \end{cases}$$

where we have used part iii) of Lemma 2 and $\widetilde{V}_x \le 1$. Applying the maximum principle, we get $Q \le 0$ in Δ, which contradicts the definition of Δ.

At last, let us show part iii). If it is not true, then

$$\mathcal{O} = \{(x,t) \in \Omega : H(x,t) < 0\} \ne \emptyset,$$

where $H(x,t) = \overline{V}(x,t;\alpha+\delta) - \overline{V}(x,t)$. Denote $F_x^\delta(x,t) = F_x(x,t;\alpha+\delta)$. It can be verified that

$$\begin{cases} -H_t - \frac{1}{2}(H_{xx} + H_x^2 + 2\overline{V}_x H_x + 2F_x^\delta H_x) + (\alpha + \delta - \frac{1}{2})H_x \\ \quad \ge -(F_x^\delta - F_x - \delta)(1 - \overline{V}_x) & \text{in } \mathcal{O}, \\ H|_{\partial \mathcal{O}} = 0. \end{cases}$$

By part iv) in Lemma 2, $F_x^\delta < F_x + \delta$, which, together with $\overline{V}_x \le 1$, gives

$$-(F_x^\delta - F_x - \delta)(1 - \overline{V}_x) \ge 0.$$

Again applying the maximum principle, we get $H \ge 0$ in $\mathcal{O}$, which is a contradiction with the definition of $\mathcal{O}$. The proof is complete. $\qquad\square$

Proof of Theorem 1

Proof of Theorem 1 According to part i) in Lemma 2 and (27),

$$\mathcal{L}_0 \overline{V} \le -(F_x + 1) + \alpha < 0, \quad \text{for } \alpha \le 0.$$

Applying the strong maximum principle, we infer $\overline{V} < 0$ in Ω for $\alpha \le 0$. Part i) then follows.

If $\alpha \ge 1$, part i) in Lemma 2 leads to

$$F_x - (\alpha - 1) \le 0.$$

So, $\overline{V} = 0$ is a solution to (27), which implies part iii).

It remains to show part ii). Since $\overline{V}_x \ge 0$, we can define a boundary

$$x_b^*(t) = \inf\{x \in (0,\infty) : \overline{V}(x,t) = 0\}, \quad \text{for any } t \in [0,T).$$

Due to $\overline{V}_t \ge 0$, we infer that $x_b^*(t)$ is monotonically decreasing in t. Let us prove $x_b^*(t) > 0$ for all t. If not, then there exists a $t_0 < T$, such that $x_b^*(t) = 0$ for all $t \in [t_0, T)$. This leads to $\overline{V}(x,t) = 0$, in $(0,+\infty) \times [t_0, T)$. By (27), we have $\mathcal{L}_0 \overline{V} + F_x - (\alpha - 1) \le 0$ in $(0,+\infty) \times [t_0, T)$, namely,

$$F_x \le \alpha - 1 \quad \text{in } (0,+\infty) \times [t_0, T),$$

which is a contradiction with $F_x(0,t) = 0, \forall t > 0$. Further, it can be shown that $x_b^*(t) < \infty$ in terms of the standard argument of Brezis and Friedman in [1] [cf. also the proof of Lemma 4.2 in [2]], where $\lim_{x\to+\infty} F_x(x,t) = -1$ will be used. In addition, due to the monotonicity of $x_b^*(t)$, we deduce that $\{x = 0\} \notin BR$. So, (19) follows.

To show $\lim_{t\to T^-} x_b^*(t) = 0$, let us assume the contrary, i.e., $\lim_{t\to T^-} x_b^*(t) = x_0 > 0$. Then, we have

$$\mathscr{L}_0\overline{V} + F_x - (\alpha - 1) = 0, \quad \forall\, 0 < x < x_0,\ 0 \le t < T,$$

which, combined with $\overline{V}(x, T) = 0$ for all x, gives

$$\overline{V}_t|_{t=T} = F_x|_{t=0} - (\alpha - 1) = -\alpha < 0, \quad \forall\, 0 < x < x_0.$$

This conflicts with $\overline{V}_t \ge 0$.

At last, we need to prove (20), which leads us to consider a stationary problem of (9). Here, we prove the case of $\frac{1}{2} < \alpha < 1$, while the case of $0 < \alpha \le \frac{1}{2}$ can be done similarly.

Noting that $\lim_{T-t\to+\infty} \Psi(x, t) = 0$, if $\frac{1}{2} < \alpha < 1$, it follows that $\lim_{T-t\to+\infty} V(x, t) = 0$, which is not desired. Thus, we define

$$\Psi^\infty(x) \doteq \lim_{t\to\infty} \sqrt{2\pi}(T - t)^{\frac{3}{2}} e^{\frac{(\alpha-\frac{1}{2})^2}{2}(T-t)} \Psi(x, t), \quad \text{and}$$

$$V^\infty(x) \doteq \lim_{t\to\infty} \sqrt{2\pi}(T - t)^{\frac{3}{2}} e^{\frac{(\alpha-\frac{1}{2})^2}{2}(T-t)} V(x, t).$$

A direct calculation shows

$$\Psi^\infty(x) = \frac{2\left(x + \frac{2}{3-2\alpha}\right)}{\left(\alpha - \frac{1}{2}\right)^2\left(\frac{3}{2} - \alpha\right)} e^{-(\frac{3}{2}-\alpha)x}. \tag{31}$$

According to (9), we see $V^\infty(x)$ satisfies,

$$\begin{cases} -\frac{1}{2}V_{xx}^\infty + (\alpha - \frac{1}{2})V_x^\infty - \frac{1}{2}(\alpha - \frac{1}{2})^2 V^\infty = 0, & 0 < x < x_\infty, \\ V_x^\infty(0) = 0, \quad V^\infty(x_\infty) = \Psi^\infty(x_\infty), \quad V_x^\infty(x_\infty) = \Psi_x^\infty(x_\infty), \end{cases} \tag{32}$$

where x_∞ is the free boundary.

It is easy to check that

$$V^\infty(x) = \begin{cases} \frac{2}{\left(\alpha-\frac{1}{2}\right)^3}\left(\frac{2}{2\alpha-1} - x\right)e^{\frac{2\alpha-1}{2}x - x_\infty}, & \text{if } 0 \le x < x_\infty, \\ \Psi^\infty(x), & \text{if } x \ge x_\infty, \end{cases}$$

satisfies (32), where $x_\infty = \frac{8(1-\alpha)}{(3-2\alpha)(2\alpha-1)}$. The proof is complete. □

Proof of Theorem 2

Proof of Theorem 2 The proof is similar to Theorem 1. The only difference is that now we have $\overline{v}_x \le 0$ instead of $0 \le \overline{V}_x \le 1$, where $\overline{v}(x, t) = \log\frac{v(x,t)}{\psi(x,t)}$. Thus, we can define the free boundary as

$$x_b^*(t) = \sup\{x \in (-\infty, 0) : \overline{v}(x, t) = 0\}, \quad \text{for any } t \in [0, T). \tag{33}$$

The existence follows immediately. Now, we turn to the asymptotic behavior of $x_b^*(t)$.

Note that $\lim_{t \to +\infty} \psi(x, t) = +\infty$, if $0 < \alpha < 1$. So, we denote

$$\psi^\infty(x) \doteq \lim_{T-t \to \infty} e^{(\alpha-1)(T-t)} \psi(x, t) = \frac{3 - 2\alpha}{2(1 - \alpha)}, \quad \text{and}$$

$$v^\infty(x) \doteq \lim_{T-t \to \infty} e^{(\alpha-1)(T-t)} v(x, t).$$

According to (9), it is easy to see $v^\infty(x)$ satisfies

$$\begin{cases} -\frac{1}{2} v_{xx}^\infty + (\alpha - \frac{1}{2}) v_x^\infty - (\alpha - 1) v^\infty = 0, & x_\infty < x < 0, \\ v_x^\infty(0) = 0, \quad v_x^\infty(x_\infty) = 0, \quad v^\infty(x_\infty) = \frac{3-2\alpha}{2(1-\alpha)}, \end{cases} \tag{34}$$

where x_∞ is the free boundary.

The general solution to (34) is

$$v^\infty(x) = Ae^x + Be^{2(\alpha-1)x}, \quad x_\infty \le x \le 0. \tag{35}$$

Due to $v_x^\infty(0) = 0$, (35) can be rewritten as

$$v^\infty(x) = A\left(e^x - \frac{1}{2(\alpha - 1)} e^{2(\alpha-1)x}\right), \quad x_\infty \le x \le 0, \tag{36}$$

which results in

$$v_x^\infty(x) = A(e^x - e^{2(\alpha-1)x}).$$

Due to $v_x^\infty(x_\infty) = 0$, we obtain $A = 0$ and $v^\infty(x) = 0$ for all x, which contradicts $v^\infty(x_\infty) = \frac{3-2\alpha}{2(1-\alpha)}$. Thus, there is no finite free boundary x_∞, i.e., $x_\infty = -\infty$. Since $\alpha < 1$, and $0 \le \lim_{x \to -\infty} v^\infty(x) \le \frac{3-2\alpha}{2(1-\alpha)}$, we deduce $A = 0$ by (36), i.e., $v^\infty(x) \equiv 0 < \psi^\infty(x)$. The proof is complete. $\qquad\square$

Proof of Theorem 3

Similar to Theorem 1, we can deal with the cases of $\alpha \le 0$ and $\alpha \ge 1$ easily. However, the case of $0 < \alpha < 1$ is more challenging because we no longer have the monotonicity of $\overline{U}$ w.r.t. t. To overcome the difficulty, we introduce an auxiliary problem:

$$\begin{cases} \mathcal{L}_0 \overline{U}^* + F_x - (\alpha - 1) = 0, & \text{in } \Omega^+, \\ \overline{U}_x^*(0, t) = 0, \quad \overline{U}^*(x, T) = 0. \end{cases} \tag{37}$$

Lemma 3 *Let $\overline{U}^*(x, t)$ be the solution to (37). Then for any $t \in [0, T)$,*

$$\overline{U}^*(0, t) > 0, \quad \text{if } \alpha > \frac{1}{2},$$

$$\overline{U}^*(0, t) = 0, \quad \text{if } \alpha = \frac{1}{2},$$

$$\overline{U}^*(0, t) < 0, \quad \text{if } \alpha < \frac{1}{2}.$$

Proof $\overline{U}^*(x,t) = \log(\frac{U^*(x,t)}{\Psi(x,t)})$ is the solution to (37), where $U^*(x,t)$ is the value function associated with a simple strategy: holding the stock until expiry T, i.e. $U^*(x,t) = \mathbb{E}(\frac{S_T}{M_T} \mid \log \frac{M_t}{S_t} = x)$. According to Shiryaev, Xu and Zhou in [8], Lemma 3 automatically follows at $x = 0$. □

Proposition 2 *Problem* (28) *has a unique solution* $\overline{U}(x,t) \in W_p^{2,1}(\Omega_N^+)$, $1 < p < +\infty$, *where* Ω_N^+ *is any bounded set in* Ω^+. *Moreover, for any* $(x,t) \in \Omega^+$,

i) $0 \le \overline{U}_x \le 1$;

ii) $\overline{U}(x,t;\alpha) \le \overline{U}(x,t;\alpha+\delta)$ *for* $\delta > 0$, *where* $\overline{U}(x,t;\beta)$ *denotes* $\overline{U}(x,t)$ *corresponding to the goodness index* β;

iii) $\overline{U}(x,t) = \overline{U}^*(x,t) > 0$ *for* $\alpha \ge \frac{1}{2}$. *And, for any* $t \in [0,T)$,

$$\overline{U}(0,t) > 0, \quad \text{if } \alpha > \frac{1}{2}, \tag{38}$$

$$\overline{U}(0,t) = 0, \quad \text{if } \alpha = \frac{1}{2}, \tag{39}$$

$$\overline{U}(0,t) = 0, \quad \text{if } \alpha < \frac{1}{2}. \tag{40}$$

Proof The proofs of part i) and ii) are the same as that of Proposition 1. Now let us prove part iii).

It is easy to see $\overline{U}_x^*(x,t) > 0$ in Ω^+ by virtue of $F_{xx} \le 0$ and the strong maximum principle. Combining with Lemma 3, we infer $\overline{U}^*(x,t) > 0$ in Ω^+ when $\alpha \ge \frac{1}{2}$. So, $\overline{U}^*(x,t)$ must be the solution to (28), which yields $\overline{U}(x,t) = \overline{U}^*(x,t)$ for $\alpha \ge \frac{1}{2}$. Then (38) and (39) follow. To show (40), clearly we have $\overline{U}(0,t) \ge 0$. Thanks to part ii) and (39), we infer $\overline{U}(0,t) \le 0$ for $\alpha < \frac{1}{2}$, which leads to (40). This completes the proof. □

Now, we are going to prove Theorem 3.

Proof of Theorem 3 Part i) and ii) follow by part iii) of Proposition 2. The proof of part iv) is similar to that of part i) in Theorem 1. Now let us prove part iii). Thanks to (40), we immediately get $\{x = 0\} \subset SR$. Combining with $\overline{U}_x \ge 0$, we can define

$$x_s^*(t) = \sup\{x \in [0,+\infty) : \overline{U}(x,t) = 0\}, \quad \text{for any } t \in [0,T).$$

We only need to show that $x_s^*(t) < \infty$. Let $x_b^*(t)$ be the free boundary as given in part ii) of Theorem 1. Due to Lemma 1, we infer $x_s^*(t) \le x_b^*(t)$, which, combined with $x_b^*(t) < \infty$, yields the desired result.

At last, let us prove $\lim_{T-t \to \infty} x_s^*(t) = \frac{1}{2\alpha-1} \log(1 - (2\alpha - 1)^2)$. Again, consider the stationary problem of (15). Denote

$$\Psi^\infty(x) = \lim_{T-t \to \infty} \Psi(x,t) \quad \text{and} \quad U^\infty(x) = \lim_{T-t \to \infty} U(x,t).$$

It is easy to see that $\Psi^\infty(x) = e^{-x} + \frac{1}{2(\alpha-1)}e^{2(\alpha-1)x}$, if $0 < \alpha < \frac{1}{2}$, and $U^\infty(x)$ satisfies

$$-\frac{1}{2}U^{\infty}_{xx} + \left(\alpha - \frac{1}{2}\right)U^{\infty}_x = 0, \quad \forall x > x_{\infty}, \tag{41}$$

$$U^{\infty}_x(x_{\infty}) = \Psi^{\infty}_x(x_{\infty}), \qquad U^{\infty}(x_{\infty}) = \Psi^{\infty}(x_{\infty}), \tag{42}$$

where x_{∞} is the free boundary.

The general solution to (41) is

$$U^{\infty}(x) = A + Be^{(2\alpha-1)x}.$$

Since $\lim_{x\to\infty} U^{\infty}(x) = 0$, we see $A = 0$. Thus, it is easy to get $U^{\infty}(x) = (e^{-x_{\infty}} + \frac{1}{2(\alpha-1)}e^{2(\alpha-1)x_{\infty}})e^{(2\alpha-1)(x-x_{\infty})}$ from (42), where $x_{\infty} = \frac{1}{2\alpha-1}\log(1 - (2\alpha - 1)^2)$. The proof is complete. $\qquad\qquad\square$

References

1. Brezis, H., Friedman, A.: Estimates on the support of solutions of parabolic variational inequalities. Ill. J. Math. **20**, 82–97 (1976)
2. Dai, M., Kwok, Y.K., Wu, L.: Optimal shouting policies of options with strike reset rights. Math. Finance **14**(3), 383–401 (2004)
3. Dai, M., Zhong, Y.: Optimal stock selling/buying strategy with reference to the ultimate average. Working Paper, National University of Singapore (2008)
4. du Toit, J., Peskir, G.: Selling a stock at the ultimate maximum. Ann. Appl. Probab. **19**, 983–1014 (2009)
5. Friedman, A.: Variational Principles and Free Boundary Problems. Wiley, New York (1982)
6. Shiryaev, A.: Quickest detection problems in the technical analysis of the financial data. Mathematical Finance – Bachelier Congress 2000, pp. 487–521. Springer, Berlin (2000)
7. Shiryaev, A., Xu, Z., Zhou, X.: Response to comment on "Thou shalt buy and hold". Quant. Finance **8**, 761–762 (2008a)
8. Shiryaev, A., Xu, Z., Zhou, X.: Thou shalt buy and hold. Quant. Finance **8**, 765–776 (2008b)

Continuity Theorems in Boundary Crossing Problems for Diffusion Processes

Konstantin A. Borovkov, Andrew N. Downes,
and Alexander A. Novikov

Abstract Computing the probability for a given diffusion process to stay under a particular boundary is crucial in many important applications including pricing financial barrier options and defaultable bonds. It is a rather tedious task that, in the general case, requires the use of some approximation methodology. One possible approach to this problem is to approximate given (general curvilinear) boundaries with some other boundaries of a form enabling one to relatively easily compute the boundary crossing probability. We discuss results on the accuracy of such approximations for both the Brownian motion process and general time-homogeneous diffusions and also some contiguous topics.

1 Introduction

The main motivation for the research presented in this paper is the question of how to price general barrier options. These are path dependent options whose payoff depends on whether or not the price of the underlying asset crosses a certain barrier during the option's lifetime. There are four basic types of barrier options whose names begin with the self-explanatory "down-and-out", "down-and-in", "up-and-out" and "up-and-in". Thus, an up-and-out option gives the holder the right but not

K.A. Borovkov · A.N. Downes
Dept. of Mathematics and Statistics, Univ. of Melbourne, Parkville 3010, Australia

K.A. Borovkov
e-mail: borovkov@unimelb.edu.au

A.N. Downes
e-mail: andrew.downes@anz.com

A.A. Novikov (✉)
Dept. of Mathematical Sciences, The University of Technology, PO Box 123, Broadway, Sydney, NSW 2007, Australia
e-mail: alex.novikov@uts.edu.au

C. Chiarella, A. Novikov (eds.), *Contemporary Quantitative Finance*,
DOI 10.1007/978-3-642-03479-4_17, © Springer-Verlag Berlin Heidelberg 2010

the obligation to buy (in the case of a call) or sell (in the case of a put) shares of an underlying asset at a pre-determined strike price so long as the price of that asset did not go above a pre-determined barrier during the option lifetime, whereas an up-and-in option is worthless unless the price of the asset exceeds the barrier at some time during the life of the contract. Down-and-out/in barrier options are defined in a similar way, but for lower barriers. Of course, one can also consider two-sided barriers, when, say, the condition for validity of the option is that the price of the underlying stays between two given boundaries during a specified time interval (such options are referred to as "kick-out" ones).

Simple single barrier options are popular attractive alternatives to the respective vanilla options, for at least two reasons: they are obviously cheaper and also have simple closed-from pricing formulae. In the basic Black-Scholes framework, Merton priced a down-and-out call option with a "flat" barrier in his seminal paper [12]. The classic paper providing analytic pricing formulae for such barriers is [17], a popular monograph reference being [11]. In fact, options of this type have become so common that one can even find online barrier option calculators[1] on the Web.

However, already in the case of the Black-Scholes model with variable (deterministic) interest rates and/or volatility term structure, no closed formulae for barrier options are available in the general case. Indeed, assume that the bank account process is non-random and has the form

$$B_t = \exp\left\{\int_0^t r_s ds\right\}, \quad t \in [0, T],$$

where $r_t > 0$ is a deterministic function of time (the spot interest rate). Under the assumptions of a Black-Scholes type diffusion model with a term structure that stipulates a variable deterministic volatility $\sigma_t > 0$ with $\sigma^2 := \frac{1}{T}\int_0^T \sigma_t^2 dt < \infty$, the time t price of the underlying asset S_t has the following dynamics under the risk-neutral measure:

$$S_t = S_0 \exp\left\{\int_0^t [r_s - \sigma_s^2/2] ds + \int_0^t \sigma_s dW_s\right\}, \quad t \in [0, T], \tag{1}$$

where $\{W_t\}$ is the standard Wiener process.

As is well known, under the no-arbitrage assumption, the fair price of an option (on an underlying asset with the price process $\{S_t\}$) with maturity T and payoff X is given by $\mathbf{E}(X/B_T)$, where $\mathbf{E}$ denotes expectation with respect to a risk-neutral measure $\mathbf{P}$ (see e.g. Sect. 6.2.2 in [3]). Now consider a kick-out barrier call option with strike K_T and time-dependent lower/upper barriers $G_\pm(t)$ such that $G_-(t) < G_+(t)$, $t \le T$. In this case, the payoff function is given by

$$X = (S_T - K_T)\mathbf{1}_{\{S_T > K_T; \; G_-(t) < S_t < G_+(t), \; t \in [0,T]\}}, \tag{2}$$

$\mathbf{1}_A$ denoting the indicator of the event A.

---

[1] See e.g. http://www.sitmo.com/live/OptionBarrier.html. Cited 13 Aug 2009.

It is obvious that, using the "natural time change" (the new "time" being $t' :=$ $\sigma^{-2} \int_0^t \sigma_s^2 ds$), one can make the volatility constant over the time interval $[0, T]$. For simplicity's sake, let us just assume that we have already done that, thus replacing (1) with the (natural time scale) dynamics

$$S_t = S_0 \exp\left\{ \int_0^t [r_s - \sigma^2/2] ds + \sigma W_t \right\}, \quad t \in [0, T] \tag{3}$$

(we retain the same notation for the time variable t and also for the respectively transformed interest rate r_t, barrier functions $G_\pm(t)$ and the Wiener process $\{W_t\}$; the interested reader could easily do all the calculations by him/herself).

Using the same standard change-of-measure technique that leads to the Black-Scholes formula for the price of a vanilla call, one can show that the fair price of the above double-barrier call option (2) under model (3) is given by

$$S_0 P_0 - K_T \exp\left\{ -\int_0^T r_s \, ds \right\} P_1, \tag{4}$$

where

$$P_0 = \mathbf{P}(g_-(t) - \sigma t < W_t < g_+(t) - \sigma t, \; t \in [0, T]; \; W_T > h - \sigma T),$$

$$P_1 = \mathbf{P}(g_-(t) < W_t < g_+(t), \; t \in [0, T]; \; W_T > h),$$

$$h = \frac{1}{\sigma}\left[\ln(K_T/S_0) + \frac{\sigma^2 T}{2} - \int_0^T r_s \, ds \right],$$

and

$$g_\pm(t) = \frac{1}{\sigma}\left[\ln(G_\pm(t)/S_0) + \frac{\sigma^2 t}{2} - \int_0^t r_s \, ds \right], \quad t \in [0, T] \tag{5}$$

(one can easily prove this statement using Girsanov's transformation; for details in the case of a one-sided barrier, see e.g. [15]).

Thus, (4) is a natural generalization of the Black-Scholes formula having the same structure as the latter, the only difference being that while pricing a European call it suffices to find the values of two normal probabilities, to price a barrier call one should simply find the values of two probabilities of the form

$$P(g_-, g_+; B) := \mathbf{P}(g_-(t) < W_t < g_+(t), \quad t \in [0, T]; \; W_T \in B), \tag{6}$$

B being a Borel set (more specifically, a half-line in this case). Unfortunately, even in the one-sided case (when either $g_- \equiv -\infty$ or $g_+ \equiv +\infty$), there are no closed form expressions for probabilities of the form (6) in the general case.

Of course, one could use the Kolmogorov-Petrovskii theorem according to which one has $P(g_-, g_+; B) = u(0, 0)$, where $u = u(t, x)$ is the solution to the parabolic PDE

$$\frac{\partial u}{\partial t} + \frac{1}{2}\frac{\partial^2 u}{\partial x^2} = 0, \quad 0 < t < T, \; g_-(t) < x < g_+(t),$$

subject to the boundary conditions

$$u(t,x) = \begin{cases} 0 & \text{if } (t,x) \in L_- \cup L_+ \cup L^-, \\ 1 & \text{if } (t,x) \in L^+, \end{cases}$$

where $L_\pm = \{(t,x) : 0 < t < T, \ x = g_\pm(t)\}$, $L^- = \{(T,x) : g_-(T) \le x \le h\}$, and $L^+ = \{(T,x) : h < x \le g_+(T)\}$. Alternatively, one can use a "brute-force" crude Monte Carlo approach (or its antithetic variation), simulating a large number of entire trajectories $\{W_t\}_{0 \le t \le T}$ (of course, using fine enough discrete time lattices for that purpose) and counting the frequencies at which the events in P_i occur.

Both approaches have been used by practitioners, but neither of them proved to be satisfactory due to their low efficiency. A possible alternative approach is to approximate (6) by $P(f_-, f_+; B)$, where the boundaries $f_\pm$ are close in some sense to $g_\pm$ and such that fast computation of $P(f_-, f_+; B)$ is feasible, and then to obtain a bound for the rate of the approximation to justify its use.

The simplest special case when (6) is easy to find is when $g_- \equiv -\infty$, g_+ is linear: conditioning on W_T, one can just use the well-known linear boundary crossing probability by the Brownian bridge process (see (8) below), which immediately yields the above-mentioned standard formulae for flat barrier options in the standard Black-Scholes framework. Another special case in which there are closed-form expressions for one-sided boundary crossing probabilities is that of the so-called "generalized Daniels' boundaries", see e.g. [7, 13]. None of these classes is wide enough to be used to approximate general (nice enough) boundaries, but one can use the following observation [15]: combining the total probability formula and the Markov property of the Wiener process, one obtains that

$$P(f_-, f_+; B) = \mathbf{E}\left[\mathbf{1}_{\{W_T \in B\}} \prod_{i=0}^{n-1} p_i(f_-, f_+ | W_{t_i}, W_{t_{i+1}}) \right], \qquad (7)$$

where we put

$$p_i(f_-, f_+ | x_i, x_{i+1}) := \mathbf{P}(f_-(s) < W_s < f_+(s), \ s \in [t_i, t_{i+1}] | W_{t_i} = x_i, W_{t_{i+1}} = x_{i+1}).$$

Note that, in the special case of $B = \mathbf{R}$, this representation for

$$P(f_-, f_+) := P(f_-, f_+; \mathbf{R})$$

also appeared as Theorem 2 in [16].

When $f_-(s) = -\infty$, $s \in [t_i, t_{i+1}]$, and $f_+(s)$ is linear on this interval, the last probability has the following simple form used in [21]:

$$p_i(-\infty, f_+ | x_i, x_{i+1}) = 1 - \exp\left\{ -\frac{2(f_+(t_i) - x_i)(f_+(t_{i+1}) - x_{i+1})}{t_{i+1} - t_i} \right\} \qquad (8)$$

(this is a well-known expression for the linear boundary crossing probability by the Brownian bridge process, see e.g. p. 63 in [4]). Thus, if f_+ is piece-wise linear with nodes located at times t_i, then the RHS of (7) becomes a linear combination of Gaussian integrals of exponential functions, that can be transformed (changing the underlying measure) into the respective combination of values of multivariate normal cumulative distribution functions. In the case of two-sided linear boundaries $f_\pm$ that are linear on $[t_i, t_{i+1}]$, the probability $p_i(f_-, f_+ | x, y)$ is given by a rapidly convergent infinite series of exponential functions (for more detail and numerical examples, see e.g. [14] or [16]).

Thus, in the cases of one-sided or two-sided piecewise linear boundaries, one will simply have to compute the respective n-fold Gaussian integrals, which can be done rather quickly for small (or even moderate) values of n. It remains to establish how well the probabilities $P(f_-, f_+; B)$ with, say, suitably chosen piece-wise linear $f_\pm$ will approximate the desired value of (6). A similar question can be posed in the case when one replaces W_t in (6) with an arbitrary diffusion process.

One of the first results of that kind was obtained in [16] (see also further references mentioned in that paper). The authors, under the assumptions that the boundaries $g_\pm$ are twice continuously differentiable with $g_\pm''(0) \neq 0$ and $g_\pm''(t) = 0$ at most at finitely many points $t \in (0, T]$, proposed a special rule for choosing a sequence of "optimal partitions" $t_0^{(n)} = 0 < t_1^{(n)} < \cdots < t_n^{(n)} = T$ of $[0, T]$ (generally speaking, depending on the boundaries) with the following property: if $g_\pm^{(n)}$ are piecewise linear boundaries with nodes at $(t_i^{(n)}, g_\pm(t_i^{(n)}))$, $i = 0, 1, \ldots, n$, then for

$$\Delta_n := |P(g_-, g_+) - P(g_-^{(n)}, g_+^{(n)})|$$

one has the asymptotic bound

$$\limsup_{n \to \infty} n^2 \Delta_n \leq A, \tag{9}$$

where the constant A depends on both the shape of the boundaries $g_\pm$ and the rule used to form the partitions $\{t_i^{(n)}\}_{0 \leq i \leq n}$ (through a couple of integrals that could actually be computed—at least, numerically). Unfortunately, this asymptotic bound is of little use when n is relatively small (such that the computation of $P(g_-^{(n)}, g_+^{(n)})$ is still feasible), while conditions on the boundaries seem to be excessive. As it turns out, one can obtain an exact bound for Δ_n just in terms on the uniform distance between the boundaries $g_\pm$ and $g_\pm^{(n)}$, respectively, which immediately leads to explicit upper bounds of the form cn^{-2} for, say, piece-wise C^2-boundaries and (almost) uniform lattices $\{t_i^{(n)}\}$ (see Corollary 1 below).

This paper presents a number of recent and new results obtained in the outlined direction. In Sect. 2 we deal with approximation rates in the case of the Brownian motion process and basically show that $P(g_-, g_+; B)$ is a locally Lipschitz function of $(g_-(\cdot), g_+(\cdot))$ (in uniform metric), giving explicit bounds for the Lipschitz constant. Then we discuss an extension of the obtained results to the case of general time-homogeneous diffusions in Sect. 3. Section 4 deals with the differentiability of the boundary (non-)crossing probability $P(-\infty, g_+; B)$ viewed as a functional of the boundary g_+.

2 Approximation Rates in the Brownian Motion Case

In this section we discuss the approximation rates in the case of the Brownian motion process, to outline the basic ideas to be used in more general situations as well. By Lip_K we will denote the class of Lipschitz functions on $[0, T]$ with the constant $K \in (0, \infty)$, so that $g \in \text{Lip}_K$ iff

$$|g(t + h) - g(t)| \leq Kh, \quad 0 \leq t < t + h \leq T,$$

and by $\|\cdot\|$ the uniform norm of a (bounded) function on $[0, T]$: $\|g\| = \sup_{0 \leq t \leq T} |g(t)|$. The following result was obtained in [7].

Theorem 1 *Assume that $g_\pm \in \text{Lip}_K$ and that, for some functions $f_\pm$ on $[0, T]$, one has $\|g_\pm - f_\pm\| \leq \varepsilon$ for an $\varepsilon > 0$. Then*

$$|P(-\infty, g_+) - P(-\infty, f_+)| \leq (2.5K + 2T^{-1/2})\varepsilon \qquad (10)$$

and

$$|P(g_-, g_+) - P(f_-, f_+)| \leq (5K + 4T^{-1/2})\varepsilon. \qquad (11)$$

The same bounds will also hold for the differences

$$|P(-\infty, g_+; B) - P(-\infty, f_+; B)| \quad \text{and} \quad |P(g_-, g_+; B) - P(f_-, f_+; B)|. \qquad (12)$$

Remark 1 Note that bounds (10) and (11) are quite sharp: the coefficient of ε on the right-hand side on (10) cannot be less than $2K + \sqrt{2/\pi}\,T^{-1/2}$. This is obvious from the well-known explicit formula for $P(-\infty, g_+)$ in the case of the straight line boundary $g_+(t) = \varepsilon + Kt$ (formula 1.1.4 on p. 197 in [4]) and the fact that $P(-\infty, f_+) = 0$ when $f_+(t) = Kt$.

Next we will formulate our improvement of (9).

Corollary 1 *Let $g_\pm \in C^1[0, T]$, $K = \max\{\|g'_-\|, \|g'_+\|\}$, and let $g'_\pm$ be absolutely continuous satisfying $|g''_\pm| \leq \gamma < \infty$ a.e. For a partition $0 = t_0 < t_1 < \cdots < t_n = T$ of $[0, T]$ of rank $\delta = \max_{0 < i \leq n} |t_i - t_{i-1}|$ and $f_\pm$ piecewise linear functions with nodes at the points $(t_i, g_\pm(t_i))$, one has*

$$|P(-\infty, g_+) - P(-\infty, f_+)| \leq (0.313K + 0.25T^{-1/2})\gamma\delta^2 \qquad (13)$$

and

$$|P(g_-, g_+) - P(f_-, f_+)| \leq (0.625K + 0.5T^{-1/2})\gamma\delta^2. \qquad (14)$$

The same bounds hold for the differences (12).

In particular, in the case of uniform partitions $t_i = iT/n$, $0 \leq i \leq n$, and $g_\pm^{(n)}$ the respective piecewise linear approximations to $g_\pm$, one has $\delta = T/n$ and hence

$\Delta_n \leq C n^{-2}$ with $C = (0.625K + 0.5T^{-1/2})\gamma T^2$ (instead of the asymptotic bound (9)).

The corollary is an immediate consequence of Theorem 1 and the elementary bound $\|f - g\| \leq \gamma \delta^2/8$ that holds true under the above assumptions (see e.g. Sect. 2 in [7]).

The proof of Theorem 1 is based on the simple observations that

$$P(g_- + \varepsilon, g_+ - \varepsilon) \leq P(f_-, f_+) \leq P(g_- - \varepsilon, g_+ + \varepsilon) \tag{15}$$

and

$$0 \leq P(g_- - \varepsilon, g_+ + \varepsilon) - P(g_-, g_+)$$
$$= [P(g_- - \varepsilon, g_+ + \varepsilon) - P(g_-, g_+ + \varepsilon)] + [P(g_-, g_+ + \varepsilon) - P(g_-, g_+)], \tag{16}$$

where, assuming for simplicity that $T = 1$, one has

$$P(g_-, g_+ + \varepsilon) - P(g_-, g_+)$$
$$= \mathbf{P}\left(0 \leq \sup_{0 \leq t \leq 1} (W_t - g_+(t)) < \varepsilon, \inf_{0 \leq t \leq 1} (W_t - g_-(t)) > 0\right)$$
$$\leq \mathbf{P}\left(0 \leq \sup_{0 \leq t \leq 1} (W_t - g_+(t)) < \varepsilon\right) = P(-\infty, g_+ + \varepsilon) - P(-\infty, g_+)$$
$$=: D_\varepsilon(g_+) \leq D_\varepsilon := \sup_{f \in \mathrm{Lip}_K} D_\varepsilon(f). \tag{17}$$

As the same argument applies to the first term on the RHS of (16) as well, we get

$$0 \leq P(g_- - \varepsilon, g_+ + \varepsilon) - P(g_-, g_+) \leq 2D_\varepsilon.$$

The difference $P(g_-, g_+) - P(g_- + \varepsilon, g_+ - \varepsilon)$ can be bounded in the same way. Together with (15) this implies that

$$|P(g_-, g_+) - P(f_-, f_+)| \leq 2D_\varepsilon.$$

The same argument shows that $|P(g_-, g_+; B) - P(f_-, f_+; B)| \leq 2D_\varepsilon$ as well. In the case of one-sided boundaries, an even simpler argument yields

$$|P(-\infty, g_+) - P(-\infty, f_+)| \leq D_\varepsilon.$$

To bound $D_\varepsilon(g)$ for a given $g \in \mathrm{Lip}_K$, set $\tau := \inf\{t > 0 : W_t > g(t)\}$ and observe that

$$D_\varepsilon(g) = \mathbf{P}\left(0 \leq \sup_{0 \leq t \leq 1} (W_t - g(t)) < \varepsilon\right)$$
$$= \int_0^1 \mathbf{P}(\tau \in dt)\, \mathbf{P}\left(\sup_{t \leq s \leq 1} (W_s - g(s)) < \varepsilon \,\Big|\, W_t = g(t)\right)$$

$$\leq \int_0^1 \mathbf{P}(\tau \in dt)\mathbf{P}\Big(\sup_{0\leq s\leq 1-t}(W_s - K^+ s) < \varepsilon\Big). \tag{18}$$

The last probability is known in explicit form (see e.g. formula 1.1.4 on p. 197 in [4]), and it is easy to see that it does not exceed $\sqrt{2/(\pi(1-t))}\varepsilon + 2K^+\varepsilon$. So it remains to obtain an upper bound for the density $p(t)$ of τ. This can be done by introducing (for a fixed $t \in [0,1]$) the linear boundary

$$g_t(s) := g(t) + K(t-s), \quad 0 \leq s \leq 1,$$

and its crossing time

$$\tau_t := \inf\{s > 0 : W_s > g_t(s)\} = \inf\{s > 0 : W_s + Ks > g(t) + Kt\}.$$

Obviously,

$$\mathbf{P}(\tau \in (t, t+h)) \leq \mathbf{P}(\tau_t \in (t, t+h)), \quad 0 \leq t < t+h \leq 1, \tag{19}$$

and hence we can bound $p(t)$ by the value of the density of τ_t (which is well-known, see e.g. formula 2.0.2 on p. 223 in [4]) at the point t. The rest of the proof is careful estimation of the integrals arising (for more detail, see Sect. 2 in [7]).

3 Approximation Rates for General Diffusions

In this section we will deal with general time-homogeneous diffusions governed by stochastic differential equations of the form

$$dU_t = v(U_t)dt + \sigma(U_t)dW_t,$$

where $\sigma(y)$ is differentiable and non-zero inside the diffusion interval (that is, the smallest interval $I \subseteq \mathbf{R}$ such that $U_t \in I$ almost surely). Interest to extending the results of Sect. 2 to this case is also motivated by barrier options' pricing, but in more general frameworks than the Black-Scholes one (e.g. when using Bessel processes to model the stock price dynamics).

As is well-known (see e.g. [20, p. 161]), by putting (for some y_0 from the diffusion interval of $\{U_t\}$)

$$F(y) := \int_{y_0}^y \frac{du}{\sigma(u)}, \qquad X_t := F(U_t),$$

one transforms the process $\{U_t\}$ to one with a unit diffusion coefficient and a drift coefficient $\mu(y)$ given by the composition $\mu(y) = (v/\sigma - \sigma'/2) \circ F^{-1}(y)$. So from here on we work with the transformed diffusion process $\{X_t\}$ governed by

$$dX_t = \mu(X_t)dt + dW_t, \qquad X_0 = x \tag{20}$$

(by $\mathbf{P}_x$ we will denote probabilities conditional on the process in question starting at x); of course, any boundaries considered must also be transformed accordingly) and assume without loss of generality that $T = 1$.

To apply the same approach as used in Sect. 2 in the case of the Brownian motion, we should have two components available: suitable bounds for the linear boundary crossing probability by the respective diffusions and for their first crossing density for our boundaries $g(t)$. It turns out that one can obtain the above by using "reference processes" that have unit diffusion coefficients and admit such bounds of simple enough form.

We will consider two such reference processes, depending on the diffusion interval of $\{X_t\}$, for which we will assume one of the two possibilities:

[A] The diffusion interval of $\{X_t\}$ is the whole real line $\mathbf{R}$.
[B] The diffusion interval of $\{X_t\}$ is $(0, \infty)$.

In case [A], we take $\{W_t\}$ to be the reference process, whereas in case [B] we take a Bessel process $\{R_t\}$ of dimension $d = 3, 4, \ldots$ Recall that this process gives the Euclidean distance from the d-dimensional Brownian motion to the origin: given that $R_0 = x \geq 0$, one can define R_t, $t > 0$, by letting

$$R_t := \sqrt{\left(x + W_t^{(1)}\right)^2 + \left(W_t^{(2)}\right)^2 + \cdots + \left(W_t^{(d)}\right)^2},$$

where the $\{W_t^{(i)}\}$ are independent standard Brownian motions, $i = 1, \ldots, d$. As is well known (see e.g. [18, p. 445]), $\{R_t\}$ satisfies the SDE

$$dR_t = \frac{d-1}{2} \frac{1}{R_t} dt + dW_t, \tag{21}$$

that can be used to define Bessel processes of non-integer "dimensions" d. The process has a transition density function $p_R(t, y, z)$ given by

$$p_R(t, y, z) = z \left(\frac{z}{y}\right)^{\eta} t^{-1} e^{-(y^2 + z^2)/2t} I_{\eta}\left(\frac{yz}{t}\right), \tag{22}$$

where $\eta = d/2 - 1$ and $I_{\eta}(z)$ is the modified Bessel function of the first kind. For further information, see e.g. [18].

An important step in bounding the first crossing density of a boundary $g(t)$ is estimating linear boundary crossing probabilities for diffusion bridges. This is one of the arguments where we use reference processes. One can give the following bounds for the above-mentioned probabilities: denoting by $p(t, x, y)$ and $q(t, x, y)$ the transition densities of the original process $\{X_t\}$ and the corresponding reference process, respectively, and setting $\overline{g}(t) := \max_{0 \leq s \leq t} g(s)$, one has, in case [A],

$$\mathbf{P}_x\left(\sup_{0 \leq s \leq t} (X_s - g(s)) < 0 \middle| X_t = z\right)$$

$$\leq \frac{q(t, x, z)}{p(t, x, z)} e^{G(z) - G(x) - (t/2)M(t)} \mathbf{P}_x\left(\sup_{0 \leq s \leq t} (W_s - g(s)) < 0 \middle| W_t = z\right) \tag{23}$$

with

$$G(y) := \int_{y_0}^{y} \mu(z)dz, \qquad M(t) := \inf_{-\infty < y \leq \overline{g}(t)} (\mu'(y) + \mu^2(y)), \qquad (24)$$

and, in case [B],

$$\mathbf{P}_x \left(\sup_{0 \leq s \leq t} (X_s - g(s)) < 0 \Big| X_t = z \right)$$

$$\leq \frac{q(t,x,z)}{p(t,x,z)} e^{G(z)-G(x)-(t/2)M(t)} \mathbf{P}_x \left(\sup_{0 \leq s \leq t} (R_s - g(s)) < 0 \Big| R_t = z \right),$$

with

$$G(y) := \int_{y_0}^{y} \left(\mu(z) - \frac{d-1}{2z} \right) dz,$$

$$M(t) := \inf_{0 < y \leq \overline{g}(t)} \left(\mu'(y) - \frac{(d-1)(d-3)}{4y^2} + \mu^2(y) \right).$$

These results are based on the approach from [1] using which we obtain, for example, the following representation in case [B]. Let $\mathbf{P}_x^z$ denote the law of $\{X_s\}$ starting at $X_0 = x$ and pinned by $X_t = z$ (in the respective space of continuous functions). Then, for any $A \in \sigma(X_u : u \leq t)$,

$$\mathbf{P}_x^z(A) = \frac{q(t,x,z)}{p(t,x,z)} e^{G(z)-G(x)} \widehat{\mathbf{E}}_x^z [e^{-(1/2)N(t)} \mathbf{1}_{\{X \in A\}}],$$

where $\widehat{\mathbf{E}}_x^z$ denotes expectation with respect to the conditional law of the Bessel processes $\{R_s\}$ of dimension $d \geq 3$ on $[0,t]$, starting at $R_0 = x$ and finishing at $R_t = z$, and

$$N(t) := \int_0^t \left(\mu'(X_u) - \frac{(d-1)(d-3)}{4X_u^2} + \mu^2(X_u) \right) du.$$

The above bounds are instrumental in proving the following theorem from [8]. To simplify its statement, we first set $\overline{x} := x$ in case [A] and $\overline{x} := -x$ in case [B].

Theorem 2 *Let $\{X_t\}$ be a diffusion process satisfying (20) with $\mu(y)$ differentiable inside the diffusion interval. Let $g \in \mathrm{Lip}_K$ be a function on $[0,1]$ such that $g(0) > x$. Then the first passage time τ of the process $\{X_t\}$ of the boundary $g(t)$, $t \in (0,1)$, has a density $p_\tau(t)$ satisfying*

$$p_\tau(t) \leq B(g,t) := \frac{1}{t}(g(t) + Kt - \overline{x})q(t,x,g(t))e^{G(g(t))-G(x)-(t/2)M(t)}. \qquad (25)$$

In case [A], one can obtain similar upper bounds for lower boundaries' first crossing time densities and also lower bounds for the densities.

Developing the approach presented in Sect. 2 above and using Theorem 2, along with a number of other auxiliary results, leads to sharp bounds for approximation rates similar to the ones presented in Theorem 1. Set

$$D_\varepsilon(g) := P(-\infty, g+\varepsilon) - P(-\infty, g), \qquad D_\varepsilon^-(g) := P(g-\varepsilon, \infty) - P(g, \infty),$$

and introduce

$$\underline{g}(t) := \min_{0 \le s \le t} g(s), \qquad L_t := \inf_{\underline{g}(t) \le y < \infty} \left(\mu'(y) + \mu^2(y)\right),$$

$$C(g,t) := \frac{1}{t}(x - g(t) + Kt)q(t, x, g(t))e^{G(\underline{g}(t)) - G(x) - (t/2)L(t)}.$$

Theorem 3 *Let $g_\pm \in \mathrm{Lip}_K$ and functions $f_\pm$ on $[0,1]$ be such that $\|g_\pm - f_\pm\| \le \varepsilon$ for some $\varepsilon > 0$. For $t_0 \in [0,1)$, let*

$$\overline{B}_{t_0}(g) := \sup_{t_0 \le t \le 1} B(g,t) \quad and \quad \overline{C}_{t_0}(g) := \sup_{t_0 \le t \le 1} C(g,t),$$

$$\underline{\mu} := \inf_{\ell < x \le g(0) + K + \varepsilon} \mu(x) \quad and \quad \overline{\mu} := \sup_{g(0) - K - \varepsilon \le x < \infty} \mu(x),$$

and set $K^* := \max\{0, K - \underline{\mu}\}$ and $\hat{K} := \max\{0, K + \overline{\mu}\}$. If $\underline{\mu} > -\infty$, then

$$|P(-\infty, g_+) - P(-\infty, f_+)| \le \max\{D_\varepsilon(g_+), D_\varepsilon(g_+ - \varepsilon)\},$$

where

$$D_\varepsilon(g) \le \left(\sqrt{\frac{2}{\pi}}\left(\frac{1}{\sqrt{1-t_0}} + 2\overline{B}(g)\sqrt{1-t_0}\right) + 2K^*\right)\varepsilon. \tag{26}$$

In case [A], *if we also have $\overline{\mu} < \infty$, then*

$$|P(g_-, g_+) - P(f_-, f_+)| \le \max\{D_\varepsilon(g_+) + D_\varepsilon^-(g_-), D_\varepsilon(g_+ - \varepsilon) + D_\varepsilon^-(g_- + \varepsilon)\},$$

where

$$D_\varepsilon^-(g) \le \left(\sqrt{\frac{2}{\pi}}\left(\frac{1}{\sqrt{1-t_0}} + 2\overline{C}(g)\sqrt{1-t_0}\right) + 2\hat{K}\right)\varepsilon. \tag{27}$$

An analog to Corollary 1 holds as well, so that we again have convergence rate $O(n^{-2})$ in case of smooth enough boundaries and uniform n-partitions.

To illustrate the above results, consider an Ornstein-Uhlenbeck process $\{X_t\}$ driven by

$$dX_t = -\theta(X_t - \mu)dt + dW_t$$

with $\theta > 0$, $\mu \in \mathbf{R}$ and a boundary of the form

$$g(t) = \mu + Ae^{-\theta t} + Be^{\theta t}, \tag{28}$$

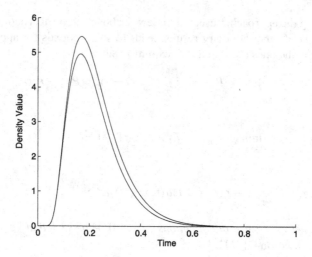

Fig. 1 The first passage time density of the boundary (28) and its upper bound for an Ornstein-Uhlenbeck process

where $x < \mu + A + B$. In this case, the first passage time density is given by

$$p_\tau(t) = 2\theta \frac{|A + B - x + \mu|}{e^{\theta t} - e^{-\theta t}} p(t, x, g(t)),$$ (29)

where

$$p(t, x, z) = \left(\frac{\theta e^{2\theta t}}{\pi(e^{2\theta t} - 1)}\right)^{1/2} \exp\left(\frac{\theta\left((z - \mu)e^{\theta t} - (x - \mu)\right)^2}{1 - e^{2\theta t}}\right)$$

is the transition density of the process $\{X_t\}$.

For the parameter choice $\theta = 0.5$, $\mu = 2$, $x = 1$, $A = 1$ and $B = -1$, Fig. 1 displays the true first passage time density (29) compared to its upper bound (25).

Theorem 3 with $t_0 \approx 0.28$ gives the bound $|P(-\infty, g) - P(-\infty, f)| \leq 1.88\varepsilon$ when $\|f - g\| \leq \varepsilon \leq 0.05$.

4 Differentiability of the Boundary Crossing Probabilities

In this section we continue treatment of the processes of the form (20).

Part of the derivation of the results of Sect. 3 was an argument establishing the upper bounds of the form (23) for linear boundary crossing probabilities by diffusion bridges. In fact, these bounds were needed in situations when z approaches $g(t)$. It turns out that, under mild assumptions (see Theorem 2 in [6]), the asymptotic behaviour of such probabilities is linear: there exits a continuous function $f(t, x)$

such that

$$\mathbf{P}_x\left(\sup_{0\le s\le t}(X_s - g(s)) < 0 \,\Big|\, X_t = z\right) = (f(t,x) + o(1))(g(t) - z), \quad z \uparrow g(t)$$

(30)

(thus, for a Brownian motion process $\{W_t\}$ and a linear boundary g, one has $f(t,x) = 2(g(0) - x)/t$, cf. (8)). When this is the case, the following interesting representation for the first crossing time of the density holds true [6].

Theorem 4 *Assume that, for a boundary $g \in \mathrm{Lip}_K$ with $g(0) > x$, one has (30). Then the first crossing time τ of the boundary has a density given by*

$$p_\tau(t) = \frac{1}{2} f(t,x) p(t,x,g(t)), \quad t > 0.$$

(31)

Note that if the boundary g satisfies Lipschitz condition in an interval $(a, b) \subset \mathbf{R}^+$ only and (30) holds in that interval as well, then one can claim that (31) holds in that interval.

Of the related results one can mention here the representation

$$p_\tau(t) = b(t)p(t,0,g(t)), \quad b(t) := \lim_{s\uparrow\uparrow t}\frac{1}{t-s}\mathbf{E}_0\big[\mathbf{1}_{\{\tau\ge s\}}(g(s) - W_s)\,|\,W_t = g(t)\big],$$

which was established in [9] under the assumption that $g(s)$ is continuous in $[0, t]$ and left differentiable at t. Thus, with our Theorem 4 we have that, for the Brownian motion,

$$b(t) = \frac{1}{2}\lim_{z\uparrow g(t)}\frac{1}{g(t) - z}\mathbf{P}_0\left(\sup_{0\le s\le t}(W_s - g(s)) < 0 \,\Big|\, W_t = z\right).$$

Theorem 4 may be viewed as an alternative expression of the results of [9] in the case of standard Brownian motion and an extension thereof to general diffusion processes.

Representation (31) could also be viewed as an extension (to the case of general diffusion processes and curvilinear boundaries) of the well-known Kendall's identity for spectrally-negative Lévy processes $\{X_s\}$: if $g(s) \equiv y = \mathrm{const}$, $X_0 = x < y$ and $\{X_s\}$ has a transition density, then (see e.g. [5] and references therein)

$$p_\tau(t) = \frac{y - x}{t} p(t,x,y), \quad t > 0.$$

Theorem 4 is proved using rather tedious methods employed in deriving the results of theorems from Sect. 3. However, under additional regularity conditions (basically, μ and g should be C^2 functions, both μ and μ' being bounded) one can establish (31) in a much simpler way. A representation formula for the hazard rate of τ that can be shown to be equivalent to our (31) was stated in Lemma 1 of [19] without proof, but with a reference to a still unpublished manuscript on "Some inequalities for one-dimensional conditioned diffusions and boundary hitting times" by G. O. Roberts (1993), from which the respective portion is reproduced below (with its author's permission).

The above-mentioned additional conditions must, in particular, ensure that the transition taboo density

$$\pi(t, x, z) := \mathbf{P}_x(\tau > t, X_t \in dz)/dz$$

satisfies in $\{(t, z) : t > 0, z < g(t)\}$ the forward Kolmogorov equation

$$\frac{\partial}{\partial t} \pi(t, x, z) = -\frac{\partial}{\partial z} \mu(z)\pi(t, x, z) + \frac{1}{2}\frac{\partial^2}{\partial z^2} \pi(t, x, z).$$

Then, assuming that all the manipulations used in the calculation below are justified and using the relations $\lim_{z \uparrow g(t)} \pi(t, x, z) = \lim_{z \to -\infty} \pi(t, x, z) = 0$, one can write:

$$p_\tau(t) = -\frac{\partial}{\partial t} \mathbf{P}_x(\tau > t) = -\frac{\partial}{\partial t} \int_{-\infty}^{g(t)} \pi(t, x, z)dz = -\int_{-\infty}^{g(t)} \frac{\partial}{\partial t} \pi(t, x, z)dz$$

$$= \int_{-\infty}^{g(t)} \left[\frac{\partial}{\partial z} \mu(z)\pi(t, x, z) - \frac{1}{2}\frac{\partial^2}{\partial z^2} \pi(t, x, z)\right]dz = -\frac{1}{2} \lim_{z \uparrow g(t)} \frac{\partial}{\partial z} \pi(t, x, z).$$

Assuming that the last limit exists and observing that the left-hand side of our condition (30) coincides with $\pi(t, x, z)/p(t, x, z)$, we see from (30) that one should have $\frac{\partial}{\partial z} \pi(t, x, z) \to -f(t, x)p(t, x, g(t))$ as $z \uparrow g(t)$, which leads to (31).

Further developing the technique used when proving Theorem 4 we can derive an interesting result that could be interpreted as refinement of the assertions of Theorem 3. Namely, assuming again for simplicity that $T = 1$ and denoting by

$$P(g) := \mathbf{P}_x\left(\sup_{0 \le s \le 1} (X_s - g(s)) < 0\right)$$

the probability that our diffusion $\{X_s\}$ does not cross the boundary $g \in \text{Lip}_K$ during the time interval $[0, 1]$, Theorem 3 claims that the functional $P(g)$ is "locally Lipschitz" in the uniform norm:

$$|P(g + h) - P(g)| \le C(g) \sup_{0 \le t \le 1} |h(t)|.$$

The next theorem from [6] proves that $P(g)$ is actually Gâteaux differentiable and, moreover, gives a representation for the derivative in terms of the Brownian meander process $\{W_s^\oplus\}_{0 \le s \le 1}$. Recall that the process can be defined as follows: letting $\tau = \sup\{t \le 1 : W_t = 0\}$ be the last zero of the Brownian motion in $[0, 1]$, we set

$$W_s^\oplus = (1 - \tau)^{-1/2}|W_{\tau + (1-\tau)s}|, \quad 0 \le s \le 1.$$

This is a continuous non-homogeneous Markov process that appears as the limit in a number of conditional functional central limit theorems [2] (see also [10] and further references therein).

Theorem 5 *Let $\{X_s\}$ be a non-explosive diffusion satisfying (20) with diffusion interval* $\mathbf{R}$ *and with* $\mu \in C^1$ *such that (20) has a unique strong solution and such that there exists a function* $\overline{Q}(y)$ *satisfying*

$$\mu'(y) + \mu^2(y) \geq -\overline{Q}(y), \quad y \in \mathbf{R}, \qquad \limsup_{y \to -\infty} \frac{\overline{Q}(y)}{y^2} < 1.$$

Assume that $g(t)$ and $h(t)$, $0 \leq t \leq 1$, are twice continuously differentiable. Then there exists the limit

$$\lim_{\varepsilon \to 0} \varepsilon^{-1}[P(g + \varepsilon h) - P(g)] = \int_0^1 h(v)\psi(v)p_\tau(v)dv,$$

where

$$\psi(v) = \left(\frac{2}{\pi(1-v)}\right)^{1/2} \mathbf{E} \exp\{G(g(1) - \sqrt{1-v}W_1^\oplus)$$
$$- G(g(v)) + \sqrt{1-v}W_1^\oplus g'(1) + \overline{N}_{1-v}(1-v)\},$$

$\{W_s^\oplus\}_{0 \leq s \leq 1}$ *is the Brownian meander, G was defined in (24), $g_{0,u}(s) := g(1 - u + s)$, $0 \leq s + u \leq 1$, and we set*

$$\overline{N}_u(t) := -\frac{1}{2}\int_0^t \left[\mu'(-\sqrt{t}W_{s/t}^\oplus + g_{0,u}(s)) + \mu^2(-\sqrt{t}W_{s/t}^\oplus + g_{0,u}(s))\right]ds$$
$$- \sqrt{t}\int_0^t g_{0,u}''(s)W_{s/t}^\oplus ds - \frac{1}{2}\int_0^t (g_{0,u}'(s))^2 ds.$$

The proof of this theorem consists of four steps. For the first three steps, we assume $h(t) \geq 0$, $0 \leq t \leq 1$. In the first step, we observe that the difference $P(g + \varepsilon h) - P(g)$ can be written as an integral by conditioning on the first crossing time τ of g. In the second step we transform the integrand (using Girsanov's theorem, transforming the trajectory space and then using Girsanov's theorem once again) so it is written as the product of an expectation of a functional of the Brownian meander and a well-known fixed level non-crossing probability for the Brownian motion. In the third step, we calculate the limit of the ratio of the thus obtained expression to ε as $\varepsilon \to 0$. This requires Theorem 2.1 from [10] and involves careful treatment near the right end point of the integration interval. Finally, in the fourth step, we show how to extend the result to general h which are twice continuously differentiable (for more detail, see [6]).

To illustrate the above result, consider the case of the Wiener process and linear functions $g(t) = a_1 + b_1 t$ and $h(t) = a_2 + b_2 t$, $0 \leq t \leq 1$, where $b_1, a_2, b_2 \in \mathbf{R}$ and, without loss of generality, $a_1 > 0$. Then the assertion of Theorem 5 reduces to

$$\lim_{\varepsilon \to 0} \varepsilon^{-1}[P(g + \varepsilon h) - P(g)]$$

$$= \sqrt{\frac{2}{\pi}} \int_0^1 \frac{a_2 + b_2(1-t)}{\sqrt{t}} \mathbf{P}_x(1 - \tau \in dt) \mathbf{E} \, e^{\sqrt{t} b_1 W_1^\oplus - \frac{1}{2} b_1^2 t}. \tag{32}$$

On the other hand, we know (see e.g. (1.1.4) on p. 250 in [4]) that, for the boundary $a + bt$ with $a, b > 0$,

$$P(a + bt) = \overline{\Phi}(-a - b) - e^{-2ba} \Phi(-a + b), \tag{33}$$

where $\Phi(x)$ is the standard normal distribution function, $\overline{\Phi}(x) = 1 - \Phi(x)$,

$$p_\tau(t) = \frac{a}{\sqrt{2\pi} t^{3/2}} \exp\left\{ -\frac{(a + bt)^2}{2t} \right\}, \tag{34}$$

while we have (see e.g. (1.1) in [10])

$$\mathbf{P}(W_1^\oplus \in dy) = y e^{-y^2/2} dy, \quad y > 0. \tag{35}$$

In the special case $b_1 = b_2 = 0$, one can evaluate both sides of (32) and find that they each give $(2/\pi)^{1/2} a_2 e^{-a_1^2/2}$. In the general case, we use (35) to evaluate the required Laplace transform:

$$\mathbf{E} \, e^{\lambda W_1^\oplus} = 1 + \sqrt{2\pi} \lambda e^{\lambda^2/2} \overline{\Phi}(-\lambda), \quad \lambda \in \mathbf{R}. \tag{36}$$

Using (33), (34) and (36), we see that in this case (32) is equivalent to the following curious identity:

$$a_2 \sqrt{\frac{2}{\pi}} e^{-(a_1 + b_1)^2/2} + 2(a_2 b_1 + a_1 b_2) e^{-2a_1 b_1} \Phi(b_1 - a_1)$$

$$= \frac{a_1}{\pi} \int_0^1 \frac{a_2 + b_2(1-t)}{\sqrt{t}(1-t)^{3/2}} \exp\left\{ -\frac{(a_1 + b_1(1-t))^2}{2(1-t)} - \frac{b_1^2 t}{2} \right\}$$

$$\times \left(1 + \sqrt{2\pi t} b_1 e^{t b_1^2/2} \overline{\Phi}(-\sqrt{t} b_1) \right) dt. \tag{37}$$

For given values of a_1, a_2, b_1 and b_2, one can verify the identity by numerically evaluating the integral on its right-hand side — that will also confirm that (32) holds for these values. Thus, using the values $a_1 = a_2 = b_1 = b_2 = 1$, both sides of (37) give 0.379. Alternatively, for the values $a_1 = 1$, $a_2 = -0.5$, $b_1 = -1$, $b_2 = 2$ (in which case the "increment" $h(t)$ assumes values of both signs on [0, 1]), both sides of (37) give 0.442.

Acknowledgement This research was supported by the ARC Centre of Excellence for Mathematics and Statistics of Complex Systems (MASCOS) and ARC Discovery Grant DP0880693.

References

1. Baldi, P., Caramellino, L.: Asymptotics of hitting probabilities for general one-dimensional pinned diffusions. Ann. Appl. Probab. **12**, 1071–1095 (2002)

2. Belkin, B.: An invariance principle for conditioned recurrent random walk attracted to a stable law. Z. Wahrscheinlichkeitstheor. Verw. Geb. **21**, 45–64 (1972)
3. Bingham, N.H., Kiesel, R.: Risk-Neutral Valuation. Springer, London (1998)
4. Borodin, A.N., Salminen, P.: Handbook of Brownian Motion – Facts and Formulae, 2nd edn. Birkhäuser, Basel (2002)
5. Borovkov, K., Burq, Z.: Kendall's identity for the first crossing time revisited. Electron. Commun. Probab. **6**, 91–94 (2001)
6. Borovkov, K., Downes, A.N.: On boundary crossing probabilities for diffusion processes. Stoch. Proc. Appl. [To appear] (2009)
7. Borovkov, K., Novikov, A.: Explicit bounds for approximation rates of boundary crossing probabilities for the Wiener process. J. Appl. Probab. **42**, 82–92 (2005)
8. Downes, A.N., Borovkov, K.: First passage densities and boundary crossing probabilities for diffusion processes. Methodol. Comput. Appl. Probab. **10**, 621–644 (2008)
9. Durbin, J.: The first-passage density of a continuous Gaussian process to a general boundary. J. Appl. Probab. **22**, 99–122 (1985)
10. Durrett, R.T., Iglehart, D.L., Miller, D.R.: Weak convergence to Brownian meander and Brownian excursion. Ann. Probab. **5**, 117–129 (1977)
11. Haug, E.G.: The Complete Guide to Option Pricing Formulas, 2nd edn. McGraw-Hill, New York (2006) [For the current list of corrections to the book, see http://www.espenhaug.com/Corrections2ndEdtion.pdf. Cited 8 Oct 2009]
12. Merton, R.: Theory of rational option pricing. Bell J. Econ. Manag. Sci. **4**, 141–184 (1973)
13. Di Nardo, E., Nobile, A.G., Pirozzi, E., Ricciardi, L.M.: A computational approach to first-passage-time problems for Gauss-Markov processes. Adv. Appl. Probab. **33**, 453–482 (2001)
14. Novikov, A.A., Frishling, V., Kordzakhia, N.: Approximations of boundary crossing probabilities for a Brownian motion. J. Appl. Probab. **36**, 1019–1030 (1999)
15. Novikov, A.A., Frishling, V., Kordzakhia, N.: Time-dependent barrier options and boundary crossing probabilities. Georgian Math. J. **10**, 325–334 (2003)
16. Potzelberger, K., Wang, L.: Boundary crossing probability for Brownian motion. J. Appl. Probab. **38**, 152–164 (2001)
17. Reiner, E., Rubinstein, M.: Breaking down the barriers. Risk **4**(8), 28–35 (1991)
18. Revuz, D., Yor, M.: Continuous Martingales and Brownian Motion, 3rd edn. Springer, Berlin (1999)
19. Roberts, G.O., Shortland, C.F.: The hazard rate tangent approximation for boundary hitting times. Ann. Appl. Probab. **5**, 446–460 (1995)
20. Rogers, L.C.G.: Smooth transition densities for one-dimensional diffusions. Bull. London Math. Soc. **17**, 157–161 (1985)
21. Wang, L., Pötzelberger, K.: Boundary crossing probability for Brownian motion and general boundaries. J. Appl. Probab. **34**, 54–65 (1997)

Binomial Models for Interest Rates

John van der Hoek

Abstract Recombining binomial tree models are very convenient for derivative pricing. But traditionally trinomial tree approximations are often used and have become a standard way of approximating continuous time interest rate models. We provide a methodology for using binomial models rather than trinomial models for such approximation and provide comparison with these trinomial approaches. We will demonstrate some advantages of these binomial models over the popular trinomial tree models.

We believe our approach is easier to apply and will perform better computationally and will help to improve some of the methods used by other researchers promoting lattice methods. The probabilities produced by our methodology are never negative for any discretization step length. The binomial model constructed by our method also preserves properties of the continuous time model like mean reversion.

1 The Methodology

The details of the general approximation methodology was originally presented in van der Hoek in [16]. There we presented a novel way to approximate a diffusion by a recombining binomial tree model. The method is obtained by approximating a procedure to find a weak solution of a stochastic differential equation. We indicated some theory which provided analysis that the method does indeed provide an approximation. We presented a list of examples of one dimensional diffusions and an illustrative two dimensional example. Here we will now review the theory and then provide applications to interest rate modeling.

In derivative pricing and in risk analysis, it is the distribution of an underlying process that is important, not the path wise properties of its solution. It is for this

J. van der Hoek (✉)
School of Mathematics and Statistics, University of South Australia, City West Campus, GPO Box 2471, Adelaide, South Australia 5001, Australia
e-mail: john.vanderhoek@unisa.edu.au

C. Chiarella, A. Novikov (eds.), *Contemporary Quantitative Finance*,
DOI 10.1007/978-3-642-03479-4_18, © Springer-Verlag Berlin Heidelberg 2010

reason that we focus on providing a discretization of a weak solution. For the strong solution we seek X with inputs (X_0, B) in (1) and (2) below, while for the weak solution we seek the pair (X, B) for a given input X_0. We start with giving a method to construct weak solutions, and then provide and approximation of it. The weak solution concept is frequently employed in stochastic control problems. It seems that weak solution construction given below is not widely known and may be novel in its own right.

We shall first provide a recombining binomial tree model approximation for the solution of the stochastic differential equation:

$$dX(t) = \mu(t, X(t))dt + \sigma(t, X(t))dB(t) \tag{1}$$

$$X(0) = X_0 \tag{2}$$

on some probability space $(\Omega, \mathscr{F}, P)$, where B is standard one-dimensional Brownian motion. We will study these equations over a time interval $[0, T]$.

We shall then provide examples of two dimensional models and provide an appendix for trinomial approximations which could be extended to multinomial models.

1.1 The Weak Solution in Continuous Time

We construct a weak solution to (1) and (2) as follows:

Step 1:

On a probability space $(\Omega, \mathscr{F}, \overline{P})$ let $\overline{B}$ be standard one-dimensional Brownian motion and suppose that

$$X(t) = \phi(t, \overline{B}(t)) \tag{3}$$

where ϕ solves the differential equation

$$\frac{\partial \phi}{\partial z}(t, z) = \sigma(t, \phi(t, z)) \tag{4}$$

$$\phi(0, 0) = X_0 \tag{5}$$

then

$$dX(t) = \overline{m}(t, \overline{B}(t))dt + \sigma(t, X(t))d\overline{B}(t) \tag{6}$$

where

$$\overline{m}(t, \overline{B}(t)) = \frac{\partial \phi}{\partial t}(t, \overline{B}(t)) + \frac{1}{2}\sigma(t, \phi(t, \overline{B}(t)))\frac{\partial \sigma}{\partial z}(t, \phi(t, \overline{B}(t))). \tag{7}$$

These statements follow from Itô's Lemma provided that a solution of (4) and (5) is smooth enough. These conditions can be checked in any application.

In other words, for the Brownian motion $\overline{B}$ we have given the correct form to the volatility term in (1).

Step 2:

We now make a change of probabilities to adjust the drift in (6) to coincide with that in (1). This can be achieved by setting:

$$\frac{dP}{d\overline{P}}\bigg|_{\mathscr{F}_T} = \Lambda_T = \exp\left[\int_0^T \psi(u)d\overline{B}(u) - \frac{1}{2}\int_0^T \psi(u)^2 du\right] \qquad (8)$$

for suitable ψ satisfying the Novikov condition, say, when

$$B(t) = \overline{B}(t) - \int_0^t \psi(u)du \qquad (9)$$

is standard one-dimensional Brownian motion under P and $\{\mathscr{F}_t\}$ is the filtration generated by $\overline{B}$.

Under P, (6) becomes

$$dX(t) = \overline{m}(t, \overline{B}(t))dt + \sigma(t, X(t))[dB(t) + \psi(t)dt] \qquad (10)$$

and we choose ψ so that

$$\mu(t, X(t)) = \overline{m}(t, \overline{B}(t)) + \sigma(t, X(t))\psi(t) \qquad (11)$$

or

$$\psi(t) = \frac{\mu(t, \phi(t, \overline{B}(t))) - m(t, \overline{B}(t))}{\sigma(t, \phi(t, \overline{B}(t)))} \equiv \Psi(t, \overline{B}(t)). \qquad (12)$$

Thus under P, X given in (3) provides a (weak) solution of (1).

In each application of this construction, technical conditions may need to be checked so that all steps in the construction are valid.

1.2 The Approximation of the Weak Solution

Here and below, we will use the notation for binomial models used in van der Hoek and Elliott in [17].

Let N be a positive integer and let $\Delta t = \frac{T}{N}$. We then define:

$$X(0, 0) = X_0 \qquad (13)$$

$$X(n, j) = \phi(n\Delta t, (2j - n)\sqrt{\Delta t}) \qquad (14)$$

for $j = 0, 1, \ldots, n$ and $n = 0, 1, \ldots, N$.

From (n, j) (time n and state j) we can move to either $(n+1, j+1)$ or $(n+1, j)$. In this way we obtain a recombining binomial tree of values for X. If $(n, j) \to (n+1, j+1)$ and $(n, j) \to (n+1, j)$ occur with equal probability, then X in (13) and (14) provides a numerical approximation to (6). For this we refer the reader to [13].

We now assign new probabilities $p(n, j)$ to $(n, j) \to (n+1, j+1)$ and $1 - p(n, j)$ to $(n, j) \to (n+1, j)$ so that X in (13) and (14) provides a numerical approximation to (1). We now motivate the formulas for the $p(n, j)$. The details of the convergence are again provided by Nelson and Ramaswamy in [13].

Let us set for $t < s$

$$\Lambda_{t,s} = \exp\left[\int_t^s \psi(u)d\overline{B}(u) - \frac{1}{2}\int_t^s \psi(u)^2 du\right] \tag{15}$$

and let Y be $\mathscr{F}_s$ measurable. Then using E for expectations under P and $\overline{E}$ for expectations under $\overline{P}$

$$E[Y|\mathscr{F}_t] = \frac{\overline{E}[\Lambda_{0,T}Y|\mathscr{F}_t]}{\overline{E}[\Lambda_{0,T}|\mathscr{F}_t]}$$

$$= \frac{\overline{E}[\Lambda_{t,s}Y|\mathscr{F}_t]}{\overline{E}[\Lambda_{t,s}|\mathscr{F}_t]}$$

One now applies this calculation with $s = t + \Delta t$ and with $Y = Y_+ = \mathbf{I}[\overline{B}(t + \Delta t) - \overline{B}(t) = \sqrt{\Delta t}]$ and $Y = Y_- = \mathbf{I}[\overline{B}(t + \Delta t) - \overline{B}(t) = -\sqrt{\Delta t}]$. We use $\mathbf{I}(A)(\omega) = 0$ or 1 depending on whether $\omega \in A$ or not.

Of course in this we are employing the approximation

$$\overline{B}(t + \Delta t) - \overline{B}(t) = \pm\sqrt{\Delta t}$$

with equal probabilities under $\overline{P}$. We are led to the approximation:

$$E[Y_+|\mathscr{F}_t] \approx \frac{\frac{1}{2}\exp\left[\psi(t)\sqrt{\Delta t} - \frac{1}{2}\psi(t)^2\Delta t\right]}{\frac{1}{2}\exp\left[\psi(t)\sqrt{\Delta t} - \frac{1}{2}\psi(t)^2\Delta t\right] + \frac{1}{2}\exp\left[-\psi(t)\sqrt{\Delta t} - \frac{1}{2}\psi(t)^2\Delta t\right]}$$

$$= \frac{\frac{1}{2}\exp\left[\psi(t)\sqrt{\Delta t}\right]}{\frac{1}{2}\exp\left[\psi(t)\sqrt{\Delta t}\right] + \frac{1}{2}\exp\left[-\psi(t)\sqrt{\Delta t}\right]}$$

$$= \frac{1}{2} + \frac{1}{2}\tanh\left[\psi(t)\sqrt{\Delta t}\right] \tag{16}$$

and likewise

$$E[Y_-|\mathscr{F}_t] \approx \frac{1}{2} - \frac{1}{2}\tanh\left[\psi(t)\sqrt{\Delta t}\right]. \tag{17}$$

These heuristic calculations lead to our choices for the $p(n, j)$. We use (12) to set

$$p(n, j) = \frac{1}{2} + \frac{1}{2}\tanh\left[\Psi(n\Delta t, (2j - n)\sqrt{\Delta t})\sqrt{\Delta t}\right]. \tag{18}$$

When $N \to \infty$ the results of [13] show that X in (13) and (14) converges to a solution to (1).

We can also apply this analysis to a system of d stochastic differential equations driven by d-dimensional Brownian motion using analogous arguments. This is illustrated in Sect. 4 with $d = 2$.

Let us also note that without restrictions on the size of Δt, the probabilities satisfy $0 \le p(n, j) \le 1$. This already shows an important advantage of this approach over the moment matching methods which do not share this property.

While our focus will be on binomial recombining models, it is easy to modify this procedure to construct multinomial approximations. We will provide brief details for the trinomial case, by way of example, in the Appendix.

2 The Black and Scholes Model

This is the simplest example. We have $X = S$, a stock price process. Write $S(n, j)$ instead of $X(n, j)$ and S for $X(0, 0)$.

We have $\mu(t, x) = \mu x$ and $\sigma(t, x) = \sigma x$. Then $\phi(t, z) = S \exp(\sigma z)$ and we set

$$S(n, j) = S \exp\left[(2j - n)\sigma\sqrt{\Delta t}\right]$$
$$= S u^j d^{n-j}$$

where

$$u = \exp\left[\sigma\sqrt{\Delta t}\right], \qquad d = \exp\left[-\sigma\sqrt{\Delta t}\right]$$
$$\psi(t) = \frac{\mu}{\sigma} - \frac{1}{2}\sigma$$

and

$$p(n, j) = \frac{1}{2} + \frac{1}{2}\tanh\left[\left(\frac{\mu}{\sigma} - \frac{1}{2}\sigma\right)\sqrt{\Delta t}\right].$$

Without going into detailed error analysis, we now make some brief remarks connected with moment matching.

If we are modeling S in the risk-neutral world, then $\mu = r$ a constant risk free rate of interest. In the Cox, Ross and Rubinstein binomial approximation (see [5]), the probabilities used were (for $0 < \Delta t < (\sigma/r)^2$)

$$\pi(n, j) = \frac{\exp[r\Delta t] - \exp\left[-\sigma\sqrt{\Delta t}\right]}{\exp\left[\sigma\sqrt{\Delta t}\right] - \exp\left[-\sigma\sqrt{\Delta t}\right]} \equiv p$$

while we are using instead, without restrictions on Δt,

$$p(n, j) = \frac{1}{2} + \frac{1}{2}\tanh\left[\left(\frac{r}{\sigma} - \frac{1}{2}\sigma\right)\sqrt{\Delta t}\right] \equiv \pi.$$

In the continuous set up we can calculate risk neutral moments

$$\mathbf{E}[S(t+\Delta t)^m \mid \mathscr{F}_t] = S(t)^m \exp\left(\left(r - \frac{1}{2}\sigma^2\right)m\Delta t + \frac{1}{2}m^2\sigma^2\Delta t\right) \equiv S(t)^m F_c(m, \Delta t).$$

This expression could be compared with (thinking of t as $n\Delta t$)

$$S(n, j)^m\left[\pi \exp\left[m\sigma\sqrt{\Delta t}\right] + (1 - \pi)\exp\left[-m\sigma\sqrt{\Delta t}\right]\right] \equiv S(n, j)^m F_\pi(m, \Delta t)$$

and

$$S(n, j)^m\left[p \exp\left[m\sigma\sqrt{\Delta t}\right] + (1 - p)\exp\left[-m\sigma\sqrt{\Delta t}\right]\right] \equiv S(n, j)^m F_p(m, \Delta t).$$

Then

$$|F_c(m, \Delta t) - F_\pi(m, \Delta t)| = C_\pi(m)(\Delta t) + O((\Delta t)^2)$$

with $C_\pi(m) = 0$ for $m = 1$ but $C_\pi(m) \neq 0$ if $m > 1$, while

$$|F_c(m, \Delta t) - F_p(m, \Delta t)| = C_p(m)(\Delta t)^2 + O((\Delta t)^3)$$

for all m.

Despite this, The CRR and our method perform similarly for a standard European call option. We illustrate this with results for $T = 1$ and $\Delta t = T/N$ with N taking various values. We take $\mu = r = 0.05$, $\sigma = 0.20$, strike $K = 100$, initial stock price $S(0) = 95$ and 100. The exact Black–Scholes (BS) values are also provided.

N	$S = 95$		$S = 100$	
	CRR	vdH	CRR	vdH
100	7.5195	7.5201	10.4298	10.4306
500	7.51	7.5101	10.4464	10.4466
1000	7.5095	7.5095	10.4485	10.4486
2000	7.5112	7.5113	10.4495	10.4496
3000	7.5103	7.5103	10.4499	10.4499
4000	7.5107	7.5107	10.4501	10.4501
5000	7.5106	7.5106	10.4502	10.4502
10000	7.5109	7.5109	10.4504	10.4504
BS	7.5109		10.4506	

3 One Factor Interest Rate Models

We illustrate our approach with applications to some well known examples.

3.1 The Hull and White One-factor Model

We consider first the Hull-White (see [7]) model ($X = r$ here)

$$dr(t) = (\theta(t) - ar(t))dt + \sigma dB(t) \qquad (19)$$

which models the short rate $r(t)$ in the risk neutral world. The $\theta(t)$ is determined to match the initial term structure, to match the correct values for the zero-coupon bond prices $P(0, t)$. In fact these zero-coupon bond prices can be expressed explicitly

$$P(t, T) = E\left[\exp\left(-\int_t^T r(u)du\right)\right] = \exp(A(t, T) + B(t, T)r(t))$$

where

$$B(t, T) = -\frac{1}{a}(1 - e^{-a(T-t)})$$

$$A(t, T) = \int_t^T \theta(u) B(u, T) du + \frac{1}{2} \int_t^T \sigma^2 B(u, T)^2 du$$

and using forward prices $f(s, t)$ defined though

$$P(t, T) = \exp\left(-\int_t^T f(t, u)du\right)$$

from which

$$f(t, T) = -\frac{\partial}{\partial T} \log P(t, T).$$

We then have the well known result

$$\theta(t) = af(0, t) + \frac{\partial f}{\partial t}(0, t) - \frac{\sigma^2}{2a^2}(1 - e^{-2at})$$

for which we would require a smooth (twice continuous differentiable) interpolation of zero-coupon bond prices $P(0, t)$.

Given values for $\theta(t)$ we can now build a binomial recombining tree for interest rates $r(n, j)$ as follows:

$$\phi(t, z) = r(0) + \sigma z$$

$$\psi(t) = \frac{\theta(t) - ar(t)}{\sigma}$$

$$r(n, j) = r(0) + \sigma(2j - n)\sqrt{\Delta t}$$

$$p(n, j) = \frac{1}{2} + \frac{1}{2} \tanh\left[\frac{\theta(n\Delta t) - ar(n, j)}{\sigma}\sqrt{\Delta t}\right].$$

We note that $p(n, j) > \frac{1}{2}$ when $\theta(n\Delta t) > ar(n, j)$ and $p(n, j) < \frac{1}{2}$ when $\theta(n\Delta t) < ar(n, j)$. The binomial model then displays a mean-reversion of interest rates to $\theta(t)/a$.

However Hull-White proceed in an alternate manner. They first construct a trinomial tree for the process:

$$d\tilde{r}(t) = -a\tilde{r}dt + \sigma dB(t)$$

with $\tilde{r}(0) = 0$. We can approximate this process with a binomial tree using

$$\tilde{r}(n, j) = \sigma(2j - n)\sqrt{\Delta t}$$

$$p(n, j) = \frac{1}{2} + \frac{1}{2} \tanh\left[-\frac{a}{\sigma}\tilde{r}(n, j)\sqrt{\Delta t}\right]$$

$$= \frac{1}{2} + \frac{1}{2} \tanh[-a(2j - n)\Delta t].$$

A deterministic function $\alpha(t)$ is then constructed so that $r(t) = \tilde{r}(t) + \alpha(t)$. In fact

$$\frac{d\alpha}{dt}(t) + a\,\alpha(t) = \theta(t).$$

Let us write α_n for $\alpha(n\Delta t)$, and set

$$r(n, j) = \alpha_n + \tilde{r}(n, j) = \alpha_n + \sigma(2j - n)\sqrt{\Delta t}.$$

We now compute the α_n as did Hull-White so that we price the zero-coupon bonds with this model to agree with $P(0, n\Delta t)$ for all n.

We introduce state prices $\lambda(n, j)$ for nodes (n, j). See [17, pp. 69–71] for a discussion and computation. Then if we know the time n state prices we find α_n from

$$\sum_{j=0}^{n} \lambda(n, j)\exp(-r(n, j)\Delta t) = P(0, (n + 1)\Delta t)$$

by

$$\alpha_n = -\frac{1}{\Delta t}\log\left[\frac{P(0, (n + 1)\Delta t)}{\sum_{j=0}^{n}\lambda(n, j)\exp(-\tilde{r}(n, j)\Delta t)}\right].$$

Of course $\lambda(0, 0) = 1$ and once we have α_n we have $r(n, j)$ and then the $\lambda(n + 1, j)$ are determined from Jamshidian [12] forward induction:

$$\lambda(n + 1, j) = (1 - p(n, j))\beta(n, j)\lambda(n, j) + p(n, j - 1)\beta(n, j - 1)\lambda(n, j - 1)$$

where $\beta(n, j) = \exp(-r(n, j)\Delta t)$ and we always set $\lambda(n, j) = 0$ is $j < 0$ or $j > n$.

We now present a numerical example and compare our results with those of Hull-White in [7]. They used a model for yields

$$y(t) = 0.08 - 0.05\exp(-0.18\,t)$$

so that the zero coupon bond prices are given by

$$P(0, t) = \exp(-t\,y(t)).$$

For example,

t	$y(t)$	$P(0, t)$
1 year	0.038236	0.962485
2 year	0.045116	0.913719
3 year	0.050863	0.858484

Hull-White then calibrate the model

$$dr(t) = (\theta(t) - 0.1\,r(t))dt + 0.01\,dB(t)$$

so $a = 0.1$ and $\sigma = 0.01$. As in Hull-White, we provide illustrations with $\Delta t = 1$. Using $\lambda(0, 0) = 1$ and $\tilde{r}(0, 0) = 0$,

$$\alpha_0 = -\frac{1}{\Delta t} \log\left[\frac{P(0, \Delta t)}{\lambda(0, 0) \exp(-\tilde{r}(0, 0)\Delta t)}\right] = 0.038236.$$

Then $r(0, 0) = 0.038236$. We then proceed with the binomial model rather that the Hull-White trinomial model. Using our formulas $p(0, 0) = 0.5$, $\beta(0, 0) = \exp(-r(0, 0)\Delta t) = 0.962485$ from which we compute $\lambda(1, 0) = \lambda(1, 1) = 0.481242$. From our formulas, $\tilde{r}(1, 0) = -\sigma\sqrt{\Delta t} = -0.01$ and $\tilde{r}(1, 1) = \sigma\sqrt{\Delta t} = 0.01$. Then

$$\alpha_1 = -\frac{1}{\Delta t} \log\left[\frac{P(0, 2\Delta t)}{\lambda(1, 0) \exp(-\tilde{r}(1, 0)\Delta t) + \lambda(1, 1) \exp(-\tilde{r}(1, 1)\Delta t)}\right] = 0.052045.$$

Hull-White reported $\alpha_1 = 0.0520$ and so we have agreement. Then $r(1, 0) = 0.042045$ and $r(1, 1) = 0.062045$. We can then compute $\beta(1, 0)$ and $\beta(1, 1)$ and $\lambda(2, 0)$, $\lambda(2, 1)$ and $\lambda(2, 2)$ and $\alpha_2 = 0.062636$, and so forth. In tabular form, we give some of the answers. For $\lambda(n, j)$ we have

	$n = 0$	$n = 1$	$n = 2$	$n = 3$
$j = 0$	1.0000	0.4812	0.2077	0.0799
$j = 1$		0.4812	0.5024	0.3551
$j = 2$			0.2036	0.3482
$j = 1$				0.0752

and for interest rates $r(n, j)$ and the risk-neutral probabilities $p(n, j)$, we have

$r(0, 0) =$	0.0382	$r(1, 0) =$	0.0420	$r(2, 0) =$	0.0425
$p(0, 0) =$	0.5000	$p(1, 0) =$	0.5498	$p(2, 0) =$	0.5987
		$r(1, 1) =$	0.0620	$r(2, 1) =$	0.0625
		$p(1, 1) =$	0.4502	$p(2, 1) =$	0.5000
				$r(2, 2) =$	0.0825
				$p(2, 2) =$	0.4013

Once that α_n values have been computed in this manner to match the initial term structure, we have a set of values for $(r(n, j), p(n, j))$ on the recombining binomial tree and this can then be used to price a variety of derivatives in the usual way.

3.2 A Black-Karasinsky Type Model

Hull-White in [9] considered interest models of the form

$$df(r(t)) = (\theta(t) - af(r(t)))dt + \sigma dB(t)$$

to include the Black and Karasinski [1] model where $f(r) = \log r$. Pelsser in his
PhD thesis (see [14]) used $f(r) = \sqrt{r}$ (see [11]).

Thus $r(t) = g(X(t))$ where g is the inverse of f and

$$dX(t) = (\theta(t) - aX(t))dt + \sigma dB(t).$$

Again, $\theta(t)$ is to be chosen so that we can match observed zero coupon bond prices
from market data. Now there is no formula for $\theta(t)$ as in the previous example.
However we propose to model X as in the previous example

$$X(n, j) = \alpha_n + \sigma(2j - n)\sqrt{\Delta t}$$

$$p(n, j) = \frac{1}{2} + \frac{1}{2}\tanh[-a(2j - n)\Delta t].$$

Now α_n is found numerically by solving

$$\sum_{j=0}^{n} \lambda(n, j)\exp\big(g\big(\alpha_n + \sigma(2j - n)\sqrt{\Delta t}\big)\Delta t\big) = P(0, (n + 1)\Delta t).$$

Hull-White solve this equation by the Newton-Raphson procedure, but as the left
hand side is monotone in α_n in most applications, interval division algorithm may
be easier to program.

If a and σ are functions of t, we may easily modify these methods.

3.3 Cox, Ingersoll and Ross Model

The Cox, Ingersoll and Ross (see [4]) model can be approximated as follows. We
use $X(t) = r(t)$ again in the general theory. The model is

$$dr(t) = (a - br(t))dt + \sigma\sqrt{r(t)}dB(t).$$

We have $\mu(t, x) = a - bx$ and $\sigma(t, x) = \sigma\sqrt{x}$. Then

$$\phi(t, z) \equiv \phi(z) = \begin{cases} \left[\sqrt{r(0)} + \frac{1}{2}\sigma z\right]^2 & \text{if } \sqrt{r(0)} + \frac{1}{2}\sigma z \geq 0 \\ 0 & \text{otherwise} \end{cases}$$

$$r(n, j) = \phi((2j - n)\sqrt{\Delta t})$$

$$\psi(t) = \left(\frac{a}{\sigma} - \frac{\sigma}{4}\right)\frac{1}{\sqrt{r(t)}} - \frac{b}{\sigma}\sqrt{r(t)}$$

$$p(n, j) = \frac{1}{2} + \frac{1}{2}\tanh\left\{\left[\left(\frac{a}{\sigma} - \frac{\sigma}{4}\right)\frac{1}{\sqrt{r(n, j)}} - \frac{b}{\sigma}\sqrt{r(n, j)}\right]\sqrt{\Delta t}\right\}$$

and we note that as $r(n, j) \to 0+$

$$p(n, j) \to \begin{cases} 1 & \text{if } a > \frac{1}{4}\sigma^2 \\ 0 & \text{if } a < \frac{1}{4}\sigma^2 \end{cases}$$

which is why we often assume the first case ($\sigma^2 < 4a$) in this model.

Valuing bond and derivative prices have traditionally used the solution of partial differential equations and their approximation. This often leads to trinomial tree methods if explicit finite difference approximations are used. See for example, [8]. As an illustration we use our binomial method to value bond prices with the same model data as was used in this paper. Let us use

$$dr(t) = 0.4\,(0.1 - r(t))dt + 0.06\sqrt{r(t)}\,dB(t)$$

so $a = 0.04$, $b = 0.4$, $\sigma = 0.06$. We will consider $r(0) = 0.06, 0.08, 0.10, 0.12, 0.14$ and maturities $T = 5, 10, 15, 20$ years. Hull and White use $\Delta t = 0.05$ years and this is used in vdH1. For vdH2 we use $\Delta t = 0.005$. As an exact solution for bond prices exists for this model, the various numerical results are compared with the exact values.

We now give the MATLAB code used to illustrate the simplicity of using our method.

```
function[price,lattice] = cirv(r0, a, b, sig, T, N)
%price = cirv(r0, a, b, sig, T, N)
%This function calculates the zero coupon bond prices
%in CIR model: dr(t) = (a- b r(t))dt + sig sqrt(r(t)) dB(t)
%This function calculates P(0,T) using an N step tree.

deltaT = T./N; lattice = zeros(N+1, 1);
%this declares the space needed for lattice.

for j = 0:N
   lattice(j+1) = 1;
end
% This sets V(N,j) = 1 for j = 0,1, ..., N.
% Initially lattice(j+1) = V(N,j)

for n = N-1:-1:0
   for j = 0:n
      r(j+1) = (max(sqrt(r0)+0.5*sig.*(2*j-n).*sqrt(deltaT),...
               eps))^2;
      %r(j+1) is r(n,j) in this loop

      ps(j+1) = (((a/sig)-(sig/4))/sqrt(r(j+1))) - ...
               (b/sig)*sqrt(r(j+1));
      p(j+1) = 0.5 + 0.5*tanh(ps(j+1).*sqrt(deltaT));
      %p(j+1) is p(n,j) in this loop

      lattice(j+1) = exp(-r(j+1).*deltaT).*(lattice(j+1) + ...
                  p(j+1).*(lattice(j+2)-lattice(j+1)));
      %lattice(j+1) is V(n,j) in this loop

   end
end price = lattice(1)
```

The zero coupon bond prices values are given for 100 maturity value. HW are the Hull and White values given in [8] and we have used the exact values computed there using the exact formula in [4].

The model with $\sigma(t, x) = \sigma x^\beta$ with $0 < \beta < 1$ is treated in a similar way. Many other one-factor interest rate models can be treated in a similar way.

		6%	8%	10%	12%	14%
5 years	HW	66.31	63.53	60.86	58.30	55.85
	vdH1	66.3502	63.4725	60.7814	58.204	55.7254
	vdH2	66.2458	63.4541	60.7821	58.2226	55.7698
	Exact	66.24	63.45	60.78	58.23	55.78
10 years	HW	40.92	38.98	37.14	35.38	33.71
	vdH1	40.8773	38.9075	37.0477	35.2764	33.5821
	vdH2	40.8353	38.8947	37.0478	35.2885	33.6121
	Exact	40.83	38.89	37.05	35.29	33.62
15 years	HW	25.02	23.82	22.68	21.59	20.55
	vdH1	24.9654	23.7447	22.5931	21.4972	20.4495
	vdH2	24.939	23.7364	22.5928	21.5043	20.4677
	Exact	24.94	23.74	22.59	21.51	20.47
20 years	HW	15.28	14.54	13.85	13.18	12.55
	vdH1	15.2288	14.4828	13.7791	13.1095	12.4694
	vdH2	15.2124	14.4776	13.7787	13.1136	12.4803
	Exact	15.21	14.48	13.78	13.11	12.48

4 Two Factor Interest Rate Models

We refer the reader to [16] for the discussion of a two factor Schwartz and Smith (see [15]) model and notation.

4.1 The Brennan-Schwartz Two-factor Model

We consider the Brennan and Schwartz (see [2, 3]) model for short and long rates:

$$dr(t) = (a_1 + b_1(L(t) - r(t)))dt + \sigma_1 r(t)dB_1(t) \qquad (20)$$

$$dL(t) = L(t)(a_2 + b_2 r(t) + c_2 L(t))dt + \sigma_2 L(t)dB_2(t). \qquad (21)$$

We will assume correlation $dB_1(t)dB_2(t) = \varrho dt$. We will assume in this example that $a_1, a_2, b_1, b_2, c_2, \sigma_1, \sigma_2, \rho$ are all constants.

It is convenient to write

$$B_2(t) = \varrho B_1(t) + \sqrt{1 - \varrho^2} B_1^*(t)$$

where B_1 and B_1^* are independent Brownian motions.

We now let $\bar{B}_1$ and $\bar{B}_1^*$ be arbitrary independent standard Brownian motions and set

$$\bar{B}_2(t) = \varrho \bar{B}_1(t) + \sqrt{1 - \varrho^2} \bar{B}_1^*(t).$$

Set

$$r(t) = r(0) \exp(\sigma_1 \bar{B}_1(t))$$

$$L(t) = L(0) \exp(\sigma_2 \bar{B}_2(t))$$

$$= L(0) \exp\left(\sigma_2\left(\varrho \bar{B}_1(t) + \sqrt{1 - \varrho^2} \bar{B}_1^*(t)\right)\right).$$

This leads to the approximations:

$$r(n, j, k) = r(0) \exp\left(\sigma_1 (2j - n)\sqrt{\Delta t}\right)$$

$$L(n, j, k) = L(0) \exp\left(\sigma_2\left[\varrho(2j - n)\sqrt{\Delta t} + \sqrt{1 - \varrho^2}(2k - n)\sqrt{\Delta t}\right]\right)$$

and now for each node (n, j, k) we must calculate four probabilities:

$$p_1(n, j, k) \quad \text{for} \quad (n, j, k) \to (n+1, j+1, k+1)$$
$$p_2(n, j, k) \quad \text{for} \quad (n, j, k) \to (n+1, j+1, k)$$
$$p_3(n, j, k) \quad \text{for} \quad (n, j, k) \to (n+1, j, k+1)$$
$$p_4(n, j, k) \quad \text{for} \quad (n, j, k) \to (n+1, j, k).$$

Note that

$$dr(t) = \frac{1}{2}\sigma_1^2 r(t)dt + \sigma_1 r(t)d\bar{B}_1(t)$$

$$dL(t) = \frac{1}{2}\sigma_2^2 L(t)dt + \sigma_2 L(t)d\bar{B}_2(t)$$

$$= \frac{1}{2}\sigma_2^2 L(t)dt + \sigma_2 L(t)\left(\varrho d\bar{B}_1(t) + \sqrt{1 - \varrho^2}d\bar{B}_1^*(t)\right).$$

We choose ψ_1 and ψ_2 so that

$$dB_1(t) = d\bar{B}_1(t) - \psi_1(t)dt$$
$$dB_1^*(t) = d\bar{B}_1^*(t) - \psi_2(t)dt$$

so that (20) and (21) hold. This leads to

$$\psi_1(t) = \frac{a_1 + b_1(L(t) - r(t)) - \frac{1}{2}\sigma_1^2 r(t)}{\sigma_1 r(t)}$$

$$\psi_2(t) = \frac{a_2 + b_2 r(t) + c_2 L(t) - \sigma_2 \varrho \psi_1(t) - \frac{1}{2}\sigma_2^2}{\sigma_2 \sqrt{1 - \varrho^2}}.$$

We then set

$$\psi_1(n, j, k) = \frac{a_1 + b_1(L(n, j, k) - r(n, j, k)) - \frac{1}{2}\sigma_1^2 r(n, j, k)}{\sigma_1 r(n, j, k)}$$

$$\psi_2(n, j, k) = \frac{a_2 + b_2 r(n, j, k) + c_2 L(n, j, k) - \sigma_2 \varrho \psi_1(n, j, k) - \frac{1}{2}\sigma_2^2}{\sigma_2 \sqrt{1 - \varrho^2}}$$

and

$$\tau_1(n, j, k) = \tanh\left[\psi_1(n, j, k)\sqrt{\Delta t}\right]$$

$$\tau_2(n, j, k) = \tanh\left[\psi_2(n, j, k)\sqrt{\Delta t}\right]$$

we use:

$$p_1(n, j, k) = \frac{1}{4}(1 + \tau_1(n, j, k))(1 + \tau_2(n, j, k))$$

$$p_2(n, j, k) = \frac{1}{4}(1 + \tau_1(n, j, k))(1 - \tau_2(n, j, k))$$

$$p_3(n, j, k) = \frac{1}{4}(1 - \tau_1(n, j, k))(1 + \tau_2(n, j, k))$$

$$p_4(n, j, k) = \frac{1}{4}(1 - \tau_1(n, j, k))(1 - \tau_2(n, j, k)).$$

It is automatic in this construction that $p_i(n, j, k) \geq 0$ for each $i = 1, 2, 3, 4$ and the probabilities sum to 1.

4.2 The Hull-White Two-factor Model

Hull-White in [10] considered the two-factor model:

$$df(r(t)) = [\theta(t) + u(t) - af(r(t))]dt + \sigma_1 dB_1(t)$$

$$du(t) = -bu(t)dt + \sigma_2 dB_2(t)$$

with $dB_1(t)dB_2(t) = \varrho dt$. In this example a, b, σ_1, σ_2 are constants. The values of $\theta(t)$ must be chosen so that the model agrees with values of $P(0, t)$ from market data. This model has a stochastic reversion level contributed by u.

This example is treated in a similar fashion to the two factor Schwartz and Smith model in [16] for an approximation to

$$d\tilde{X}(t) = [u(t) - a\tilde{X}(t)]dt + \sigma_1 dB_1(t)$$

$$du(t) = -bu(t)dt + \sigma_2 dB_2(t).$$

Then

$$X(t) = \alpha(t) + \tilde{X}(t)$$

with

$$\frac{d\alpha}{dt}(t) + a\alpha(t) = \theta(t)$$

as before. We then continue as in example 3.1 and 3.2 above.

5 Conclusions

We have presented a methodology for the construction of binomial recombining tree models as approximations for continuous time interest rate models. The method is based on approximating the weak solution of the continuous time model and is not based on moment matching constructions that are often used, but on approximating the Girsanov theorem. Each application will have to be studied to deal with any of its own technical issues. We also note that the risk neutral probabilities constructed by our method will always satisfy $0 \leq p(n, j) \leq 1$ without restrictions on the discretization time step Δt. Moment matching usually require some restriction on the discretization time step size for their validity. For this reason our methods are easier to apply and to program. Valuing American style interest rate derivative are now straight forward, and do not require the numerical solution of free boundary value problems for partial differential equations. It is useful to compare our methods for two dimensional examples with that proposed in [6].

We admit that our methodology is difficult to apply in models where the process X involves some form of feedback and in multidimensional problems with the volatility structure depending on more than one process. This would be the case with the Heston stochastic volatility model. Further approximations are needed to adapt our method to such problems, and is the subject of ongoing research.

Appendix

Trinomial trees use the approximation

$$\bar{B}(t + \Delta t) - \bar{B}(t) = \begin{cases} \sqrt{\frac{3}{2}\Delta t} & \text{with probability } \bar{p} = \frac{1}{3} \\ 0 & \text{with probability } 1 - \bar{p} - \bar{q} = \frac{1}{3} \\ -\sqrt{\frac{3}{2}\Delta t} & \text{with probability } \bar{q} = \frac{1}{3}. \end{cases}$$

It is easily verified that this approximation matches the first two moments of the increment of the Brownian motion.

We label the nodes on a trinomial tree with (n, j) as in binomial trees, but for each $n \geq 0$, j takes integer values between $-n$ and $+n$. Thus we have

$$X(0, 0) = X_0 \tag{22}$$

$$X(n, j) = \phi\left(n\Delta t, j\sqrt{\frac{3}{2}\Delta t}\right). \tag{23}$$

We then move from (n, j) to $(n + 1, j + 1)$, $(n + 1, j)$ or to $(n + 1, j - 1)$. We then obtain a recombining trinomial tree. If these transitions occur with equal probabilities, then X in (22) and (23) provides a numerical approximation to (6).

If we assign new probabilities $p(n, j)$ to $(n, j) \to (n + 1, j + 1)$ and $q(n, j)$ to $(n, j) \to (n + 1, j - 1)$ and $1 - p(n, j) - q(n, j)$ to $(n, j) \to (n + 1, j)$ with

$$p(n, j) = \frac{\exp\left(\Psi\left(n\Delta t, j\sqrt{\tfrac{3}{2}\Delta t}\right)\sqrt{\tfrac{3}{2}\Delta t}\right)}{\exp\left(-\Psi\left(n\Delta t, j\sqrt{\tfrac{3}{2}\Delta t}\right)\sqrt{\tfrac{3}{2}\Delta t}\right) + 1 + \exp\left(\Psi\left(n\Delta t, j\sqrt{\tfrac{3}{2}\Delta t}\right)\sqrt{\tfrac{3}{2}\Delta t}\right)}$$

$$q(n, j) = \frac{\exp\left(-\Psi\left(n\Delta t, j\sqrt{\tfrac{3}{2}\Delta t}\right)\sqrt{\tfrac{3}{2}\Delta t}\right)}{\exp\left(-\Psi\left(n\Delta t, j\sqrt{\tfrac{3}{2}\Delta t}\right)\sqrt{\tfrac{3}{2}\Delta t}\right) + 1 + \exp\left(\Psi\left(n\Delta t, j\sqrt{\tfrac{3}{2}\Delta t}\right)\sqrt{\tfrac{3}{2}\Delta t}\right)}.$$

Then $p(n, j) \geq 0$, $q(n, j) \geq 0$ and $1 - p(n, j) - q(n, j) \geq 0$ and with these new probabilities X in (22) and (23) provides a numerical approximation to (1).

References

1. Black, F., Karasinski, P.: Bond and option pricing when the short rates are lognormal. Financ. Anal. J., July–August, 52–59 (1991)
2. Brennan, M.J., Schwartz, E.S.: An equilibrium model of bond pricing and a test of market efficiency. J. Financ. Quant. Anal. **17**, 301–329 (1982)
3. Brennan, M.J., Schwartz, E.S.: Alternative methods for valuing debt options. J. Finance **4**, 118–138 (1983)
4. Cox, J.C., Ingersol, J.E., Ross, R.A.: An equilibrium characterization theory of the term structure. Econometrica **53**, 385–407 (1985)
5. Cox, J.C., Ross, S.A., Rubinstein, M.: Option pricing: A simplified approach. J. Financ. Econ. **7**, 229–263 (1979)
6. Hahn, W.J., Dyer, J.S.: A discrete-time approach for valuing real options with underlying mean-reverting stochastic processes. McCombs School of Business, The University of Texas at Austin (2004) and Eur. J. Oper. Res. **84**, 534–548 (2007)
7. Hull, J., White, A.: Pricing interest rate derivative securities. Rev. Financ. Stud. **3**, 573–592 (1990)
8. Hull, J., White, A.: Valuing derivative securities using the explicit finite difference method. J. Financ. Quant. Anal. **25**, 87–100 (1990)
9. Hull, J., White, A.: Numerical methods for implementing term structure models I: Single-factor models. J. Deriv., Fall, 7–16 (1994)
10. Hull, J., White, A.: Numerical methods for implementing term structure models II: Two-factor models. J. Deriv., Winter, 37–48 (1994)
11. James, J., Webber, N.: Interest Rate Modelling. Wiley, New York (2000)
12. Jamshidian, F.: Forward induction and construction of yield curves in diffusion models. J. of Fixed Income, June, 62–74 (1991)
13. Nelson, D.B., Ramaswamy, K.: Simple binomial processes and diffusion approximations in financial models. Rev. Financ. Stud. **3**, 393–430 (1990)
14. Pelsser, A.: Efficient Methods for Valuing Interest Rate Derivatives. Springer, Berlin (2000)
15. Schwartz, E., Smith, J.E.: Short-term variations and long-term dynamics in commodity prices. Manag. Sci. **46**, 893–911 (2000)
16. van der Hoek, J.: Recombining binomial tree approximations for diffusions. In: Bensoussan, A., Zhang, Q. (eds.) Mathematical Modeling and Numerical Methods in Finance, pp. 361–368. Elsevier, Amsterdam (2009)
17. van der Hoek, J., Elliott, R.J.: Binomial Models in Finance. Springer, Berlin (2005)
18. Vasicek, O.: A theory of the term structure of interest rates. J. Financ. Econ. **5**, 177–188 (1977)

Lognormal Forward Market Model (LFM) Volatility Function Approximation

In-Hwan Chung, Tim Dun, and Erik Schlögl

Abstract In the lognormal forward Market model (LFM) framework, the specification for time-deterministic instantaneous volatility functions for state variable forward rates is required. In reality, only a discrete number of forward rates is observable in the market. For this reason, traders routinely construct time-deterministic volatility functions for these forward rates based on the tenor structure given by these rates. In any practical implementation, however, it is of considerable importance that volatility functions can also be evaluated for forward rates not matching the implied tenor structure. Following the deterministic arbitrage-free interpolation scheme introduced by Schlögl in (Advances in Finance and Stochastics: Essays in Honour of Dieter Sondermann. Springer, Berlin 2002) in the LFM, this paper, firstly, derives an approximate analytical formula for the volatility function of a forward rate not matching the original tenor structure. Secondly, the result is extended to a swap rate volatility function under the lognormal forward rate assumption. Finally, a modified Black's market formula is derived for the cases of the payoff dates differing from the original tenor structure, employing the LFM convexity adjustment formula. Implications for Monte Carlo simulation of non-tenor rates are also discussed. Most importantly in this context, the present paper introduces a simulation method avoiding the unrealistic behaviour of implied volatilities for interpolated caplets documented in (Schlögl, Advances in Finance and Stochastics: Essays in Honour of Dieter Sondermann. Springer, Berlin 2002).

I.-H. Chung
ANZ, 2/20 Martin Place, Sydney NSW 2000, Australia
e-mail: inhwan.chung@anz.com

T. Dun
Westpac Banking Corporation, 275 Kent Street, Sydney NSW 2000, Australia
e-mail: tdun@westpac.com.au

E. Schlögl (✉)
Quantitative Finance Research Centre, University of Technology, Sydney, Broadway NSW 2007, Australia
e-mail: erik.schlogl@uts.edu.au

C. Chiarella, A. Novikov (eds.), *Contemporary Quantitative Finance*,
DOI 10.1007/978-3-642-03479-4_19, © Springer-Verlag Berlin Heidelberg 2010

1 Introduction

Since its publication, the Black/Scholes formula has undoubtedly become widely accepted by practitioners. Black's formula, the fixed income market version of the Black/Scholes formula, has played a significant role in the interest rate derivatives market as well. The LFM[1] embeds Black's formula for caplets in a consistent term structure model, and thus is very attractive for market practitioners. The use of Black's formula for caplets and, in close approximation, for swaptions facilitates the all-important calibration to at-the-money implied volatilities observed in the market, while a consistent model allows for the pricing of more exotic instruments with path-dependency or Bermudan exercise. However, many interest rate derivatives cannot be priced analytically. In addition, to incorporate correlation effects between forward rate dynamics, the driving diffusion is quite often multi-dimensional. Furthermore, the joint dynamics of forward rates in the LFM are Markovian only in a large number of state variables at best. For these reasons, Monte Carlo simulation (MC) is typically the numerical method of choice in the LFM.

Another issue is that although Brace, Gatarek and Musiela in [6] show that the LFM can be extended to continuous tenor using all δ-compounded forward rates, this is impractical. Firstly, their continuous-tenor model is Markovian only in an infinite-dimensional state variable — the continuous term structure of interest rates. Secondly, in reality, only a finite number of forward rates are observable in the market. Typically, a tenor structure comprises three-month-spaced tenor dates. A volatility function calibrated given this tenor structure, consequently, only provides volatility information for forward rates relevant to these tenor dates. In other words, there is no direct and unique way of getting volatility information from calibrated volatility functions for forward rates with start and end dates[2] not matching the tenor dates. To resolve this problem, one needs an interpolation scheme.

Since pricing by MC is quite slow, as far as European style caps and swaptions are concerned, a Black's type formula is sought. For caps not conforming to the tenor structure, and for swaptions in general, at first glance this sort of analytical formula seems infeasible. However, under some assumptions that will be discussed later, a quite accurate approximate analytical formula for European-style caps and swaptions can be derived. A number of authors such as Brace [2, 3], Brace Gatarek and Musiela [6], Andersen and Andreasen [1], Hull and White [9], and Jäckel and Rebonato [11] present formulae of this type. Unfortunately, they are applicable only when the instruments are standard.[3] Note that non-standard features can occur even for instruments, which match the standard tenor structure at their inception. For

---

[1] See [6, 12] and [10].

[2] A forward rate is defined using two bond prices. The earlier maturity of the two marks the start date and the latter the end date. An exact definition will be given below.

[3] The term "standard" means that all payment schedules are matching the tenor dates of a given discrete-tenor model. Accordingly, "non-standard" represents the case where all or some of the schedules are non-standard.

example, suppose a trader calibrates time-deterministic volatility functions for standard forward rates based on a pre-determined tenor structure using observable market instrument prices today. Based on the volatility functions, the trader quotes a standard instrument price, say, a plain vanilla European swaption price and writes the option. However, the next morning, the trader re-calibrates the volatility functions, and finds that the plain vanilla instrument sold yesterday is not standard anymore since all the schedules of that instrument are shifted by one day relative to the tenor dates on which the new calibration is based. As time goes on this problem grows progressively worse — given a three-month tenor structure, dates for existing instruments may be out of sync by up to a month and a half. Already for this reason alone, an approximate formula accepting non-standard features is needed.

The purpose of this paper firstly is to provide such a formula in the discrete tenor LFM framework. Following the deterministic interpolation scheme introduced in [16] in the LFM, this paper derives an approximate analytical formula for the volatility function of a non-tenor forward rate. Secondly, the result is extended to a swap rate volatility function under the lognormal forward rate assumption. We generalise in such ways as to permit

- Differing (and completely generic) reset and LIBOR start, end, and payoff dates; and
- Correct day count fractions.

Secondly, for analytical valuation as well as Monte Carlo simulation, we propose a method of "dead rate simulation" to avoid the undesirable effect that naïve interpolation has on caplet implied volatilities. Without dead rate simulation, interpolated caplet implied volatilities between standard tenor points are substantially lower than those for the standard tenor dates, an effect which is generally considered unrealistic and unacceptable by practitioners. We will illustrate how "dead rate simulation" resolves this issue.

In this context, one should note that our choice of interpolation scheme, combined with "dead rate simulation," is by no means the only way of tackling the problem. Tang and Li in [17] describe a generic method of ensuring that an interpolation scheme is consistent with the absence of arbitrage by preserving the relevant martingale properties, though they do not specifically address avoiding the implied drop in interpolated caplet volatilities. Brace in [4] also proposes methods for interpolating rates for maturities between node points in the tenor grid, and discusses "dead rate simulation" for these methods. However, Brace notes that these methods are not strictly arbitrage-free, though "in practice they work fairly accurately." Our particular choice is motivated by its relative simplicity and its straightforward interpretation, as in [16], as an arbitrage-free method of interpolation by day count fractions.

The structure of the paper is as follows. In Sect. 2, generic forward rates are defined and a brief exposition of the interpolation employed is given, followed by a detailed caplet volatility formula derivation in Sect. 3. There we present an instantaneous volatility function for a generic forward rate in terms of the volatility function for standard forward rates using the interpolation scheme. From this instantaneous volatility function, we propose an approximate Black's volatility for a generic caplet. Convexity adjustment and dead rate simulation techniques are explored and a modified Black's caplet formula[4] is derived. Next, in Sect. 4, an instantaneous volatility function for a generic swap rate is derived using an instantaneous volatility function for standard forward rates, and thus an approximation for Black's swaption volatility is derived. The performance of the formulae are tested against MC and the results are presented in Sect. 5.

2 Generic Forward Rates

2.1 Base Definitions

Assume a tenor structure $\{T_k \mid k = 0, 1, \ldots, N + 1\}$ such that $T_{k+1} = T_k + \delta_k$ and $T_0 = 0$. The δ_k can be arbitrary. However, a typical value is approximately three months or six months, or one year represented in terms of year fractions. Using that tenor structure, we can define a forward rate as

$$K(t, T_k, T_{k+1}) = \frac{1}{\delta_k}\left(\frac{P(t, T_k)}{P(t, T_{k+1})} - 1\right) \tag{1}$$

where $P(t, T)$ is the time t price of a zero coupon bond (ZCB) maturing at time T. This definition states that a forward rate is specified by two bond prices. The maturity of the shorter bond determines the start date and that of the longer one represents the end date of the accrual period of the forward rate. Since the start and end dates all match the tenor dates, we call such forward rates *standard* and the tenor *base tenor*. Unfortunately, in practice, most of the required forward rates are not standard.

As is the usual practice in the LFM literature, we also define the tenor structure locator function $\eta(t) = \{k : T_{k-1} \leq t < T_k\}$.

2.2 Generic Forward Rate Definition

Consider a generic forward rate with the following configuration:

- forward start date in years: T^s; and
- forward end date in years: T^e

Denote by $K(t, T^s, T^e)$ the forward rate setting at date t for the accrual period between dates T^s and T^e as given by

$$K(t, T^s, T^e) = \frac{1}{\delta}\left(\frac{P(t, T^s)}{P(t, T^e)} - 1\right) \tag{2}$$

---

[4] In Appendix 7.2.

where $\delta = T^e - T^s$. Since there is no guarantee that each pair of T^s and T^e for $K(t, T^s, T^e)$ matches the given tenor dates, we call these rates *non-standard forward rates*.

When the reset date t is equal to T^s, $K(t, T^s, T^e)$ becomes a spot rate covering the (T^s, T^e) period. In case of $t < T^s$, the forward rate $K(t, T^s, T^e)$ is determined by the yield curve realised at t. In case of t being greater than T^s, $K(t, T^s, T^e)$ is sometimes called a *dead rate* because it is already realised.

$P(t, T^s)$ and $P(t, T^e)$ denote the bond prices at time t maturing at non-tenor dates T^s and T^e, respectively. From the time t yield curve, one needs to calculate these bond prices with some form of interpolation strategy. What interpolation strategy one would choose would be a matter of preference, though certain no-arbitrage consistency conditions need to be taken into account. In the next section, we will discuss one of the available interpolation methods.

2.3 Interpolation by Day-Count Fractions

Following the arbitrage-free *interpolation by day-count fractions* approach, we can write the ratio of a zero coupon bond maturing at a non-tenor date to a ZCB maturing at the next tenor date as follows:

$$\frac{P(t, T)}{P(t, T_{\eta(T)})} = 1 + (T_{\eta(T)} - T) K(t, T_{\eta(T)-1}, T_{\eta(T)}).$$

Consistent application of this interpolation method is arbitrage free, see [16]. Henceforth, we drop the end date in the notation of $K(t,,)$ as far as start and end dates match tenor dates.

Applying this to the start and end dates, we have

$$\frac{P(t, T^s)}{P(t, T_{\eta(T^s)})} = 1 + \delta_s K(t, T_{\eta(T^s)-1}) \tag{3}$$

$$\frac{P(t, T^e)}{P(t, T_{\eta(T^e)})} = 1 + \delta_e K(t, T_{\eta(T^e)-1}) \tag{4}$$

for

$$\delta_s = T_{\eta(T^s)} - T^s$$
$$\delta_e = T_{\eta(T^e)} - T^e.$$

Consider now the ratio

$$\frac{P(t, T_{\eta(T^s)})}{P(t, T_{\eta(T^e)})}$$

which appears in the definition of the generic forward rate. Using the interpolation strategy, we can rewrite the ratio as

$$\frac{P(t, T^s)}{P(t, T^e)} = \frac{P(t, T^s)}{P(t, T_{\eta(T^s)})} \frac{P(t, T_{\eta(T^s)})}{P(t, T_{\eta(T^e)})} \frac{P(t, T_{\eta(T^e)})}{P(t, T^e)}. \tag{5}$$

The middle term is given by

$$\frac{P(t, T_{\eta(T^s)})}{P(t, T_{\eta(T^e)})} = \prod_{i=\eta(T^s)}^{\eta(T^e)-1} (1 + \delta_i K(t, T_i)). \tag{6}$$

Thus, from (3), (4), and (6), the ZCB ratio becomes

$$\frac{P(t, T^s)}{P(t, T^e)} = \frac{1 + \delta_s K(t, T_{\eta(T^s)-1})}{1 + \delta_e K(t, T_{\eta(T^e)-1})} \prod_{i=\eta(T^s)}^{\eta(T^e)-1} (1 + \delta_i K(t, T_i)). \tag{7}$$

Then (2) can rewritten as

$$K(t, T^s, T^e) = \frac{1}{\delta} \left(\frac{1 + \delta_s K(t, T_{\eta(T^s)-1})}{1 + \delta_e K(t, T_{\eta(T^e)-1})} \prod_{i=\eta(T^s)}^{\eta(T^e)-1} (1 + \delta_i K(t, T_i)) - 1 \right).$$

3 Caplet Volatility

This section consists of four subsections. In the first subsection, we are going to derive the instantaneous volatility function for a non-standard forward process $\frac{P(t,T^s)}{P(t,T^e)}$ in terms of the standard volatilities. This will be done via the derivation of the forward process stochastic differential equation (SDE). We recognise the forward process as a product of two functions appearing in (7).

Since the drift terms in the SDE can be eliminated by an appropriate change of measure, we are only interested in the volatility terms. Thus, for notational simplicity, we will omit drift terms in the SDEs presented below. Similarly, we simply use dW_t to represent a Wiener increment under any probability measure appropriate to the context.

The second subsection will transform the non-standard forward process SDE into the non-standard forward rate SDE, followed by the non-standard caplet volatility approximation. The remaining part will deal with Black's caplet formula and relevant issues arising from the early reset or late payoff of the caplet.

3.1 Ito Calculus

3.1.1 Step 1

Let

$$X_t = \prod_{i=\eta(T^s)}^{\eta(T^e)-1} (1 + \delta_i K(t, T_i))$$

$$dX_t = (\ldots)dt + X_t v_X dW_t \tag{8}$$

and

$$Y_t = \frac{1 + \delta_s K(t, T_{\eta(T^s)-1})}{1 + \delta_e K(t, T_{\eta(T^e)-1})}$$

$$dY_t = (\ldots)dt + Y_t \nu_Y dW_t.$$

(9)

Thus ν_X and ν_Y represent the relative volatility functions for X_t and Y_t, respectively. Then, we have

$$d\left(\frac{P(t, T^s)}{P(t, T^e)}\right) = d(X_t Y_t)$$

$$= X_t dY_t + Y_t dX_t + (\ldots)dt$$

$$= X_t Y_t (\nu_Y + \nu_X)dW_t + (\ldots)dt.$$

(10)

3.1.2 Step 2: Derivation of ν_X

Given a discrete-tenor lognormal LIBOR Market Model, a standard forward rate is assumed to be an exponential martingale under the appropriate probability measure and we can write the forward rate dynamics as

$$dK(t, T_i) = K(t, T_i)\gamma(t, T_i)dW_t^{T_{i+1}}$$

(11)

under that forward measure relevant to the bond price maturing at T_{i+1} as the numeraire asset. The $\gamma(t, T_i)$ represents the volatility function for $K(t, T_i)$. In general, W_t is a d-dimensional vector-valued Brownian motion and γ a vector-valued function, i.e. (11) is to be interpreted as

$$dK(t, T_i) = K(t, T_i)\sum_{j=1}^{d}\gamma_j(t, T_i)dW_t^{(j)}$$

where $W_t^{(j)}$ is the j-th component of the vector-valued Brownian motion (in (11) the j-th component of $W_t^{T_{i+1}}$). Under any equivalent probability measure,

$$dK(t, T_i) = K(t, T_i)\gamma(t, T_i)dW_t + (\ldots)dt$$

where dW_t is Wiener increment under that measure. Now, we apply Ito's lemma to X_t:

$$dX_t = \sum_{i=\eta(T^s)}^{\eta(T^e)-1} \frac{\partial X_t}{\partial K(t, T_i)} dK(t, T_i)$$

$$+ \frac{1}{2} \sum_{i=\eta(T^s)}^{\eta(T^e)-1} \sum_{j=\eta(T^s)}^{\eta(T^e)-1} \frac{\partial^2 X_t}{\partial K(t, T_i)\partial K(t, T_j)} d\langle K(t, T_i), K(t, T_j)\rangle$$

$$= \sum_{i=\eta(T^s)}^{\eta(T^e)-1} \frac{\partial X_t}{\partial K(t, T_i)} dK(t, T_i) + (\ldots)dt.$$

(12)

Since

$$\frac{\partial X_t}{\partial K(t, T_i)} = \frac{\delta_i X_t}{1 + \delta_i K(t, T_i)}$$

we obtain

$$dX_t = \sum_{i=\eta(T^s)}^{\eta(T^e)-1} \frac{\delta_i X_t}{1 + \delta_i K(t, T_i)} dK(t, T_i) + (\ldots)dt$$

$$= \sum_{i=\eta(T^s)}^{\eta(T^e)-1} \frac{\delta_i X_t}{1 + \delta_i K(t, T_i)} K(t, T_i)\gamma(t, T_i)dW_t + (\ldots)dt$$

$$= X_t \sum_{i=\eta(T^s)}^{\eta(T^e)-1} \mu(t, T_i)\gamma(t, T_i)dW_t + (\ldots)dt \tag{13}$$

where

$$\mu(t, T_i) = \frac{\delta_i K(t, T_i)}{1 + \delta_i K(t, T_i)}.$$

Thus, we have

$$\nu_X = \sum_{i=\eta(T^s)}^{\eta(T^e)-1} \mu(t, T_i)\gamma(t, T_i). \tag{14}$$

3.1.3 Step 3: Derivation of ν_Y

From

$$Y_t = \frac{1 + \delta_s K(t, T_{\eta(T^s)-1})}{1 + \delta_e K(t, T_{\eta(T^e)-1})},$$

denote the numerator by S_t and the denominator by E_t. Then

$$dS_t = \delta_s K(t, T_{\eta(T^s)-1})\gamma(t, T_{\eta(T^s)-1})dW_t + (\ldots)dt$$
$$dE_t = \delta_e K(t, T_{\eta(T^e)-1})\gamma(t, T_{\eta(T^e)-1})dW_t + (\ldots)dt.$$

Thus, using the result (37) in the Appendix:

$$dY_t = d\left(\frac{S_t}{E_t}\right)$$
$$= Y_t \left[\frac{\delta_s K(t, T_{\eta(T^s)-1})\gamma(t, T_{\eta(T^s)-1})}{1 + \delta_s K(t, T_{\eta(T^s)-1})} - \frac{\delta_e K(t, T_{\eta(T^e)-1})\gamma(t, T_{\eta(T^e)-1})}{1 + \delta_e K(t, T_{\eta(T^e)-1})}\right]dW_t$$
$$+ (\ldots)dt$$
$$= Y_t \left(\mu_s(t, T_{\eta(T^s)-1})\gamma(t, T_{\eta(T^s)-1}) - \mu_e(t, T_{\eta(T^e)-1})\gamma(t, T_{\eta(T^e)-1})\right)dW_t$$
$$+ (\ldots)dt \tag{15}$$

where

$$\mu_s(t, T_{\eta(T^s)-1}) = \frac{\delta_s K(t, T_{\eta(T^s)-1})}{1 + \delta_s K(t, T_{\eta(T^s)-1})}$$

$$\mu_e(t, T_{\eta(T^e)-1}) = \frac{\delta_e K(t, T_{\eta(T^e)-1})}{1 + \delta_e K(t, T_{\eta(T^e)-1})}.$$

3.1.4 Step 4

Combining everything together, we have

$$v_X = \sum_{i=\eta(T^s)}^{\eta(T^e)-1} \mu(t, T_i)\gamma(t, T_i)$$

$$v_Y = \mu_s(t, T_{\eta(T^s)-1})\gamma(t, T_{\eta(T^s)-1}) - \mu_e(t, T_{\eta(T^e)-1})\gamma(t, T_{\eta(T^e)-1})$$

and

$$d\left(\frac{P(t, T^s)}{P(t, T^e)}\right)$$

$$= X_t Y_t (v_Y + v_X) dW_t + (\ldots)dt$$

$$= X_t Y_t \left[\mu_s(t, T_{\eta(T^s)-1})\gamma(t, T_{\eta(T^s)-1}) - \mu_e(t, T_{\eta(T^e)-1})\gamma(t, T_{\eta(T^e)-1}) \right.$$

$$\left. + \sum_{i=\eta(T^s)}^{\eta(T^e)-1} \mu(t, T_i)\gamma(t, T_i) \right] dW_t + (\ldots)dt$$

$$= \frac{P(t, T^s)}{P(t, T^e)} \left[\mu_s(t, T_{\eta(T^s)-1})\gamma(t, T_{\eta(T^s)-1}) - \mu_e(t, T_{\eta(T^e)-1})\gamma(t, T_{\eta(T^e)-1}) \right.$$

$$\left. + \sum_{i=\eta(T^s)}^{\eta(T^e)-1} \mu(t, T_i)\gamma(t, T_i) \right] dW_t + (\ldots)dt. \tag{16}$$

Now, we have the volatility for the non-standard forward process, i.e. the expression inside the square bracket.

3.2 Non-standard Forward SDE

To simplify the notation of the SDE accounting for the diffusion dynamics for the non-standard forward process, $\frac{P_t, T^s)}{P(t, T^e)}$, we introduce $\Theta(t, T_i)$, defined by

$$\Theta(t, T_{\eta(T^s)-1}) = \mu_s(t, T_{\eta(T^s)-1})$$

$$\Theta(t, T_i) = \mu(t, T_i), \quad i = \eta(T^s), \ldots, \eta(T^e) - 2$$

$$\Theta(t, T_{\eta(T^e)-1}) = \mu(t, T_{\eta(T^e)-1}) - \mu_e(t, T_{\eta(T^e)-1})$$

so that the SDE becomes

$$d\left(\frac{P(t,T^s)}{P(t,T^e)}\right) = \frac{P(t,T^s)}{P(t,T^e)}\left(\sum_{i=\eta(T^s)-1}^{\eta(T^e)-1}\Theta(t,T_i)\gamma(t,T_i)\right)dW_t + (\ldots)dt. \quad (17)$$

From (2), the SDE for the non-standard forward rate is

$$dK(t,T^s,T^e) = \frac{1}{\delta}d\left(\frac{P(t,T^s)}{P(t,T^e)}\right)$$

$$= \frac{1}{\delta}\frac{P(t,T^s)}{P(t,T^e)}\left(\sum_{i=\eta(T^s)-1}^{\eta(T^e)-1}\Theta(t,T_i)\gamma(t,T_i)\right)dW_t + (\ldots)dt$$

$$= \frac{1}{\delta}(1+\delta K(t,T^s,T^e))\left(\sum_{i=\eta(T^s)-1}^{\eta(T^e)-1}\Theta(t,T_i)\gamma(t,T_i)\right)dW_t + (\ldots)dt,$$

$$(18)$$

or equivalently,

$$\frac{dK(t,T^s,T^e)}{K(t,T^s,T^e)} = \frac{(1+\delta K(t,T^s,T^e))}{\delta K(t,T^s,T^e)}\left(\sum_{i=\eta(T^s)-1}^{\eta(T^e)-1}\Theta(t,T_i)\gamma(t,T_i)\right)dW_t + (\ldots)dt$$

$$= \frac{\left(\sum_{i=\eta(T^s)-1}^{\eta(T^e)-1}\Theta(t,T_i)\gamma(t,T_i)\right)}{\mu^k(t,T^s,T^e)}dW_t + (\ldots)dt \quad (19)$$

where

$$\mu^k(t,T^s,T^e) = \frac{\delta K(t,T^s,T^e)}{(1+\delta K(t,T^s,T^e))}. \quad (20)$$

By defining

$$A_i(t) = \frac{\Theta(t,T_i)}{\mu^k(t,T^s,T^e)},$$

we can re-write (19) as

$$\frac{dK(t,T^s,T^e)}{K(t,T^s,T^e)} = \sum_{i=\eta(T^s)-1}^{\eta(T^e)-1}A_i(t)\gamma(t,T_i)dW_t + (\ldots)dt. \quad (21)$$

3.3 Volatility Approximation

The SDE (21) does not lend itself to analytical solution due to the dependence of the relative volatility on the level of interest rates. To get around this difficulty, we follow Brace in [2] and apply a Wiener Chaos approximation of order zero, "freezing" the level-dependence at its initial value, i.e.

$$\left\| \sum_{i=\eta(T^s)-1}^{\eta(T^e)-1} A_i(s)\gamma(s,T_i) \right\|^2 \approx \left\| \sum_{i=\eta(T^s)-1}^{\eta(T^e)-1} A_i(0)\gamma(s,T_i) \right\|^2$$

$$= \left\| \sum_{i=\eta(T^s)-1}^{\eta(T^e)-1} A_i\gamma(s,T_i) \right\|^2 .$$

Since $A_i(0)$ no longer depends on the time variable s, we shorten notation from $A_i(s)$ to A_i. Under this approximation, Black's formula applies to options on $K(t,T^s,T^e)$, $0 \le t \le T_f$, and the implied volatility can be represented as

$$\sigma^2_{K(t,T^s,T^e)}(T_f) \approx \int_t^{T_f} \left\| \sum_{i=\eta(T^s)-1}^{\eta(T^e)-1} A_i\gamma(s,T_i) \right\|^2 ds. \qquad (22)$$

Some authors, such as Jäckel and Rebonato in [11], and Hull and White in [9], justify this approximation on the grounds that the frozen coefficients do not change much. The practicality of this approximation will be tested in Sect. 5. Simplifying the integral in (22), we have

$$\int_t^{T_f} \left\| \sum_{i=\eta(T^s)-1}^{\eta(T^e)-1} A_i\gamma(s,T_i) \right\|^2 ds$$

$$= \int_t^{T_f} \left(\sum_{i=\eta(T^s)-1}^{\eta(T^e)-1} A_i\gamma(s,T_i) \right) \left(\sum_{j=\eta(T^s)-1}^{\eta(T^e)-1} A_j\gamma(s,T_j) \right) ds$$

$$= \sum_{i=\eta(T^s)-1}^{\eta(T^e)-1} \sum_{j=\eta(T^s)-1}^{\eta(T^e)-1} A_i A_j \int_t^{T_f} \gamma(s,T_i)\gamma(s,T_j)ds$$

$$= \sum_{i=\eta(T^s)-1}^{\eta(T^e)-1} \sum_{j=\eta(T^s)-1}^{\eta(T^e)-1} A_i A_j \lambda_{ij}(t,T_f)$$

$$\text{where} \quad \lambda_{ij}(t,T_f) = \int_t^{T_f} \gamma(s,T_i)\gamma(s,T_j)ds.$$

Consequently,

$$\sigma_{K(t,T^s,T^e)} \approx \sqrt{\sum_{i=\eta(T^s)-1}^{\eta(T^e)-1} \sum_{j=\eta(T^s)-1}^{\eta(T^e)-1} A_i A_j \lambda_{ij}(t,T_f)}. \qquad (23)$$

3.4 Dead Rate Volatility

For the calculation of λ_{ij} in (23), we need to allow for cases where T_f is beyond $T_{\eta(T^s)-1}$. For example, suppose a tenor structure of $T_0 = 0$, $T_1 = 0.25$, $T_2 = 0.5$,

$T_3 = 0.75$, $T_4 = 1.0$ and a non-standard forward rate $K(T^f, T^s, T^e)$ with $T^f = 0.3$, $T^s = 0.3$ and $T^e = 0.7$. In this case, K is represented in terms of $K(\cdot, T_1)$ and $K(\cdot, T_2)$. The reset date, T^f, is beyond the start date of the relevant standard forward rate, $K(\cdot, T_1)$, used in the interpolation. In practice this rate does not evolve beyond T_1. Thus, after T_1, it is termed a *dead rate* and it would seem natural to set its volatility $\gamma(s, T_1)$ beyond T_1 to be zero.

$$\lambda_{1,2}(0, T_f) = \int_0^{0.25} \gamma(s, T_1)\gamma(s, T_2)ds + \int_{0.25}^{0.3} \gamma(s, T_1)\gamma(s, T_2)ds$$

$$= \int_0^{0.25} \gamma(s, T_1)\gamma(s, T_2)ds + 0.$$

Consequently, caplets based on non-standard rates will have lower implied volatilities than one would expect based on, say, simple linear interpolation of the implied volatilities for standard rates.[5] To prevent such a drop in volatility, we introduce *dead rate volatility*, letting a rate $K(t, T_i)$ continue to evolve beyond its reset date T_i up until its maturity date T_{i+1}. Over the period $T_i \le t < T_{i+1}$, this gives us the additional freedom to choose the volatility $\gamma(t, T_i)$. For example, it can be chosen to be constant and equal to $\gamma(T_i, T_i)$. This dead rate has no physical significance in itself, but is needed for the interpolation of short term zero coupon bonds.

Based on this, we define λ_{ij} as follows. Consider the function $\lambda_{1,2}(t, T)$ where we assume $T_1 \le T_2$ (this is not a restrictive assumption, as the function is symmetric). We therefore have three cases, depending on whether the reset date T is below, between, or above the times T_1 and T_2. Treating the cases in turn, starting from

$$\lambda_{1,2}(t, T) = \int_t^T \gamma(s, T_1)\gamma(s, T_2)ds,$$

we have

- Case 1, $T \le T_1 \le T_2$ (the "routine" case)

$$\lambda_{1,2}(t, T) = \int_t^T \gamma(s, T_1)\gamma(s, T_2)ds$$

- Case 2, $T_1 < T \le T_2$

$$\lambda_{1,2}(t, T) = \int_t^{T_1} \gamma(s, T_1)\gamma(s, T_2)ds + \gamma(T_1, T_1)\int_{T_1}^T \gamma(s, T_2)ds$$

and
- Case 3, $T_1 \le T_2 < T$

---

[5] See [16] for an example of this effect.

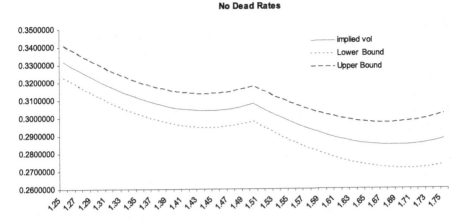

Fig. 1 Monte Carlo estimates of 40000 paths for caplets starting from 1.25 yr to 1.75 yrs without dead rates simulation: Implied volatilities with two-standard deviation bounds are plotted. Clearly volatility drops appear as in [16]

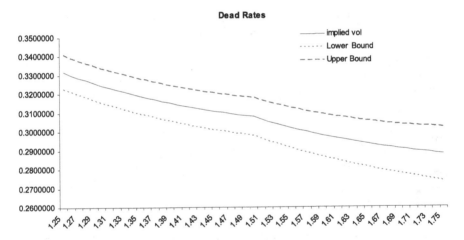

Fig. 2 Monte Carlo estimates of 40000 paths for caplets starting from 1.25 yr to 1.75 yrs with dead rates simulation: Implied volatilities with two-standard deviation bounds are plotted. Most of the volatility drops disappear

$$\lambda_{1,2}(t, T) = \int_t^{T_1} \gamma(s, T_1)\gamma(s, T_2)ds + \gamma(T_1, T_1)\int_{T_1}^{T_2} \gamma(s, T_2)ds$$
$$+ \gamma(T_1, T_1)\gamma(T_2, T_2)(T - T_2).$$

To visualise the benefits of using dead rate volatility, consider Figs. 1 to 7. Figure 1 and Figs. 3 to 5 are based on Monte Carlo prices for "broken date" caplets (i.e. caplets that fall between the original maturity grid of the model), generated without

No Dead Rates

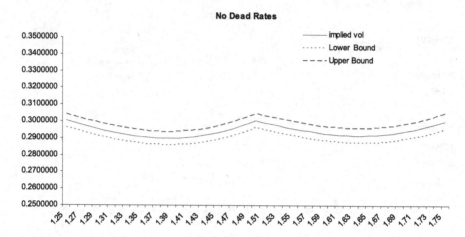

Fig. 3 Monte Carlo estimates of 40000 paths for caplets starting from 1.25 yr to 1.75 yrs without dead rates simulation: Using one factor model with flat yield of 5% and flat vol of 30%, implied volatilities with two-standard deviation bounds are plotted. Clearly volatility drops appear as in [16]

No Dead Rates

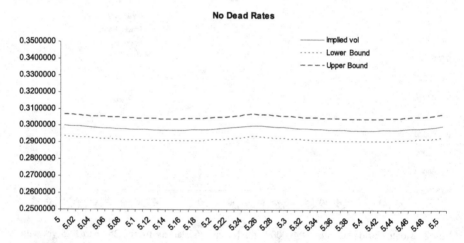

Fig. 4 Monte Carlo estimates of 40000 paths for caplets starting from 5 yr to 5.5 yrs without dead rates simulation: Using one factor model with flat yield of 5% and flat vol of 30%, implied volatilities with two-standard deviation bounds are plotted. Volatility drops start to fade out due to the relatively shorter life of the dead rate compared to that of the caplet

using dead rate volatilities, while the others are generated with dead rate volatilities. For Figs. 1 to 2 the model was calibrated to market data as of 31/March/2009 from Reuters (initial interest rate term structure and at-the-money volatilities). In order to isolate the "volatility drop" effect, the other figures are based on a one factor LFM with flat initial yield curve of 5% and a flat term structure of at-the-money caplet

No Dead Rates

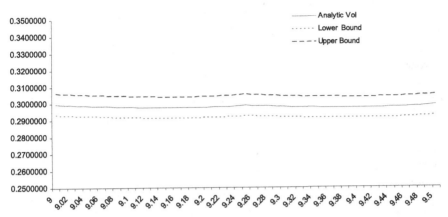

Fig. 5 Monte Carlo estimates of 40000 paths for caplets starting from 9 yr to 9.5 yrs without dead rates simulation: Using one factor model with flat yield of 5% and flat vol of 30%, implied volatilities with two-standard deviation bounds are plotted. Volatility drops almost disappear due to very small proportion of the life for the dead rate compared to that for the caplet

Dead Rates

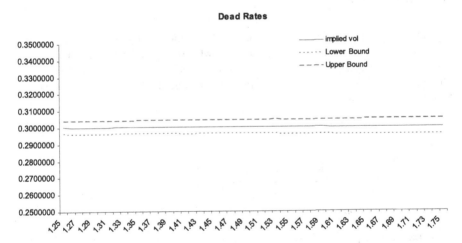

Fig. 6 Monte Carlo estimates of 40000 paths for caplets starting from 1.25 yr to 1.75 yrs with dead rates simulation: Using one factor model with flat yield of 5% and flat vol of 30%, implied volatilities with two-standard deviation bounds are plotted. Clearly volatility drops disappear

volatilities of 30%. Without dead rate volatility, the *volatility drops* are quite evident in Figs. 1 and 3. However, as the caplet expiries get longer, the effects fade out as in Fig. 4 and finally almost disappear as in Fig. 5, as the dead rates contribute proportionally less to the variance of the terminal distribution of the rate underlying the caplet. When dead rate volatilities are introduced, the volatility drops almost disappear regardless of the expiries of caplets.

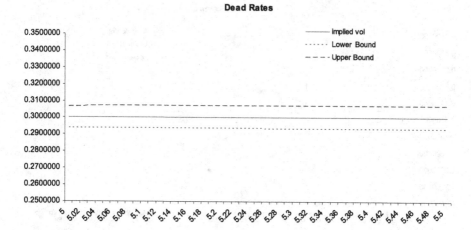

Fig. 7 Monte Carlo estimates of 40000 paths for caplets starting from 5 yr to 5.5 yrs with dead rates simulation: Using one factor model with flat yield of 5% and flat vol of 30%, implied volatilities with two-standard deviation bounds are plotted. Clearly volatility drops disappear

3.5 Caplet Volatility

The above calculations lead to fairly reasonable values for the non-standard caplet volatilities in the sense that they are similar to what one would expect from naïve interpolation of the standard caplet implied volatilities.[6] This volatility can be entered directly into the Black's caplet formula in order to get a price, where the equation with unit notional is

$$Caplet(t) = \delta^k P(t, T^p)(K(t, T^s, T^e)\Phi(d_1) - \kappa\Phi(d_2)) \qquad (24)$$

where

$$d_1 = \frac{\ln\left(\frac{K(t,T^s,T^e)}{\kappa}\right) + \frac{1}{2}\sigma_K^2(t, T^s, T^e)(T_f - t)}{\sigma_K(t, T^s, T^e)\sqrt{T_f - t}}$$

$$d_2 = d_1 - \sigma_K(t, T^s, T^e)\sqrt{T_f - t},$$

δ^k is the accrual period of the caplet, κ is the caplet strike, T^p is the payoff date, and $P(t, T^p)$ is the initial bond price maturing at T^p.

There are, of course, several approximations being made here and the formula is not consistent with an arbitrage-free model in the strict sense of the theory. For the formula to be strictly correct, $K(t, T^s, T^e)$ has to be exponential martingale under the T^p forward measure, which is true only when $T^p = T^e$. When T^p is very close to T^e, (24) may work well. However, if T^p is substantially different from T^e, a convexity adjustment has to be made, i.e., we need to modify the Black's caplet

---

[6]In fact, any desired arbitrage-free interpolation of standard caplet implied volatilities can be achieved by an appropriate choice of dead rate volatility.

formula to take into account the drift under the measure relevant to the payoff date. This modified Black's formula is given by

$$Caplet(t) = \delta^k P(t, T^p) \left[K(t, T^s, T^e) e^{\tilde{\mu}} \Phi \left(\frac{\ln\left(\frac{K(s, T^s, T^e)}{\kappa}\right) + \tilde{\mu} + \frac{1}{2}\sigma^2}{\sigma} \right) \right.$$
$$\left. - \kappa \Phi \left(\frac{\ln\left(\frac{K(s, T^s, T^e)}{\kappa}\right) + \tilde{\mu} - \frac{1}{2}\sigma^2}{\sigma} \right) \right], \tag{25}$$

where

$$\tilde{\mu} = - \int_t^{T_f} \sum_{i=\eta(T^e)-1}^{\eta(T^p)-1} \frac{\delta_i K(t, T_i, T_{i+1})}{1 + \delta_i K(t, T_i, T_{i+1})} \gamma(s, T_i)\gamma^*(s)ds$$

$$\sigma^2 = \int_t^{T_f} \|\gamma^*(s)\|^2 ds.$$

$\gamma^*(s)$ denotes the instantaneous volatility for the forward rate $K(s, T^s, T^e)$ and $\delta_i = T_{i+1} - T_i$, $\Phi()$ is a cumulative normal density function. A detailed derivation of the modified Black's formula is presented in the Appendix. If we regard $K(t, T^s, T^e) e^{\tilde{\mu}}$ as the convexity adjusted rate, then we still have the original Black's formula intact, where the initial rate is replaced by the adjusted rate.

3.6 LFM Convexity Adjustment

Employing Girsanov's Theorem,

$$dW_t^{T^p} = dW_t^{T^e} + \bar{\sigma}(t, T^e, T^p)dt, \tag{26}$$

where $\bar{\sigma}(t, T^e, T^p)$ is the volatility of the ratio of zero coupon bonds

$$\frac{P(t, T^e)}{P(t, T^p)}.$$

As noted in Sects. 3.2 and 3.3, this volatility can be expressed in the following approximate form

- for $T^e \leq T^p$

$$\bar{\sigma}(t, T^e, T^p) \approx \sum_{i=\eta(T^e)-1}^{\eta(T^p)-1} \frac{\delta^k K(t, T^s, T^e)A_i}{(1 + \delta^k K(t, T^s, T^e))} \gamma(s, T_i) \tag{27}$$

- for $T^e \geq T^p$,

$$\bar{\sigma}(t, T^e, T^p) \approx - \sum_{i=\eta(T^p)-1}^{\eta(T^e)-1} \frac{\delta^k K(t, T^s, T^e)A_i}{(1 + \delta^k K(t, T^s, T^e))} \gamma(s, T_i) \tag{28}$$

with the A_i as in Sect. 3.3. Now, from (21) and (26), the SDE for the forward rate with $T^e \leq T^p$ is

$$dK^f(t, T^s, T^e) = K(t, T^s, T^e) \sum_{i=\eta(T^s)-1}^{\eta(T^e)-1} A_i \gamma(t, T_i) dW_t^{T^e}$$

$$\approx K(t, T^s, T^e) \sum_{i=\eta(T^s)-1}^{\eta(T^e)-1} A_i \gamma(t, T_i) \left(dW_t^{T^P} \right.$$

$$\left. - \sum_{i=\eta(T^e)-1}^{\eta(T^P)-1} \frac{\delta^k K(t, T^s, T^e) A_i}{(1+\delta^k K(t, T^s, T^e))} \gamma(t, T_i) dt \right)$$

with the approximate solution (following the notation introduced in Sect. 3.3)

$$\frac{K(T^f, T^s, T^e)}{K(0, T^s, T^e)}$$

$$\approx \exp\left[-\int_0^{T^f} \left(\sum_{i=\eta(T^s)-1}^{\eta(T^e)-1} A_i \gamma(t, T_i) \sum_{j=\eta(T^e)-1}^{\eta(T^P)-1} \frac{\delta^k K(0, T^s, T^e) A_j}{(1+\delta^k K(0, T^s, T^e))} \gamma(t, T_j) \right) dt \right.$$

$$\left. - \frac{1}{2} \int_0^{T^f} \left\| \sum_{i=\eta(T^s)-1}^{\eta(T^e)-1} A_i \gamma(t, T_i) \right\|^2 dt + \int_0^{T^f} \left(\sum_{i=\eta(T^s)-1}^{\eta(T^e)-1} A_i \gamma(t, T_i) \right) dW_t^{T^P} \right]$$

$$= \exp\left[\frac{-\delta^k K(0, T^s, T^e)}{(1+\delta^k K(0, T^s, T^e))} \sum_{i=\eta(T^s)-1}^{\eta(T^e)-1} \sum_{j=\eta(T^e)-1}^{\eta(T^P)-1} A_i A_j \lambda_{ij}(0, T^f) \right.$$

$$\left. - \frac{1}{2} \int_0^{T^f} \left\| \sum_{i=\eta(T^s)-1}^{\eta(T^e)-1} A_i \gamma(t, T_i) \right\|^2 dt + \int_0^{T^f} \left(\sum_{i=\eta(T^s)-1}^{\eta(T^e)-1} A_i \gamma(t, T_i) \right) dW_t^{T^P} \right].$$

Thus the expected forward rate under this measure is given by

$$E^{T^P}[K(T^f, T^s, T^e)]$$

$$\approx K(0, T^s, T^e) \exp\left[-\frac{\delta^k K(0, T^s, T^e)}{(1+\delta^k K(0, T^s, T^e))} \sum_{i=\eta(T^s)-1}^{\eta(T^e)-1} \sum_{j=\eta(T^e)-1}^{\eta(T^P)-1} A_i A_j \lambda_{ij}(0, T_f) \right],$$

i.e. we set $\tilde{\mu}$ in (25) to

$$-\frac{\delta^k K(0, T^s, T^e)}{(1+\delta^k K(0, T^s, T^e))} \sum_{i=\eta(T^s)-1}^{\eta(T^e)-1} \sum_{j=\eta(T^e)-1}^{\eta(T^P)-1} A_i A_j \lambda_{ij}(0, T_f).$$

Note that the modified Black's formula with convexity adjustment holds only approximately, by "freezing coefficients" as in Sect. 3.3.

4 Swaption Volatility Calculation

The purpose of this section is to provide an approximate formula for swaption volatility that permits the calculation of European swaption prices analytically using

Black's swaption formula. It is obvious that swap rates cannot be lognormal while forward rates are lognormal and vice versa. Fortunately, however, it is also well known that the probability distribution of swap rates is very close to lognormal under the forward rate lognormality assumption. Thus, we can rely on the Black's swaption formula, even though we are in LFM framework. In this section, the approximate formula for calculating European swaption volatilities within the LFM model will be derived. This is again an application of the "frozen coefficient" approximation, and the basic formula can be traced back to Brace, Gatarek and Musiela in [6] and Brace, Dun and Barton in [5]. We generalise this method by introducing non-standard configurations such as:

- payoff, reset, start and end dates may differ from base tenor dates;
- reset dates may differ from start dates;
- payoff dates may differ from end dates;
- correct day count fractions;
- accrual periods differ from forward definition periods;
- variable notionals;
- variable strikes (i.e. variable fixed legs of the underlying swap).

4.1 Definition

Consider a swap with the following notation: For $i = 1, \ldots, n$,

- accrual periods[7] in year fraction: δ_i^b;
- payoff date in fractions of a year: T_i^p;
- reset date in fractions of a year: T_i^f;
- start date in fractions of a year: T_i^s;
- end date in fractions of a year: T_i^e;
- variable notionals Q_i, and
- variable strikes κ_i.

Similarly to (2), we denote a forward rate[8] $K(t, T_i^s, T_i^e)$ covering a period from T_i^s to T_i^e as

$$K(t, T_i^s) = \frac{1}{\delta_i^k}\left(\frac{P(t, T_i^s)}{P(t, T_i^e)} - 1\right) \qquad (29)$$

where $\delta_i = T_i^e - T_i^s$. Once again, the underlying period (T_i^s, T_i^e) may differ from the standard tenors chosen for the LFM.

---

[7]It can be different from δ_i that is the period from the start to the end of i-th forward rate.

[8]$K(t, T_i^s, T_i^e)$ is the full notation. For a lighter notational burden, we drop the end date T_i^e until stated otherwise.

The value of a swap (payer swap) is determined by considering the discounted expectations of the n future cash flows, taken under the appropriate forward measure:

$$Pswap(t) = \sum_{i=1}^{n} Q_i \delta_i^b P(t, T_i^p) \left[E_{T_i^p} \{ K(T_i^f, T_i^s) | \mathcal{F}_t \} - \kappa_i \right].$$

In the case that the payoff date T_i^p and end date T_i^e coincide in the payer swap formula, the expectation $E_{T_i^p}\{K(T_i^f, T_i^s)|\mathcal{F}_t\}$ reduces to $K(t, T_i^s)$. In all other cases, the convexity adjustment presented in Sect. 3.6 is required. The magnitude of the adjustment typically is a few basis points. By interpreting $K(t, T_i^s)$ as the convexity adjusted forward rate where required, we can write

$$Pswap(t) = \sum_{i=1}^{n} Q_i \delta_i^b P(t, T_i^p)(K(t, T_i^s) - \kappa_i).$$

We deal with the case of variable strikes by introducing fixed rate notionals Q_i' defined as

$$Q_i' = Q_i \frac{\kappa_i}{\kappa}$$

where κ is the "average" strike, chosen more or less arbitrarily. This definition allows us to express the swap in terms of the single strike as

$$Pswap(t) = \sum_{i=1}^{n} \delta_i^b P(t, T_i^p)(Q_i K(t, T_i^s) - Q_i' \kappa). \qquad (30)$$

A swap rate $\varphi(t, T^s, T^e)$ is the value of the strike κ which sets the net present value of the swap to zero and is given by

$$\varphi(t, T_\varphi^s, T_\varphi^e) = \frac{\sum_{i=1}^{n} \delta_i^b P(t, T_i^p) Q_i K(t, T_i^s)}{\sum_{i=1}^{n} Q_i' \delta_i^b P(t, T_i^p)} \qquad (31)$$

where T_φ^s represents the swap start date and T_φ^e the swap end date. Dividing the numerator and the denominator by $P(t, T_\varphi^s)$ in order to express (31) in terms of forward bond prices, that is

$$F_T(t, T^*) = \frac{P(t, T^*)}{P(t, T)} \quad \text{for } \leq T^* \leq T,$$

(31) can be written as

$$\varphi(t, T_\varphi^s, T_\varphi^e) = \frac{\sum_{i=1}^{n} \frac{Q_i \delta_i^b}{\delta_i^k F_{T_i^p}(t, T_\varphi^s)} (F_{T_i^e}(t, T_i^s) - 1)}{\sum_{i=1}^{n} \frac{Q_i' \delta_i^b}{F_{T_i^p}(t, T^s)}}. \qquad (32)$$

Thus, the swap value can be expressed in terms of the swap rate as

$$Pswap(t) = \sum_{i=1}^{n} Q_i' \delta_i^b P(t, T_i^p)(\varphi(t, T_\varphi^s, T_\varphi^e) - \kappa).$$

4.2 Approximate Swaption Volatility Formula

Consider standardised LFM tenor dates T_i such that $0 = T_0 < T_1 < \cdots < T_{n+1}$, and $T_{i+1} - T_i = \delta_i$, $i = 0, \ldots, n$. The standard forward rates defined as

$$K(t, T_i) = \frac{1}{\delta_i} \left(\frac{P(t, T_i)}{P(t, T_{i+1})} - 1 \right)$$

have volatility given by the LFM volatility function of $\gamma(t, T_i)$, and follow the SDE

$$dK(t, T_i) = K(t, T_i)\gamma(t, T_i)dW_t^{T_{i+1}}.$$

We set $\frac{P(t,T^s)}{P(t,T^e)}$ as $F_{T^e}(t, T^s)$. Thus, (17) can be rewritten as

$$dF_{T^e}(t, T^s) = F_{T^e}(t, T^s)\sigma_{T^e}(t, T^s)dW_t + (\ldots)dt$$

$$\text{where } \sigma_{T^e}(t, T^s) \approx \sum_{i=\eta(T^s)-1}^{\eta(T^e)-1} A_i \gamma(t, T_i).$$

with the A_i again "frozen coefficients" as in Sect. 3.3. Using Ito's lemma and ignoring drift terms, for any Ito process X,

$$d\left(\frac{1}{X}\right) = -\frac{dX}{X^2} + (\ldots)dt.$$

Thus

$$d\left(\frac{1}{F_{T^e}(t, T^s)}\right) = -\frac{\sigma_{T^e}(t, T^s)}{F_{T^e}(t, T^s)}dW_t + (\ldots)dt \quad \text{where } T^s \leq T^e.$$

The SDE for the numerator of the swap rate is thus given by

$$d\sum_{i=1}^{n} \frac{Q_i \delta_i^b}{\delta_i^k} \left(\frac{F_{T_i^e}(t, T_i^s)}{F_{T_i^p}(t, T_\varphi^s)} - \frac{1}{F_{T_i^p}(t, T_\varphi^s)} \right)$$

$$= \sum_{i=1}^{n} \frac{Q_i \delta_i^b}{\delta_i^k} \left(d\left(\frac{F_{T_i^e}(t, T_i^s)}{F_{T_i^p}(t, T_\varphi^s)} \right) - d\left(\frac{1}{F_{T_i^p}(t, T_\varphi^s)} \right) \right)$$

$$= \sum_{i=1}^{n} \frac{Q_i \delta_i^b}{\delta_i^k} \left[\frac{F_{T_i^e}(t, T_i^s)}{F_{T_i^p}(t, T_\varphi^s)} (\sigma_{T_i^e}(t, T_i^s) - \sigma_{T_i^p}(t, T_\varphi^s)) + \frac{\sigma_{T_i^p}(t, T_\varphi^s)}{F_{T_i^p}(t, T_\varphi^s)} \right] dW_t + (\ldots)dt$$

$$= \sum_{i=1}^{n} \frac{Q_i \delta_i^b}{\delta_i^k F_{T_i^p}(t, T_\varphi^s)} \left[F_{T_i^e}(t, T_i^s)(\sigma_{T_i^e}(t, T_i^s) - \sigma_{T_i^p}(t, T_\varphi^s)) + \sigma_{T_i^p}(t, T_\varphi^s) \right] dW_t$$

$$+ (\ldots)dt. \tag{33}$$

The SDE for the denominator of the swap rate is given by

$$d \sum_{i=1}^{n} \frac{Q_i' \delta_i^b}{F_{T_i^p}(t, T_\varphi^s)} = - \sum_{i=1}^{n} Q_i' \delta_i^b \frac{\sigma_{T_i^p}(t, T_\varphi^s)}{F_{T_i^p}(t, T_\varphi^s)} dW_t + (\ldots) dt. \tag{34}$$

Now, we are ready to obtain the volatility function for the swap rate such that

$$\frac{d\varphi(t, T_\varphi^s, T_\varphi^e)}{\varphi(t, T_\varphi^s, T_\varphi^e)} = (\ldots) dt + \gamma_s(t, T_\varphi^s, T_\varphi^e) dW_t,$$

where $\gamma_s(t, T_\varphi^s, T_\varphi^e)$ denotes instantaneous volatility of $\varphi(t, T_\varphi^s, T_\varphi^e)$. Set

$$\varphi(t, T_\varphi^s, T_\varphi^e) = \frac{X}{Y},$$

for

$$X = \sum_{i=1}^{n} \frac{Q_i \delta_i^b}{\delta_i^k} \frac{F_{T_i^e}(t, T_i^s) - 1}{F_{T_i^p}(t, T_\varphi^s)}$$

$$Y = \sum_{i=1}^{n} \frac{Q_i' \delta_i^b}{F_{T_i^p}(t, T_\varphi^s)}.$$

We apply the identity[9]

$$\frac{d\left(\frac{X}{Y}\right)}{\frac{X}{Y}} = (\ldots) dt + (\sigma_X - \sigma_Y) dW_t,$$

for

$$\frac{dX}{X} = (\ldots) dt + \sigma_X dW_t$$

$$\frac{dY}{Y} = (\ldots) dt + \sigma_Y dW_t.$$

Note that the SDEs are under the same measure via measure transform. From (33) and (34), the volatility terms are expressed by

$$X\sigma_X = \sum_{i=1}^{n} \frac{Q_i \delta_i^b}{\delta_i^k F_{T_i^p}(t, T_\varphi^s)} \left[F_{T_i^e}(t, T_i^s) \left(\sigma_{T_i^e}(t, T_i^s) - \sigma_{T_i^p}(t, T_\varphi^s) \right) + \sigma_{T_i^p}(t, T_\varphi^s) \right]$$

$$= \sum_{i=1}^{n} \frac{Q_i \delta_i^b}{\delta_i^k F_{T_i^p}(t, T_\varphi^s)} \left[F_{T_i^e}(t, T_i^s) \sigma_{T_i^e}(t, T_i^s) - \sigma_{T_i^p}(t, T_\varphi^s)(F_{T_i^e}(t, T_i^s) - 1) \right]$$

$$= \sum_{i=1}^{n} \frac{Q_i \delta_i^b (F_{T_i^e}(t, T_i^s) - 1)}{\delta_i^k F_{T_i^p}(t, T_\varphi^s)} \left[\frac{F_{T_i^e}(t, T_i^s) \sigma_{T_i^e}(t, T_i^s)}{(F_{T_i^e}(t, T_i^s) - 1)} - \sigma_{T_i^p}(t, T_\varphi^s) \right], \tag{35}$$

$$Y\sigma_Y = - \sum_{i=1}^{n} Q_i' \delta_i^b \frac{\sigma_{T_i^p}(t, T_\varphi^s)}{F_{T_i^p}(t, T_\varphi^s)}.$$

---

[9]See Appendix A.1.

By setting

$$w_i(t) = \frac{c_i P(t, T_i^p) K(t, T_i^s)}{\sum\limits_{j=1}^{n} c_j P(t, T_j^p) K(t, T_j^s)}, \quad c_i = Q_i \delta_i^b, \ c_i' = Q_i' \delta_i^b,$$

$$u_i(t) = \frac{c_i' P(t, T_i^p)}{\sum\limits_{j=1}^{n} c_j' P(t, T_j^p)}, \qquad \mu_i(t) = \frac{\delta_i^k K(t, T_i^s)}{1 + \delta_i^k K(t, T_i^s)}$$

we can write

$$\sigma_X - \sigma_Y = \frac{\sum\limits_{i=1}^{n} \frac{Q_i \delta_i^b (F_{T_i^e}(t, T_i^s) - 1)}{\delta_i F_{T_i^p}(t, T_\varphi^s)} \left[\frac{F_{T_i^e}(t, T_i^s) \sigma_{T_i^e}(t, T_i^s)}{(F_{T_i^e}(t, T_i^s) - 1)} - \sigma_{T_i^p}(t, T_\varphi^s) \right]}{\sum\limits_{j=1}^{n} \frac{Q_j \delta_j^b}{\delta_j (F_{T_j^p}(t, T_\varphi^s))} (F_{T_j^e}(t, T_j^s) - 1)}$$

$$+ \frac{\sum\limits_{i=1}^{n} \frac{Q_i' \delta_i^b \sigma_{T_i^p}(t, T_\varphi^s)}{F_{T_i^p}(t, T_\varphi^s)}}{\sum\limits_{i=1}^{n} \frac{Q_i' \delta_i^b}{F_{T_i^p}(t, T_\varphi^s)}}$$

$$= \frac{\sum\limits_{i=1}^{n} c_i \frac{P(t, T_i^p)}{\delta_i P(t, T_\varphi^s)} \left(\frac{P(t, T_i^s)}{P(t, T_i^e)} - 1 \right) \left[\frac{\frac{P(t, T_i^s)}{P(t, T_i^e)} \sigma_{T_i^e}(t, T_i^s)}{\frac{P(t, T_i^s)}{P(t, T_i^e)} - 1} - \sigma_{T_i^p}(t, T_\varphi^s) \right]}{\sum\limits_{j=1}^{n} c_j \frac{P(t, T_j^p)}{\delta_j P(t, T_\varphi^s)} \left(\frac{P(t, T_j^s)}{P(t, T_j^e)} - 1 \right)}$$

$$+ \frac{\sum\limits_{i=1}^{n} c_i' \frac{P(t, T_i^p)}{P(t, T_\varphi^s)} \sigma_{T_i^p}(t, T_\varphi^s)}{\sum\limits_{i=1}^{n} c_i' \frac{P(t, T_i^p)}{P(t, T_\varphi^s)}}.$$

Thus,

$$\gamma_s(t, T_\varphi^s, T_\varphi^e) = \sum_{i=1}^{n} w_i(t) \left[\frac{\sigma_{T_i^e}(t, T_i^s)}{\mu_i(t)} - \sigma_{T_i^p}(t, T_\varphi^s) \right] + \sum_{i=1}^{n} u_i(t) \sigma_{T_i^p}(t, T_\varphi^s)$$

$$= \sum_{i=1}^{n} (u_i(t) - w_i(t)) \sigma_{T_i^p}(t, T_\varphi^s) + \sum_{i=1}^{n} w_i(t) \frac{\sigma_{T_i^e}(t, T_i^s)}{\mu_i(t)}. \qquad (36)$$

As in Sect. 3.2, we can write the volatilities as

$$\sigma_{T_i^p}(t, T_\varphi^s) = \sum_{j=\eta(T_\varphi^s)-1}^{\eta(T_i^p)-1} A_j(t) \gamma(t, T_j)$$

$$\sigma_{T_i^e}(t, T_i^s) = \sum_{j=\eta(T_i^s)-1}^{\eta(T_i^e)-1} B_j(t)\gamma(t, T_j)$$

for some coefficients $A_j(t)$ and $B_j(t)$ and so we can rewrite the swap rate volatility in (36) as

$$\gamma_s(t, T_\varphi^s, T_\varphi^e) = \sum_{i=1}^{n} \frac{w_i(t)}{\mu_i(t)} \sum_{j=\eta(T_i^s)-1}^{\eta(T_i^e)-1} B_j(t)\gamma(t, T_j)$$

$$+ \sum_{i=1}^{n} (u_i(t) - w_i(t)) \sum_{j=\eta(T_\varphi^s)-1}^{\eta(T_i^P)-1} A_j(t)\gamma(t, T_j).$$

Let $l_m = \max(\eta(T_n^e) - 1, \eta(T_n^P) - 1)$, and $f_m = \eta(T_\varphi^s) - 1$. Since $\gamma_s(t, T_\varphi^s, T_\varphi^e)$ is represented in terms of standard rate volatilities, we can denote

$$\gamma_s(t, T_\varphi^s, T_\varphi^e) = \sum_{i=f_m}^{l_m} C_i(t)\gamma(t, T_i),$$

where the coefficients $C_i(t)$ can be calculated iteratively.

By freezing $C_i(t)$ to $C_i(0)$ and denoting it as C_i,[10] the approximate Black swaption implied volatility is represented as

$$\sigma^2(T_{\exp}, T_\varphi^s, T_\varphi^e)T_{\exp} \approx \int_0^{T_{\exp}} \left\| \sum_{i=f_m}^{l_m} C_i\gamma(s, T_i) \right\|^2 ds,$$

with the swaption expiring at date $T_{\exp}$. Expanding the square summation, we have

$$\int_0^{T_{\exp}} \left\| \sum_{i=f_m}^{l_m} C_i\gamma(s, T_i) \right\|^2 ds = \int_0^{T_{\exp}} \left(\sum_{i=f_m}^{l_m} C_i\gamma(s, T_i) \right)\left(\sum_{j=f_m}^{l_m} C_j\gamma(s, T_j) \right) ds$$

$$= \sum_{i=f_m}^{l_m} \sum_{j=f_m}^{l_m} C_i C_j \int_0^{T_{\exp}} \gamma(s, T_i)\gamma(s, T_j) ds$$

$$= \sum_{i=f_m}^{l_m} \sum_{j=f_m}^{l_m} C_i C_j \lambda_{i,j}(0, T_{\exp}).$$

The payer swaption price is thus given by

$$Pswapn(T_{\exp}, T_\varphi^s, T_\varphi^e) = \sum_{i=1}^{n} Q_i' \delta_i^b P(t, T_i^P)(\varphi(0, T_\varphi^s, T_\varphi^e)\Phi(d_1) - \kappa\Phi(d_2))$$

---

[10]We are using the same approximation method as in Sect. 3.3. This is again a Wiener Chaos expansion of order zero.

$$\text{with} \quad d_1 = \frac{\ln \frac{\varphi(0, T^s, T^e)}{\kappa} + \frac{1}{2}\sigma^2(T_{\exp}, T_\varphi^s, T_\varphi^e)T_{\exp}}{\sigma(T_{\exp}, T_\varphi^s, T_\varphi^e)\sqrt{T_{\exp}}}$$

$$d_2 = d_1 - \sigma(T_{\exp}, T_\varphi^s, T_\varphi^e)\sqrt{T_{\exp}}.$$

5 Testing the Approximate Formulae

In this section, the performance of the approximate formulae is tested and compared to Monte Carlo (MC) simulations under several scenarios. The following scenarios are used:

Case	Instrument	Coupon	Start	End	Length	n	ATM	ITM	OTM
1	caplet	3m	1	1.24	0.25	13	ATM	2.25%	4.59%
2	caplet	3m	5	5.24	0.25	13	ATM	4.19%	6.55%
3	swaption	3m	2	3.5	5	16	ATM	4.31%	6.58%
4	swaption	3m	3.5	5	10	16	ATM	4.59%	6.61%
5	swaption	1m	3.5	5	0.833	16	ATM	4%	6.50%

"Coupon" means the length of the coupon period. Cases 1 to 4 have 3-month coupon periods while only Case 5 is based on a 1-month coupon period. The heading "n" shows how many instruments are included in each case to cover the "broken dates" between the dates of the standard three-month tenor grid. For Case 1, for example, thirteen caplets are included to cover rest dates between the "start" date one year out and the "end" date 1.24 years out. This the first caplet tested in Case 1 covers the accrual period $(1, 1.25)$, the second covers $(1.02, 1.27)$, the third covers $(1.04, 1.29)$, and so on until the last, which covers $(1.24, 1.49)$. Each case has three different sub-scenarios with respect to the strikes (at-the-money, in-the-money and out-of-the-money).

Per each sub-scenario of each case, the analytical volatilities from the approximate formulae will be compared against two standard error bounds of the Monte Carlo estimates. The performance is deemed to be acceptable if the pricing error (in terms of basis points difference) is economically insignificant. For the comparison to make sense, the 95% confidence intervals around the MC estimates must be sufficient tight, i.e. the standard deviations of the MC estimates must be less than half of the maximum error, which is considered economically insignificant. For plain vanilla interest rate derivative instruments, a typical bid- ask spread quoted by traders is about one vega, which is 1% difference of volatilities in absolute terms. Thus, the criterion for "successful" performance in the test in term of implied volatilities would be that the maximum errors between the MC bounds and approximate implied volatilities are less than 0.5 percentage points.

5.1 Assumptions and Descriptions for Performance Tests

- For all tests, standard tenor dates are based on 3-month forward rates. All periods and dates are measured in fractions of a year.
- The tenor horizon is 15 years, and thus there are 60 tenor periods;
- The instantaneous volatility function is pre-calibrated using the Pedersen approach in [14]. It is time-deterministic and piece-wise constant during each tenor period. The AUD yield curve, AUD cap curve and AUD swaption volatility matrix as of 31/March/2009 are used. They are all mid values of bid and ask quotes from Reuters.
- A historical correlation matrix is used.[11]
- Three factors are assumed.
- For uniform random number generation, ran2 in [15] is used in conjunction with the Box/Muller method to generate normal variates.

The procedure for the comparisons is as follows:

- For each instrument, obtain a Monte Carlo estimate from one million paths for the price of each instrument;
- Calculate the standard deviation (SD) of the Monte Carlo estimate using the formula

$$SD = \sqrt{\frac{\frac{1}{N-1} \sum_{i=1}^{N} x_i^2 - \left(\frac{1}{N} \sum_{i=1}^{N} x_i\right)^2}{N}};$$

- Calculate a two SD interval (95% confidence intervals), upper bound and lower bound, around the Monte Carlo price;
- By inverting the corresponding Black's formula using a root finding algorithm, obtain the volatilities implied by the two bounds (volatility bounds);
- Calculate the corresponding approximate volatility (caplet vols or swaption vols) using the formulae in previous sections and compare with the volatility bounds.

The test results are plotted in Figs. 8 to 19. The correlation matrix and swaption matrix used in the test are depicted in Figs. 23 and 24, respectively. For Case 1 and Case 2, the approximate analytic implied volatilities are all within the MC bounds. However, for Case 3 and Case 4, in some of the sub-scenarios the approximate volatilities are slightly outside the bounds, though still quite accurate — note that the bounds achieved with one million MC simulations are substantially tighter than the 0.5 percentage points around the MC estimate set as the threshold of "economic significance." As discussed in Sect. 4, swap rates are not exactly lognormal under the lognormality assumption on forward LIBOR. The results here show that the errors from this source are not economically significant.

---

[11] The justification for the use of historical correlation matrix can be found in [7] — essentially, the choice of correlation matrix has little impact on the derivative instruments considered here.

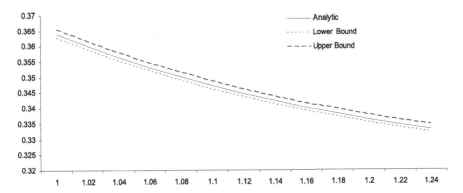

Fig. 8 Case 1 ATM, 3-month caplets expiring from 1 year to 1.24 years with two standard deviation bounds of 1 million paths; "Analytic" represents approximate closed-form volatilities

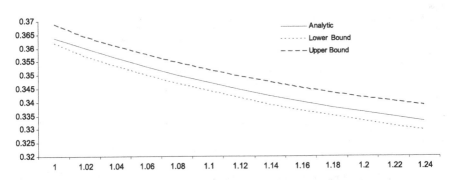

Fig. 9 Case 1 ITM, 3-month caplets expiring from 1 year to 1.24 years, strikes = 2.25%, with two standard deviation bounds of 1 million paths; "Analytic" represents approximate closed-form volatilities

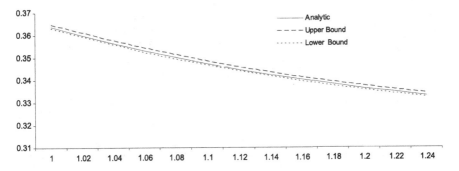

Fig. 10 Case 1 OTM, 3-month caplets expiring from 1 year to 1.24 years, strikes = 4.59%, with two standard deviation bounds of 1 million paths; "Analytic" represents approximate closed-form volatilities

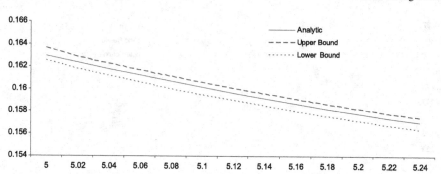

Fig. 11 Case 2 ATM, 3-month caplets expiring from 5 years to 5.24 years with two standard deviation bounds of 1 million paths; "Analytic" represents approximate closed-form volatilities

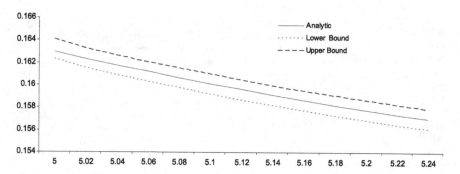

Fig. 12 Case 2 ITM, 3-month caplets expiring from 5 years to 5.24 years, strikes = 2.25%, with two standard deviation bounds of 1 million paths; "Analytic" represents approximate closed-form volatilities

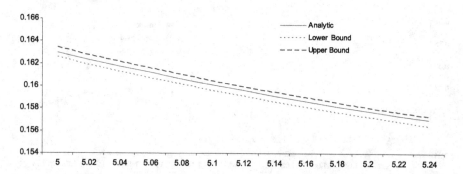

Fig. 13 Case 2 OTM, 3-month caplets expiring from 5 years to 5.24 years, strike = 6.55% with two standard deviation bounds of 1 million paths; "Analytic" represents approximate closed-form volatilities

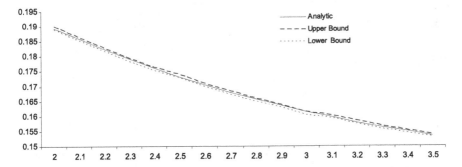

Fig. 14 Case 3 ATM, 5-year European swaptions expiring from 2 years to 3.5 years with two standard deviation bounds of 1 million paths; "Analytic" represents approximate closed-form volatilities

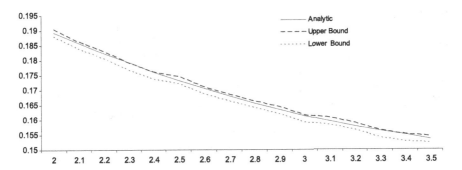

Fig. 15 Case 3 ITM, 5-year European swaptions expiring from 2 years to 3.5 years, strikes = 4.21%, with two standard deviation bounds of 1 million paths; "Analytic" represents approximate closed-form volatilities

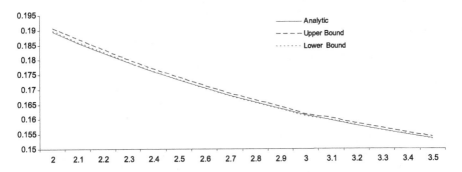

Fig. 16 Case 3 OTM, 5-year European swaptions expiring from 2 years to 3.5 years, strikes = 6.58%, with two standard deviation bounds of 1 million paths; "Analytic" represents approximate closed-form volatilities

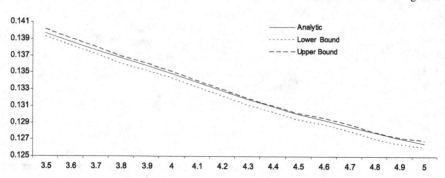

Fig. 17 Case 4 ATM, 10-year European swaptions expiring from 3.5 years to 5 years with two standard deviation bounds of 1 million paths; "Analytic" represents approximate closed-form volatilities

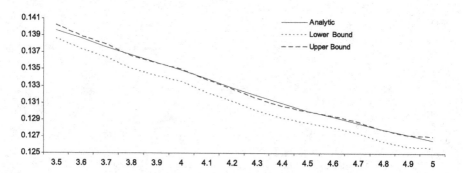

Fig. 18 Case 4 ITM, 10-year European swaptions expiring from 3.5 years to 5 years, strikes = 4.59%, with two standard deviation bounds of 1 million paths; "Analytic" represents approximate closed-form volatilities

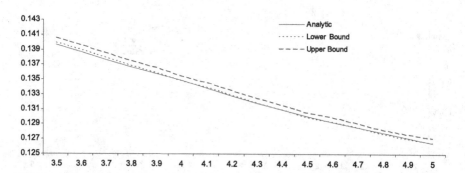

Fig. 19 Case 4 OTM, 10-year European swaptions expiring from 3.5 years to 5 years, strikes = 6.61%, with two standard deviation bounds of 1 million paths; "Analytic" represents approximate closed-form volatilities

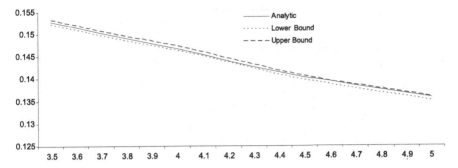

Fig. 20 Case 5 ATM, 1-month coupon 5-year European swaptions expiring from 3.5 years to 5 years, with two standard deviation bounds of 1 million paths; "Analytic" represents approximate closed-form volatilities

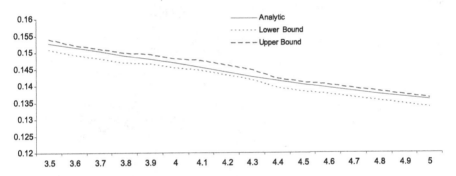

Fig. 21 Case 5 ITM, 1-month coupon 5-year European swaptions expiring from 3.5 years to 5 years, strikes = 4%, with two standard deviation bounds of 1 million paths; "Analytic" represents approximate closed-form volatilities

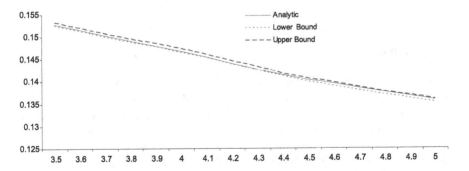

Fig. 22 Case 5 OTM, 1-month coupon 5-year European swaptions expiring from 3.5 years to 5 years, strikes = 6.5%, with two standard deviation bounds of 1 million paths; "Analytic" represents approximate closed-form volatilities

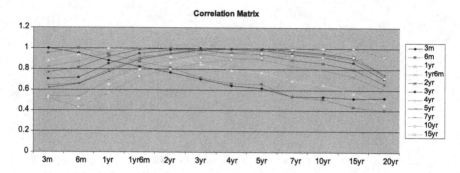

Fig. 23 Historical correlation matrix for 3-month forward LIBORs

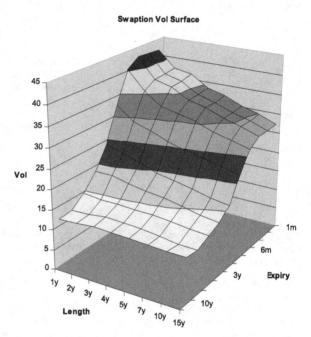

Fig. 24 AUD Swaption
volatility matrix (source:
Reuters, 31/March/2009, mid
of bid and ask quotes)

6 Conclusion

In the present paper, several problems arising in the practical implementation of
the Lognormal Forward Rate Market Model (LFM) are addressed. Firstly, actual
payment dates of interest rate derivatives rarely match the tenor grid of the LFM
exactly. The formulae derived here allow one to explicitly take this mismatch into
account. When verified against Monte Carlo simulation in an LFM calibrated to actual market data, the approximating assumptions did not lead to any economically
significant errors. Secondly, in general there also may be a mismatch between the
payment date of an interest rate derivative and the end of the accrual period of the
underlying rate. In this case, a convexity correction becomes necessary, and this cor-

rection is made specific in the context of the fully general payment dates considered in this paper. Thirdly, we have shown how one can avoid the unrealistic implied volatilities for "broken date" caplets resulting from interest rate interpolation. This is achieved by maintaining a non-zero volatility for the modelled rates underlying the interpolation, until the *end* of their respective accrual periods rather than the beginning; though beyond the beginning of their accrual period these rates only have meaning for the purpose of interest rate interpolation.

Appendix

7.1 Quotient SDE

Let

$$C_t = \frac{A_t}{B_t}$$
$$dA_t = \mu_a dt + \gamma_a dW_t$$
$$dB_t = \mu_b dt + \gamma_b dW_t.$$

Then,

$$d\left(\frac{A_t}{B_t}\right) = \frac{\partial C_t}{\partial A_t} dA_t + \frac{\partial C_t}{\partial B_t} dB_t + \frac{1}{2}\frac{\partial^2 C_t}{\partial A_t^2}(dA_t)^2 + \frac{1}{2}\frac{\partial^2 C_t}{\partial B_t^2}(dB_t)^2 + \frac{\partial^2 C_t}{\partial A_t \partial B_t} dA_t dB_t$$

where

$$\frac{\partial C_t}{\partial A_t} = \frac{1}{B_t}, \qquad \frac{\partial C_t}{\partial B_t} = -\frac{A_t}{B_t^2}, \qquad \frac{\partial^2 C_t}{\partial A_t^2} = 0,$$

$$\frac{\partial^2 C_t}{\partial B_t^2} = 2\frac{A_t}{B_t^3}, \qquad \frac{\partial^2 C_t}{\partial A_t \partial B_t} = -\frac{1}{B_t^2}.$$

Thus,

$$dC_t = \frac{1}{B_t}(\mu_a dt + \gamma_a dW_t) - \frac{A_t}{B_t^2}(\mu_b dt + \gamma_b dW_t) + \frac{A_t}{B_t^3}\gamma_b^2 dt - \frac{1}{B_t^2}\gamma_a\gamma_b dt$$

$$= \left(\frac{\gamma_a}{B_t} - \frac{\gamma_b A_t}{B_t^2}\right) dW_t + (\ldots)dt$$

$$= C_t\left(\frac{\gamma_a}{A_t} - \frac{\gamma_b}{B_t}\right) dW_t + (\ldots)dt. \tag{37}$$

7.2 Modified Black's Formula

When the payoff date does not match the forward end date, we cannot use the standard Black's formula to price a caplet. Consider a caplet:

$$Cpl(0) = E\left(\frac{1}{\beta(T^P)}\left[(K(T_f, T^s, T^e) - \kappa)^+\right]\right)\delta$$

$$= P(0, T^P)\delta E_{T^P}\left[(K(T_f, T^s, T^e) - \kappa)^+\right]. \tag{38}$$

If the payoff date does not match the forward end date, $K(T_f, T^s, T^e)$ is not a martingale under the $\mathbf{P}_{T^P}$ measure, and the standard Black's formula does not apply. Rather, we need to modify Black's formula to take into account the drift of $K(\cdot, T^s, T^e)$ under the T^P measure. For the reader's convenience, the well-known solution to the problem will be presented in two steps: Step 1 is for the derivation of the modified Black's formula, while Step 2 calculates the drift of a non-standard forward rate.

7.2.1 Step 1

Let

$$K(T_f, T^s, T^e) = K_0 e^z$$

where

$$K_0 = K(0, T^s, T^e)$$

and

$$z \sim \Phi(\mu, \sigma^2)$$

is a normally distributed random variable with mean μ and variance σ^2. The forward rate $K(T_f, T^s, T^e)$ has a drift under the measure $\mathbf{P}_{T^P}$, where $T^P \neq T^e$. Then

$$I = E_{T^P}\left[(K(T_f, T^s, T^e) - \kappa)^+|\mathcal{F}_0\right]$$

$$= \int_{-\infty}^{\infty} \frac{1}{\sqrt{2\pi}\sigma}(K(T_f, T^s, T^e) - \kappa)^+ e^{-\frac{(z-\mu)^2}{2\sigma^2}} dz$$

or, setting $L = \ln(\frac{\kappa}{K_0})$,

$$I = \int_L^{\infty} \frac{1}{\sqrt{2\pi}\sigma}(K(T_f, T^s, T^e) - \kappa) e^{-\frac{(z-\mu)^2}{2\sigma^2}} dz.$$

Normalizing z, define:

$$y = \frac{z - \mu}{\sigma}.$$

Then $\frac{dy}{dz} = \frac{1}{\sigma}$ and by changing the variable from z to y

$$I = \int_{\frac{L-\mu}{\sigma}}^{\infty} \frac{1}{\sqrt{2\pi}}(K_t e^{\mu+\sigma y} - \kappa) e^{-\frac{y^2}{2}} dy$$

$$= \int_{\frac{L-\mu}{\sigma}}^{\infty} \frac{1}{\sqrt{2\pi}} K_t e^{\mu+\sigma y-\frac{1}{2}y^2} dy - \int_{\frac{L-\mu}{\sigma}}^{\infty} \kappa \frac{1}{\sqrt{2\pi}} e^{-\frac{y^2}{2}} dy$$

$$= \int_{\frac{L-\mu}{\sigma}}^{\infty} \frac{1}{\sqrt{2\pi}} K_t \, e^{\mu + \frac{\sigma^2}{2} - \frac{y^2 - 2\sigma y + \sigma^2}{2}} \, dy - \int_{\frac{L-\mu}{\sigma}}^{\infty} \kappa \frac{1}{\sqrt{2\pi}} e^{-\frac{y^2}{2}} \, dy$$

$$= K_t \, e^{\mu + \frac{\sigma^2}{2}} \int_{\frac{L-\mu}{\sigma}}^{\infty} \frac{1}{\sqrt{2\pi}} e^{-\frac{(y-\sigma)^2}{2}} \, dy - \int_{\frac{L-\mu}{\sigma}}^{\infty} \kappa \frac{1}{\sqrt{2\pi}} e^{-\frac{y^2}{2}} \, dy.$$

Using the symmetry property of the normal distribution:

$$I = K_t \, e^{\mu + \frac{\sigma^2}{2}} \, \Phi\left(\frac{\ln\left(\frac{K_t}{\kappa}\right) + \mu + \sigma^2}{\sigma} \right) - \kappa \Phi\left(\frac{\ln\left(\frac{K_t}{\kappa}\right) + \mu}{\sigma} \right)$$

where $\Phi(\cdot)$ is the cumulative distribution function of the standard normal distribution. Finally,

$$Cpl(t) = P(t, T^P)\delta \left[K_t \, e^{\mu + \frac{\sigma^2}{2}} \, \Phi\left(\frac{\ln\left(\frac{K_t}{\kappa}\right) + \mu + \sigma^2}{\sigma} \right) \right.$$
$$\left. - \kappa \Phi\left(\frac{\ln\left(\frac{K_t}{\kappa}\right) + \mu}{\sigma} \right) \right], \tag{39}$$

where δ is the length of the accrual period in fractions of a year.

7.2.2 Step 2

The forward rate $K(\,\cdot\,, T^s, T^e)$ is martingale under the corresponding measure $\mathbf{P}_{T^e}$. The change of the probability measure from $\mathbf{P}_{T^e}$ to $\mathbf{P}_{T^P}$ give rise to a drift. Using Girsanov's Theorem,

$$dW_t^{T^P} = dW_t^{T^e} + \int_{T^e}^{T^P} \sigma(t, s) ds \, dt \quad \text{for } 0 \leq t \leq T_f.$$

The discrete tenor version of the integral term is

$$\int_{T^e}^{T^P} \sigma(t, s) ds = \sum_{i=\eta(T^e)-1}^{\eta(T^P)-1} \frac{\delta_i K(0, T_i, T_{i+1})}{1 + \delta_i K(0, T_i, T_{i+1})} \gamma(t, T_i).$$

Then

$$K(T_f, T^s, T^e) = K_0 \exp\left\{ -\int_0^{T_f} \sum_{i=\eta(T^e)-1}^{\eta(T^P)-1} \frac{\delta_i K(0, T_i, T_{i+1})}{1 + \delta_i K(0, T_i, T_{i+1})} \gamma(s, T_i) \gamma^*(s) ds \right.$$
$$\left. - \frac{1}{2} \int_t^{T_f} \|\gamma^*(s)\|^2 ds + \int_0^{T_f} \gamma^*(s) dW_s^{T^P} \right\}$$

and thus

$$\mu = -\int_0^{T_f} \sum_{i=\eta(T^e)-1}^{\eta(T^P)-1} \frac{\delta_i K(0, T_i, T_{i+1})}{1 + \delta_i K(0, T_i, T_{i+1})} \gamma(s, T_i) \gamma^*(s) ds - \frac{1}{2} \int_0^{T_f} \|\gamma^*(s)\|^2 ds$$

where $\gamma^*(s)$ denotes the instantaneous volatility for the forward rate $K(s, T^s, T^e)$. If we set $\mu = \tilde{\mu} - \frac{1}{2}\sigma^2$,

$$\tilde{\mu} = -\int_0^{T_f} \sum_{i=\eta(T^e)-1}^{\eta(T^p)-1} \frac{\delta_i K(0, T_i, T_{i+1})}{1 + \delta_i K(0, T_i, T_{i+1})} \gamma(s, T_i)\gamma^*(s)ds$$

$$\sigma^2 = \int_0^{T_f} \|\gamma^*(s)\|^2 ds$$

then (39) can be written as

$$Cpl(0) = P(0, T^p)\delta\left[K_t\, e^{\tilde{\mu}}\Phi\left(\frac{\ln\left(\frac{K_0}{\kappa}\right) + \tilde{\mu} + \frac{1}{2}\sigma^2}{\sigma}\right) - \kappa\Phi\left(\frac{\ln\left(\frac{K_0}{\kappa}\right) + \tilde{\mu} - \frac{1}{2}\sigma^2}{\sigma}\right)\right].$$
(40)

The calculations for $\tilde{\mu}$ and σ^2 can be found in Sect. 3.6, where

$$\tilde{\mu} = -\sum_{i=\eta(T^s)-1}^{\eta(T^e)-1} \sum_{\eta(T^e)-1}^{\eta(T^p)-1} \left(\frac{\delta_i K(0, T_i, T_{i+1})A_i}{1 + \delta_i K(0, T_i, T_{i+1})}\lambda_{ij}(0, T_f)\right)$$

$$\sigma^2 = \sum_{i=\eta(T^s)-1}^{\eta(T^e)-1} \sum_{i=\eta(T^s)-1}^{\eta(T^e)-1} A_i A_j \lambda_{ij}(0, T_f).$$

References

1. Andersen, L., Andreasen, J.: Volatility skews and extensions of the libor market model. Appl. Math. Finance **7**(1), 1–32 (2000)
2. Brace, A.: Dual swap and swaption formulae in forward models. FMMA NOTES (1996)
3. Brace, A.: Rank-2 swaption formulae. FMMA NOTES (1997)
4. Brace, A.: Engineering BGM. Chapman & Hall, London (2008)
5. Brace, A., Dun, T., Barton, G.: Towards a central interest rate model. In: Jouini, E., Cvitanić, J., Musiela, M. (eds.) Option Pricing, Interest Rates and Risk Management. Cambridge University Press, London (2001)
6. Brace, A., Gatarek, D., Musiela, M.: The market model of interest rate dynamics. Math. Finance **7**(2), 127–155 (1997)
7. Choy, B., Dun, T., Schlögl, E.: Correlating market models. Risk **September**, 124–129 (2004)
8. Heath, D., Jarrow, R., Morton, A.: Bond pricing and the term structure of interest rates: A new methodology. Econometrica **60**(1), 77–105 (1992)
9. Hull, J., White, A.: Forward rate volatilities. Swap rate volatilities, and the implementation of the LIBOR market model. J. Fixed Income **10**(2), 46–62 (2000)
10. Jamshidian, F.: LIBOR and swap market models and measures. Finance Stoch. **1**(4), 293–330 (1997)
11. Jäckel, P., Rebonato, R.: Linking caplet and swaption volatilities in a BGM/J framework. http://www.quarchome.org (2000)
12. Miltersen, K., Sandmann, K., Sondermann, D.: Closed form solutions for term structure derivatives with lognormal interest rates. J. Finance **52**(1), 409–30 (1997)
13. Musiela, M., Rutkowski, M.: Continuous time term structure models: A forward measure approach. Finance Stoch. **1**, 261–291 (1997)

14. Pedersen, M.: Calibrating LIBOR market models. Working paper, Financial Research Department, Simcorp A/S (1998)
15. Press, W.H., Teukolsky, S.A., Vetterling, W.T., Flannery, B.P.: Numerical Recipes: The Art of Scientific Computing, 3rd edn. Cambridge University Press, London (2002)
16. Schlögl, E.: Arbitrage-free interpolation in models of market observable interest rates. In: Sandmann K., Schönbucher P.J. (eds.) Advances in Finance and Stochastics: Essays in Honour of Dieter Sondermann, pp. 197–218. Springer, Berlin (2002)
17. Tang, Y., Li, B.: Quantitative Analysis, Derivatives Modeling, and Trading Strategies. World Scientific, Singapore (2007)

Maximum Likelihood Estimation for Integrated Diffusion Processes

Fernando Baltazar-Larios and Michael Sørensen

Abstract We propose a method for obtaining maximum likelihood estimates of parameters in diffusion models when the data is a discrete time sample of the integral of the process, while no direct observations of the process itself are available. The data are, moreover, assumed to be contaminated by measurement errors. Integrated volatility is an example of this type of observations. Another example is ice-core data on oxygen isotopes used to investigate paleo-temperatures. The data can be viewed as incomplete observations of a model with a tractable likelihood function. Therefore we propose a simulated EM-algorithm to obtain maximum likelihood estimates of the parameters in the diffusion model. As part of the algorithm, we use a recent simple method for approximate simulation of diffusion bridges. In simulation studies for the Ornstein-Uhlenbeck process and the CIR process the proposed method works well.

1 Introduction

We consider estimation for a general one-dimensional diffusion process. Likelihood based estimation (including Bayesian) for discretely observed diffusion processes has been investigated by [1–3, 9, 19–21, 34–37]. Martingale estimating functions for discretely observed diffusions are reviewed in [38] and [40].

In this paper we consider maximum likelihood estimation in the situation where we do not observe the diffusion process itself directly, but instead observe integrals

F. Baltazar-Larios
Instituto de Investigación en Matemáticas Aplicadas y en Sistemas, Universidad Nacional Autónoma de México, A.P. 20-726, 01000 Mexico, D.F., Mexico
e-mail: sheva7.fernando@gmail.com

M. Sørensen (✉)
Department of Mathematical Sciences, University of Copenhagen, Universitetsparken 5, 2100 Copenhagen, Denmark
e-mail: michael@math.ku.dk

C. Chiarella, A. Novikov (eds.), *Contemporary Quantitative Finance*,
DOI 10.1007/978-3-642-03479-4_20, © Springer-Verlag Berlin Heidelberg 2010

of the process over disjoint time-intervals. These observations are, moreover, assumed to be contaminated by measurement errors. Integrated diffusion processes play an important role in finance as models for realized volatility, see e.g. [4–6, 11]. These processes are also used for modeling purposes in fields of engineering and the sciences. An example is provided by the records of the concentration of oxygen isotopes in ice-core data from Greenland and Antarctica, see e.g. [17]. Such data are used to investigate the paleo-climate.

The likelihood function for a discretely sampled integrated diffusion with observation error is in almost all cases not explicitly available. Moreover, the integrated process is not a Markov process, so there is no easily calculated martingales. Therefore martingale estimating functions are not a feasible alternative, but prediction-based estimating function can be applied, see [39]. In the present paper, we note instead that the data can be viewed as incomplete observations from a model with a tractable likelihood function. The full data set is a continuous time record of the diffusion process and the observation errors. We can therefore find maximum likelihood estimates by applying the Expectation-Maximization (EM) algorithm, see [16] and [32]. To do so we need to calculate the conditional expectation of the log-likelihood function for the full model given the observations. We do this by simulating sample paths of the diffusion process given the data using ideas from [13]. An essential step in doing this is to simulate a part of a sample path given the rest, which corresponds to simulation a diffusion bridge. This is done by applying the method for approximate diffusion bridge simulation recently proposed by [10].

Parametric inference for integrated diffusion process has been considered by [11, 18, 22, 24, 25]. Nonparametric inference has been considered in [14].

In Sect. 2 the model of an integrated diffusion process with measurement error and its assumptions are presented. Section 3 contains the calculation of the likelihood function with full diffusion observation and the EM-algorithm. In Sect. 4 we present a method for simulation of a diffusion process conditional on integrals observed with measurement error. In Sect. 5 we consider the Ornstein-Uhlenbeck process in detail and a simulation study for this model is reported. A similar investigation for the CIR/square root process is presented in Sect. 6, where also stochastic volatility models are briefly discussed. Some concluding remarks are given in Sect. 7.

2 Model and Data

We consider likelihood estimation for the general one-dimensional diffusion process $X = \{X_t\}_{t \geq 0}$ given by the stochastic differential equation

$$dX_t = b(X_t; \psi)dt + \sigma(X_t; \psi)dW_t \qquad (1)$$

where $W = \{W_t\}$ is a standard Wiener process, and where the drift and diffusion coefficients depend on an unknown p-dimensional parameter ψ belonging to the parameter set $\Psi \subseteq R^p$. We assume that the solution X is an ergodic, stationary diffusion with invariant measure with density function $v_\psi(x)$ ($X_0 \sim v_\psi$ is independent of W). We also assume that the stochastic differential equation has a unique weak solution, i.e. a solution exists and all solutions have identical finite-dimensional distributions; see e.g. [28]. It is well-known that sufficient conditions for these assumptions can be expressed in terms of the so-called scale function and speed measure; see e.g. [29].

In this paper we consider the situation where the process X has not been observed directly. Instead the data are integrals of X_t over intervals $[t_{i-1}, t_i]$ observed with measurement error, i.e.

$$Y_i = \int_{t_{i-1}}^{t_i} X_s ds + Z_i, \quad i = 1, \ldots, n, \tag{2}$$

where $Z_i \sim N(0, \tau^2)$, $i = 1, \ldots, n$ are mutually independent and independent of X. We assume that $t_0 = 0$, so the total interval of observation is $[0, t_n]$. Note that the variance of the measurement error, τ^2, is an extra unknown parameter. Thus we need to estimate the $p + 1$-dimensional parameter $\theta = (\psi, \tau^2)$.

Conditionally on the sample path of X, the observations Y_i, $i = 1, \ldots, n$ are independent and normal distributed:

$$Y_i \mid X_t : t \in [0, t_n] \sim N\left(\int_{t_{i-1}}^{t_i} X_s ds, \tau^2\right). \tag{3}$$

We assume that the coefficients of the stochastic differential equation (1) satisfy the following conditions which we need in the following sections.

Condition 1 *The drift and diffusion coefficients of (1), $b(x; \psi)$ and $\sigma(x; \psi)$ satisfy that for all $\psi \in \Psi$*

- *$b(x; \psi)$ is continuously differentiable w.r.t. x*
- *$\sigma(x; \psi)$ is twice continuously differentiable w.r.t. x*
- *$\sigma(x; \psi) > 0$ for all x in the state space of X.*

3 The Likelihood Function and the EM-Algorithm

We can think of the data set $Y = (Y_1, \ldots, Y_n)$ as an incomplete observation of a full data set given by the sample path X_t, $t \in [0, t_n]$ and the measurement errors $Z_1, \ldots, Z_n$, or equivalently X_t, $t \in [0, t_n]$ and $Y = (Y_1, \ldots, Y_n)$. Therefore likelihood based estimation can be done by means of the EM-algorithm or MCMC-methods. In this paper we concentrate on the EM-algorithm. We need to find the likelihood function for the full data set and the conditional expectation of this full log-likelihood function given the observations $Y = (Y_1, \ldots, Y_n)$.

3.1 Likelihood with Full Diffusion Observation

The full observation of a diffusion sample path in the time interval $[0, t_n]$ is an element in the space C of continuous functions from $[0, t_n]$ to $\mathbb{R}$. We equip this space with the usual σ-algebra, $\mathscr{C}$, generated by the cylinder sets, and consider the probability measures induced on $(C, \mathscr{C})$ by the solutions to (1). These measures are in general singular because the diffusion coefficient depends on the parameter ψ. In order to obtain a likelihood function, we use the standard 1-1 transformation

$$h(x; \psi) = \int_{x^*}^{x} \frac{1}{\sigma(u; \psi)} du, \tag{4}$$

where x^* is some arbitrary element of the state space of X. By this parameter dependent transformation, we obtain a diffusion process with unit diffusion coefficient. Specifically, we obtain (by Ito's formula) that

$$U_t = h(X_t; \psi)$$

satisfies the stochastic differential equation

$$dU_t = \mu(U_t; \psi)dt + dW_t, \tag{5}$$

with

$$\mu(u; \psi) = \frac{b(h^{-1}(u; \psi); \psi)}{\sigma(h^{-1}(u; \psi); \psi)} - \frac{\sigma'(h^{-1}(u; \psi); \psi)}{2},$$

where σ' denotes the derivative of σ w.r.t. x. In (5) the diffusion coefficient does not depend on the parameters, so the probability measures induced on $(C, \mathscr{C})$ by the solution to (5) are equivalent and the likelihood function can be found.

We can express the observations Y_i in terms of the process U. By inserting $X_s = h^{-1}(U_s; \psi)$ in (2), we find that

$$Y_i = \int_{t_{i-1}}^{t_i} h^{-1}(U_s; \psi)ds + Z_i, \quad i = 1, \ldots, n.$$

Therefore we will think of the full dataset as U_t, $t \in [0, t_n]$ and $Y = (Y_1, \ldots, Y_n)$. Since conditionally on the sample path of U the observations Y_i, $i, \ldots, n$ are independent, we have that the likelihood of Y conditional on the sample path of U in $[0, t_n]$ is

$$L(Y_1, \ldots, Y_n \mid U_t, t \in [0, t_n]) = \prod_{i=1}^{n} \phi \left(Y_i; \int_{t_{i-1}}^{t_i} h^{-1}(U_s; \psi)ds, \tau^2 \right) \tag{6}$$

where $\phi(u; a_1, a_2)$ denotes the density of the normal distribution with mean a_1 and variance a_2 evaluated at u.

Let P_ψ be the probability measure induced by $U = \{U_t\}_{t \in [0, t_n]}$ on $(C, \mathscr{C})$, i.e. the probability measure with respect to which the coordinate process has the same distribution as U, and let Q be the Wiener measure on $(C, \mathscr{C})$. We assume that the coefficient μ satisfies conditions ensuring that the Girsanov theorem holds so that we have the Radon-Nykodym derivative

$$\frac{dP_\psi}{dQ}(B) = \exp \left\{ \int_0^{t_n} \mu(B_t; \psi)dB_t - \frac{1}{2} \int_0^{t_n} \mu^2(B_t; \psi)dt \right\}; \tag{7}$$

see e.g. [27, 31, 33].

The evaluation of $\frac{dP_\psi}{dQ}$ is difficult because of the Ito integral term. To simplify the likelihood function, we apply the transformation

$$a(x; \psi) = \int^x \mu(u; \psi) du$$

(any antiderivative of μ), which under Condition 1 is twice continuously differentiable. By Ito's formula

$$\int_0^{t_n} \mu(B_t) dB_t = a(B_{t_n}; \psi) - a(B_0; \psi) - \frac{1}{2} \int_0^{t_n} \mu'(B_t; \psi) dt,$$

where μ' denotes the derivative of $\mu(u; \psi)$ w.r.t. u. We can now write the likelihood function (7) as

$$\frac{dP_\psi}{dQ}(B) = \exp\left\{ a(B_{t_n}; \psi) - a(B_0; \psi) - \frac{1}{2} \int_0^{t_n} [\mu(B_t; \psi)^2 + \mu'(B_t; \psi)] dt \right\}.$$

By combining this expression and (6), we see that the log-likelihood function for θ based on the full data set $U_t, t \in [0, t_n]$ and $Y = (Y_1, \ldots, Y_n)$ is given by

$$\log L(\theta; Y_1, \ldots, Y_n, U_t, t \in [0, t_n])$$

$$= \sum_{i=1}^n \log \phi\left(Y_i; \int_{t_{i-1}}^{t_i} h^{-1}(U_s; \psi) ds, \tau^2 \right)$$

$$+ a(U_{t_n}; \psi) - a(U_0; \psi) - \frac{1}{2} \int_0^{t_n} (\mu(U_t; \psi)^2 + \mu'(U_t; \psi)) dt. \qquad (8)$$

3.2 EM Algorithm

We can now apply the EM-algorithm to the full log-likelihood function (8) to obtain the maximum likelihood estimate of the parameter θ.

As the initial value for the algorithm, let $\hat{\theta}$ be any value of the parameter vector $\theta = (\psi, \tau^2) \in \Psi \times (0, \infty)$. Then the EM-algorithm works as follow.

1. **E-STEP.**

   Generate M sample paths of the diffusion process X, $X^{(k)}, k = 1, \ldots, M$, conditional on the observations $Y_1, \ldots, Y_n$ using the parameter value $\hat{\theta} = (\hat{\psi}, \hat{\tau}^2)$, and calculate

   $$g(\theta) = \frac{1}{M - M_0} \sum_{k=M_0+1}^M \log L(\theta; Y_1, \ldots, Y_n, h(X_t^{(k)}; \hat{\psi}), t \in [0, t_n]),$$

   for a suitable burn-in period M_0 and M sufficiently large.

2. **M-STEP.**

$\hat{\theta} = \text{argmax} g(\theta)$.

3. *Go to 1.*

To implement this algorithm, the main issue is how to generate sample paths of X conditionally on $Y_1, \ldots, Y_n$, where the relation between the Y_is and X is given by (2). The algorithm must produce a sequence $X^{(k)}$, $k = 1, \ldots, M$, that is sufficiently mixing to ensure that $g(\theta)$ approximates the conditional expectation of the full log-likelihood function (8) given the data. This problem is discussed at the next section.

4 Conditional Diffusion Process Simulation

In this section we present a method for generating a sample from

$$\{X_t; t \in [0, t_n]\} | (Y_1, \ldots, Y_n)$$

for a given value of the parameter vector θ, i.e. for simulating the diffusion X conditional on the observations $Y = (Y_1, \ldots, Y_n)$ of integrals of X over subintervals $[t_{j-1}, t_j]$, $j = 1, \ldots, n$ perturbed by measurement errors. This can be done by means of a Metropolis-Hastings algorithm, see e.g. [12] or [23]. However, if the sample path in the entire time interval $[0, t_n]$ is updated in one step, the rejection probability is typically very large. Therefore it is more efficient to randomly divide the time interval into subintervals and update the sample path in each of the subintervals conditional on the rest of the sample path. This corresponds to simulating a (conditional) diffusion bridge in each subinterval (except the end-intervals). The method outlined in this section is a modification of the method in [13], where we use the algorithm for approximate diffusion bridge simulation proposed by [10].

In the following the parameter value $\theta = (\psi, \tau^2)$ is fixed.

Algorithm 1

1. *Generate an initial unrestricted stationary sample path, $\{X_t^{(0)} : t \in [0, t_n]\}$, of the diffusion given by* (1) *using for instance the Milstein scheme or one of the other methods in* [30].
2. *Set $l = 1$.*
3. *Generate a sample path $\{X_t^{(l)} : t \in [0, t_n]\}$ conditional on Y by updating the subsets of the sample path:*
   a. *Randomly split the time interval from 0 to t_n in K blocks, and write these subsampling times as*

   $$0 = \tau_0 \leq \tau_1 \leq \cdots \leq \tau_K = t_n,$$

   where each τ_i is one of the end-points of the integration intervals, t_j, $j = 0, \ldots, n$. Let $Y_{\{k\}}$ denote the collection of all observations Y_j for which $\tau_{k-1} < t_j \leq \tau_k$.

b. *Draw* $X_0^{(l)}$ *from the stationary distribution,* ν_ψ, *and simulate the conditional subpath*

$$\{X_t^{(l)} : t \in [\tau_{k-1}, \tau_k]\} \mid Y_{\{k\}}, X_{\tau_{k-1}}^{(l)}, X_{\tau_k}^{(l-1)} \tag{9}$$

for $k = 1, \ldots, K - 1$. *Finally, simulate a sample path from*

$$\{X_t^{(l)} : t \in [\tau_{K-1}, \tau_K]\} \mid Y_{\{K\}}, X_{\tau_{K-1}}^{(l)}.$$

4. $l = l + 1$.
5. *Go to* 3.

To implement of this algorithm, the main issue is how to sample variables of the type (9), which is a non-linear diffusion bridge. We use the method for approximate diffusion bridge simulation proposed by [10]. The idea (in the case of a diffusion bridge in the time interval [0, 1]) is to let one diffusion process move forward from time zero out of one given point, a, until it meets another diffusion process that independently moves backwards from time one out of another given point, b. Conditional on the event that the two diffusions intersect, the process constructed in this way is an approximation to a realization of a diffusion bridge between a and b. The diffusions can be simulated by means of simple procedures like the Euler scheme or the Milstein scheme, see [30]. The method is therefore very easy to implement. The resulting sample path is an approximation to a diffusion bridge in the sense that it has the distribution of a diffusion bridge from a to b conditional on the event that the bridge is hit by an independent diffusion with stochastic differential equation (1) and initial distribution with density $p_1(b, \cdot)$. Simulation studies in [10] indicate that the approximation is very good for bridges between points that are likely to appear on a sample path of the diffusion, which is the type of bridges that are relevant to this paper.

Alternative methods that provide exact diffusion bridges have been proposed by [7] and [8]. When the drift and diffusion coefficients satisfy certain boundedness conditions, this algorithm is relatively simple, but under weaker condition it is more complex. A simulation study in [10] indicates that for the method which we use here, the CPU-time is linear in the length of the interval where the diffusion bridge is defined, whereas for the method in [7], the CPU time increases exponentially with the interval length. This is an advantage of the method in [10] in the present context. MCMC algorithms for simulation of diffusion bridges were proposed by [13, 19, 37].

To generate the random subintervals in step 3 (a) of Algorithm 1, we use the following algorithm, where the number of integration subintervals $[t_{j-1}, t_j]$ included in one of the random subintervals is a Poisson distributed random number plus 1. The draws in the algorithm are independent. First choose the expectation of the Poisson distribution, $\lambda \geq 1$.

Algorithm 2

1. *Draw* $k_1 \sim Poisson(\lambda) + 1$: *if* $k_1 \geq n$ *set* $k_1 = n$, $K = 1$ *and stop, otherwise set* $i = 2$.

2. *Draw $k_i \sim Poisson(\lambda) + 1$, if $\sum_{j=1}^{i} k_j \geq n$ set $k_i = n$, $K = i$ and stop, else set $i = i + 1$ and repeat 2.*

Finally define $\tau_i = t_{k_i}$, $i = 1, \ldots, K$.

We have discussed how to simulate diffusion bridges, but we need diffusion bridges conditional on the data Y. Sample paths of the conditional bridges (9) can be obtained by the following Metropolis-Hastings algorithm. By a (t, a, s, b)-bridge, we mean a diffusion bridge in the time interval $[t, s]$ with $X_t = a$ and $X_s = b$. After a burn-in period the algorithm will output samples from a $(\tau_{k-1}, a, \tau_k, b)$-bridge conditional on $Y_{\{k\}}$, the data in $(\tau_{k-1}, \tau_k]$. To formulate the algorithm we need to specify that the end-point τ_{k-1} is equal to t_j, and that there are n_k observations in the interval $(\tau_{k-1}, \tau_k]$, namely, $Y_{j+1}, \ldots, Y_{j+n_k}$.

Algorithm 3

1. *Simulate a $(\tau_{k-1}, a, \tau_k, b)$-bridge, $X^{(0)}$, and set $l = 1$.*
2. *Propose a new sample paths by simulating a $(\tau_{k-1}, a, \tau_k, b)$-bridge, $X^{(l)}$.*
3. *Accept the proposed diffusion bridge with probability*

$$\min\left(1, \prod_{i=1}^{n_k} \frac{\phi\left(Y_{j+i}; \int_{t_{j+i-1}}^{t_{j+i}} X_s^{(l)} ds, \tau^2\right)}{\phi\left(Y_{j+i}; \int_{t_{j+i-1}}^{t_{j+i}} X_s^{(l-1)} ds, \tau^2\right)}\right).$$

   Otherwise set $X^{(l)} = X^{(l-1)}$.
4. *Set $l = l + 1$ and go to 2.*

As previously, $\phi(x; \mu, \tau^2)$ denotes the density function of the normal distribution with mean μ and variance τ^2.

5 The Ornstein-Uhlenbeck Process: A Simulation Study

In this section we apply the method developed above to the Ornstein-Uhlenbeck process, which is a solution of the stochastic differential equation

$$dX_t = -\alpha X_t dt + \sigma dW_t, \tag{10}$$

where $\alpha > 0$ and $\sigma > 0$ are unknown parameters to be estimated, and W is a standard Wiener process. We investigate the bias of the estimators in a simulation study.

5.1 The Likelihood and the EM-algorithm

The transformation (4) is here given by

$$h(x; \sigma) = \frac{x}{\sigma},$$

so $h^{-1}(x; \sigma) = \sigma x$. Hence $U_t = h(X_t; \sigma) = X_t/\sigma$, solves the stochastic differential equation

$$dU_t = -\alpha U_t dt + dW_t.$$

We have $\mu(u; \alpha, \sigma) = -\alpha u$, so

$$a(u; \alpha, \sigma) = -\frac{1}{2}\alpha u^2.$$

Thus the full log-likelihood function (8) is given by

$$\log L(\theta; Y_1, \ldots, Y_n, U_t, t \in [0, t_n])$$

$$= \sum_{i=1}^{n} \log \phi\left(Y_i; \sigma \int_{t_{i-1}}^{t_i} U_s ds, \tau^2\right) + \frac{\alpha}{2}(U_0^2 - U_{t_n}^2 + t_n) - \frac{\alpha^2}{2}\int_0^{t_n} U_t^2 dt, \quad (11)$$

where $\theta = (\alpha, \sigma, \tau^2)$.

Now, the EM algorithm works as follow.

E-STEP

The objective function $g(\theta)$ is for the Ornstein-Uhlenbeck process given by

$$g(\theta) = -\frac{1}{2\tau^2(M - M_0)} \sum_{k=M_0+1}^{M} \sum_{i=1}^{n}\left(Y_i - \sigma \int_{t_{i-1}}^{t_i} U_t^{(k)} dt\right)^2 - \frac{n}{2}\log(2\pi\tau^2) + \frac{\alpha}{2}t_n$$

$$+ \frac{\alpha}{2(M - M_0)} \sum_{k=M_0+1}^{M}((U_0^{(k)})^2 - (U_{t_n}^{(k)})^2)$$

$$- \frac{\alpha^2}{2(M - M_0)} \sum_{k=M_0+1}^{M}\int_0^{t_n}(U_t^{(k)})^2 dt.$$

Here $U_t^{(k)} = X_t^{(k)}/\hat{\sigma}$, where $X_t^{(k)}$ is the k-th sample path of the process X simulated conditionally on the data Y using the Algorithms 1–3 with the parameter value obtained in the previous step $(\hat{\alpha}, \hat{\sigma}, \hat{\tau}^2)$.

M-STEP

The maximum $\hat{\theta}$ is obtained as the solution to the following system of equations

$$\frac{\partial g(\theta)}{\partial \alpha} = \frac{1}{2}t_n + \frac{\sum_{k=M_0+1}^{M}[(U_0^{(k)})^2 - (U_{t_n}^{(k)})^2]}{2(M - M_0)} - \frac{\alpha \sum_{k=M_0+1}^{M}\int_0^{t_n}(U_t^{(k)})^2 dt}{M - M_0}$$

$$= 0, \quad (12)$$

$$\frac{\partial g(\theta)}{\partial \sigma} = \frac{\sum_{k=M_0+1}^{M}\sum_{i=1}^{n}\left(Y_i - \sigma \int_{t_{i-1}}^{t_i} U_t^{(k)} dt\right)\left(\int_{t_{i-1}}^{t_i} U_t^{(k)} dt\right)}{\tau^2(M - M_0)} = 0 \quad (13)$$

and

$$\frac{\partial g(\theta)}{\partial \tau^2} = \frac{\sum_{k=M_0+1}^{M} \sum_{i=1}^{n} \left(Y_i - \sigma \int_{t_{i-1}}^{t_i} U_t^{(k)} dt\right)^2}{2\tau^4(M - M_0)} - \frac{n}{2\tau^2} = 0. \tag{14}$$

From (12) we have

$$\hat{\alpha} = \frac{t_n(M - M_0) + \sum_{k=M_0+1}^{M} [(U_0^{(k)})^2 - (U_{t_n}^{(k)})^2]}{2 \sum_{k=M_0+1}^{M} \int_0^{t_n} (U_t^{(k)})^2 dt},$$

and from (13)

$$\hat{\sigma} = \frac{\sum_{k=M_0+1}^{M} \sum_{i=1}^{n} Y_i \int_{t_{i-1}}^{t_i} U_t^{(k)} dt}{\sum_{k=M_0+1}^{M} \sum_{i=1}^{n} \left(\int_{t_{i-1}}^{t_i} U_t^{(k)} dt\right)^2}. \tag{15}$$

Now inserting $\hat{\sigma}$ given by (15) in (14) we obtain

$$\hat{\tau}^2 =$$

$$\frac{(M - M_0)(\sum_{i=1}^{n} Y_i^2)[\sum_{k=M_0+1}^{M} \sum_{i=1}^{n} \left(\int_{t_{i-1}}^{t_i} U_t^{(k)} dt\right)^2] - [\sum_{k=M_0+1}^{M} \sum_{i=1}^{n} Y_i \int_{t_{i-1}}^{t_i} U_t^{(k)} dt]^2}{n(M - M_0) \sum_{k=M_0+1}^{M} \sum_{i=1}^{n} \left(\int_{t_{i-1}}^{t_i} U_t^{(k)} dt\right)^2}.$$

The Hessian matrix of $g(\theta)$ evaluated at $\hat{\theta}$ is negative define, so $\hat{\theta}$ is maximum.

5.2 A Simulation Study

In this section we present the result of a small simulation study, in which we simulated 1000 datasets and for each of them obtained estimates by means of the EM-algorithm proposed in the present paper. Each data set was obtained by simulating a sample path of length 1500 with initial distribution $X_0 \sim N(0, \sigma^2/(2\alpha))$, and then calculating data Y_i, $i = 1, \ldots, 1500$ by (2) with $t_i = i$, $i = 0, \ldots, n$. The parameter values were $\alpha = 0.1$, $\sigma = 0.5$ and $\tau^2 = 1.25$.

The EM-algorithm was run with $M = 10000$ and $M_0 = 1000$ and for three different values of λ, namely $\lambda = 10, 20, 30$. The average of the estimates obtained for the 1000 dataset are given in Table 1. The bias is small, and is overall most satisfactory for $\lambda = 20$.

Table 1 Average of parameter estimates obtained from 1000 simulated datasets with parameter values $\alpha = 0.1$, $\sigma = 0.5$ and $\tau^2 = 1.25$

λ	α	σ	τ^2
10	0.106	0.523	1.229
20	0.101	0.507	1.235
30	0.084	0.458	1.252

6 The CIR Process and a Stochastic Volatility Model

In this section we apply our method to the CIR process, which solves

$$dX_t = (\alpha - \beta X_t)dt + \sigma\sqrt{X_t}dW_t, \tag{16}$$

where $X_0 > 0$, $\alpha > 0$, $\beta > 0$, $\sigma > 0$ and W is a standard Brownian motion. If $2\alpha > \sigma^2$, the CIR process is strictly positive. Otherwise it can reach the boundary 0 in finite time with positive probability, but if the boundary is made instantaneously reflecting, the process stays non-negative. In both cases, the stationary distribution is $\Gamma(2\alpha/\sigma^2, \sigma^2/\beta)$.

The CIR process plays important roles in financial mathematics, where it has been used to describe the evolution of interest rates [15], and to model the volatility in the Heston model, [26]. In the latter model, the dynamics of the logarithm of the price, P_t, of a financial asset is given by

$$dP_t = (\kappa + \nu X_t)dt + \sqrt{X_t}dB_t,$$

where the volatility process X is given by (16), and where the standard Brownian motion B may possibly be correlated with the Brownian motion W in (16). If high frequency observations of the asset price are available at the time points $j\delta$, $j = 0, \ldots, N$, then the integrated volatility over longer time intervals of length $\Delta = m\delta$ (e.g. hours or days)

$$\int_{(i-1)\Delta}^{i\Delta} X_t dt,$$

$i = 1, \ldots, n = [N/m]$, can be estimated by the quadratic variation/realized volatility

$$V_i = \sum_{j=(i-1)m+1}^{im} (P_{j\delta} - P_{(j-1)\delta})^2$$

$i = 1, \ldots, n$. We can therefore estimate the parameters α, β, σ in the volatility process (16) by treating the realized volatilities V_i, $i = 1, \ldots, n$, as observations Y_i of the type (2) with X given by (16) and $t_i = i\Delta$.

In the simulation study below, we investigate the bias of the estimators obtained by our methods for data Y_i, $i = 1, \ldots, n$ of the type (2) with X given by the CIR process (16).

6.1 The Likelihood and the EM-algorithm

We apply the transformation (4), which for the CIR process is

$$h(x; \sigma) = \frac{2\sqrt{x}}{\sigma},$$

with $h^{-1}(x;\sigma) = \sigma^2 x^2/4$. Hence $U_t = h(X_t;\sigma) = 2\sqrt{X_t}/\sigma$, solves the stochastic differential equation

$$dU_t = \mu(U_t;\alpha,\beta,\sigma)dt + dW_t,$$

where

$$\mu(u;\alpha,\beta,\sigma) = \frac{4\alpha - \beta\sigma^2 u^2 - \sigma^2}{2\sigma^2 u},$$

so that

$$a(u;\alpha,\beta,\sigma) = \log(u)\left(\frac{2\alpha}{\sigma^2} - \frac{1}{2}\right) - \frac{\beta u^2}{4}.$$

For the CIR model the full log-likelihood function (8) is given by

$$\log L(\theta; Y_1,\ldots,Y_n, U_t, t \in [0,t_n])$$

$$= \sum_{i=1}^{n} \log\phi\left(Y_i; \tfrac{1}{4}\sigma^2 \int_{t_{i-1}}^{t_i}(U_s)^2 ds, \tau^2\right) + \left(\frac{2\alpha}{\sigma^2} - \frac{1}{2}\right)\log\left(\frac{U_{t_n}}{U_0}\right) + \frac{\beta}{4}(U_0^2 - U_{t_n}^2)$$

$$+ \frac{\alpha\beta t_n}{\sigma^2} - \frac{\beta^2}{8}\int_0^{t_n} U_t^2 dt + \left(\frac{2\alpha}{\sigma^2} - \frac{2\alpha^2}{\sigma^4} - \frac{3}{8}\right)\int_0^{t_n} U_t^{-2} dt,$$

where $\theta = (\alpha,\beta,\sigma,\tau^2)$.

Now, we can specify the EM algorithm.

E-STEP

The objective function $g(\theta)$ is for the CIR process given by

$$g(\theta) = -\frac{1}{2\tau^2(M-M_0)} \sum_{k=M_0+1}^{M} \sum_{i=1}^{n}\left(Y_i - \tfrac{1}{4}\sigma^2 \int_{t_{i-1}}^{t_i}(U_t^{(k)})^2 dt\right)^2 - \frac{n}{2}\log(2\pi\tau^2)$$

$$+ \frac{\alpha\beta t_n}{\sigma^2} + \frac{2\alpha\sigma^{-2} - \frac{1}{2}}{M-M_0}\sum_{k=M_0+1}^{M} \log\left(\frac{U_{t_n}^{(k)}}{U_0^{(k)}}\right)$$

$$+ \frac{\beta}{4(M-M_0)}\sum_{k=M_0+1}^{M}\left((U_0^{(k)})^2 - (U_{t_n}^{(k)})^2\right)$$

$$+ \frac{2\alpha\sigma^2 - 2\alpha^2}{\sigma^4(M-M_0)}\sum_{k=M_0+1}^{M}\int_0^{t_n}(U_t^{(k)})^{-2} dt$$

$$- \frac{3}{8(M-M_0)}\sum_{k=M_0+1}^{M}\int_0^{t_n}(U_t^{(k)})^{-2} dt - \frac{\beta^2}{8(M-M_0)}\sum_{k=M_0+1}^{M}\int_0^{t_n}(U_t^{(k)})^2 dt.$$

Here $U_t^{(k)} = 2\sqrt{X_t^{(k)}}/\hat{\sigma}$, where $X_t^{(k)}$ is the k-th sample path of the process X simulated conditionally on the data Y using the Algorithms 1–3 with the parameter value obtained in the previous step $(\hat{\alpha}, \hat{\beta}, \hat{\sigma}, \hat{\tau}^2)$.

M-STEP

The maximum $\hat{\theta}$ is obtained as the solution to the following system of equations

$$\frac{\partial g(\theta)}{\partial \alpha} = \frac{2}{\sigma^2(M - M_0)} \sum_{k=M_0+1}^{M} \log\left(\frac{U_{t_n}^{(k)}}{U_0^{(k)}}\right) + \frac{\beta t_n}{\sigma^2}$$

$$+ \frac{2\sigma^2 - 4\alpha}{\sigma^4(M - M_0)} \sum_{k=M_0+1}^{M} \int_0^{t_n} \left(U_t^{(k)}\right)^{-2} dt = 0,$$

$$\frac{\partial g(\theta)}{\partial \beta} = \frac{1}{4(M - M_0)} \sum_{k=M_0+1}^{M} \left((U_0^{(k)})^2 - (U_{t_n}^{(k)})^2\right) + \frac{\alpha t_n}{\sigma^2}$$

$$- \frac{\beta}{4(M - M_0)} \sum_{k=M_0+1}^{M} \int_0^{t_n} \left(U_t^{(k)}\right)^2 dt = 0,$$

$$\frac{\partial g(\theta)}{\partial \sigma} = \frac{\sigma}{2\tau^2(M - M_0)} \sum_{k=M_0+1}^{M} \sum_{i=1}^{n} \left(Y_i - \tfrac{1}{4}\sigma^2 \int_{t_{i-1}}^{t_i} \left(U_t^{(k)}\right)^2 dt\right) \int_{t_{i-1}}^{t_i} \left(U_t^{(k)}\right)^2 dt$$

$$- \frac{2\alpha\beta t_n}{\sigma^3} - \frac{4\alpha}{\sigma^3(M - M_0)} \sum_{k=M_0+1}^{M} \log\left(\frac{U_{t_n}^{(k)}}{U_0^{(k)}}\right)$$

$$+ \frac{4\alpha(2\alpha - \sigma^2)}{\sigma^5(M - M_0)} \sum_{k=M_0+1}^{M} \int_0^{t_n} \left(U_t^{(k)}\right)^{-2} dt = 0,$$

and

$$\frac{\partial g(\theta)}{\partial \tau^2} = \frac{1}{2\tau^2(M - M_0)} \sum_{k=M_0+1}^{M} \sum_{i=1}^{n} \left(Y_i - \tfrac{1}{4}\sigma^2 \int_{t_{i-1}}^{t_i} \left(U_t^{(k)}\right)^2 dt\right)^2 - \frac{n}{2\tau^2} = 0.$$

The solution $\hat{\theta}$ is given by

$$\hat{\alpha} = \frac{C_5(2C_2C_4 + 2C_1C_4 + C_3 t_n)}{C_6(C_4C_2 - t_n^2)}$$

$$\hat{\beta} = \frac{C_3C_2 + 2t_n(C_2 + C_1)}{C_4C_2 - t_n^2}$$

$$\hat{\sigma} = \sqrt{4C_5/C_6}$$

$$\hat{\tau}^2 = \frac{C_7C_6 - C_5^2}{n(M - M_0)C_6},$$

where the values of the constants C_i are

$$C_1 = \frac{1}{M - M_0} \sum_{k=M_0+1}^{M} \log\left(U_{t_n}^{(k)} / U_0^{(k)}\right)$$

$$C_2 = \sum_{k=M_0+1}^{M} \int_{t_0}^{t_n} \left(U_t^{(k)}\right)^{-2} dt$$

$$C_3 = \sum_{k=M_0+1}^{M} \left(\left(U_0^{(k)}\right)^2 - \left(U_{t_n}^{(k)}\right)^2\right)$$

$$C_4 = \sum_{k=M_0+1}^{M} \int_0^{t_n} \left(U_t^{(k)}\right)^2 dt$$

$$C_5 = \sum_{k=M_0+1}^{M} \sum_{i=1}^{n} Y_i \int_{t_{i-1}}^{t_i} \left(U_t^{(k)}\right)^2 dt$$

$$C_6 = \sum_{k=M_0+1}^{M} \sum_{i=1}^{n} \left(\int_{t_{i-1}}^{t_i} \left(U_t^{(k)}\right)^2 dt\right)^2$$

$$C_7 = \sum_{k=M_0+1}^{M} \sum_{i=1}^{n} Y_i.$$

6.2 A Simulation Study

Here we present a simulation study for the integrated CIR-model. We simulated 1500 datasets, and for each of them obtained estimates by means of our EM-algorithm. Each data set was obtained by simulating a sample path of length 1500 with initial distribution $X_0 \sim \Gamma(2\alpha/\sigma^2, \sigma^2/\beta)$, and then calculating data Y_i, $i = 1, \ldots, 1500$ by (2) with $t_i = i$, $i = 0, \ldots, n$. The parameter values were $\alpha = 0.5$, $\beta = 0.2$, $\sigma = 0.5$ and $\tau^2 = 1.25$.

The EM-algorithm was run with $M = 10000$ and $M_0 = 1000$ for three values of λ. The average of the estimates obtained for the 1500 dataset are given in Table 2. Also for the CIR model the bias is small.

	λ	α	β	σ	τ^2
Table 2 Average of parameter estimates obtained from 1500 simulated datasets with parameter values $\alpha = 0.1$, $\beta = 0.2$, $\sigma = 0.5$ and $\tau^2 = 1.25$	30	0.4802	0.2056	0.4787	1.2432
	20	0.4727	0.2043	0.4698	1.2406
	10	0.4587	0.1965	0.4609	1.2287

7 Concluding Remarks

We have presented an EM-algorithm for obtaining maximum likelihood estimates of parameters in diffusion models when the data is a discrete time sample of the integral of the diffusion process contaminated by measurement errors, while no direct observations of the process itself are available. This was done by viewing the data as an incomplete observation, where the full data set includes a continuous time record of the diffusion process.

It is not difficult to generalize the method presented in this paper to the situation, where the diffusion process is integrated w.r.t. a more general measure than the Lebesgue measure considered in this paper. This would allow analysis of e.g. weighted averages of diffusion processes, and discrete time observation would be a particular case. Note also that a Gibbs sampler could easily be set up in close analogy to the EM-algorithm used in the present paper. This would be much closer to the approach in [13].

Acknowledgements The research of Michael Sørensen was supported by the Danish Center for Accounting and Finance funded by the Danish Social Science Research Council and by the Center for Research in Econometric Analysis of Time Series funded by the Danish National Research Foundation. The support to Fernando Baltazar-Larios from Consejo Nacional de Ciencia y Tecnología is gratefully acknowledged. This support financed a one year stay at the University of Copenhagen.

References

1. Aït-Sahalia, Y.: Maximum likelihood estimation of discretely sampled diffusions: A closed-form approximation approach. Econometrica **70**, 223–262 (2002)
2. Aït-Sahalia, Y.: Closed-form likelihood expansions for multivariate diffusions. Ann. Stat. **36**, 906–937 (2008)
3. Aït-Sahalia, Y., Mykland, P.: The effects of random and discrete sampling when estimating continuous-time diffusions. Econometrica **71**, 483–549 (2003)
4. Andersen, T.G., Bollerslev, T., Diebold, F.X., Ebens, H.: The distribution of realized stock return volatility. J. Financ. Econ. **61**, 43–76 (2001)
5. Andersen, T.G., Bollerslev, T., Diebold, F.X., Labys, P.: The distribution of exchange rate volatility. J. Am. Stat. Assoc. **96**, 42–55 (2001)
6. Barndorff-Nielsen, O.E., Shephard, N.: Econometric analysis of realized volatility and its use in estimating stochastic volatility models. J. R. Stat. Soc., Ser. B Stat. Methodol. **64**, 253–280 (2002)
7. Beskos, A., Papaspiliopoulos, O., Roberts, G.O.: Retrospective exact simulation of diffusion sample paths with applications. Bernoulli **12**, 1077–1098 (2006)
8. Beskos, A., Papaspiliopoulos, O., Roberts, G.O.: A factorization of diffusion measure and finite sample path constructions. Methodol. Comput. Appl. Probab. **10**, 85–104 (2008).
9. Beskos, A., Papaspiliopoulos, O., Roberts, G.O., Fearnhead, P.: Exact and computationally efficient likelihood-based estimation for discretely observed diffusion processes. J. R. Stat. Soc., Ser. B Stat. Methodol. **68**, 333–382 (2006)
10. Bladt, M., Sørensen, M.: Simple simulation of diffusion bridges with application to likelihood inference for diffusions. Working paper, Dept. Math. Sciences, Univ. of Copenhagen (2009)

11. Bollerslev, T., Zhou, H.: Estimating stochastic volatility diffusion using conditional moments of integrated volatility. J. Econom. **109**, 33–65 (2002)
12. Chib, S., Greenberg, E.: Understanding the Metropolis-Hastings algorithm. Am. Stat. **49**, 327–335 (1995)
13. Chib, S., Pitt, M.K., Shephard, N.: Likelihood based inference for diffusion driven state space models. Working paper (2006)
14. Comte, F., Genon-Catalot, V., Rozenholc, Y.: Nonparametric adaptive estimation for integrated diffusions. Stoch. Process. Appl. **119**, 811–834 (2009)
15. Cox, J.C., Ingersoll, Jr., J.E., Ross, S.A.: A theory of the term structure of interest rates. Econometrica **53**, 385–407 (1985)
16. Dempster, A.P., Laird, N.M., Rubin, D.B.: Maximum likelihood from incomplete data via the EM algorithm (with discussion). J. R. Stat. Soc., Ser. B Stat. Methodol. **39**, 1–38 (1977)
17. Ditlevsen, P.D., Ditlevsen, S., Andersen, K.K.: The fast climate fluctuations during the stadial and interstadial climate states. Ann. Glaciol. **35**, 457–462 (2002)
18. Ditlevsen, S., Sørensen, M.: Inference for observations of integrated diffusion processes. Scand. J. Stat. **31**, 417–429 (2004)
19. Durham, G.B., Gallant, A.R.: Numerical techniques for maximum likelihood estimation of continuous-time diffusion processes. J. Bus. Econ. Stat. **20**, 297–338 (2002)
20. Elerian, O., Chib, S., Shephard, N.: Likelihood inference for discretely observed non-linear diffusions. Econometrica **69**, 959–993 (2001)
21. Eraker, B.: MCMC analysis of diffusion models with application to finance. J. Bus. Econ. Stat. **19**, 177–191 (2001)
22. Forman, J.L., Sørensen, M.: The Pearson diffusions: A class of statistically tractable diffusion processes. Scand. J. Stat. **35**, 438–465 (2008)
23. Gilks, W.R., Richardson, S., Spiegelhalter, D.J.: Markov Chain Monte Carlo in Practice. Chapman & Hall, London (1996)
24. Gloter, A.: Parameter estimation for a discrete sampling of an integrated Ornstein-Uhlenbeck process. Statistics **35**, 225–243 (2000)
25. Gloter, A.: Parameter estimation for a discretely observed integrated diffusion process. Scand. J. Stat. **33**, 83–104 (2006)
26. Heston, S.L.: A closed-form solution for options with stochastic volatility with applications to bond and currency options. Rev. Financ. Stud. **6**, 327–343 (1993)
27. Jacod, J., Shiryaev, A.N.: Limit Theorems for Stochastic Processes. Springer, New York (1987)
28. Karatzas, I., Shreve, S.E.: Brownian Motion and Stochastic Calculus. Springer, New York (1991)
29. Karlin, S., Taylor, H.M.: A Second Course in Stochastic Processes. Academic Press, Orlando (1981)
30. Kloeden, P.E., Platen, E.: Numerical Solution of Stochastic Differential Equations. 3rd revised printing. Springer, New York (1999)
31. Liptser, R.S., Shiryaev, A.N.: Statistics of Random Processes. Springer, New York (1977)
32. McLachlan, G.J., Krishnan, T.: The EM Algorithm and Extensions. Wiley, New York (1997)
33. Øksendal, B.: Stochastic Differential Equations. Springer, New York (1998)
34. Ozaki, T.: Non-linear time series models and dynamical systems. In: Hannan, E.J., Krishnaiah, P.R., Rao, M.M. (eds.) Handbook of Statistics, vol. 5, pp. 25–83. Elsevier, Amsterdam (1985)
35. Pedersen, A.R.: A new approach to maximum likelihood estimation for stochastic differential equations based on discrete observations. Scand. J. Stat. **22**, 55–71 (1995)
36. Poulsen, R.: Approximate maximum likelihood estimation of discretely observed diffusion processes. Working paper 29, Centre for Analytical Finance, Aarhus (1999)
37. Roberts, G.O., Stramer, O.: On inference for partially observed nonlinear diffusion models using Metropolis-Hastings algorithms. Biometrika **88**, 603–621 (2001)
38. Sørensen, M.: Estimating functions for discretely observed diffusions: A review. In: Basawa, I.V., Godambe, V.P., Taylor, R.L. (eds.) Selected Proceedings of the Symposium on Estimating Functions. *IMS Lecture Notes – Monograph Series*, vol. 32, pp. 305–325. Institute of Mathematical Statistics, Hayward (1997)

39. Sørensen, M.: Prediction-based estimating functions. Econom. J. **3**, 123–147 (2000)
40. Sørensen, M.: Estimating functions for diffusion-type processes. In: Kessler, M., Lindner, A., Sørensen, M. (eds.) Statistical Methods for Stochastic Differential Equations. Chapman & Hall, London (2010, forthcoming)